NINTH EDITION

BIOLOGICAL INVESTIGATIONS

Form, Function, Diversity & Process

Warren D. Dolphin
David Vleck, James T. Colbert, and Linda M. Westgate

Iowa State University

BIOLOGICAL INVESTIGATIONS: FORM, FUNCTION, DIVERSITY, AND PROCESS, NINTH EDITION

Published by McGraw-Hill, a business unit of The McGraw-Hill Companies, Inc., 1221 Avenue of the Americas, New York, NY 10020. Copyright © 2011 by The McGraw-Hill Companies, Inc. All rights reserved. Previous editions © 2008, 2005, and 2002. No part of this publication may be reproduced or distributed in any form or by any means, or stored in a database or retrieval system, without the prior written consent of The McGraw-Hill Companies, Inc., including, but not limited to, in any network or other electronic storage or transmission, or broadcast for distance learning.

Some ancillaries, including electronic and print components, may not be available to customers outside the United States.

 This book is printed on recycled, acid-free paper containing 10% postconsumer waste.

2 3 4 5 6 7 8 9 0 QDB/QDB 1 0 9 8 7 6 5 4 3 2 1

ISBN 978-0-07-338305-7
MHID 0-07-338305-8

Vice President & Editor-in-Chief: *Marty Lange*
Vice President, EDP: *Kimberly Meriwether David*
Publisher: *Janice Roerig-Blong*
Director of Development: *Kristine Tibbetts*
Marketing Manager: *Chris Loewenberg*
Senior Project Manager: *Joyce Watters*
Production Supervisor: *Nicole Baumgartner*
Design Coordinator: *Brenda A. Rolwes*
Cover Designer: *Studio Montage, St. Louis, Missouri*
(USE) Cover Image: *© Getty Images/Digital Vision*
Lead Photo Research Coordinator: *Carrie K. Burger*
Photo Research: *Pam Carley/Sound Reach*
Compositor: *Laserwords Private Limited*
Typeface: *10/12 Times Roman*
Printer: *Quad/Graphics*

All credits appearing on page or at the end of the book are considered to be an extension of the copyright page.

Library of Congress Cataloging-in-Publication Data

Dolphin, Warren D.
  Biological investigations : form, function, diversity, and process/Warren D. Dolphin.
— 9th ed.
      p. cm.
  Includes bibliographical references and index.
  ISBN 978-0-07-338305-7 (alk. paper)
  1. Biology—Handbooks, manuals, etc. I. Title.
  QH310.D65 2011
  570—dc22
                                                2009049872

www.mhhe.com

Brief Contents

Preface ix

Lab Topic 1	Science: A Way of Knowing 1
Lab Topic 2	Using Microscopes 9
Lab Topic 3	Cellular Structure Reflects Function 19
Lab Topic 4	Membranes, Diffusion, and Osmosis 29
Lab Topic 5	Using Quantitative Techniques 39
Lab Topic 6	Visualizing Biological Molecules 53
Lab Topic 7	Determining the Properties of an Enzyme 75
Lab Topic 8	Measuring Cellular Respiration 89
Lab Topic 9	Investigating Photosynthesis 99
Lab Topic 10	Mitosis and Chromosome Number 111
Lab Topic 11	Modeling Meiosis and Determining Cross-Over Frequency 121
Lab Topic 12	Analyzing Fruit Fly Phenotypes and Genotypes 133
Interchapter	An Outline of Sterile Technique 143
Lab Topic 13	Isolating DNA and Transformation with Plasmids 145
Lab Topic 14	Modeling Processes in Evolution and Mutagenesis 155
Lab Topic 15	Investigating Bacterial Diversity 171
Lab Topic 16	Evolution of Eukarya and Diverse Protists 183
Lab Topic 17	Ancestral and Derived Characteristics of Seedless Plants 199
Lab Topic 18	Derived Characteristics of Seed Plants 213
Lab Topic 19	Investigating Fungal Diversity and Symbiotic Relationships 229
Lab Topic 20	From Basal to Bilateral Animals 247
Lab Topic 21	Lophotrochozoa: Increased Complexity 265
Lab Topic 22	Ecdysozoa: Simple and Complex 283
Lab Topic 23	Deuterostomes and the Origins of Vertebrates 301
Lab Topic 24	Investigating Plant Cells and Tissues 317
Lab Topic 25	Investigating Functional Anatomy of Vascular Plants 331
Lab Topic 26	Investigating Pollination, Development, and Dispersal 351
Interchapter	Investigating Animal Form and Function 361
Lab Topic 27	Investigating Digestive, Renal, and Reproductive Systems 363
Lab Topic 28	Investigating Circulatory Systems 377
Lab Topic 29	Investigating Gas Exchange Systems 391
Lab Topic 30	Investigating the Properties of Muscle and Skeletal Systems 401
Lab Topic 31	Investigating Nervous and Sensory Systems 417
Lab Topic 32	Investigating Early Events in Animal Development 429
Lab Topic 33	Estimating Population Size and Growth 441
Appendix A	Significant Figures and Rounding 451
Appendix B	Making Graphs 453
Appendix C	Simple Statistics 455
Appendix D	Writing Reports and Scientific Papers 461

Photo Credits 465

Contents

Preface ix

Lab Topic 1   Science: A Way of Knowing 1
Exploring Discovery-Based Science 3
Exploring Hypothesis-Based Science 4
Evaluating Published Information 5

Lab Topic 2   Using Microscopes 9
The Dissecting Microscope 10
The Compound Microscope 11
   Avoiding Hazards in Microscopy 13
Making a Slide and Viewing It 13
The Compound Microscope Image 14
Measuring with a Microscope (Optional) 15
Using an Oil Immersion Lens (Optional) 16

Lab Topic 3   Cellular Structure Reflects Function 19
Prokaryotic Cells 20
Eukaryotic Cells 21
   Protists 21
   Fungal Cells 22
   Plant Cells 22
   Animal Cells 23
Cell Ultrastructure 25

Lab Topic 4   Membranes, Diffusion, and Osmosis 29
Modeling Cell Membranes 31
Diffusion in Gels 33
Osmotic Challenges to Living Cells 33
   Visualizing the Challenge 33
   Cell Walls Osmotically Protect 34
   Expelling Water 35

Lab Topic 5   Using Quantitative Techniques 39
Measuring Mass 40
Measuring Volumes 40
Calculating Simple Statistics 44
Using a Spectrophotometer 44
   Spectrophotometer Experiments 47
      Measuring an Absorption Spectrum 47
      Constructing a Standard Curve 48
      Determining an Unknown Concentration 49

Lab Topic 6   Visualizing Biological Molecules 53
Getting Started 55
Carbohydrates 56
   Sugars 56
   Polysaccharides 58
Lipids 58
   Fatty Acids 60
   Glycerides 60
   Membranes 61
Proteins 62
   Amino Acids 62
   Peptides 63
   Protein Structure 65
   Advanced Analysis of Peroxidase Active Site 68
Nucleic Acids 70
   Nucleotides 71
   DNA 71

Lab Topic 7   Determining the Properties of an Enzyme 75
Peroxidase 77
Preparing Peroxidase 77
Standardizing Amount of Enzyme 77
Factors Affecting Peroxidase Activity 79
   Effect of Temperature 79
   Effect of pH 81
   Effect of Boiling on Peroxidase 82
   Probing the Active Site 83
Advanced Molecular Analysis 83

Lab Topic 8   Measuring Cellular Respiration 89
Respiratory Flexibility in Yeast 91
   Tests for Respiratory Products 91
   Demonstration of Respiratory Efficiency 92
Aerobic Respiration in Germinating Seeds 93
   Doing Discovery Science 93
   Measurement 94
   Analysis 95

Lab Topic 9   Investigating Photosynthesis 99
Chloroplasts 101
Photosynthetic Pigments 101

Measuring Photosynthetic Rate 104
Starch Accumulation in Leaves 107

Lab Topic 10   Mitosis and Chromosome Number 111

Mitosis in Animal Cells 113
Mitosis in Plant Cells 114
Staining Chromosomes in Dividing Cells 115
   Testing a Hypothesis 117

Lab Topic 11   Modeling Meiosis and Determining Cross-Over Frequency 121

Testing Understanding Through Modeling 123
   Modeling Meiosis with No Crossing-Over 123
   Modeling Meiosis with Crossing-Over 124
   Modeling Fertilization 125
Measuring the Frequency of Crossing-Over 126
   Background on Experimental Organism 126
   Advanced Statistical Analysis 129

Lab Topic 12   Analyzing Fruit Fly Phenotypes and Genotypes 133

Fruit Fly Life Cycle 134
Fruit Fly Anatomy 135
Giant Chromosomes (Optional) 136
Theoretical Background 137
   Alleles on Autosomes' 137
Performing a Genetic Cross 138

Interchapter   An Outline of Sterile Technique 143

Lab Topic 13   Isolating DNA and Transformation with Plasmids 145

Isolating Genomic DNA 147
Determining the Amount of DNA in Solution 148
Transformation by Plasmids 148
   Making Competent Cells 149
   Transformation Procedure 150
   Selective Growth 150
   Serial Dilution 150
   Transformation Efficiency 151

Lab Topic 14   Modeling Processes in Evolution and Mutagenesis 155

Simulating Impact of Natural Selection on
   Phenotypic Frequency 158
   Exploring the Habitat 159
   Analyze Impact of Predation 159
   Impact of Natural Selection over Generations 159
   Analysis of Natural Selection 161
Hardy-Weinberg Equilibrium Simulation 161
   Verifying the Hardy-Weinberg Equilibrium Model 162

Effect of Selection for a Dominant Allele in a Large Population 163
Effect of Small Population Size 164
Testing Combined Effect of Changes in Assumptions 165
Inducing Mutations 165

Lab Topic 15   Investigating Bacterial Diversity 171

Finding Bacteria 173
Bacterial Morphology 174
Gram Staining 176
Differential Growth 178
Bacterial Population Counts 180

Lab Topic 16   Evolution of Eukarya and Diverse Protists 183

Supergroup 1: Excavata 185
   Parabasalid Clade 185
   Euglenozoan Clade 185
   Kinetoplastid Clade 186
Supergroup 2: Chromalveolata 187
   Alveolata Clade 187
   Stramenopila Clade 189
Supergroup 3: Archaeplastida 190
   Rhodophyte (Red Algae) Clade 190
   Chlorophyte (Green Algae) Clade 191
   Plant Clade 193
Supergroup 4: Rhizaria 194
   Foraminiferan Clade 194
Supergroup 5: Unikonta 194
   Amoebozoan Clade 194
   Opisthokontan Clade 196

Lab Topic 17   Ancestral and Derived Characteritics of Seedless Plants 199

Plant Adaptations to Land 200
Algal Preadaptations to Land 201
   Stonewort Anatomy and Life Cycle 202
Bryophytes 203
   Phylum Hepatophyta (Liverworts) 203
   Phylum Bryophyta (Mosses) 205
Vascular Plants 207
   Aspects of Body Plan 207
   Phylum Pterophyta (Ferns and Horsetails) 208

Lab Topic 18   Derived Characteristics of Seed Plants 213

Gymnosperms 215
   Phylum Coniferophyta (Conifers) 216
     Sporophyte Stage of Pine 216
     Microsporangia and Male Gametophyte Stage 216
     Megasporangia and Female Gametophytes 217

Fertilization and New Sporophyte Generation 218
Angiosperms 219
    Sporophyte Floral Structure 221
        Microsporangia and Male Gametophyte 222
        Megasporangia and Female Gametophyte 223
    Fertilization 224
    Seeds and Fruits 225
    Grass Flowers 225

Lab Topic 19 Investigating Fungal Diversity and Symbiotic Relationships 229

Observation of Field Samples 232
Phylum Chytridiomycota 232
Phylum Zygomycota 234
Phylum Ascomycota: Sac Fungi 236
    Molds and Mildews 237
        Asexual Reproduction 237
    Cup Fungi 237
Phylum Basidiomycota: Club Fungi 239
Fungal Associations 241
    Lichens 241
    Mycorrhizae 243
Fun with Fungi (Optional) 244
    Identifying Unknowns 244
    Fungus Amon-gus 244

Lab Topic 20 From Basal to Bilateral Animals 247

Metazoa 250
Parazoa 250
    Phylum Porifera 250
Eumetazoa 253
Radiata 253
    Phylum Cnidaria 253
        A Polyp: Hydra 254
        A Medusa: Gonionemus 256
        A Colonial Form: Obelia 257
Bilateria 257
    Phylum Platyhelminthes 258
        Class Turbellaria 258
        Class Trematoda 260
        Class Cestoidea 261

Lab Topic 21 Lophotrochozoa: Increased Complexity 265

Clade Lophotrochozoa 267
Phylum Bryozoa (Ectoprocta) 267
Phylum Annelida 268
    Live Aquatic Oligochaetes 269
    Earthworm Dissection 270
Phylum Mollusca 275
    Class Bivalvia 276
    Class Cephalopoda 278

Lab Topic 22 Ecdysozoa: Simple and Complex 283

Phylum Nematoda 284
    Functional Anatomy 285
    Live Nematodes 287
    Rogue's Gallery 287
Phylum Arthropoda 287
    Clade Chelicerata (=Cheliceriformes) 288
        Horseshoe Crab 289
        Spider 289
    Clade Crustacea 290
        Barnacles 290
        Daphnia 291
        Crayfish 292
    Clade Hexapoda (=Insects) 296
        Functional Anatomy 297
        Keying Insects (Optional) 299

Lab Topic 23 Deuterostomes and the Origins of Vertebrates 301

Phylum Echinodermata 302
    Echinoderm Diversity 302
    Asteroidea 303
        External Anatomy 303
        Internal Anatomy 304
Phylum Chordata 306
    Subphylum Urochordata 306
        Observation of Derived Chordate Traits 306
        Adult Functional Anatomy 307
    Subphylum Cephalochordata 308
        Observation of Derived Chordate Traits 308
        Functional Anatomy 309
    Subphylum Vertebrata 310
        Fish External Anatomy 311
        Fish Functional Anatomy 313

Lab Topic 24 Investigating Plant Cells and Tissues 317

Investigating Unique Structures in Plant Cells 319
    Cell Walls 319
    Vacuoles 321
    Plastids 322
Investigating Plant Tissues 322
    Meristematic Tissue 322
    Dermal Tissue System 322
    Ground Tissue System 324
    Vascular Tissue System 326

Lab Topic 25 Investigating Functional Anatomy of Vascular Plants 331

Root Structure 334
    Whole Roots 334
    Root Histology 334

Stem Structure 339
    External Eudicot Stem Anatomy 339
    Herbaceous Eudicot Stem Histology 339
    Woody Eudicot Stem Histology 340
    Structure of Wood 342
    Monocot Stem Histology 343
Leaf Structure 343
    Whole Leaves 344
    Leaf Histology 344
    Guard Cell Response to Osmotic Stress 347
Investigating Water Movement Mechanisms 348
    Water Movement Due to Transpiration 348
    Testing Water Tension–Cohesion Hypothesis 348

Lab Topic 26 Investigating Pollination, Development, and Dispersal 351

Pollination and Seed Dispersal 353
Female Gametophyte 354
Embryo Development 354
Seed Anatomy 355
Fruits 356
Germination 357
    Research Project: Hormone Effects on Germination and Seedlings 358
    Photomorphogenesis in Seedlings 359

Interchapter Investigating Animal Form and Function 361

Lab Topic 27 Investigating Digestive, Renal, and Reproductive Systems 363

Feeding Behavior in *Hydra* 365
Mammalian Digestive System Anatomy 366
    Anatomy of the Mouth 366
    Alimentary Canal Anatomy 366
    Histology of Small Intestine 369
Mammalian Excretory System 370
    Nephron Structure 370
Mammalian Reproductive System 372
    Male System 372
    Female System 373

Lab Topic 28 Investigating Circulatory Systems 377

Open Circulatory Systems 379
Mammalian Circulatory System 379
    The Heart and Major Vessels 380
    Vessels Cranial to the Heart 380
    Vessels Caudal to the Heart 382
    Internal Heart Structure 383
    Histology of Vessels 384
    Blood 385
    Capillary Circulation 386
Measuring Blood Pressure and Heart Rate 387
    Blood Pressures under Experimental Conditions 389

Lab Topic 29 Investigating Gas Exchange Systems 391

Insect Tracheal System 392
Gills 393
Mammalian Respiratory System 394
    Gross Anatomy 394
    Microscopic Anatomy of Lung 396
    Lung Ventilation Mechanism 396
    Measuring Human Respiratory Volumes 396

Lab Topic 30 Investigating the Properties of Muscle and Skeletal Systems 401

Muscular System 402
    Microscopic Anatomy of Muscle 402
    Fetal Pig Superficial Muscles (Optional) 405
    Physiology of Muscle 406
        Locating a Motor Point 406
        Determining Threshold and Observing Motor Unit Recruitment 407
        Temporal Summation 408
Skeletal Systems 409
    Hydrostatic Skeletons 409
    Exoskeletons 410
    Endoskeletons 410
        Bone Structure 411
        Comparative Vertebrate Endoskeletons 411

Lab Topic 31 Investigating Nervous and Sensory Systems 417

Histology of the Nervous System 419
    Structure of Neurons 419
    Cross Section of Spinal Cord 419
        Spinal Reflex Experiment 420
    Neuromuscular Junctions 420
Mammalian Central Nervous System 421
    Spinal Cord 421
    Autonomic Nervous System 422
    Brain Anatomy 422
Mammalian Sensory Systems 424
    Eye Functional Anatomy 424
        Gross Anatomy of the Eye 424
        Afterimages Experiment 425
    Ear Functional Anatomy 426
        Histology of Inner Ear 427

Lab Topic 32 Investigating Early Events in Animal Development 429

Early Sea Star Development 430
    Fertilization and Cleavage 430
    Gastrulation 432
Comparison of Radial and Spiral Cleavage 432
Experimental Embryology with Sea Urchins (Optional) 433

Obtaining Gametes 433
Observing Fertilization 434
Observing Cleavage 434
Amphibian Development 435
Cleavage 435
Gastrulation 436
Neurulation 436
Avian Development (Optional) 436

Lab Topic 33 Estimating Population Size and Growth 441

Quadrat Sampling 442
Mark and Recapture 445
Population Growth 446
Exploring Threatened and Endangered Species Lists 448

Appendix A Significant Figures and Rounding 451

What Are Significant Figures? 451
What Is Rounding? 451

Appendix B Making Graphs 453

Line Graphs 453
Derivative Graphs 453
Histograms 454

Appendix C Simple Statistics 455

Dealing with Measurement Data 455
Comparing Count Data: Dealing with Variability 458

Appendix D Writing Reports and Scientific Papers 461

Format 461
General Comments on Style 463
Further Readings 464

Photo Credits 465

Preface

Designed to be used with all majors-level general biology textbooks, the lab topics included in this substantially revised lab manual are investigative, using both discovery-based and hypothesis-based modes of inquiry. Students experimentally investigate topics; use critical thinking skills to summarize, predict and test ideas; observe structure and diversity; and engage in hands-on learning. Students are often asked, "what evidence do you have that. . ." to encourage independent thinking. By emphasizing investigative, quantitative, and comparative approaches to the topics, the authors illustrate how the biological sciences are integrative, yet unique. As stated in its subtitle, the manual addresses the fundamental biological principles taught in introductory classes while emphasizing three evolutionary themes: form reflects function; unity despite diversity; and the adaptive processes of life. This manual is an excellent choice for colleges and universities that want their students to experience the breadth of modern biology.

For this edition, three colleagues from Iowa State University joined the senior author in making revisions. Warren Dolphin is now University Professor, Emeritus, of Genetics, Cell, and Developmental Biology; David Vleck is Associate Scientist of Ecology, Evolution and Organismal Biology; James Colbert is Associate Professor of Ecology, Evolution, and Organismal Biology and Coordinator of the Undergraduate Biology Program; and Linda Westgate is the Biology Teaching Lab Coordinator. Together, representing over 60 years of experience in teaching introductory biology labs to large classes, their insights and diverse backgrounds assure that this new edition meets the needs of both instructors and students.

Laboratory Preparation Guide

A guide, available online at *www.mhhe.com/labcentral*, provides directions on scheduling and preparing materials, lesson plans, teaching tips, and multiple-choice questions for quizzes and practical exams. Contact your McGraw-Hill sales representative for access.

Changes to the Ninth Edition

Based on our experience and comments of reviewers, nearly every topic was revised.

- Procedures were updated for clarity and currentness.
- New experiments were added to over half the topics.
- Introductions were rewritten for more than half the topics.
- Guidelines for data analysis clarified and expanded.
- Both intermediate and terminal lab summaries were added to most labs.
- Added 28 new illustrations or photographs.

Specific changes include:

Lab Topic 1   Science: A Way of Knowing
- Discovery-based science added to correlate with textbooks.
- New experiment for hypothesis-based science reduces time needed.

Lab Topic 2   Using Microscopes
- Reorganized as a quick introduction to microscopes.

Lab Topic 3   Cellular Structure Reflects Function
- Streamlined observations of prokaryotic versus eukaryotic cells.
- Cellular ultrastructure section moved here.

Lab Topic 4   Membranes, Diffusion, and Osmosis
- Completely rewritten with several new experiments.
- Emphasis on osmotic challenges that cells encounter.

Lab Topic 5   Using Quantitative Techniques
- New emphasis given to precision and accuracy in measurements.

Lab Topic 6   Visualizing Biological Molecules
- Revised to use the open source Jmol Molecular Visualization program.
- Analysis of protein structure rewritten to emphasize levels of structure.

Lab Topic 7   Determining the Properties of an Enzyme
- Improved directions on making derivative graphs.

Lab Topic 8   Measuring Cellular Respiration
- Introduction rewritten to give better background on experiments.
- New experiment on respiratory flexibility in yeasts.

Lab Topic 9   Investigating Photosynthesis
- Topic moved to metabolism section from plant biology.
- New experiment added on visualizing chloroplasts.
- Analysis section rewritten.

Lab Topic 10   Mitosis and Chromosome Number
- New introduction with more modern emphasis.
- Procedures and analysis sections clarified.

Lab Topic 11   Modeling Meiosis and Determining Cross-Over Frequency
- *Ascaris* egg observations replaced with chromosome simulation activity.

Lab Topic 12   Analyzing Fruit Fly Phenotypes and Genotypes
- Simplified experiment added to determine parental genotypes from progeny phenotypes.

Lab Topic 13   Isolating DNA and Transformation with Plasmids
- Plasmid procedure rewritten to include use of jellyfish green fluorescent protein as a reporter gene and section added on transformation efficiency.

Lab Topic 14   Modeling Processes in Evolution and Mutagenesis
- Rewritten to use new BioQuest computer simulation.
- Section added to illustrate the role of camouflage in natural selection.
- Experimental induction of mutations now in *E. coli*.

Lab Topic 15   Investigating Bacterial Diversity
- New introduction, capturing the diverse biology of this prokaryote group.
- Procedures edited to clarify activities and new data tables added.

Lab Topic 16   Evolution of Eukarya and Diverse Protists
- Clade and phylum names updated based on latest phylogenetic research.
- Discovery science at its best as students investigate the diversity.

Lab Topic 17   Ancestral and Derived Characteristics of Seedless Plants
- Headings added to allow instructors greater flexibility in choosing sections to be completed.
- More accurate fern section and focus on discovery science.

Lab Topic 18   Derived Characteristics of Seed Plants
- Introduction and gymnosperm section rewritten.
- Angiosperm reproduction moved from the plant biology section to here.

Lab Topic 19   Investigating Fungal Diversity and Symbiotic Relationships
- Reorganized to promote comparisons between phyla.
- Emphasis added on fungal body plan.
- Deleted powdery mildews and added section on *Penicillium*.

Lab Topic 20   From Basal to Bilateral Animals
- Presents underlying ideas for new clades and phylogeny of animals.
- Discusses ancestral and derived characters of sponges, cnidarians, and flatworms. Nematodes no longer included.

Lab Topic 21   Lophotrochozoa: Increased Complexity
- Illustrates the basis for new animal clades.
- Organ systems in annelids and molluscs comparatively explored.

Lab Topic 22   Ecdysozoa: Simple and Complex
- Explores the basis for this new animal clade.
- Nematodes moved to this topic and dissection rewritten.
- New barnacle dissection with photo.

Lab Topic 23   Deuterostomes and the Origins of Vertebrates
- Derived characters in each group now presented.
- Echinoderm dissection rewritten with a more systems approach.
- Invertebrate chordate sections given a more comparative approach.

Lab Topic 24   Investigating Plant Cells and Tissues
- Revised to emphasize structure-function relationships.

Lab Topic 25   Investigating Functional Anatomy of Vascular Plants
- Revised to emphasize structure-function relationships.
- Leaf structure added.

Lab Topic 26   Investigating Pollination, Development, and Dispersal

- Rewritten to focus on seed development, hormonal control of germination, and plant-animal interactions.

Lab Topic 27   Investigating Digestive, Renal, and Reproductive Systems

- For dissection efficiency, digestive, renal, and reproductive systems have been grouped in this reorganization of the fetal pig dissection.
- Emphasis added on digestive processes and function of liver and pancreas.

Lab Topic 28   Investigating Circulatory Systems

- Added measurements of blood pressure/pulse and exercise response using computer interfaced data collection.
- Blood pressure protocols updated.
- Added emphasis on health relevance.

Lab Topic 29   Investigating Gas Exchange Systems

- Revised to emphasize structure-function relationships.
- Protocols updated to include widely used computer-based systems.
- Expanded coverage of gill systems.

Lab Topic 30   Investigating the Properties of Muscle and Skeletal Systems

- Revised to include structure-function comparisons of vertebrate groups.
- Computer-based muscle contraction protocols are integrated into physiology portion of lab.

Lab Topic 31   Investigating Nervous and Sensory Systems

- Revised discussions of glial cells and specialization of brain regions
- Eye and ear sections revised to focus on structural adaptation for sensitivity, acuity, and color in eyes and frequency, amplitude, and direction in ears.

Lab Topic 32   Investigating Early Events in Animal Development

- Topic moved to animal functional anatomy section.

Lab Topic 33   Estimating Population Size and Growth

- New section on bald eagle population recoveries since 1960.
- New section on investigating endangered species using Internet.

Acknowledgments

Reviewers who made constructive suggestions for previous editions include:

Olukemi Adewusi, *Ferris State University*
Linda L. Allen, *Lon Morris College*
Gordon Atkins, *Andrews University*
William Barstow, *University of Georgia*
Maria Begonia, *Jackson State University*
Paul Biebel, *Dickinson College*
Brenda Blackwelder, *Central Piedmont Community College*
Thomas Clark Bowman, *The Citadel Military College*
Natalie Bronstein, *Mercy College*
Christian Chauret, *Indiana University—Kokomo*
Mary Anne Clark, *Texas Wesleyan University*
Naomi D'Alessio, *Nova Southeastern University*
Marvin Druger, *Syracuse University*
Renata Dusenbury, *St. Augustina's College*
San Eisen, *Christian Brothers College*
Gerald Gates, *University of Redlands*
Becky Green-Marroquin, *Los Angeles Valley College*
Ida Greidanus, *Passaic County Community College*
Dana Brown Haine, *Central Piedmont Community College*
Cynthia M. Handler, *University of Delaware*
Sidney S. Herman, *Lehigh University*
Peter King, *Francis Marion University*
Margaret Krawiec, *Lehigh University*
Peter A. Lauzetta, *Kingsborough Community College (CUNY)*
Raymond Lewis, *Wheaton College*
Lewis Lutton, *Mercyhurst College*
Charles Lycan, *Tarrant County Junior College*
Lee Anne Martinez, *University of Southern Colorado*
Thomas Mertens, *Ball State University*
Karel Rogers, *Adams State College*
Stephen G. Saupe, *St. Johns University (Minnesota)*
Mary Schmall, *McDaniel College*
Nancy Segsworth, *Capilano College (British Columbia)*
Mindy Skarda, *Southwestern Community College*
Gary A. Smith, *Tarrant County Junior College*
Stacy Smith, *Lexington Community College*
Timothy A. Stabler, *Indiana University Northwest*
Daryl Sweeney, *University of Illinois*
Conrad Toepfer, *Milikin University*
Linda R. Van Thiel, *Wayne State University*
Miryam Wahrman, *William Paterson University of New Jersey*
Jan Whitson, *Concordia University*
Lise Wilson, *Siena College*
Lynne Zeman, *Kirkwood Community College*
William J. Zimmerman, *University of Michigan–Dearborn*

Reviewers of the eighth edition, who provided many thoughtful suggestions, were:

Gordon Atkins, *Andrews University*

Jennifer R. Chase, *Northwest Nazarene University*
F. Paul Doerder, *Cleveland State University*
Jim R. Goetze, *Laredo Community College*
Chad Heins, *Bethany Lutheran College*
Jeffrey Hughes, *Millikin University*
James R. Jacob, *Tompkins Cortland Community College*
Paul Jarrell, *Pasadena City College*
Marcy Kelly, *Pace University*
Jamee Nixon, *Northwest Nazarene University*
Trenton Smith, *Simpson University*
Ron Strohmeyer, *Northwest Nazarene University*
Randall Tracy, *Worcester State College*

	Brooker et al. 2010	Raven et al. 2010	Mader 2009	Hoefnagels 2010	Campbell, Reece 2008	Freeman 2007	Sadava et al. 2010	Starr et al. 2008
	Biology, 2e	Biology, 9e	Biology, 10e	Biology: Concepts & Investigations, 1e	Biology, 8e	Biological Science, 3e	Life: The Science of Biology, 9e	Biology: Unity and Diversity, 12e
1 Science: A Way of Knowing	1	1	1	1	1	1	1	1
2 Using Microscopes	4	4	4	3	6	7	5	4
3 Cellular Structure Reflects Function	4	4	4	3	6	7	5	40, 41
4 Membranes, Diffusion and Osmosis	5	5	5	3	7	6	6	5
5 Using Quantitative Techniques								
6 Visualizing Biological Molecules	2, 3	2, 3	3	2	4, 5	3, 4, 5, 6	2, 3, 4	2, 3
7 Determining the Properties of an Enzyme	6	6	6	4	5, 8	3	8	6
8 Measuring Cellular Respiration	7	7	8	6	9	9	9	8
9 Investigating Photosynthesis	8	8	7	5	10	10	10	7
10 Mitosis and Chromosome Number	15	10	9	8	12	11	11	9
11 Modeling Meiosis and Determining Cross-Over Frequency	15	11	10	9	13	12	11	10
12 Analyzing Fruit Fly Phenotypes and Genotypes	16	12	11	10	14	13	12	11
13 Isolating DNA and Transformation with Plasmids	11, 18	14, 17	12, 14	12	16, 20	14, 19	13, 15	13, 16
14 Modeling Processes in Evolution and Mutagenesis	14, 24	20	15, 16	13	23	24, 25	15, 21	18
15 Investigating Bacterial Diversity	27	28	20	18	27	28	26	21
16 Evolution of Eukarya and Diverse Protists	28	29	21	19	28	29	27	22
17 Ancestral and Derived Characteristics of Seedless Plants	29	30	23	20	29	30	28	23
18 Derived Characteristics of Seed Plants	30	31	23	20	30	30	29	23
19 Investigating Fungal Diversity and Symbiotic Relationships	31	32	22	21	31	31	30	24
20 From Basal to Bilateral Animals	32,33	33, 34	28	22	32,33	32	31	25
21 Lophotrochozoa: Increased Complexity	33	35	28	22	33	33	32	25
22 Ecdysozoa: Simple and Complex	33	35	28	22	33	33	32	25
23 Deuterostomes and the Origins of Vertebrates	33,34	36	29	23	33, 34	34	33	25, 26
24 Investigating Plant Cells and Tissues,	35	37,38	24	24	35	36	34	28
25 Investigating Functional Anatomy of Vascular Plants	38	39	25	25	36	37	35	29
26 Investigating Pollination, Development and Dispersal	39	43	27	26	38	40	38	30, 31
27 Investigating Digestive, Renal and Reproductive Systems	45, 49, 51	49, 52, 54	34, 36, 41	34, 35, 37	41, 44, 46	42, 43	43, 51, 52	40, 41, 42
28 Investigating Circulatory Systems	47	50	32	32	42	44	50	37
29 Investigating Gas Exchange Systems	48	15	35	33	42	44	49	39
30 Investigating the Properties of Muscle and Skeletal	44	48	39, 40	31	50	46	48	36
31 Investigating Nervous and Sensory Systems	41, 42, 43	45, 46	37, 38	28, 29	48, 49	45, 46	45, 46	33, 34
32 Investigating Events in Early Animal Development	52	55	42	37	47	48	44	43
33 Estimating Population Size and Growth	56	57	44	39	53	52	55	45

Lab Topic 1

Science: A Way of Knowing

Supplies

Resource guide available at
   www.mhhe.com/labcentral

Materials

125 ml Erlenmeyer flasks

Paper clips

Pirate-style eye patches

Photo copies of newspaper, magazine, and journal articles about biology (AIDS, rainforests, or cloning would be good examples, especially if articles were coordinated so students see same material intended for different audiences.)

Student Prelab Preparation

Before doing this lab, read the Background material and sections of the lab topic that have been assigned by your instructor.
   Find definitions for the following terms:

Dependent variable

Hypothesis

Independent variable

Scientific literature

   You should be able to describe in your own words the following concepts:

Experimental design

Scientific method

   After finishing the prelab review, write any questions you have about terms, concepts, or techniques in the margins of this lab topic. The lab experiments should help you answer these questions, or you can ask your instructor during the lab.

Objectives

1. To understand the central roles of discovery science and hypothesis testing in modern scientific methods
2. To design and conduct experiments using discovery science and hypothesis testing
3. To learn to draw conclusions from data
4. To evaluate writing samples from several sources for content and style

Background

Many dictionaries define science as a body of knowledge dealing with facts or truths concerning nature. The emphasis is on facts, and there is an implication that absolute truth is involved. Ask scientists whether this is a reasonable definition and few will agree. To them, science is a process. It involves gathering and critically examining information to increase humankind's understanding of nature. At the same time, they would add that this understanding is always considered tentative and subject to revision in light of new discoveries.

   Science is based on three fundamental ideas:

The *principle of universal causality:* when experimental conditions are replicated, identical results will be obtained regardless of when or where the work is repeated. This principle allows science to be self-analytical and self-correcting, but it requires a standard of measurement and calibration to make results comparable.

The *principle of uniformity in nature:* the future will resemble the past so what we learned yesterday applies tomorrow.

The *law of parsimony:* the simplest explanation with the fewest modifying statements is best.

   For many, science is just a refined way of using common sense in finding answers to questions. During our everyday lives, we observe things that are new to us and try to determine cause and effect relationships. We presume that relationships we learned from past experience will also apply in the future. We ask ourselves questions about daily experiences and often propose tentative explanations that we seek to confirm through additional observations. We interpret new information in light of old and make decisions about whether our hunches are right or wrong. The process of science is similar.

   The origin of today's scientific methods can be found in the logical methods of Aristotle. He advocated that three principles should be applied to any study of nature:

1. One should carefully collect observations about the natural phenomenon.
2. These observations should be studied to determine the similarities and differences; *i.e.,* a compare and

contrast approach should be used to summarize the observations.
3. A summarizing or explanatory principle should be developed.

In adapting Aristotle's principles, modern scientists follow two approaches that are often interwoven so that it is difficult to separate the importance of one from the other.

Discovery-based science is an approach where observations of nature are recorded as data sets that may be qualitative or quantitative. This approach is science because the workers use the methods of science: curiosity, careful observation, accurate recording, and meticulous descriptions that would allow others to repeat the work to test its accuracy. Sometimes called descriptive science, it seeks to answer the "what," "where," and "when" types of questions, *e.g.*, what is the anatomy of a whale's kidney, what diseases are transmitted by insects, or when did flowering plants first appear in the fossil record? These data sets may later be explored for correlations and tentative explanations, a form of inductive reasoning that proceeds from general knowledge toward specific unifying statements.

Hypothesis-based science is the second approach and it starts with observations as Aristotle suggested but goes on to test proposed answers to the "how" and "why" questions; why does the number of species seem to increase over time, how do genes express themselves, or how will a population of prey change when predators are introduced?

Some are dismissive of the importance of discovery science, but they overlook the impact that it has had, and continues to have, on modern biology. Darwin in his early years did discovery-based science almost exclusively. As the naturalist aboard the *HMS Beagle,* his job was to discover and record plants, animals, and natural environments during a five-year voyage, something he did with great energy and curiosity building a large database of what, where, and when organisms were found. Later, he spent years looking back over his collections and notes, thinking about how and why questions. Those thoughts led to his book, *Origin of the Species,* that profoundly changed how we look at the natural world. Few biologists have had as great an impact as he did.

Discovery science can also be found as the basis for our incredible knowledge of cells and genetics. Today work being done to decipher the genomes of organisms is discovery science that is detailing what genes are found where. The accumulated data are being organized into systems and classifications that are resulting in new hypotheses on evolutionary relationships. Hypothesis-based science starts with what we know whereas discovery-based science starts with what we do not know, describing what no one else may have ever seen. As you go through this lab manual, you will see that the planned activities are often a combination of these two approaches: observing, recording, analyzing, and forming hypotheses that are tested in experiments.

Hypothesis-based science warrants more discussion. It is about explaining nature using a deductive approach; if x happens in an experiment then y must be true. It seeks to unravel cause and effect in nature. After looking at a database of information, a hypothesis-testing scientist will ask a question about how something is similar to or different from something else or how two or more observations relate to each other.

After considering the questions, a scientist will state a research **hypothesis,** a tentative answer to a key question. This process consists of studying events until one feels confident in predicting future events. In forming a hypothesis, the assumptions are stated and a tentative explanation is proposed to link possible cause and effect.

A key aspect of a hypothesis is that it must be *falsifiable; i.e.,* if a critical experiment were performed and yielded certain information, the hypothesis would be declared false and discarded. If a hypothesis cannot be proven false by additional experiments and observations, it is considered to be tentatively true and useful, but not absolute truth. Possibly another experiment could prove it false, even though scientists cannot think of one at the moment. Thus, recognize that science is always tentative in its conclusions and adaptable to accomodate new information should it appear.

Science advances as a result of the rejection of false ideas tested through experiments or observations. Hypotheses that are not falsified over the years and which are useful in predicting natural phenomena are called theories or principles—for example, the principles of Mendelian genetics.

In designing experiments to test a hypothesis, predictions are made. If a hypothesis is true, predictions based on it should happen. In converting a research hypothesis into a prediction, a deductive reasoning approach is employed using if-then statements: if the hypothesis is true, then this will happen when an experimental variable is changed. The experiment is then conducted. If the response does not correspond to the prediction, the hypothesis is falsified and rejected; if the response matches the prediction, the hypothesis is supported.

The design of experiments to test hypotheses requires considerable thought! The variables must be identified, appropriate measures developed, and extraneous influences must be controlled. The **independent variable** is what the experimenter changes; it is the cause. The **dependent variable** is the effect; it should change as a result of varying the independent variable. **Control variables** are also identified and are kept constant throughout the experiment. It is reasoned that if kept constant, they cannot cause changes and confuse the interpretation of the experiment.

Once the variables are defined, decisions must be made regarding how to measure the effect. Measures may be quantitative (numerical) or qualitative (categorical). The metric system has been adopted as the international standard for quantitative science, allowing results to be compared. A decision must be made concerning the scale or level of the treatments. For example, if something is to be warmed, what will be the range of temperatures used? Most (but not all) biological material stops functioning (dies) at temperatures above 40°C and it would not be productive to test at temperatures every 10°C throughout the range 0° to 100°C. Another aspect of experimental design is the idea of replication: how many times should the experiment be repeated in order to have confidence in the results and to develop an appreciation of the variability in the response.

Once collected, experimental data are reviewed and summarized to answer the question: do the data falsify or support the hypothesis? The research conclusions then state the decision regarding the hypothesis and discuss the implications of the decision.

If the hypothesis is in a popular area of research, others may independently devise experiments to test the same hypothesis. A hypothesis that cannot be falsified, despite repeated attempts, will gradually be accepted by others as a description that is probably true and worthy of being considered as suitable background material when making new hypotheses. If, on the other hand, the data do not conform to the prediction based on the hypothesis, it is rejected.

Modern science is a collaborative activity. When a scientist reviews the work of others in journals or when scientists work in lab teams, they help one another with interpretation of data and in the design of experiments. When discovery-based data are obtained or a series of experiments completed and the results are judged to be significant, the scientist then shares this information with others. This is done by making a presentation at a scientific meeting or writing an article for a journal. In both forms of communication, the author shares observations, the data from the experiments that tested the hypothesis, and the conclusions based on the data. Thus, the information becomes public and is carefully scrutinized by peers who may find a flaw in the logic or who may accept it as a valuable contribution to the field. Thus, discussion fostered by presentation and publication creates an evaluation function that makes science self-correcting. Only robust hypotheses survive this careful scrutiny and contribute to the development of scientific theories.

LAB INSTRUCTIONS

You will simulate a discovery-based investigation by classifying biological specimens. After that you will simulate a hypothesis-based research project where you state a hypothesis, design an experiment to test it, conduct the experiment, summarize the data, and come to a conclusion about the acceptability of the hypothesis. You will also practice evaluating scientific information from various published sources.

Exploring Discovery-Based Science

A common misconception about science is that, to be "science," hypothesis testing experiments must be involved. Answering questions in science often involves doing experiments, but many scientific questions are addressed by doing "discovery science" (also called "descriptive science"). This approach involves careful observation of the natural world and the recording of those observations to produce scientific data that can then be analyzed in various ways. Areas of biological science that make extensive use of discovery science include studies of biochemical pathways, some aspects of genetics, anatomy, animal behavior, and biological diversity. The data from these studies add much to our understanding of the world. Some estimate that only 5% of the species living on the deep ocean floor have ever been seen and other estimates indicate we know the functions of only a small fraction of our genes. Much work is yet to be done before we develop a full understanding of what can be found in our biological world.

The central characteristics of a good discovery scientist are curiosity, drive, keen observational skills, and an ability to accurately record what is discovered. Observational and recording capabilities can be enhanced by using instrumentation: hydrophones to record whale sounds, chemical tests to determine what enzymes may be present in a bacterium, gene-sequencing machines to analyze genomes, or computers linked to remote sensors in lakes and streams. As you work through this lab manual, you will many times do discovery-based science as you find things that are new to you and build your knowledge base. You will use microscopes, analytical instruments, or computer interfaced data gatherers to extend your senses for both discovery science and experimental science.

1 ▶ *Data gathering* To practice discovery science, get a tray of biological specimens that may include leaves, fungi, insects and shells, bones, and other biological materials.

1. Observe each specimen carefully. Use your dissecting microscope when appropriate, and determine **how many** distinct types of organisms are represented. We will call these species.

2. Create a **two-word descriptive name** for each species.

3. Write a **brief description** for each of your "species" focusing on the unique features that characterize that species.

4. Once the descriptions are complete, analyze your descriptions looking for commonalities as well as differences and begin to group your observations. Give each grouping a name, list the "species" in the group and describe the shared characteristics.

2 ▸ *Analysis and Discussion* A reporting session should be organized to compare and contrast the collective results. After all groups have reported their data-gathering results and grouping analyses, the entire lab should discuss answers to the following questions:

1. In trays analyzed by different groups, some trays had the same types of specimens. Did you give the same "species" name to the specimens as did other groups?
2. Why is it important that biologists agree on one name for a particular species?
3. Why is it important to use standardized terminology to describe organisms?
4. Do data organizational systems make it easier to see commonalities and accentuate differences?

Exploring Hypothesis-Based Science

As biological scientists practice discovery science to describe the natural world, they begin to develop questions about how natural processes occur and why organisms have certain structures and adaptations. Tentative answers are then proposed to these questions, *i.e.*, hypotheses are formed. These hypotheses can then be tested by designing and performing experiments. For example, vertebrates have two eyes. Why is this the case? Is there an advantage when two eyes are present?

3 ▸ Work in a group of four and make a list of the potential advantages of having two eyes rather than one eye.

Forming Hypotheses Each advantage you have written down represents a potentially testable hypothesis about why vertebrates have two eyes. Some of these hypotheses could be easily tested with an experiment, while others would be more difficult to test. For example, one advantage you might have written down is "improved depth perception." Written as a hypothesis this could be phrased: "Binocular vision provides better depth perception compared to monocular vision." Discuss in your group, and **4** ▸ write down, two other ways in which this hypothesis could be rephrased.

Testing a Hypothesis The hypothesis that binocular vision improves depth perception can be tested in a simple experiment. Work in groups of four. Two individuals will be scientists and the other two will be research subjects. Place a 125 ml Erlenmeyer flask about 75 cm from the edge of the table. Place a pile of 2-inch (5 cm) paper clips at the edge of the table. The independent variable in this experiment will be the number of eyes available to do the task of placing paper clips into the flask. The dependent variable will be the number of paper clips successfully placed into the flask.

5 ▸ Select one research subject to serve as the control, *i.e.*, the subject who will attempt the task using two eyes. The experimental subject will place a "pirate-style" eye patch over one eye corresponding to the hand used to place paper clips into the flask. Thus, the experimental treatments, the independent variable will be monocular versus binocular vision. The scientists will allow each subject to place as many paper clips as possible, one at a time, into the flask in a 30-second time period. One scientist should serve as a timer, while the other oversees the experimental subjects. The subject must remain seated while placing paper clips into the flask. The number of paper clips will be recorded and the experiment will be repeated twice more. Record your data in **table 1.1**.

6 ▸ *Analysis and Discussion* Do the results of your group's experiment support or falsify the hypothesis?

Table 1.1	Number of Paper Clips Added to Flask in 30 Seconds				
Treatment	Trial 1	Trial 2	Trial 3	Average	Class Average
Monocular					
Binocular					

If your results support your hypothesis, what could you do to obtain further support for the hypothesis?

Do you think that biological scientists have previously investigated the importance of binocular vision? How would you find out?

If your results falsify your hypothesis, what could you do increase your confidence that this hypothesis is indeed false?

After you've completed collecting data, your instructor will ask you to contribute your group's data to the calculation of a class average. Enter the class averages into table 1.1. Do the class averages support or falsify the hypothesis that binocular vision improves depth perception?

Discuss with the members of your group further experiments that could be designed to test this hypothesis. Describe in writing one such experiment that you think would be possible to do.

Evaluating Published Information

(Adapted from notes prepared by Chuck Kugler at Radford University and Chris Minor when at Iowa State University.)

Daily, we are exposed to scientific information in newspapers, magazines, over the World Wide Web, and through books. How do you evaluate such information? Is a newspaper best because it is available daily or is the WWW better because no editors have changed words to fit a story in the column space? In classes, future jobs, and everyday life, you will be asked to evaluate what you read.

In this section you will learn how to evaluate a written report. First, read the following material, describing key factors to look for in a published report.

Evaluate Format

First, be suspicious of any supposedly scientific report that is not written in a style that parallels the scientific method. Discovery goals or hypotheses should be clearly identified, data presented, and the reasoning leading to the conclusions explained. The formal elements of a scientific paper are discussed in appendix D. If a report lacks these elements, be skeptical. On the other hand, reading about a discovery in the newspaper can alert you to locate the actual report in a journal.

Evaluate the Source

Several thousand journals publish information of interest to biologists. The journals range from magazines such as *National Geographic* and *Scientific American* to scholarly journals, published by professional associations, such as the *American Journal of Botany, Journal of Cell Biology,*

Genetics, Ecology, Science, etc. Magazine articles are usually written by journalists and not by scientists who did the research. They can be quite helpful in developing a general appreciation for a topic, but they are not ultimate sources of scientific information.

What makes scholarly journals so reliable is the use of a peer review system. Articles are written by scientists who did the work and sent to the journal editor, who is usually a scientist. When he/she receives the article, it is sent to other scientists who are working on similar problems and they are asked to make comments about the work. These reviews can be harsh and may criticize experimental design, methods, data quality, and conclusions as well as writing style. If a paper is considered good science despite some flaws, the reviewers' comments are returned to the author who then revises the paper before it is published. It is this peer review system that maintains the quality of the information in journals. Popular magazines such as *Time* or television shows (even those on the *Discovery* channel) have been created for entertainment purposes, lack such peer review, and are far less reliable sources of evidence.

Evaluate Style

Good scientific writing is factual and concise. It is not overly argumentative, nor should it be an appeal to the emotions. As you read any scientific report, watch for

1. Repetition: some authors believe that the more they say something, the more likely you are to believe it;
2. Exaggeration: often identifiable by the use of the words "all" or "never";
3. Emotional: the author attempts to get you to agree based on "feelings," not reason;
4. Dichotomous simplification: expressing a complex situation as if there were only two alternatives.

Evaluate the Arguments

Analyze how the author seeks to convince you that what is reported is true, significant, and applicable to science. Be on guard for the following types of rhetorical arguments:

1. Appeals to authority: citing a well-known person or organization to make a point. Recognize that authorities can be biased, be experts in fields other than the one under consideration, and be wrong.
2. Appeals to the democratic process: using the phrase "most people." Remember, only 200 years ago, most people erroneously believed in the spontaneous generation of life.
3. Use of personal incredulity: implying that you could not possibly believe something, *e.g.,* "how could something as complex as the human just evolve?"
4. Use of irrelevant arguments: statements that might be true but which are not relevant, *e.g.,* "suggesting that complex animals could have resulted by chance is like saying that a clock could result from putting gears in a box and shaking it."
5. Using straw arguments: presenting information incorrectly and then criticizing the information because it is wrong, *e.g.,* "the evolution of a wing requires 20 simultaneous mutations—an impossibility." There is no basis for saying that the evolution of a wing requires 20 mutations; it could be fewer but most likely many more, and they need not be simultaneous.
6. Arguing by analogy: using an analogy to suggest that an idea is correct or incorrect, *e.g.,* "intricate watches are made by careful designers, so complex organisms must have had a designer."

Evaluate the Evidence

Before getting too involved in interpreting trends in the data, spend a few moments thinking about the type of evidence presented. Be especially skeptical of reports that have the following flaws in their evidence:

1. Distinguish between evidence (data) and speculation (an educated guess).
2. Use of stories usually involving single events that are not the results of designed experiments, *e.g.,* "bee stings are lethal; my uncle died when he was stung."
3. Correlation used to imply cause and effect: remember that correlation is only a probability of two events occurring together. While it is interesting to speculate that one might cause the other, this is not necessarily so; *e.g.,* at the instant before a major earthquake has struck a major city, there is a high probability that someone was slamming a car door. Did the slam cause the earthquake?
4. Sample selection: in statistical studies, a large number of situations should be examined and the procedures should be free of bias. You do not choose to report only experiments that support your beliefs.
5. Misrepresentation of source: source material can be quoted out of context or badly paraphrased; *e.g.,* "Moderate drinking of red wine is correlated with lower risk of heart attack" could be misrepresented as "Drinking alcohol is good for you."

Check the Data

When data are presented, do routine checks. If percentages are involved, do you know the sample size? It is an impressive statement to say that 75% of the people surveyed preferred brand X, but it is less impressive to find out that this calculation is based on a sample size of four rather than 400 or more. When percentages are reported, be sure to check that they add up to 100. If on the eve of an election 42% of the voters were for Obama and 41% were for McCain, it would seem that Obama would win, except that 17% of the voters were unaccounted for and could have swung the election one way or the other.

Continue the habit of doing simple arithmetic checks when examining data in tables. If totals are given for columns of numbers, do some quick math to see if things check out. If not, you might not want to base major decisions on the report because you do not know what other kinds of errors went undetected!

With the advent of computer graphics, it is now rather easy to use computer programs to produce appealing graphics. However, one should not accept data based on its beauty of presentation. To illustrate this point, see **figure 1.1** for an interesting graphic that appeared in a newspaper.

Evaluate the Conclusions

Conclusions are not a summary of the data. Conclusions deal with the decision that is to be made about the hypothesis that was being tested. You should ask, "Are the data thoroughly reviewed to test the hypothesis?" Ask yourself whether there is another explanation other than what the author is telling you. Once a decision was made by the author to accept or reject the hypothesis, were the implications of the decision discussed? In some cases, the implications may be extrapolated to new situations, but over extrapolation can result in problems. For example, raising a frog's body temperature from 10° to 20°C may increase the frog's metabolic rate twofold, but this does not mean by extrapolation that raising it to 100°C will increase metabolic rate tenfold. In fact, the frog will die when its body temperature approaches 40°C.

As you look back through the newspaper, magazine, and journal articles, determine which one of these forms of publication used more of the nonscientific forms of writing.

Evaluating WWW Pages

Anyone can put information on the WWW regardless of whether the statements are true or false. There is no review process involving editors or peers. One only needs to read a few blogs to recognize this.

How do you evaluate what you have found on the Web? All of the points raised about how to critically examine published information also apply to information accessed in WWW pages. In addition, librarians at Cornell University and Iowa State University suggest that you always ask the following additional questions.

- **Authorship:** First, determine who is the author. If no author is indicated, ask why. Somebody wrote the article. Why is their name not on it? If there is a name, ask if this person is qualified to write about the subject, or is the person simply offering an opinion? In short, do you trust the author's judgment?
- **Accuracy:** Based on what you know about the topic, does this site seem to be correct or are there discrepancies? Does the author provide references or ways of verifying the information that is presented? Is contact information provided so that you can contact the author and ask questions? Check the URL domain and determine if it is a recognized authoritative site for your topic?
- **Objectivity:** Think about why the information was published and who the intended audience is. Is the information intended to inform, persuade, or sell? What are the objectives of the sponsoring WWW site? Is this just advertising or someone offering unsupported opinions? Are multiple views on an issue offered or is it one-sided?
- **Currency:** When was the page published and has it been updated? Are the links current or are they dead ends?
- **Content:** Is the site just a school page with lecture notes or a blog? Does it offer data or just opinion? How would you compare the information with that in a journal or a book? More detailed or less so?

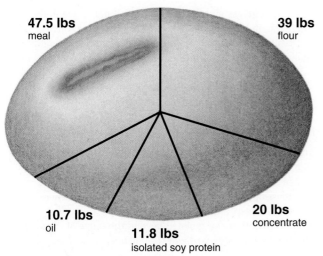

What does this chart tell you?

Do the numbers add up?

Would you use the information in this chart to make a decision?

Do you trust these data?

Figure 1.1 This pie chart was taken from a midwest newspaper. It is supposed to depict the composition of a bushel of soybeans.

Few pages will meet every one of these criteria and, ultimately, you will have to make some decisions on whether to include the pages as references for a report.

Analyzing a Published Report

8 Your instructor will have photocopies of reports from several sources (newspapers, magazines, journals, or the WWW) available for you to analyze, using the criteria just described. Read the materials quickly to get an idea of what is being described.

Now you should systematically go back through the report and analyze each section, following the directions in the Publication Evaluation Criteria that follows.

After finishing your evaluation, come to a conclusion about the report you have read. Was this a good science report containing information that you would trust?

PUBLICATION EVALUATION CRITERIA

Evaluate the Source

Are articles peer reviewed in this source?

Where does (do) the author(s) work?

Could the author(s) have vested interests?

Is there a Hypothesis?

Underline and label the hypothesis tested in the work reported. If none, indicate if it is discovery science or simply opinion.

Examine the Writing Style

Use a light-colored marker to highlight on the photocopy any passages that seem to deviate from a factual and concise style. According to the following key, write a Roman numeral next to the highlighted area indicating the type of writing style used.

  I. Use of repetition
  II. Dichotomous simplification
  III. Exaggeration
  IV. Use of emotionally charged words or statements

Examine the Arguments

Use a light-colored marker to highlight on the photocopy any arguments. According to the following key, write a number next to the highlighted area indicating the type of argument used.

  1. Appeals to authority
  2. Appeals to the democratic process
  3. Uses personal incredulity
  4. Uses irrelevant arguments
  5. Uses straw arguments
  6. Argues by analogy

Analyze the Evidence

Underline the sections of the photocopied article that present the arguments of the author. According to the following key, write a letter next to the arguments.

  A. Speculation
  B. Evidence collected using scientific methods
  C. Anecdotal evidence
  D. Correlation, not cause and effect
  E. Description of sample size and selection method
  F. Possible place to check for misrepresentation of source

Check the Data

Do all percentages given add up to 100? If not, circle in the text where the omission is located and describe the problem.

Do all numbers in columns or charts add up to the indicated totals or are there math mistakes? Circle the mistakes and indicate correction.

Are flashy graphics used to catch your attention? Write comments next to the questionable graphics.

Examine the Conclusions

Are the conclusions easy to find and clearly stated? Highlight them.

Are the conclusions based on a review of data and the hypotheses presented in the introduction? Indicate where reviews and tests are written.

Has the author extrapolated beyond the range of data collected? Indicate by writing extrapolation in the margin.

Lab Topic 2

Using Microscopes

Supplies

Resource guide available at
www.mhhe.com/labcentral

Equipment

Compound microscope (oil immersion lens optional)
Dissecting microscope

Materials

Dropper bottles with water
Dissecting needles and scissors
Ocular micrometer, if doing measurements
Stage micrometer, if doing measurements
Small colored letters from printed page
Slides and coverslips
Prepared slide
    Mammalian spinal cord smear
Various small specimens for dissecting scope (flowers, insects, lichens, *etc.*)

Student Prelab Preparation

Before doing this lab, read the Background material and sections of the lab topic assigned by your instructor.
    Find definitions for the following terms:
Brightness
Calibration
Contrast
Magnification
Resolution

    Describe in your own words the following concepts:
Light path through parts of a microscope
How to make wet-mount slide
How to calibrate an ocular micrometer

    After finishing the prelab review, write any questions you have about terms, concepts, or techniques in the margins of this lab topic. The lab experiments should help you answer these questions, or you can ask your instructor during the lab.

Objectives

1. To learn the parts of dissecting and compound microscopes and their functions
2. To investigate the optical properties of microscopes, including measuring objects
3. To understand the importance of magnification, resolution, and contrast in microscopy

Background

Because an unaided eye cannot detect anything smaller than 0.1 mm (10^{-4} meters) in diameter, cells, tissues, and many small organisms cannot be observed directly. A light microscope extends our vision a thousand times, so that objects as small as 0.2 micrometers (2×10^{-7} meters) in diameter can be seen.

More than 300 years have passed since the invention of the light microscope. The optical principles of microscopy have not changed, but manufacturing improvements now allow mass production of quality microscopes for classroom use. Your microscope is far superior to the one used by Robert Hooke, the first person to use the word "cell" in describing biological materials.

Microscopes allow us to see small objects because they resolve as well as magnify. **Magnification** simply makes things appear larger. **Resolution** is the capacity of an optical instrument to distinguish two points that are close together. High resolution allows us to see separate points (detail) while low resolution blurs close-together points into one image rather than showing them as separate ones.

The distinction between microscopic resolution and magnification can best be illustrated by an analogy. If a photograph of a newspaper is taken from across a room, the newspaper image would be small, too small for you to read the words. If the photograph were enlarged, or magnified, the image would be larger, but the print would still be unreadable. Regardless of the magnification used, the photograph would never show distinct points on the printed page. Therefore, without **resolving power,** or the ability to distinguish detail, magnification is worthless.

Modern microscopes increase both magnification and resolution. Today's light microscopes are limited to practical magnifications in the range of 1000× to 2000× and to resolving powers of 0.2 micrometers. Most student microscopes have magnification powers to 450× and resolving properties of about 0.5 micrometers. These limits are imposed by the cost of higher power lenses and the accurate alignment of lenses and light sources.

The theoretical limit for the resolving power of a microscope depends on the **wavelength** of light (the color) divided by a value called the numerical aperture of the lens system, times a constant (0.61). The numerical aperture is derived from a mathematical expression that relates the light delivered to the specimen by the condenser to the light entering the objective lens. Therefore,

$$\text{Minimum distance that can be resolved} = \frac{\text{wavelength}}{\text{numeric aperture}} \times 0.61$$

If all other factors are equal, resolving power is increased by reducing the wavelength of light used. Microscopes often have blue filters because blue light has the shortest wavelength in the visible spectrum. For example, if blue light with a wavelength of 480 nanometers is used and the numerical aperture is 2, the theoretical resolving power is 146 nanometers, or 0.146 micrometers.

Even with sufficient magnification and resolution, a specimen can be seen on a microscope slide only if there is sufficient **contrast,** the gradations from dark to light tone seen when viewing an image. Contrast is based on the differential absorption of light by parts of the specimen. Often specimens have opaque parts or contain natural pigments, such as chlorophyll, that provide contrast. But, how is it possible to view the majority of biological materials that are translucent?

Microscopists improve contrast by using stains that bind to cellular structures and absorb light to provide contrast. Some stains are specific for certain chemicals and bind only to structures composed of those chemicals. Others are nonspecific and stain all structures.

To summarize, good microscopy involves three factors: resolution, magnification, and contrast. A beginning biologist must learn to use a microscope with these factors in mind to gain access to the world that exists beyond that seen with the unaided eye.

LAB INSTRUCTIONS

In this lab topic you will be introduced to dissecting and compound microscopes, tools that you will use in most of the other lab topics in the lab manual.

The Dissecting Microscope

We start our study of microscopes by first examining a simple dissecting microscope and then move to a more complex compound microscope. Dissecting microscopes are really two microscopes that are aligned side by side in a single housing. Sometimes called stereoscopic microscopes, these microscopes allow specimens to be observed in three dimensions, a handy property when doing microscopic dissection.

1) Get a dissecting microscope from the storage cabinet. Carry it firmly with both hands, using one hand to support the base and the other to hold the column.

Parts of a Dissecting Microscope

Use **figure 2.1** to help you identify the parts of your microscope. It may not have all the features shown, or might have additional ones. However, the operational principles will be the same.

> **Stage:** This is where a specimen is placed for viewing. It may have a white, black, or clear glass plate in the center. Beneath the plate there will be a mirror or a light bulb.
>
> **Light source:** Often dissecting microscopes have two light sources: one that projects light from above (the reflected light source) and a second that projects light from below the stage (the transmitted light source). Light from above allows you to view the external features of opaque structures while light from below passes through translucent specimens revealing internal details. If the light sources are built into your microscope, there may be two light controls: one to turn the lights on and off and a rheostat to adjust intensity.
>
> **Body:** This is the heavy metal column and parts encasing the lens system of the microscope.
>
> **Ocular lens:** The ocular lenses are the lenses you look through. They are adjustable for individual differences. Gently pull the eyepieces to opposite

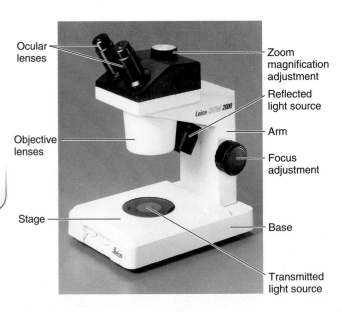

Figure 2.1 Stereoscopic dissecting microscope.

sides and the distance between them, called the **interpupillary distance,** increases. If you squeeze them together, the distance decreases. Doing this while looking through the microscope, allows you to adjust for the distances between your eyes.

Note that the ocular lens may have an adjustment ring at its base. Later you will use it to adjust for the focusing differences between your eyes.

- **Objective lenses:** These lenses are closest to the specimen, located in recesses at the bottom of the body.
- **Magnification control:** Depending on the manufacturer, magnification controls vary. Some have a knob that you twist to increase or decrease magnification. It may be on the top of the body, on the side, or as a ring at the bottom of the body. Stereomicroscopes usually have magnification capabilities in the 4X to 50X range.
- **Focusing knob:** Turning this knob raises or lowers the body (and lenses) to focus on a specimen. Turn the knob as you watch from the side to see the action.

Using a Dissecting Microscope

2▸ Set up your microscope to view a specimen with reflected light, light coming from above the stage. Turn the magnification to the lowest setting, and view any of the biological specimens available in the lab. Note that both magnification and resolution are increased, allowing you to see detail not visible to the naked eye.

Slowly move your specimen to the right and draw it toward you. Did the image move in the same directions as your specimen? _____

Customizing the Microscope for Your Eyes

Adjust the oculars for your interpupillary distance by squeezing them together or pulling them apart so that you have one field of view with both eyes open. Increase the magnification while looking at a single spot on your biological specimen. Stop at an intermediate magnification.

You will now adjust the oculars for the focusing differences between your eyes. Determine which ocular has the adjusting focus ring. Close the eye that is over the adjustable ocular but keep the other open. Use the focus adjustment knob to get a clear image of your specimen through the open eye. Now close that eye and open the other. Is the image still in focus? If not, turn the adjustment ring on the ocular to bring it into focus. The microscope in now customized for your eyes and you should not experience eye strain when using it for long periods of time. Because microscopes are shared with students in other lab periods, you should make these personal adjustments every time you get your microscope from the storage area.

Go to the supply area and get a printed page with colored letters. Cut one out and put it on the stage. Look at it first with reflected light and then with transmitted light (coming through the stage from beneath the specimen). Which gives you the best image?

Increase the magnification to its highest. Describe the colors of the ink dots you see.

This is a good example of how microscopes increase resolution. Individual dots cannot be seen with the naked eye and colors blend to yield the color of the letter. When the dots are resolved, not only do they look larger, they are distinct.

Return your dissecting scope to storage cabinet and get a compound microscope. Remember to use both hands when carrying it with one hand underneath to support the base.

The Compound Microscope

Compound microscopes are more complex than dissecting microscopes and are capable of much higher resolution and magnification, allowing biologists to study subcellular structures. Depending on its age, manufacturer, and cost, your compound microscope may not have all of the features discussed in this section.

3▸ Look over your microscope and find the parts described, referring to **figure 2.2**.

Parts of a Compound Microscope

Ocular Lens: If your microscope has one ocular, it is a **monocular** microscope. If it has two, it is **binocular.** In binocular microscopes, one ocular is adjustable to compensate for the differences between your eyes. Ocular lenses can be made with different magnifications. What magnification is stamped on your ocular lens housing?

Ocular magnification = _____

The ocular lens is actually a system of several lenses that may include a pointer and a measuring scale called an ocular micrometer. Never attempt to take an ocular lens apart.

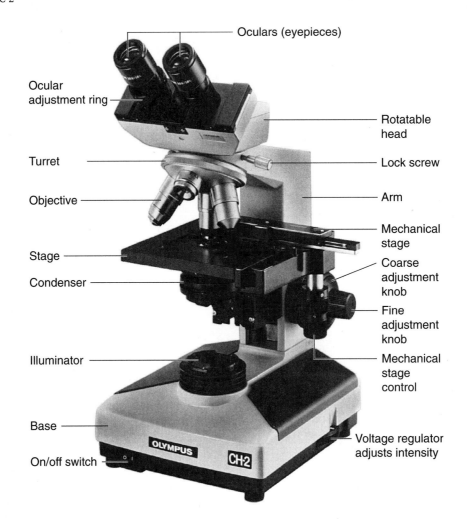

Figure 2.2 A binocular compound microscope.
Source: Courtesy of the Olympus Corporation, Lake Success, N.Y.

As on dissecting microscopes, the interpupillary distance can be adjusted. Also note that one of the oculars has a focus adjustment ring at its base, as did the dissecting microscope.

Body Tube: The body tube is the hollow housing through which light travels to the ocular. If the microscope has inclined oculars, as in figure 2.2, the body tube contains a prism to bend the light rays around the corner.

Objective Lenses: The objective lenses are a set of three to four lenses mounted on a **turret** at the bottom of the body tube. Rotate the turret and note the click as each objective comes into position. The objective gathers light from the specimen and projects it into the body tube. Magnification is stamped on each lens. Record the numbers.

Scanning (small) Lens _____

Low-power (medium) Lens _____

High-power (large) Lens _____

Oil Immersion (largest) Lens _____

Stage: The stage is the horizontal surface on which a slide is placed. It may have simple clips for holding the slide or a **mechanical stage,** a geared device for precisely moving the slide. Two knobs, either on top of or under the stage, move the mechanical stage. If your microscope has these knobs, twist them to see the movement.

Substage Condenser Lens: The substage condenser lens system, located immediately under the stage, focuses light on the specimen.

Diaphragm Control: The diaphragm is an adjustable light barrier built into the condenser. It may be either an **annular** or an **iris** type. With an annular control, a plate under the stage is rotated, placing open circles of different diameters in the light path to regulate the amount of light reaching the specimen. With the iris control, a lever projecting from one side of the condenser opens and closes the diaphragm. Which type of diaphragm does your microscope have?

Use the smallest opening that does not interfere with the field of view. The condenser and diaphragm assembly may be adjusted vertically with a knob projecting to one side. Proper adjustment often yields a greatly improved view of the specimen. Start with the condenser at its highest point.

Light Source: The light source has an off/on switch and may have adjustable lamp intensities and color filters. To prolong lamp life, use medium to low voltages whenever possible. A second diaphragm may be found in the light source. If present, experiment with it to adjust the light.

Base and Body Arm: The base and body arm are the heavy cast metal parts.

Coarse Focus Adjustment: Depending on the type of microscope, the coarse adjustment knob either raises and lowers the body tube or the stage to focus the optics on the specimen. Use the coarse adjustment only with the scanning (4×) and low-power (10×) objectives. Never use it with the high-power (40×) objective. (The reasons for this will be explained later.)

Fine Focus Adjustment: The fine adjustment changes the specimen-to-objective distance very slightly with each turn of the knob and is used for all focusing of the 40× objective. It has no noticeable effect on the focus of the scanning objective (4×), and little effect with the 10× objective.

Avoiding Hazards in Microscopy

Use care in handling your compound microscope. The following list describes common problems and how they can be avoided.

1. Microscope dropped.
   a. Carry microscope in an upright position using both hands. Hold it close to your body with one hand under the base.
   b. When placing a microscope on a table or in a cabinet, set it down gently.
2. The condensor falls out.
   Check that the set screw on the condensor body is tight.
3. Image blurred.
   a. Dirty lenses:
      Clean microscope lenses before and after use. Oils from eyelashes adhere to oculars, and wet-mount slides often encrust the objectives with salts. Use only *lens tissue* folded over at least twice to prevent skin oils from getting on the lens. Use distilled water to remove stubborn dirt. Do NOT use paper towels, facial tissue or handkerchiefs to clean objectives or oculars. They will scratch the lens.

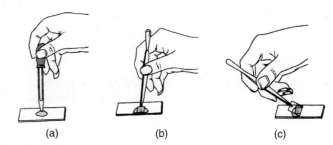

Figure 2.3 Procedure for making a wet-mount slide. (a) Place a drop of water on a clean slide. (b) Place specimen in water. (c) Place edge of coverslip against the water drop and lower coverslip onto slide.

   b. Scratched lenses:
      - Slide was removed when high-powered objective was in place, scratching lens. Remove slide only when low-power objective is in place.
      - High-power objective was pushed through the coverslip (see 4, below) and lens is dirty or scratched.
   c. The slide you are viewing is dirty. Clean slide with lens tissue and wipe dry.
4. Objective lens smashes into coverslip and slide:
   a. Always examine a slide first with the low- or medium-power objective.
   b. Never use the high-power objective to view thick specimens.
   c. Never focus downward with the coarse adjustment when using high-power objective.
5. Mechanical failure of focus mechanism:
   a. Never force an adjustment knob; this may strip gears.
   b. Never try to take a microscope apart; you need a repair manual and proper tools.

Making a Slide and Viewing It

4 ▶ **Figure 2.3** shows how to prepare a wet-mount microscope slide. Take a magazine or printed page and cut out a colored lowercase letter *e* or *a* or the number *3*, *4*, or *5*. Clean a microscope slide with a lens tissue, add a drop of water to the center, and place the letter in the drop. Add a coverslip and place the slide with the letter in its normal orientation on the microscope stage with the scanning objective in place. Always use the following steps when viewing a slide.

STEPS USED IN VIEWING A SLIDE

1. Check to be sure that the ocular lens, all objective lenses, and slide are clean.
2. Turn the light on, adjust its intensity and open the diaphragm. Center specimen over stage opening.

3. Start with the scanning objective as close to the slide as possible. While looking through the oculars, back off with the coarse adjustment knob until the specimen is in focus.
4. Readjust the light intensity so that you do not squint as you view the specimen. Center the specimen in the field of view by moving the slide. Close the iris diaphragm and, if possible, adjust the substage condenser height until the edges of the diaphragm are in focus.
5. If you have a binocular microscope, adjust the ocular lenses for the differences between your eyes. Adjust the interpupillary distance. Determine which ocular is adjustable. Close the eye over that lens and bring the specimen into sharp focus for the open eye. Open the other eye, and close the first. If the specimen is not in sharp focus, turn the adjustable ocular until the specimen is in focus. You should do this each time you get a microscope from the cabinet because other students may have adjusted it for their eyes.
6. Switch from the scanning to low-power (10×) objective. The turret should click when the objective is in place. Sharpen the focus, center the specimen in the field of view, and readjust the light intensity and diaphragm opening to get a comfortable level of illumination that does not hurt your eyes.
7. Switch to the high-power (40×) objective. When it is in place, adjust the focus with the fine focus adjustment only. If you use the coarse adjustment, you may hit the slide and damage the objective. Be sure to adjust the light intensity so that you have a comfortable level of illumination.
8. Students wonder if they should wear their glasses when using a microscope. If you are nearsighted or farsighted, you need not. The focus adjustments will compensate. If you have an astigmatism, wear your glasses.
9. If your microscope is monocular, you will tend to close one eye. Eyestrain will develop if this is continued. Learn to keep both eyes open and ignore what you see with the other eye.

The first seven steps listed are the usual procedures for using the microscope. Always start with a clean scanning objective and proceed in sequence to high power, making minor adjustments to the focus and light source. Your skill in tuning your microscope will determine what you will see on microscope slides throughout this course.

The Compound Microscope Image

The following activities are designed to familiarize you with the properties of your microscope. Use the wet-mount slide you just made.

5. *Image Orientation* Using the scanning objective, look at the slide through the microscope and then with the naked eye.

Is there a difference in the orientation of the images?

While looking through the microscope, try to move the slide so that the image moves to the left. Which way did you have to move the slide?

Try to move the image down. Which way did you have to move the slide?

How is the movement of the specimen's image, different from that observed when you did a similar experiment with your dissecting microscope?

When showing someone an interesting specimen, you can describe the location of the specimen by referring to the field of view as a clock face. (Thus, a structure of interest might be described as being at one o'clock or seven o'clock.) Some microscopes have pointers built into the ocular. In such cases, the structure can simply be moved to the end of the pointer.

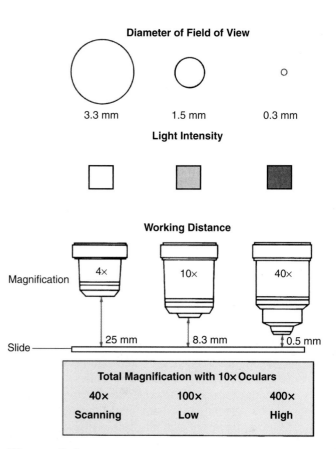

Figure 2.4 Comparison of the relative diameters of fields of view, light intensities, and working distances for three different objective magnifications.

6 ▶ *Magnification* Compound microscopes consist of two lens systems: the objective lens, which magnifies and projects a "virtual image" into the body tube, and the ocular lens, which magnifies that image further and projects the enlarged image into the eye.

The ocular lens only increases the magnification of the image and does not enhance the resolution. The objective lens magnifies and resolves. The total magnification of a microscope is the product of the magnification of the objective and the ocular lenses. If the objective lens has a magnification of 5× and the ocular 12×, then what you see is 60 times larger than the specimen.

What magnifications are possible with your microscope?

Scanning power = _____
Low power = _____
High power = _____
Oil immersion = _____ (if present)

7 ▶ *Field of View and Brightness* Look at your microscope slide first with the scanning lens, and then with the low- and high-power lens. Note that as magnification increases, the field of view decreases and brightness is reduced. Note also that the **working distance,** the distance between the slide and the objective, decreases. (This is the reason you never focus on thick specimens with a high-power objective.) These relationships are summarized in **figure 2.4**.

8 ▶ *Focal Plane and Optical Sectioning* The concept of the **focal plane** is important in microscopy. Like the eye, a microscope lens has a limited depth of focus; therefore, only part of a thick specimen is in focus at any one setting. The higher the magnification, the thinner the focal plane. In practical terms, this means that you should make constant use of the fine adjustment knob when viewing a slide with the high-power objective. If you turn the knob a quarter turn back and forth as you view a specimen, you will get an idea of the specimen's three-dimensional form. It would be possible, for example, to reconstruct the three-dimensional structure in **figure 2.5** by looking through a microscope focused at the three planes passing through the object on a slide.

9 ▶ *Image Contrast* The contrast of the image can be changed by closing the diaphragm, although this usually results in poorer resolution. Light rays are deflected from the edges of the diaphragm and enter the slide at oblique angles. Scattered light makes materials appear darker because some rays of light take longer to reach the eye than others. This can be an advantage when looking at unstained specimens. Thus, the benefits of greater contrast sometimes outweigh the loss of resolving power. Contrast is also improved by reducing light intensity or brightness. If you have to squint, reduce the light intensity. While looking through the ocular open and close the diaphragm as well as vary light intensity with the rheostat to see the effects.

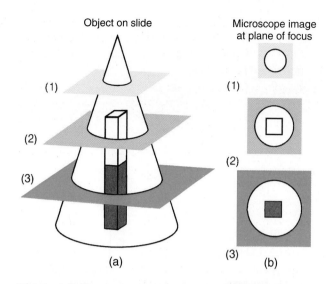

Figure 2.5 Because the depth of field for a lens is limited to a shallow plane, (a) sequentially focusing at depths (1), (2), and (3) yields (b) three different images that can be used to reconstruct the original three-dimensional structure in your mind.

Measuring with a Microscope (Optional)

Measuring microscopic structures requires a standardized **ocular micrometer.** It is a small glass disc on which are etched uniformly spaced lines in arbitrary units. The disc is inserted into an ocular, and the etched scale is superimposed on the image. If your microscope, does not have an ocular micrometer, skip to the next section.

10 ▶ The spacing between the lines on the disc must be calibrated with a very precise standard ruler called a **stage micrometer.** To calibrate an ocular micrometer, obtain a stage micrometer from the supply area. Look at it with the scanning objective.

What are the units?

What is the smallest space equal to in these units?

Follow the steps given in **figure 2.6** to calibrate your ocular micrometer.

How many units (in millimeters) on the stage micrometer equal 100 spaces on the ocular micrometer? Record in **table 2.1**. If you divide the number of stage units by 100, you will get the calibration for one ocular unit when using the scanning objective. To convert to micrometers from millimeters, multiply by 1,000. Repeat for each objective. *Be careful not to push the high-power objective through the stage micrometer. (They are expensive!)*

LAB TOPIC 2

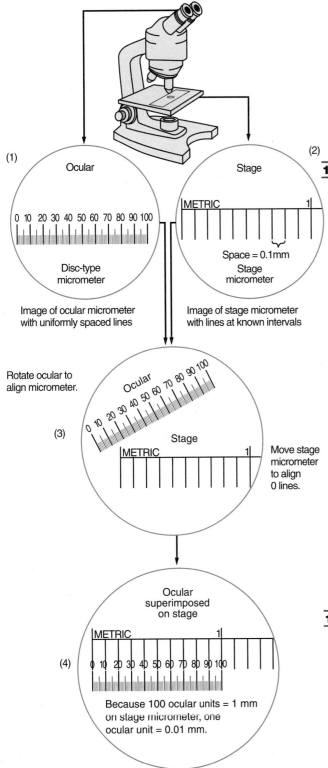

Figure 2.6 A stage micrometer is used to calibrate an ocular micrometer.

Table 2.1	Calibration of Ocular Micrometer

	Stage Units (mm)	Ocular Units	mm per Ocular Unit	Converted to μm
Scanning	___	___	___	___
Low	___	___	___	___
High	___	___	___	___

11. Now use your calibrated microscope to measure the sizes of some cells. Obtain a prepared slide of mammalian neurons in a spinal cord smear. Look at it first with the scanning objective and then with the 10×. Switch to high power and measure the diameter of the nuclei in three nerve cells. Calculate the average in micrometers.

Average nuclear diameter = _____

Using an Oil Immersion Lens (Optional)

Skip to the next section if your microscope does not have an oil immersion lens.

Oil immersion lenses allow you to approach 1,000× magnifications with an increase in resolution. However, the distance from the lens surface to the slide is very small, and it is quite easy to push the lens through the slide, breaking it and ruining the lens. Oil immersion means the space between the objective lens and a specimen is filled with oil rather than air. That increases numerical aperture, improves resolution, and allows use of higher magnification.

12.
1. Use the nerve cell slide from above. Focus first on the slide by proceeding from the scanning to high-power objectives as you did before. Now you are ready to try the oil immersion lens.

2. Do NOT touch the focus knobs or the stage knobs. Turn the turret to swing the high-power objective out of the way. Place a single drop of immersion oil on the coverslip right over where the light is coming through the stage, and rotate the **oil immersion lens** into place. The lens will actually contact the oil drop, making a column of oil from the slide surface to the lens surface.

3. Now look through the oculars and open the substage diaphragm to increase the light. The object on the slide should still be in the field of vision but will probably be out of focus. Use the fine-adjustment knob to focus clearly. ***NEVER*** *use the coarse-adjustment knob.*

4. Once you have an oil immersion lens in place, do NOT swing the 40× objective back into place.

Table 2.2	Properties of a Compound Microscope		
	Objectives		
Image Properties	Scanning	Low	High
Magnification of objective			
Total magnification of microscope			
Field of view size			
Field of view depth			
Brightness of field			
Resolving power			
Calibrated size of one ocular unit			

Because the objective focuses close to the slide, the 40× objective will get oil on it; is difficult to clean the oil from the lens surface. The 10× can be used because that lens focuses far away from the slide.

5. When you have finished using the oil immersion objective, you must clean the oil from its surface and from the slide, using lens paper. Because oil immersion lenses require extra cleanup and the danger of breaking slides is great, they usually are not used in beginning laboratories.

Lab Summary

On a separate sheet of paper, answer the following questions assigned by your instructor.

1. Define magnification and resolution. How do these properties of a microscope differ?
2. In **table 2.2**, use the words *least, intermediate, or greatest* to describe how the indicated image properties change as you change objectives on your compound microscope.
3. Describe how you would change the contrast when viewing an image.
4. Describe how you would determine the size of a specimen viewed with the 10× objective of your compound microscope.

You may want to try the critical thinking questions that apply some of the knowledge you gained in doing this lab.

Learning Biology by Writing

Write a short essay (about 150 words) describing how magnification, resolution, and contrast are important considerations in microscopy. Indicate how microscopists can increase contrast in viewing specimens.

Critical Thinking Questions

1. When looking through the oculars of a binocular compound light microscope, you see two circles of light instead of one. How would you correct this problem? If you saw no light at all, just a dark field, what correction would you make? Now, you finally have an interesting structure in view using your 10× objective lens, but, when you switch to the 40× objective lens, the structure is not in the field of view. What happened? How would you correct this?
2. What type of microscope would you use to determine the sex of a live fruit fly? If you wanted to look at the chromosomes of the fruit fly, what type of microscope would you use?
3. You are viewing a slide with a 10× objective and want to see greater detail. You rotate the 40× objective into place and it smashes the slide and gums the objective. Why did this happen?
4. When you look at a slide with your compound microscope the image is fuzzy. Name three things that could cause this.
5. After looking through a microscope for an hour, your eyes seem "tired." What two adjustments on your microscope should be checked?

Lab Topic 3

Cellular Structure Reflects Function

Supplies

Resource guide available at
www.mhhe.com/labcentral

Equipment

Compound microscopes with ocular micrometers and oil immersion objectives, if available

Optional: microtome and wax-embedded specimens for sectioning demonstration

Materials

Living organisms
    Yogurt or fresh sauerkraut from natural food store
    Mix cultures of *Anabaena* and *Gloeocapsa* together
    Culture of protozoa mixed with green algae
    Fungal hyphae (any culture)
    Elodea
Prepared slides of
    Loose connective tissue
    Neurons from cow's spinal-cord smear
    Pine secondary xylem macerate
Coverslips and slides
Razor blades and forceps
Electron micrographs from WWW or old textbooks mounted on poster board

Solutions

Methyl cellulose or *Protoslo*
Neutral red stain
India ink

Student Prelab Preparation

Before doing this lab, read the Background material and sections of the lab topic assigned by your instructor.
    Use your textbook to review the definitions of the following terms:

Algae	Colonial
Bacteria	Cyanobacteria
Cell wall	Cytoplasm
Chloroplast	Epidermis
Epithelium	Protoplast
Eukaryotic	Protists
Multicellular	Protozoa
Nucleus	Unicellular
Plasma membrane	Vacuole
Prokaryotic	Xylem

Describe in your own words the following concepts:

Cell theory

Structure reflects function

Cell compared to tissue

After finishing the prelab review, write any questions you have about terms, concepts, or techniques in the margins of this lab topic. The lab experiments should help you answer these questions, or you can ask your instructor during the lab.

Objectives

1. To observe the differences between prokaryotic and eukaryotic cell types
2. To collect evidence that cellular structure reflects function
3. To interpret subcellular structures in electron micrographs

Background

In 1665, Robert Hooke first used the word cell to refer to the basic units of life. Almost 200 years later, after other scientists had observed cells in many different organisms, Schleiden and Schwann published the cell theory. It states that the cell is the basic unit of life and that all living organisms are composed of one or more cells and the products of those cells. Today, the cell theory is accepted as a fundamental fact of biology.

If cells are the basic units of life, then the study of basic life processes is the study of cells. Today cell biologists strive to understand how cells function by using microscopes and biochemical analyses. This quest for knowledge is driven by a logical relationship; if normal organismal function is dependent on cell function, then disease and abnormal functioning can also be understood at the cellular level.

Biologists recognize two general organizational plans for cells: prokaryotic and eukaryotic. **Prokaryotic** cells are characteristic of archaeans, bacteria, and blue-green algae (called cyanobacteria). They lack a nuclear envelope, chromosomal proteins, and membranous cytoplasmic organelles. **Eukaryotic** cells have a nucleus containing chromosomes and numerous small structures made of membranes inside their cells, called cytoplasmic organelles. Organisms that have eukaryotic cells are protists, fungi, plants, and animals. Although these two types of cells are distinctly different, they also share many characteristics. Both have a plasma membrane surrounding the cell that regulates the passage of materials into and out of the cell. Both have similar enzymes, DNA as their hereditary material, and ribosomes that function in protein synthesis. The similarities suggest a common evolutionary history somewhere in the past, but the complexity of eukaryotic cells argues that they appeared later. The sequence of appearance of cell types in the fossil record supports this idea.

The cellular arrangements found in different organisms fall into three very broad categories. Many prokaryotic and eukaryotic organisms are composed of only one cell and are said to be **unicellular.** Others are **colonial,** composed of many single cells that cluster together to form what might be called a body, but all of the cells look alike and perform the same functions. None of the cells are specialized to perform only certain tasks. **Multicelluar** organisms are also composed of many cells, but the cells are specialized for different functions, usually containing different enzymes and structures. Consequently, the shapes, contents, and sizes as well as the biochemistry of specialized cells may be quite different. For example, red blood cells from a multicellular mammal are small, disk-shaped cells filled with the oxygen-binding protein hemoglobin. They pass easily through the blood vessels, carrying oxygen to body tissues. In contrast, muscle cells are large, tubular cells filled with specialized proteins, which give muscle cells their ability to contract. One cannot do the other's function because each type of cell is unique in its chemistry, cellular contents, and structure, but when working together in the whole organism their functions are integrated to yield an organism that can achieve what no cell could alone.

LAB INSTRUCTIONS

In Lab Topic 2, you learned how to use your microscope. Now you will use it to observe cells. This will introduce you to the paradox that biologists face: the unity and the diversity of living forms. Moreover, you should come to appreciate a maxim in biology: structure reflects function. After directly observing cells, you will study, electron micrographs, illustrating the amazing world of subcellular structure.

Prokaryotic Cells

All members of the Domain Bacteria and Domain Archaea have cells of the prokaryotic type. They are generally quite small and hard to see. For this lab, you will look at a few of the larger species.

Bacteria

Although we generally think of bacteria as harmful, most are beneficial. Some are used to produce food. Yogurt is made by adding bacteria in the genus *Lactobacillus* and *Streptococcus thermophilus* to milk and allowing the bacteria to anaerobically metabolize milk sugar. Lactic acid is produced and excreted by the bacteria. It curdles the milk, producing the semisolid yogurt and the bacteria die in the acid environment, thus producing a stable food. Sauerkraut is made by a similar fermentation process, starting with cabbage leaves.

1) A diluted sample of yogurt or sauerkraut was made by your instructor. Take a small drop and mix it on a slide with a drop of water. Add a coverslip and observe the slide through your microscope, progressing from the scanning to high-power objectives while adjusting the light intensity for comfortable viewing. The bacteria are small and will only be visible at high power after careful focusing.

What are the shapes of the cells? Are they unicellular or colonial? _____

Use your calibrated ocular micrometer to measure one of the cells. Sketch it below and include the dimensions.

Is any internal structure visible in the cells? _____

Cyanobacteria

All cyanobacteria are prokaryotes and most are surrounded by a gelatinous matrix. They live in soils, on moist surfaces, and in water. The common name "blue-green algae" characterizes the predominant feature of about half the organisms found in this group: they are blue because of the presence of a pigment called phycocyanin and green because they contain chlorophyll. However, some species look brown or olive because of other pigments.

2) Two cyanobacteria are available for study in this lab—*Anabaena* and *Gloeocapsa*. They have been mixed together in a single culture. Make a wet-mount slide by placing a small drop of the culture on a slide. Take a dissecting needle and dip it in India ink and touch the wet needle to the drop of blue-green algae culture. Some ink will transfer and improve the viewing. Press a coverslip down on the drop and blot away excess liquid. View the slide first with your scanning objective and then the medium and high-power objectives.

Both species are colonial but the colonies look very different.

Make a sketch of each organism below.

Anabaena colonies are filaments that contain three cell types: small spherical vegetative cells; elongate spores called akinetes with thick cell walls that surround a dormant cell that can survive harsh conditions; and large spherical heterocyst cells, which function in nitrogen fixation. Label these cells in your drawing above.

Use your calibrated ocular micrometer to measure the diameter of an akinete. Add the dimensions to your drawing.

Gloeocapsa colonies are aggregates of cells in a large gelatinous matrix. How many cells can you count in each matrix?

Label the *Gloeocapsa* colonies in your drawing above.
Look closely at the cells of both species. Can you see any internal structures such as nuclei?

Call your instructor over to verify the identification of any nuclei that you see.

Eukaryotic Cells

In contrast to the simple cells of prokaryotes, eukaryotic cells are quite complex with a lot of internal structure. Much of this complexity is not visible through light microscopes but you will observe the fine structure of eukaryotic cells at the end of this lab topic in the cell ultrastructure section. One structure, however, should be visible with

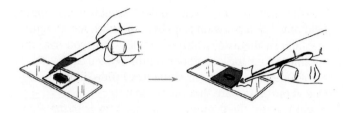

(1) Add one or two drops of stain solution to edge of coverslip.
(2) Wick the stain under by touching lens or tissue paper to the opposite side.

Figure 3.1 Method for wicking stain under the coverslip.

your microscope in all eukaryotic cells, the nucleus. In this section you will observe protist, fungal, plant, and animal cells and be able to confirm that nuclei are found in all eukaryotes. In addition, you will observe how cells have specialized structure reflecting their function.

Protists

The term protist is used to collectively refer to protozoa and algae that are classified into several clades too numerous to discuss here. See Lab Topic 16 for a more detailed treatment.

A mixed culture of protists is available in the lab. Make a wet-mount slide by placing a drop of the culture on a slide and adding a coverslip. Look at it first with the scanning objective and work your way up to high power, adjusting the light intensity for comfortable viewing. Many protists are translucent and do not offer much contrast and you might want to wick some neutral red stain under the coverslip to improve contrast (**fig. 3.1**).

Do not worry about identifying all of the "critters" that you can see on your slide. Our purpose here is to look at cells and their characteristics. One of the first things you probably noticed is that protists are larger than bacteria. As a general rule, although it is far from absolute, eukaryotic cells are larger than prokaryotic cells. Sketch one of the protist types on your slide. If you have a calibrated ocular micrometer, measure the cell and add the dimensions to your drawing.

Find a large protist, and focus up and down looking inside the cell. Is any structure visible? What cellular structure should you see in all protist cells that was not present in prokaryotes? _____

Are any of the protists moving rapidly across your field of view? If so, what is it about their cellular structure that allows them to swim so rapidly? Can you see any specializations for locomotion on the cell surface?

Can you see different structures inside different species of protists or are they all the same? _____ In eukaryotes, the cytoplasm has many types of membranous **organelles** that contain specialized enzymes to perform particular functions. In green algae in the sample, you should see green organelles, chloroplasts, in which the reactions of photosynthesis occur. How does the structure of green algae differ from that observed in blue-green algae?

Fungal Cells

Fungi are multicellular, eukaryotic organisms. When not sexually reproducing, fungal cells form hairlike hyphae consisting of single cells joined end to end. They are not specialized, and might seem to represent a colony rather than a multicellular organism. However when fungi reproduce, many of the cells specialize to form the familiar mushrooms and bracts characteristic of the group, allowing us to say they are multicellular eukaryotes.

4▶ In the lab there is a culture of hyphae, collectively called a mycelium. Take a few hyphae and mount them in a drop of neutral red stain on a slide to make a wet mount. Add a coverslip and observe through your microscope, starting with the scanning objective.

Note the **cell walls** surrounding the cell. It is made from a polysaccharide polymer called chitin that is different from the cellulose cell walls found in plants. A membrane lies just beneath the cell wall but you will not see it because it is too thin to be resolved by your microscope. Look inside the cells for any evidence of internal structure. Can you see the **nucleus**, confirming that fungi are eukaryotes? _____ Draw a few hyphae in the circle below.

If you have calibrated the ocular micrometer, measure one of the cells and record its dimensions.

Hyphae have a tremendous surface area per volume of cellular material because every cell is exposed to its environment. Can you think of a reason why this structure might promote a particular function? (Hint: fungi secrete enzymes to digest materials outside of the cells. What must happen if the fungus is to get energy from this food?)

Plant Cells

Plants and animals are also multicellular, eukaryotes, but the cells of plants differ from those in fungi and animals in several characteristics. Plant cells are always surrounded by a **cell wall** that is composed of cellulose and materials not found in fungal cell walls. Animal cells lack a cell wall. The living cell within a plant cell wall is called a **protoplast.** In the cytoplasm of some protoplasts, unique structures called **chloroplasts** are found. They carry out the complex reactions of photosynthesis. Protoplasts also usually have a large central **vacuole** filled with water and dissolved materials. As in animals, the cells of plants are specialized for particular functions and can be arranged in tissue groupings: epidermal tissues found on external surfaces; vascular tissues that conduct water, minerals and food; and ground tissues that are metabolically active.

5▶ *Photosynthetic Cells* To view cellular structures characteristic of plants, make a slide of a leaf from *Elodea,* an aquatic plant. Take a slide and place a drop of water in its center. Add one leaf to the drop and place a coverslip over the preparation. View with your compound microscope working from the scanning to the high-power objectives.

Note the thin cell walls surrounding individual protoplasts. Look closely at one cell and adjust the light intensity for best viewing. Focus up and down on the cell. This technique, called optical sectioning, allows you to see what lies above or below the plane of focus of the microscope. How many cells thick is the *Elodea* leaf? _____

Just inside the cell wall lies the cell membrane but you will not see it as a separate structure because it is too thin to be resolved by your microscope. In the center of the protoplast a large vacuole will be obvious. It too is surrounded by a non-resolvable membrane. Between the membrane surrounding the vacuole and the outer cell membrane lies the cytoplasm and nucleus. Locate the nucleus, confirming that you are looking at a eukaryotic cell.

In the cytoplasm you should easily spot several chloroplasts. How many do you see in a cell? _____

Use your calibrated ocular micrometer to measure the size of one of the cells. What are its dimensions in μm?

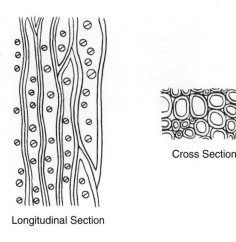

Figure 3.2 Two views of cells found in xylem tissue.

If structure reflects function, do your observations of xylem cells support the idea that they conduct fluids and support the plant? Explain.

Measure the size of its nucleus and record.

7 A redwood tree is 100 m tall. Use your measurements of cell length to calculate how many xylem cells must be stacked on end to form a conduit from the ground to the top of the tree.

Plant Cell Products Plant vascular tissues transport water, minerals, and other materials from the roots to the leaves, and also transport the products of photosynthesis from the leaves to other regions of the plant. Two basic tissues make up the plant vascular system: xylem transports water and minerals (**fig. 3.2**) and phloem transports photosynthetic products. You will look at xylem.

6 Obtain a slide of macerated pine wood from the supply area. Wood is not composed of living cells. It consists of cellulose cell walls that were made by protoplasts that then died; leaving behind "skeletons." This is why the cell theory states that living organisms are composed of cells or cell products.

Examine your slide first under scanning and then progressing to high-power. You should see elongated cells called tracheids with long, tapering end walls. The side walls of the cells are perforated by pits. Water passes to adjacent cells through the pits so that water moving from the roots to needles follows a zigzag pathway.

Note two important aspects of the cells: (1) the cell walls are thickened tubes so that they not only transport but also structurally support the plant, and (2) no **protoplast** (living cell) is visible. The protoplast of the cell functioned only to make the cell wall and then died, leaving the tubelike wall to function in water transport and support.

Sketch a few tracheids in the following blue circle. Be sure to show the pit structure in the side walls. Use your ocular micrometer to measure a tracheid and add dimensions to your illustration.

Animal Cells

In multicellular animals, cells are specialized into tissues that perform specialized functions. Four basic tissue types are found in most animals: epithelial, muscle, connective, and nerve. In this section you will look at connective and nerve tissue.

To prepare tissues from plants or animals for microscopic observation, it is necessary to instantly kill the cells with a chemical. The cells are then frozen or infiltrated with wax to make the tissue rigid. Thin sections then can be cut from this rigid block using a special cutting machine called a **microtome.** If a microtome is available in the lab, your instructor will demonstrate its use.

Tangential, longitudinal, or cross-sectional cuts may be made on embedded tissue. As **figure 3.3** indicates, the same basic structure may look different, depending on the plane of the section. After being cut, the sections are attached to slides and stained. Because the same tissue can be stained by different dyes, it is a good practice not to "memorize" tissues by color. For example, connective tissues could be stained blue on one slide and red on another.

Animal Cells and Products Our first example of animal cells will be fibroblasts in loose connective tissue (**fig 3.4**). Similar cells are also found in tendons and cartilage. Loose

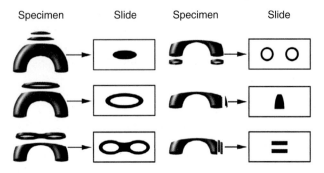

Figure 3.3 How plane of sectioning affects shape seen on slide. Sections taken through a bent tube at different levels are shown in rectangles. Where it is cut determines what you see.

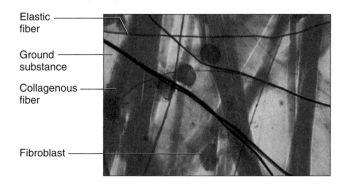

Figure 3.4 Loose connective tissue.

connective tissue attaches the skin to the body and strengthens the walls of blood vessels and organs. The fibroblasts are specialized to produce collagen, a fibrous protein, and a matrix of mucopolysaccharide gel that holds the fibers in place. Bones contain cells similar to fibroblasts. They secrete collagen fibers to form a fiber matrix and calcium salts are deposited in the matrix, cementing the fibers in place. Bones and loose connective tissue provide another example of multicellular organisms being composed of cells or cell products.

8▸ Obtain a slide of loose connective tissue from the supply area and look at it first using the scanning lens of your microscope, increasing magnification as needed. Locate the fibroblasts among the collagen fibers. What structure found in fungal and plant cells is missing? _____

Look inside the cell. What structure confirms that you are looking at a eukaryotic cell?

Draw a sample of what you see in the circle below. Use your ocular micrometer to measure a fibroblast and add the dimensions to your drawing.

Animal Cell Structure Reflects Function Nerve cells, called neurons, are cells specialized to transmit messages from one part of the body to another. In mammals, most nerve cell bodies reside in the spinal cord or brain, and cytoplasmic extensions pass out to muscles or to sensory receptors (**fig. 3.5**). If you have sensory receptors in your toes and cell bodies at the base of your spine, how long must the cytoplasmic extensions of the neurons be? _____

9▸ Obtain a slide of a neuron prepared by smearing a section of a cow's spinal cord on a slide. Observe it under low power with your compound microscope. The large cells are neurons and the smaller ones are supporting cells called glial cells. Note the neuron's cell body (soma) containing the nucleus. Extending from the soma are cytoplasmic extensions. Those that conduct impulses away from the cell body are called axons and those that conduct impulses toward the cell body are called dendrites. However, you cannot look at a slide and determine which way the impulses travel. Sketch a neuron below.

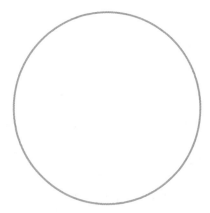

10▸ Use your ocular micrometer to measure the longest dimension of the somas of five neurons. Calculate the average length from your measurements.

Cell #	Length in μm
1	_____
2	_____
3	_____
4	_____
5	_____
Average length =	_____

CELLULAR STRUCTURE REFLECTS FUNCTION

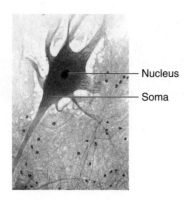

Figure 3.5 Nerve Cell.

Is a neuron larger, smaller, or the same size as the fibroblast that you measured when you looked at connective tissue?

Compared to other cells you have studied, how does the structure of a neuron reflect its function?

Using Your Knowledge

11. On a table in the laboratory, your instructor may have set up five microscopes. Each has a slide of an unknown prokaryote or eukaryote cell type. Identify the type of cell at the end of the pointer in each. In **table 3.1** list your reasons for naming each type.

Cell Ultrastructure

A fundamental problem in using light microscopes to study cells is limited resolving power. Objects smaller than 0.2 μm cannot be seen. Many structures inside eukaryotic cells are smaller.

Electron Microscopy

About 70 years ago, biologists realized that a microscope could be built to use electrons rather than light as an illumination source. The theoretical resolution limit for such an instrument would be 100,000× greater than that of the light microscope because electrons have a wavelength of 0.005 nm, a hundred thousand times smaller than that of visible light.

A transmission electron microscope (TEM) is shown in **figure 3.6**. The central column is the microscope proper, and all the rest is electronic equipment, vacuum pumps, and plumbing used to control the microscope.

The TEM is essentially a vertical television tube with the electron gun at the top and the fluorescent screen at the bottom. Electrons leave a hot filament at the top, are accelerated by high voltage, and pass down the tube to the screen. The tube must be in a vacuum so that the electrons can pass without interference.

As the electrons strike the screen, a glow appears. If a biological specimen is treated and placed in the column, it deflects electrons away from the screen, and an image of the electron-opaque and electron-transmitting sections of the specimen appear on the screen as shadows and light areas, respectively. When this image is recorded on photographic film, a picture like that in **figure 3.7** is obtained. Clearly, this technique reveals detail inside a cell that cannot be seen with a compound microscope.

To view, a specimen in a TEM, it must be carefully treated and prepared. This treatment includes fixing the tissue to preserve structure. After the cells are fixed and the structure stabilized, it is infiltrated and embedded in hard epoxy plastic. The plastic then may be sliced thinly, cutting the embedded cells, much as a butcher slices a loaf of salami. These sections are stained with heavy metal salts and finally viewed in the microscope.

Keep in mind that transmission electron micrographs are pictures of thinly sectioned material. If a structure does not appear in a particular picture, it does not mean that the structure does not occur in that cell. When the section was cut, the structure simply may not have been included in that section, in much the same way that a peppercorn does not occur in every slice from a salami loaf. This is especially important to remember when you are trying to count structures in cells.

Table 3.1	Unknown Identification	
	Cell/Specimen Type	**Reason**
1.		
2.		
3.		
4.		
5.		

Figure 3.6 A transmission electron microscope.

Interpreting TEMs

Figures 3.7 and 3.8 are transmission electron micrographs of animal and plant cells. Your instructor may have additional photos taken through a transmission electron microscope. Your task is to work with a partner to interpret these images.

Eukaryotic cellular structure is based on membranes that form compartments within the cells. These compartments are called **organelles** and are where enzymes for special functions are found.

12 Using your textbook from lectures as a resource, identify the organelles and structures in the photographs. Place a check in table 3.2 when you have identified them and add a brief description of their functions.

What structures are found in plant cells but are absent from animal cells?

If plant and animal cells have some of the same structures, does this mean that both types of cells can perform many of the same functions? _____

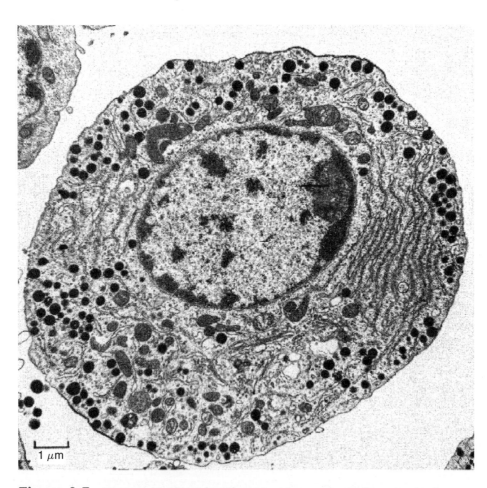

Figure 3.7 Transmission electron microscope image of a pituitary gland cell.

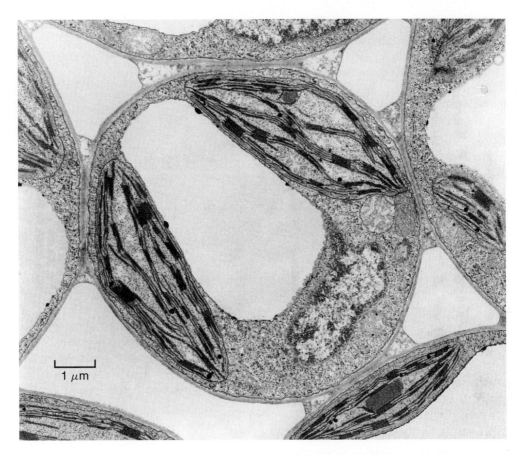

Figure 3.8 Transmission electron micrograph of a plant parenchyma cell from a *Coleus* leaf, magnification 10,000× as printed.

Do you expect that bacterial cells would have a similar or a different structural basis? Why?

The magnifications for the photographs are often given in the figure legends. If you have a millimeter ruler, you can determine the actual size of the organelle in a cell, by using the following formula:

$$\text{Structure size} = \frac{\text{Measured size of structure (in mm)}}{\text{Magnification of photo}}$$

Table 3.2	Work Table for Electron Micrographs of Eukaryotic Cells			
Organelle	**Size**	**Animal Cell**	**Plant Cell**	**Function**
Cell wall				
Plasma membrane				
Cytoplasm				
Nucleus				
Chromatin				
Nucleolus				
Endoplasmic reticulum				
Golgi apparatus				
Ribosomes				
Lysosomes				
Mitochondria				
Chloroplasts				

Another way to determine size is to use the size bar at the lower left in each of the figures in this lab manual. The length of this line represents the size of 1 μm when it is magnified the same amount as the structures in the photo. To use the size bar, measure its length in millimeters and, using the following formula, calculate how may micrometers equal a millimeter at this magnification:

$$\mu m/mm = \frac{1 \mu m}{\text{bar length in mm}}$$

What value did you obtain? _____

13 Now measure in mm some of the structures in the figures and multiply by this conversion factor to determine the actual size of each organelle. Add this information to table 3.2.

Table 3.3 Dimensions of Cells in μm

Cell Type	Describe Shape	Dimensions
Lactobacillus		
Anabaena		
Protozoan		
Alga		
Fungus		
Elodea		
Xylem		
Fibroblast		
Neuron		

Lab Summary

14 Your study of cells has ranged across all types of organisms. It has provided you with evidence to support the cell theory. In this section, organize your observations to illustrate these principles.

On a separate sheet of paper, answer the following questions as assigned by your instructor.

1. What evidence do you have from your lab observations to support the following statements?
   a. Prokaryotic cells are fundamentally different from eukaryotic cells;
   b. All organisms are either unicellular, colonial or multicellular;
   c. Organisms ranging from bacteria to animals are composed of cells and cell products;
   d. Multicellular organisms have tissues specialized for different functions.

   If you do not think you have any evidence to support these statements, what would convince you of their validity? What would convince you that the statements are false?

2. How do plant cells differ from animal cells? Among the cells that you have observed, what evidence, if any, do you find that the structure of a cell correlates with its function?

3. Comparing what you saw inside eukaryotic cells to the electron micrographs in the cell ultrastructure section, what evidence can you cite that what we see is limited by the resolution of the microscope used?

4. You measured many different cells. Summarize those measurements in **table 3.3**.

Learning Biology by Writing

Based on your observations, write a essay on the theme "form reflects function" at the cellular level. Cite four examples that you observed.

Critical Thinking Questions

1. Why is it logical to think that the first living cells on earth had a prokaryotic cell plan and that eukaryotic cells developed later?

2. Long before electron microscopes were invented, biologists reasoned that every cell had a surface membrane. If you cannot see membranes with a light microscope, what observations allowed them to say this?

3. Cells in the pancreas produce many proteins that function as digestive enzymes and others as hormones. How should the internal structure of pancreatic cells reflect these functions?

Lab Topic 4

Membranes, Diffusion, and Osmosis

Supplies

Resource guide available at
www.mhhe.com/labcentral

Equipment
Compound microscopes
Balance (0.1 g sensitivity)

Materials
Slides and coverslips
Dropper bottles
Markers
Dialysis tubing precut to 15 cm
1 ml pipettes
Test tubes and rack
Tes-Tape glucose strips (drugstore)
Cork borer
250 ml beaker
Petri plates poured 5 mm deep with 1% plain agar
Live organisms
   Elodea
   Paramecium caudatum
   Red blood cells from fish, bird, or animal

Solutions
1% soluble starch in 1% Na_2SO_4
25% dextrose made up in Lugol's solution
I_2KI solution (5 g I_2: 10 g KI: 85 ml H_2O) Lugol's
2% $BaCl_2$
20% Na_2SO_4
25% sucrose made in Lugol's solution
0.9% NaCl
3.0% NaCl
1.0% agar

Student Prelab Preparation

Before doing this lab, read the Background material and sections of the lab topic assigned by the instructor.

Use your textbook to review the definitions of the following terms:

Active transport
Contractile vacuole
Cell Membrane
Diffusion
Hypertonic
Hypotonic

Isotonic
Kinetic energy
Osmoregulation
Osmosis
Plasmolysis
Turgor

Describe in your own words the following concepts:
Facilitated diffusion
Osmotic pressure
Selectively permeable membrane

After finishing the prelab review, write any questions you have about terms, concepts, or techniques in the margins of this lab topic. The lab experiments should help you answer these questions, or you can ask your instructor for help during the lab.

Objectives

1. To determine if diffusion and osmosis both occur through selectively permeable membranes
2. To observe diffusion in a gel
3. To test how animal, plant, and protist cells respond to osmotic challenges

Background

Polar and ionic molecules dissolve in water because water molecules surround each molecule as a **hydration shell.** It prevents the dissolved materials from aggregating and coming out of solution. These hydrated, dissolved molecules are in constant motion because of their kinetic energy. They continuously collide and rebound to travel in different directions. The result of this is seen when a few crystals of a colored substance are added to calm water. The color gradually spreads from the dissolving crystals, until the color is uniformly distributed throughout the solution. Molecular movement does not cease when uniformity is reached. It simply can no longer be observed with the naked eye. Molecular motion continues with dye (and water) molecules moving in random motion from one place in the container to another, exchanging places on a one-for-one basis in a dynamic equilibrium. This process of molecular movement from a region of high concentration to low concentration due to thermal motion is called **diffusion.**

The rate of diffusion is dependent on four factors: (1) the cross-sectional area through which the molecules can move; (2) the concentration differences between two regions; (3) the temperature, which influences the speed of random molecular motion; and (4) the molecular mass of the substance involved. Fick's law describes the relationship among these variables and can be stated as follows: the rate of diffusion increases as the differences in concentration increase; the cross-sectional area becomes larger; and the temperature increases, but decreases as the size of the involved molecules increases.

All cells are surrounded by cell membranes that act as dynamic boundaries between a cell's cytoplasm and the surrounding environment. Composed of two layers of phospholipids, membranes are a selective hydrophobic barrier allowing small, uncharged molecules to pass through while excluding others. Larger molecules such as proteins, sugars, and amino acids cannot pass through the barrier without assistance, nor can many ions such as Na^+, K^+, Mg^{++}, Ca^{++}, or Cl^-. Often transport proteins are embedded in and span the membrane. These act as selective channels through which large organic molecules and ions may diffuse. Because passage is assisted by a protein, this phenomenon is called **facilitated diffusion.** The transport proteins may be directional, allowing some materials to diffuse only inward, others only outward, and yet others in both directions. Without such transport aids, materials necessary for normal cell functioning would not be able to cross membranes. By regulating what proteins are in their membranes and by metabolically regulating the functioning of those proteins, cells can control what substances enter and leave their cytoplasm. The result is a regulated stream of molecular traffic with some substances excluded or retained and other substances crossing the membrane, each at its own rate. Both diffusion and facilitated transport are passive processes driven by thermal motion and concentration gradients, requiring no direct energy from a cell.

Because membranes block, slow, or accelerate the passage of specific substances, they are described as being **selectively permeable.** As a result, concentrations of dissolved solutes differ between the cytoplasm and the extracellular fluid bathing a cell. Consider glucose in our bodies. After a meal, it is present in high concentrations in the blood and extracellular fluids. Glucose enters our cells by binding with a membrane transport protein that assists it across the membrane. Glucose that enters is almost immediately metabolized, so intracellular glucose concentration remains below that in the extracellular fluids, and glucose continues to diffuse inward.

Diffusion may occur directly through the phospholipids layer as well as through the embedded proteins. Nonpolar, small molecules such as oxygen and ethanol can pass through the spaces between the phospholipid molecules. For example, in organisms actively carrying on photosynthesis, oxygen is at higher concentrations inside the cell than in the environment. Consequently, oxygen diffuses out of the cell. However, oxygen's story gets more interesting for when it gets dark, photosynthesis stops. Now the direction of oxygen diffusion reverses because the cellular levels of oxygen drop below those of the environment due to oxygen consumption in aerobic metabolism.

For many substances, favorable diffusion gradients do not exist. For example, sodium ions are found at higher concentrations outside mammalian cells, yet the net movement of sodium is from inside to outside the cell. For such materials, cellular energy must be expended to transport the molecules across the cell membrane. Called **active transport,** certain proteins in the cell membrane bind with the substances to be transported and use metabolic energy to drive the "pumping" of an ion or molecule into or out of a cell.

For years, biologists were puzzled as to how water could move so quickly across cell membranes. The hydrophobic barrier should allow only slow passage, if any. The answer came with the discovery of a group of membrane proteins called **aquaporins.** These function as specific channels for the passage of water through membranes. First found in mammals, they are now known to occur in prokaryotes and all eukaryotes

Water, whether inside or outside a cell, exists in two states. Some water molecules are free to diffuse, to move randomly by thermal motion. Others are less mobile because they are bound in hydration shells around dissolved solutes. When two solutions differ in their total solute concentration, they also differ in their concentrations of free water molecules. The solution with the greatest solute concentration has the least amount of free water. If the two solutions are separated by a membrane that is not permeable to solutes, an interesting situation occurs. Given the random thermal motion of only the free water molecules, a net movement of water occurs from the side with the least solute to the side with the most. This diffusion of water through a membrane separating solutions of differing concentration is called **osmosis.** A simple cell like an amoeba crawling on the bottom of a freshwater pond will take up water by osmosis because the solutes in its cytoplasm are more concentrated than in the surrounding water. If that same amoeba is placed in a concentrated salt solution, the osmotic gradient will be reversed. Water will now flow outward. As water is lost, the internal concentration of solutes increases until it equals that of the bathing solution, producing a steady state with no net gain or loss of water.

Tonicity is a root word used to describe the relative concentrations of nonpermeating solutes on either side of a selectively permeable membrane. Three prefixes are added to the root: *iso-* meaning equal, *hypo-* meaning less than, and *hyper-* meaning more than. An **isotonic** solution is one where the total concentration of solutes is equal on both sides of a selective membrane. A cell placed in an isotonic solution neither gains nor loses water by osmosis. A **hypotonic** solution is one where the total solute concentration is lower outside the cell than inside. A cell placed in a hypotonic environment gains water by osmosis. A **hypertonic** solution is one where the solutes are more concentrated outside. A cell placed in a hypertonic solution loses water.

Different environments pose a range of osmotic challenges to cells which have been met by a number of different strategies. Terrestrial organisms are always losing water by evaporation of body fluids. They try to maintain body fluid solute concentrations isotonic to their cells by seeking water or eliminating solutes, but during periods of dehydration their body fluids can become hypertonic to their cells, leading to water loss and death. Freshwater organisms are always hypertonic to their environment (it is hypotonic relative to them). They constantly gain water by osmosis that can disrupt cell functioning and even cause the cells to burst. Multicellular freshwater animals cope by excreting water. Some single-celled organisms have special cellular organelles that extrude any osmotically gained water. Marine organisms often regulate the solute concentrations in their cells so they remain isotonic to their environment and neither gain nor lose water. Bacteria, algae, plants, and fungi have cell walls that are splendid osmoregulatory devices. When a cell that is surrounded by a rigid cell wall takes on water, the hydrostatic pressure inside the cell increases as the cytoplasm presses against the wall. With increasing pressure, the kinetic energy of the free water molecules in the cytoplasm increases so that every molecule of water osmotically entering a walled cell is matched by a water molecule leaving as a result of increased molecular motion. The pressure at which this occurs is called the **osmotic pressure** or **turgor pressure** of the cell.

Lab Instructions

You will observe diffusion and the properties of selectively permeable artificial membranes. Start the experiment "Modeling Cell Membranes" first so that sufficient time can elapse to see results. You will formulate and test experimental hypotheses regarding how living cells react to hypo- and hypertonic environments.

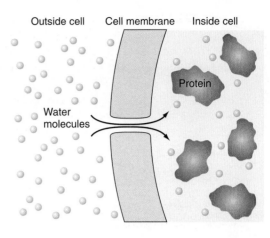

Figure 4.1 Model of osmosis through a differentially permeable membrane. Water molecules can pass through small pores in the membrane but larger protein molecules cannot.

starches which typically have molecular weights above 12,000 to pass (**fig. 4.1**).

In this section you are going to make a bag from dialysis tubing and fill it with a solution containing ions and glucose to act as model of a cell. You will immerse the "cell" in a beaker containing a solution of starch and ions to represent the cell's environment. Chemical tests will allow you to determine which materials can cross the membrane. You do not know the molecular weight cutoff for the tubing you are using, but the results of this experiment will give you some hints.

1 ▶ *Procedure* Obtain a 15 cm section of dialysis tubing that has been soaked in distilled water. Tie or fold and clip one end of the tubing to form a leakproof bag. Half fill the bag with a solution of 1% starch in 20% Na_2SO_4. Also add a 1 ml sample of the same solution into each of two test tubes labeled "**Inside Start**" for later analysis.

Now tie the bag closed with a leakproof seal. Wash the bag with distilled water, blot it on a paper towel, weigh it to the nearest 0.1 g, and record the Mass in **table 4.1**, in the Inside Cell Start column.

Place the bag in a 250 ml beaker containing 25% glucose in I_2KI solution. Place 1 ml samples of the fluid from

Modeling Cell Membranes

Dialysis tubing is an artificial material that can be used to model processes that occur across a cell's membrane. Made from cellulose, the size of pores in the membrane can be controlled during the manufacturing process. Consequently, a membrane can be produced that will allow small molecules to pass through but will block larger ones. For example, dialysis tubing with a molecular weight cut off of 1,000 would allow H_2O (MW 18) or KI (MW 166) to pass through readily but would not allow proteins or

Table 4.1	Results of Osmosis/Diffusion Experiment with Dialysis Tubing			
	Inside Cell		Outside Cell	
	Start	End	Start	End
Mass			not needed for outside	
I_2KI				
Na_2SO_4				
Glucose				
Starch				

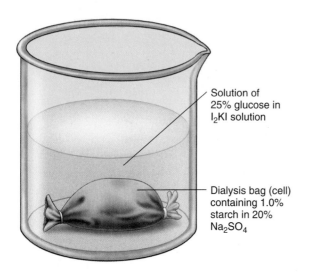

Figure 4.2 Initial conditions for osmosis and diffusion experiment, showing composition of solutions inside and outside dialysis bag.

the beaker in two test tubes labeled **"Outside Start"** for later analysis. The starting conditions are summarized in **figure 4.2**. This experiment will run for approximately an hour. Go on to the other experiments starting at arrow 4 while this experiment runs in the background.

At 15-minute intervals, swirl the beaker containing the bag or place the beaker on a slowly turning magnetic stirrer.

2▶Analysis After your model cell has been in the starch solution for at least one hour, you are ready to stop the experiment and take final samples for chemical analysis.

First, look at the bag in the beaker and note any visible changes. There should be an obvious one. What is it?

I_2KI is a chemical test for the presence of starch, reacting with it to form a bluish chemical complex. At the beginning, I_2KI was inside the bag and starch was outside. Where is the blue complex located? What does this tell you about the permeability of the dialysis membrane and the direction of movement of the iodine and potassium iodide ions? Were they following a concentration gradient?

Were the starch molecules able to enter the bag? How can you tell?

Take two 1 ml samples from the beaker for further chemical tests and place them in tubes labeled **"Outside End."**

Remove the bag from the beaker, blot it and weigh it to the nearest 0.1 g. Record the mass in table 4.1 in the "Inside Cell End" column. Compare it to the starting mass. What might have caused this change?

Open the bag and pour its contents into a beaker. Take two 1 ml samples for further analysis and put them in tubes labeled **"Inside End."**

You will now analyze the four samples taken during your experiment for the presence of Na_2SO_4 and glucose. The tests are:

Sulfate ions: Add a few drops of 2% $BaCl_2$ to one *inside* and one *outside start* sample. Repeat for *inside* and *outside end* samples. If SO_4^{-2} ions are present, a white precipitate of $BaSO_4$ will form. Record the results in table 4.1 as + when SO_4^{-2} is present or − when absent.

Glucose: Take four glucose Tes-Tape strips from the jar and dip them one at a time into the remaining tubes. If glucose is present in the sample, the strip will turn green to blue. Record the results as a + when glucose is present, or − when absent, in the appropriate columns in table 4.1.

3▶Now it is time to look back at your results and analyze them in terms of the selective permeability of the artificial membrane, osmosis, and diffusion because all three factors influenced your results.

In **table 4.2**, use the chemical formulas to calculate the molecular masses for the substances tested and enter the values in the second column. From your experimental data, indicate which molecules passed through the membrane.

Is there a correlation between permeability and molecular weight? _____
Describe how could you test this idea in a second experiment.

What evidence do you have that water was able to pass through the dialysis membrane? How would the

Table 4.2 Summary of Membrane Permeabilities

Dissolved Substance	Molecular Mass	Permeable (Y/N)
H_2O		
Sulfate SO_4^{-2}		
Glucose ($C_6H_{12}O_6$)		
I_2		
KI		
Starch ($C_6H_{12}O_6$)$_{1,000}$		

evidence be stronger if the experiment were run overnight rather than for an hour?

Which side of the membrane had the greatest total solute concentration? _____

Which had the greatest solvent concentration? _____

Which side of the membrane was hypertonic to the other? _____

Which side was hypotonic? _____

Using the concepts of free water molecules and water molecules bound in hydration shells, explain why water diffused in one direction.

How would your results change if we added a 20% starch solution mixed in a 15% protein solution to the dialysis bag in place of the mixture you used?

Diffusion in Gels

4 The cytoplasm of a cell is more like a gel than it is like an aqueous solution. Does diffusion occur in gels? To answer this question, do the following experiment.

Diffusion in highly viscous solutions can be demonstrated using agar as a gelling agent. Obtain a petri plate containing 1% agar poured to a depth of 5 mm. Use a 5 mm cork borer to make two wells about 2 cm apart. Add a solution of sodium sulfate to one and barium chloride to the other. Place the plate on black paper and observe at intervals during the lab. When the ions of these two soluble compounds diffuse through the water in the agar and meet, they chemically react with each other to form a white precipitate, barium sulfate.

$$Na_2SO_4 + BaCl_2 \rightarrow BaSO_4(ppt) + 2NaCl$$

Record your results as a time diary below.

Time	Description

According to Fick's law, substances with large molecular weights (MW) diffuse slower than those with small ones. The MW of barium = 137 and sulfate's is 96. Do your results confirm or falsify this principle? _____

If you repeated this experiment using two dishes, one kept at a high temperature and the other at a low temperature, what difference would you expect to see in the results?

Osmotic Challenges to Living Cells

Visualizing the Challenge

Vertebrates as well as many invertebrates regulate the composition of the body fluids bathing their cells so that the two are isotonic. They do so by selectively eating and drinking, by synthesizing proteins such as albumin that increase the tonicity of the extracellular fluids, and by selectively eliminating salts and water through excretion. Under dire circumstances the balance may be upset due to dehydration or overhydration, but most of the time a cell's environment is constant, and water is neither gained nor lost. Consequently, many cells in multicellular animals lack mechanisms to cope with osmotic challenges. This presents an opportunity to investigate how an unprotected animal cell responds to such a challenge.

5 ▸ **Baseline Observations** In the lab are living vertebrate red blood cells (RBCs) diluted with isotonic saline (0.9%). Place a drop of the cell suspension on a microscope slide, add a coverslip and observe with your compound microscope. RBCs are disc-shaped blood cells with slightly concave tops and bottoms. When you look at one through a microscope, the centers appear lighter in color because that is where the cells are the thinnest. Note the smooth edges.

Experimental Observations While looking through your microscope, osmotically challenge the RBCs by adding a drop of distilled water near one edge of the coverslip to make their environment hypotonic. Use paper toweling or blotting paper to draw the water under the coverslip while watching the cells' reaction. See figure 3.1 for blotting method. Describe what happens to the size of the RBCs below and enter a summary word to **table 4.3**.

Wash and dry your slide. Make a new slide from the RBC stock tube. Again observe the normal RBC shape and size. You are going to osmotically challenge them by adding 3% NaCl to the edge of the coverslip, wicking it under as you did previously. What will happen to the RBCs as the saline concentration increases? Write this as a testable hypothesis below.

Do the experiment to test your hypothesis and briefly describe the results.

Table 4.3	Cell Response to Osmotic Challenges	
Medium	RBC	*Elodea*
Isotonic	Normal	Turgid
Hypotonic		
Hypertonic		

Summarize your observations by writing one or two words describing the RBCs reaction in table 4.3.

Cell Walls Osmotically Protect

The cells of plants, algae, bacteria, and fungi are surrounded by rigid cell walls. They are essentially porous boxes surrounding the cytoplasm. They do not function as barriers to diffusion. Salts, sugars, water, and most other dissolved substances easily pass through cell walls. Pressed against the inner surface of the cell wall is the plasma membrane of the living cell and it regulates what passes into or out of the cytoplasm. However, the cell wall boxes have an important osmoregulatory function which may be the reason that they are found in so many organisms. In this section, you will investigate the osmoregulatory function of plant cell walls using the aquatic plant, *Elodea*.

6 ▸ **Baseline Observations** Remove a leaf from the aquatic plant *Elodea*, cut about a third from the tip and mount it in a drop of water on a microscope slide.

View with your compound microscope and identify individual cells. In the center of the cell is a large vacuole that displaces the cytoplasm to the periphery, close to the surrounding cell wall. Several chloroplasts and a nucleus should be visible in the cytoplasm.

Note the position of the cytoplasm in relation to the cell walls. When the cells are in a hypotonic environment, the cytoplasm is closely pressed against the cell wall. Cells are said to be **turgid** when in this condition.

Sketch a turgid cell below. While looking and sketching, you may see cytoplasmic streaming, movement of the cytoplasm in a circular pattern within the cell. If you think in terms of molecular mixing, how might this benefit the cell?

Experimental Observations You will now subject the *Elodea* cells to an osmotic challenge by creating a hypertonic environment. Place a drop of 25% sucrose on the slide next to the coverslip. Touch a tissue to the opposite side of the coverslip and wick the sucrose under it to bathe the leaf in concentrated sucrose.

Use your microscope to observe the cells, looking at the position of the cytoplasm relative to the cell wall. When the cytoplasm pulls away from the cell wall due to water loss from the central vacuole in hypertonic solutions, **plasmolysis** is occurring. Draw a plasmolysed cell next to the normal cell drawn earlier and enter your observation into table 4.3.

What do you think will happen if you placed these cells in plain water, a hypotonic environment? Formulate this prediction as a hypothesis and write it below.

Test the hypothesis by wicking distilled water under the coverslip. Record your observations in table 4.3. Based on this observation must you accept or reject your hypothesis? _____

7 ***Analysis*** Using the concepts of free versus bound water and osmosis, discuss with your lab partner why both red blood cells and *Elodea* responded similarly to hypertonic media. In the hypotonic media, there was a huge difference in their responses. What was it and why did the *Elodea* cells survive while the red blood cells were destroyed? What do you think would happen if you repeated the experiment with fungal hyphal cells or bacteria that also have cell walls?

Turgor pressure is important in supporting upright plants. You undoubtedly have seen houseplants droop when deprived of water only to return upright when watered. When deprived of water, the plant's cells undergo plasmolysis allowing the cell walls to bend because they are no longer supported by hydrostatic pressure of a full vacuole in the cytoplasm, in much the same way that a firehose loses its rigidity when disconnected from a hydrant. When water is again available, the vacuole fills and the drooping branches straighten.

Expelling Water

Paramecium is a unicellular organism living in fresh water. Its environment is hypotonic to its cytoplasm so the organism continuously gains water. Although *Paramecium* is surrounded by a semirigid pellicle (protein outer covering of cell membrane), it lacks the strong cell wall you saw in plants and would expand, possibly bursting, were it not for an interesting subcellular structure that removes water from the cell. Called the **contractile vacuole** (CV), it consists of a central membranous vesicle with membrane canals radiating from it. The canals collect excess water from the cytoplasm and convey it to the vesicle. Once it reaches a certain size, the vesicle collapses as it voids the water through a pore that develops on the cell's surface. The pore then closes and the vesicle refills to start the next cycle, acting like a sump pump to maintain the cell's volume and osmotic concentration of solutes. Contractile vacuoles are found in several protists and in the cells of freshwater sponges.

You will observe the contractile vacuole in *Paramecium* and then do an experiment that demonstrates its role in osmoregulation.

8 ***Baseline Observations*** Learning to recognize the CV for this part of the experiment is going to be difficult and requires patience, but the results are worth the effort.

Paramecia move rather quickly by coordinated beating of cilia on their surfaces. The best results will be obtained when you can reduce their range of movement. To achieve this, shred a small piece (1 or 2 mm square) of lens paper or part of a cotton ball onto a microscope slide. Look at the culture of *Paramecium* and determine where the cells seem to be, swimming to one side or clustered at the bottom. Take your sample from where they are clustered without disturbing the culture tube. Add the drop of

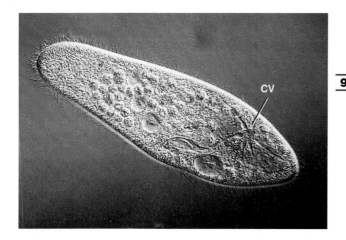

Figure 4.3 Photomicrograph of *Paramecium caudatum* showing one of its two contractive vacuoles (CV). In life, these empty and fill in a regular cycle.

Paramecium culture to the center of the shreddings and add a coverslip. Wick off excess fluid by touching the edge of the coverslip with a paper towel or blotting paper.

Locate a *Paramecium* with your medium-power objective, looking for one trapped in the fiber matrix. Switch to high power and locate the two CVs, one at either end of the cell. These will be seen as clear areas (**fig. 4.3**) that seem to appear and disappear 6 to 10 times per minute, although the rate may be more variable. Do the two vesicles contract at the same time? _____

After identifying the CVs, switch to medium power to determine if you can see the pulsations at lower magnification. If so, do all subsequent work at that magnification as it will be easier to cope with the organism's swimming movements.

Count the number of pulsations per minute for one vacuole and record your observation as trial 1 in **table 4.4**. Make and record two more measurements and then average the three.

Experimental Observations When CVs were first observed, researchers must have wondered about their function. If you put yourself in their shoes, think about how you could demonstrate that they function in osmoregulation. Shouldn't the pulsation rates change in response to osmotic challenges from its environment? By making a slide with distilled water or 3% NaCl, you can offer osmotic challenges and observe the effects on pulsation rate. If the rate changes, then it suggests an osmoregulatory function for the CV. Formulate a hypothesis to be tested by doing such an experiment and write it in table 4.4.

9> Start your experiment by making a new slide. However, before adding *Paramecium,* first place a drop of distilled water on the fiber matrix. Then add a small drop of culture so that the total solutes in the culture are less than half of what they were in the baseline observations. Make a prediction about what will happen to the CV pulsation rate compared to your baseline observations. Now make three measurements as before and record in table 4.4.

Now change the osmotic environment by adding a drop of 3% NaCl to the edge of your coverslip on the slide. Wick it under the coverslip with paper toweling or blotting paper. Make a mental prediction about what will happen to pulsation rate. Make three measurements of pulsation rate and add them to table 4.4.

10> **Analysis** If you have not done so, calculate averages for your sets of observations. Review your hypothesis in table 4.4 and organize your reasons for accepting or rejecting the hypothesis. State your conclusion and reasons on the last line of table 4.4.

Based on your data, you should now be able to describe the tonicity of the culture medium relative to the tonicity of the paramecium's cytoplasm. Using the terms, *hypotonic, isotonic,* and *hypertonic,* enter the correct term in the last column of table 4.4.

11> One last optional calculation would be to estimate the actual volume of water being pumped by the contractile vacuoles in a single *Paramecium*. If we assume that the vesicles are spheres and that the both vesicles are the same size, then we need only to calculate the volume of the sphere, multiply by 2, and multiply again by the rate of pulsations per minute.

For estimation purposes in this lab, we will say that the diameter of a CV is about 8 micrometers (1 $\mu m = 10^{-6}$ m). Its volume is:

$$V = \frac{4\pi (radius)^3}{3}$$

Table 4.4	Contractile Vacuole Pulsation Rates in *Paramecium*				
My hypothesis:					
	Trial 1	Trial 2	Trial 3	Average	Solution Tonicity Relative to Cell
Baseline pulsations					
In distilled water					
In 3% NaCl					
Accept or reject hypothesis? Summarize reasons, briefly.					

Calculate the volume of water pumped by a *Paramecium* in a minute using the units μm. Assuming there were about a million (= 10^6 paramecia) in the culture tube at the start of the experiment, how many liters would pass through their bodies in a year? A liter is equal to 10^{15} μm^3.

Lab Summary

12 On a separate sheet of paper, answer the following questions as assigned by your instructor.

1. What data do you have to support the conclusions that (1) dialysis membranes are selectively permeable and that (2) osmosis and diffusion can occur at the same time through a dialysis membrane?
2. What evidence do you have from your experiments to indicate that molecular weight influences the rate of diffusion?
3. What evidence do you have that cells are hypertonic to pure water?
4. What evidence to you have that suggests cell walls function in osmoregulation?
5. Explain why *Paramecium* continually gains water from its environment. Describe the evidence you have from your experiments that argues the water expulsion vacuole is an osmoregulatory structure?
6. Explain osmosis in terms of selectively permeable membranes, kinetic molecular energy, and bound versus free water molecules.

You may want to try the critical thinking questions that apply some of the knowledge you gained in doing this lab.

INTERNET SOURCES

Choose your favorite type of organism: bacteria, protist, plant, fungus, or animal. Use your WWW browser to access Google and inquire about aquaporins for the group. For example, typing in "aquaporins in fungi" gets you immediately into the scientific literature with the latest articles written about discoveries.

LEARNING BIOLOGY BY WRITING

Prepare a brief report describing the evidence that you have showing that membranes are selectively permeable. Start with dialysis membranes as models of cell membranes, discussing your data showing how they allow some molecules to cross the membrane in either direction while blocking the passage of others. Discuss how dialysis membranes differ from cell membranes but why they offer some insight into the phenomena of osmosis, diffusion, and selective permeability. Present the evidence that you collected showing that osmosis occurs across cell membranes of unprotected cells, walled cells, and protists and discuss how different organisms have evolved different strategies to counteract the osmotic challenges they face in their environments.

CRITICAL THINKING QUESTIONS

1. If a person's blood volume drops due to injury or severe dehydration, why do doctors administer isotonic saline intravenously instead of pure water?
2. How would an increase in temperature affect the rate of diffusion of gases? How might it affect osmosis? Explain.
3. How would increasing the concentration of a protein in your simultaneous osmosis and diffusion experiment affect the results?
4. When preparing sauerkraut, shredded cabbage is layered with coarse salt crystals. What effect would this have on the cabbage?
5. What osmotic regulatory challenges would a fish living in freshwater have versus a fish living in salt water?
6. If fruits are crushed, cooked, and then stored at room temperature, they spoil. If high concentrations of sugar are added as in making jams, their shelf life is greatly extended. Explain this in terms of the osmotic challenges facing invading bacteria and fungi.
7. Based on the techniques used in this lab, devise an experiment that would allow you to determine what concentration of saline was isotonic to the cytoplasm of the cells found in a freshwater sponge.
8. Cystic fibrosis is caused by a defect in the CFTR gene that makes a protein controlling the movement of chloride ions and water in and out of the cells that make mucus in the human body. The result is a thick mucus that can clog the lungs and interfere with the functioning of the digestive system. Use your WWW browser to access Google and enter "membrane transport disorders" to learn about other membrane related diseases. Why do mutations that affect genes end up interfering with normal membrane functions?

Lab Topic 5

Using Quantitative Techniques

SUPPLIES

Resource guide available at
www.mhhe.com/labcentral

Equipment
Spectrophotometer
Balances (minimum sensitivity 0.01 g, 0.001 g preferred)

Materials
5 ml pipettes, TD style, or micropipetters
Suction devices for pipettes
Test tubes and racks
50 ml beakers
Spectrophotometer cuvettes

Solutions
Distilled water
Bromophenol blue standard solution (0.02 mg/ml)
Bromophenol blue solutions for unknowns

STUDENT PRELAB PREPARATION

Before doing this lab, you should read over the Background material and sections scheduled by your instructor.

Use a dictionary for the definitions of the following terms:

Absorbance	Milliliter
Balance	Nanometer
Histogram	Spectrophotometer
Mean	Standard deviation
Micropipet	Wavelength

Describe in your own words the following concepts:

Beer's law

Absorption spectrum

Standard curve

Review appendices A, B, and the first part of C

After finishing your prelab review, write any questions you have about terms, concepts, or techniques in the margins of the pages of this lab topic. The lab experiments should help you answer these questions, or you can ask your instructor during the lab.

OBJECTIVES

1. To calibrate instruments for measuring mass and volume
2. To use a spectrophotometer to measure light absorbance of colored solutions
3. To construct a standard curve to determine the concentration of dye in an unknown
4. To analyze data using simple statistical techniques

BACKGROUND

For any quantitative measurement, you need to ask two fundamental questions. How **accurate** is the measurement, and how **precise** is the measurement? Accuracy means how close the measurement is to the true value. Precision, on the other hand, refers to exactness and reproducibility. Measuring a height to the nearest millimeter is a more precise measurement than measuring it to the nearest centimeter.

A measurement can be accurate, precise, both, or neither. If you weigh exactly 70 kg, a scale that reports your weight as 70.0 kg every time is both accurate and precise. If it claims you weigh 69.0 kg every time, it is still precise, but considerably less accurate—the reported weight is always 1 kg low. It is obviously important to avoid confusing precision with accuracy. If repeated weighings range from 69 to 71 kg but average 70 kg, the scale is accurate but not very precise. The average of many readings will be close to the true value, but you can trust a single reading only to ± 1 kg. If weights ranged from 68 to 69 kg, the scale would be neither accurate nor precise.

To assess the accuracy of a measuring device, we use it to measure a known standard, and see how close the measurement is to the true value. In the United States, the National Bureau of Standards supplies certified calibration standards for this purpose. If the device is not accurate, we can use such a standard to **calibrate** it. Calibration means adjusting the device or correcting the measured value to match the true value as closely as possible. Where accuracy is important, you should always check the calibration of your measuring device.

Precision can never be better than the smallest unit you can read from your measuring device, and may be worse. To assess precision, make repeated measurements of the same quantity, and see how close together those measurements are. Do not record measurement values more exactly than justified by precision. For example, if repeated measurements of a distance vary by a centimeter, do not report the distance to the nearest millimeter.

In modern labs where many instruments have digital read-out devices and handheld calculators are used for arithmetic, you will often see a number expressed to three or more decimal places. One should be skeptical before accepting such numbers and should ask a few questions. Some examples might be:

1. Were instruments calibrated with standards that had an accuracy to three decimal places?
2. Did all instruments used in preparing the sample yielding this data have the same precision as the instrument I am reading?
3. Is it necessary to have great precision in the measurement or will a value with fewer decimal places be sufficient?

The implications of precision in measurements and laboratory calculations are discussed further in appendix A. Read this so that you will know how to do arithmetic and rounding with values that have different levels of precision. You should apply these rules in the calculations that are called for in this lab.

LAB INSTRUCTIONS

This lab introduces you to skills that you will need in later experiments. Weighing, pipetting, and using spectrophotometers are performed daily in modern biological research labs and are needed skills. In addition, this lab introduces data and statistical analyses which are commonly used in experimental biology.

Measuring Mass

At the start of any quantitative work, one should be certain that the instruments being used are **calibrated.** To do this, the instrument is used to measure a **standard** to determine if the readings it gives are accurate. In this lab activity, you will first calibrate a balance using a reference standard of known mass and then use the balance as a working standard to determine the accuracy of devices used to measure volumes.

① Electronic balances are now the standard in laboratories. Depending on the quality of the balance, it may have sensitivity in the range 0.1 to 0.001 g, meaning it can detect differences in mass up to its stated sensitivity. What is the sensitivity of the balance you will use? _____

The sensitivity sets the upper limit on the number of significant figures you will use in subsequent calculations. See appendix A.

Turn on your electronic balance and follow your instructor's directions on how to **zero** it. Zeroing compensates for variations in electronic circuits and any changes in the mass of the pans.

Place a standard weight (10 g to 50 g) on the pan and record the balance reading. Remove the weight, check the zero, and replace the weight. Repeat to get three different readings. Enter your data below.

Standard's Mass	Measured Mass
Av =	Av =

Is the balance accurate? Explain your answer.

Is the balance's precision as good as its claimed sensitivity? _____

If the average reading is not the same as the stated mass of the standard, your instructor will give you directions on how to adjust the balance so that it is calibrated to the standard.

Measuring Volumes

Your instructor will demonstrate how to measure fluid volumes using several pieces of glassware that differ in their precision, and often accuracy. In this section, you will work with two: a 10-ml graduated cylinder and a pipette (or micropipetter). Below are brief instructions on how to measure and deliver volumes with these devices.

Using Graduated Cylinders Graduated cylinders typically come in capacities ranging from 10 ml to over a liter. The cylinder is graduated in units that vary and you should always check to be sure you understand the precision of the scale, usually noted on the cylinder as ± value.

When aqueous solutions are added to a cylinder, the surface of the solution is not straight across. Instead it is concave because water molecules are attracted more to the cylinder wall than they are to each other. This curvature is called the **meniscus.** When reading a volume in a graduated cylinder, one should always read from the bottom of the meniscus, not the top.

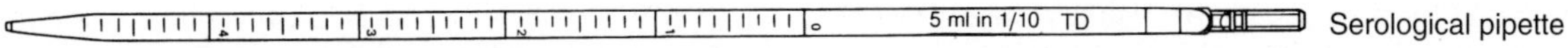

Figure 5.1 One type of glass pipette used in laboratories. The serological pipette must drain dry to dispense the calibrated volume. Pipettes should have the letters TC or TD printed on them. TC means "To Contain" and such pipettes should have the last remaining drop blown out of them. TD means "To Deliver" and such pipettes should be allowed to drain only, leaving a small amount of fluid in the tip at the end when the full contents of the pipette are delivered.

When reading the meniscus, you should also be aware of a second source of error, **parallax.** It occurs when reading an instrument that has a scale and pointer that are not aligned perpendicular to your eyes. If you ever look at the speedometer from the passenger's seat of a car, you experience a parallax error because it appears that the car is going slower than its actual speed as a result of the viewing angle. When reading a graduated cylinder (or a pipette), the meniscus must be at the same level as your eyes to avoid this problem.

To measure a volume of fluid with a graduated cylinder, fill it with the needed solution until the bottom of the meniscus is even with the appropriate calibrated marking on the wall of the cylinder when viewed at eye level. A certain amount of trial and error is involved to achieve the volume you want to use. To deliver the fluid, invert the cylinder and let it drain for three seconds, touching off any drop retained at the open edge.

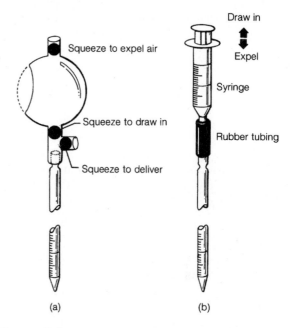

Figure 5.2 Suction devices used to fill glass pipettes: (a) rubber bulb with valves, (b) syringe with rubber connector. Other devices may be demonstrated by your instructor.

> **CAUTION**
> If you use glass pipettes, never draw chemical or biohazardous solutions into a pipette by mouth suction, as you would with a straw. You could accidentally ingest a poison or pathogen. Use a suction device (fig. 5.2.)

Using Glass Pipettes A glass pipette is a long calibrated glass tube (**fig. 5.1**). It is filled by using a syringe or valved rubber bulb attached to the end of the pipette (**fig. 5.2**). To use a pipette, immerse the tip in the appropriate fluid and draw the fluid up beyond the zero mark using the suction device. Look at the surface of the fluid in the pipette and note the **meniscus,** the concave upper surface. Hold the pipette vertically at eye level and allow the fluid to drain until the bottom of the meniscus touches the zero line.

To deliver a volume less than the total volume, let the fluid drain until the bottom of the meniscus touches the line corresponding to the desired volume. If the total volume is to be delivered, you must know if you have a TC-or TD-style pipette (fig. 5.1 legend).

Using Micropipetters Using an automatic micropipetter (**fig. 5.3**) is a bit more involved, and proper technique will

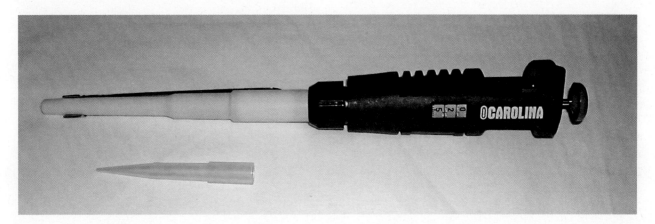

Figure 5.3 A micropipetter has a dial that allows one to set the volume to be delivered, and a disposable tip that is discarded after use. Depressing a plunger button on the top allows you to fill or empty the fluid in the tip.

be demonstrated by your instructor. Usually the following steps are involved.

1. Select a pipetter that dispenses volumes in the range you want.
2. Note the units and dial in the volume you wish to dispense.
3. Put a disposable tip on the pipette with a twisting motion. Make certain it is firmly seated.
4. Depress the plunger to the first stop and, while holding it in that position, submerge the tip in the fluid to be drawn up. Slowly release the plunger to draw in the fluid.
5. Without blotting, put the tip into the tube to receive the fluid. Press the plunger to dispense the fluid, continuing to press it past the first stop to the second, thus expelling any clinging fluid. Touch the tip to the side wall to remove any drops at the tip.
6. If a new solution is to be pipetted, press the tip ejection lever and put on a new tip to eliminate cross-contamination of solutions. The same tip can be used over again when dispensing the same fluid, even though you may change the volume.

Checking Volumetric Accuracies

Your calibrated balance can now be used as a reference tool to check the accuracy and repeatability (precision) of pipettes and graduated cylinders. This is done by dispensing measured volumes of water and checking their weight using the standardized balance. Because the density of water at room temperature is 0.998 g/ml, the expected mass of the volume of water that was delivered can be calculated using this relationship:

Expected mass = density × volume

If the expected mass is not the same as the mass measured with a calibrated balance, there is something wrong in how the volumes were delivered or measured. Either the scales on the instruments were off or technique was poor.

2 Experiment Working with a lab partner, obtain a clean dry cup, 10-ml graduated cylinder, and a pipette with a suction device or a micropipetter. You will also need a cup of water.

Look at the graduated cylinder and note the precision of the volume measurements.

What are the smallest units? _____

Now look at the pipette or micropipetter and note the same. _____

Which is more precise? _____

Your instructor will give you a slip of paper with a number on it between 2 and 10, indicating the number of milliliters of water you should add in each trial as you work through the following directions.

Place the empty cup on the balance pan and use the tare control to adjust the reading to 0. **Taring** zeroes and electronically adjusts the balance to compensate for the mass of the empty cup so that it is not included in subsequent readings.

Now use the graduated cylinder to deliver your assigned amount of water to the cup while it is on the pan and the balance is on. Remember to read water volume at the bottom of the meniscus in the graduated cylinder and hold it at eye level to eliminate parallax. Record the mass of water in the beaker in **table 5.1**.

Tare the balance to compensate for the water just added. Repeat adding your assigned volume of water a second time, recording the measurement as trial 2. Repeat the taring and measurements until you have a total of eight measurements.

Table 5.1 Masses from Repetitive Dispensing of Water

Graduated Cylinder		Pipette/Micropipetter	
Volume to Be Delivered: (ml)	Mass of Water Delivered (g)	Volume to Be Delivered: (ml)	Mass of Water Delivered (g)
Trial 1		Trial 1	
Trial 2		Trial 2	
Trial 3		Trial 3	
Trial 4		Trial 4	
Trial 5		Trial 5	
Trial 6		Trial 6	
Trial 7		Trial 7	
Trial 8		Trial 8	
Mean **mass** delivered		Mean **mass** delivered	
Mean **volume** delivered		Mean **volume** delivered	
Range of observations		Range of observations	
Mean % error		Mean % error	
Standard deviation		Standard deviation	

3 Discard the water and dry the cup. Now you will use a pipette or micropipetter to do the experiment again. Remember the difference in precision between the measuring devices.

Place the cup back on the balance and tare its mass. Add your assigned volume of water and record the results as trial 1. Repeat, adding water with a pipette and taring the balance until you have results from eight trials. Record these results in table 5.1.

4 *Analysis* Calculate the mean mass of water delivered by each device. Enter the values in table 5.1. Look over the values in each series and determine for each, the lowest and highest value recorded. These encompass the **range** of observations and are a measure of variability in the results. Enter the ranges in table 5.1.

Use the mass of water divided by its density (0.998 g/ml at 23°C) as the standard to check the accuracy of the volume measuring methods.

Using the density of water, calculate mean volume of water delivered in each series. Note this is your first opportunity to make a judgment about the use of significant figures. (See appendix A.)

How many ml of water did you expect to deliver each time? _____

How many ml were delivered on average with the graduated cylinder? _____

How many on average with the pipetter? _____

In comparing the two methods of measuring volume, which method was more accurate?

Which method was more precise?

Does this correspond to the precision of the device or to the fact that one method required judgment and coordination while the other did not? Can you tell?

A value called **% error**, sometimes called **experimental error**, is often calculated to indicate the accuracy of measurements. The equation for calculating % error is:

$$\% \text{ error} = \frac{(\text{measured value} - \text{expected value}) \times 100}{\text{expected value}}$$

5 Calculate the % error for the averages of both series of measurements and enter in table 5.1.

If your % errors are not zero, think about how you performed the experiment. Is there anything you consistently did that might have produced an error or is there a problem with the insruments? List your ideas below.

If you can rule out errors from your techniques, you are left with the conclusion that the pipettes or graduated cylinder were not accurately calibrated. The scales do not indicate true values. Before leaping to this conclusion, the experiment should be done a second or third time to see if the same results are obtained with the same devices. If the results are consistent and the error is large, exceeding the accuracy that your research requires, then there are two options. A correction factor could be applied to all volumes being measured with that cylinder or pipette. For example if the average error is +10%, then decrease the volume to be delivered by 10%. Alternatively, repair or replace the device.

Calculating Simple Statistics

Background

The measurements you have made are not true values but are simply estimations of a true value called the **expected value.** Some estimations obtained by the class are lower than the expected value, while others are higher. When it is important for scientists to obtain a true measurement, they repeat the measurement several times and calculate a **mean** (average) value as you did.

Chance errors in data sets cancel each other out when means are calculated—that is, a value that is too high due to chance error is balanced by a value that is low for the same reason. The **range** of observations gives some sense of the variability in measurement. However, range is not a very good estimator of variability because it can be artificially inflated by one or two outlying values. Consider the following two sets of hypothetical data:

	Set A	Set B
	30	30
	29	40
	31	20
	28	32
	32	31
	30	30
	29	31
	31	26
Σ (= sum)	240	240
N (= number of observations)	8	8
Mean	30	30
Range	28 – 32(±2)	20 – 40(±10)

The data sets have the same mean, but one pair of values in set B created an extremely broad range despite having the same means. Because of this problem and the need to convey information about the amount of variability in a set of measurements, scientists use prescribed calculations to obtain variability estimators called **variance** and **standard deviation.** (Appendix C contains a more thorough discussion of statistics.) These estimators are obtained by expressing all values in a data set as plus or minus variations from the mean. Variance and standard deviation are calculated as follows:

$$\text{Variance} = \frac{\Sigma(\text{measured value for each sample} - \text{mean})^2}{(N-1) \text{ one less than number of observations}}$$

$$\text{Standard deviation} = \pm\sqrt{\text{variance}}$$

6 Use the following work table to calculate the standard deviation of the hypothetical set of measured values given in data Set A.

Measured Value	Measured Minus Mean	(Measured Minus Mean)^2
15		
20		
25		
16		
20		
24		
10		
20		
30		
Σ _____		Σ _____
N _____		N – 1 _____
Mean _____		Variance _____
		SD _____

Do your answers agree with those of other students? Check your arithmetic if they do not.

Statistical Analysis of Your Data

7 Having practiced how to calculate a standard deviation, now apply this concept to your water data. Calculate the standard deviation for both sets of your water mass data and enter the values on the last line of table 5.1. Standard deviations are often reported with the signs ± in front of the number. For data sets that include more than about six measurements, about 2/3 of the observations should lie within ±1 standard deviation of the average. See appendix C for further discussion.

Standard deviation summarizes the variability in repetitive measurements. Automatically calculate it any time you calculate a mean involving six or more measurements. Good experimental technique should always yield a small standard deviation.

What does the standard deviation suggest about the precision of the two methods used to measure volume?

Students who have experience with spreadsheet programs realize that such software contains functions for simple statistical analyses as do many hand calculators. You need only enter the data set into a spreadsheet and call up the functions AVERAGE and STANDARD DEVIATION.

Using a Spectrophotometer

Background

Spectrophotometry is a method used to identify and quantify colored solutions based on their light absorbing properties. Many organic molecules absorb radiant energy because of the nature of their chemical bonds. Light-absorbing organic molecules often have a system of single and double bonds between adjacent carbons or carbon and nitrogen. For example, proteins and nucleic acids absorb ultraviolet light in the wavelength interval 240 to 300 nanometers (nm), pigments and dyes absorb visible light (about 400 to 770 nm), and many organic compounds absorb infrared energy (above 770 nm). Our perception of color is related to the ability of pigment molecules in our eyes' cone cells to absorb light energy. If an object appears to be red, it contains molecules whose chemical bonds absorb blue or green light while reflecting or transmitting red light to the red sensitive cones of our eyes.

A **spectrophotometer** is an instrument designed to detect the amount of light energy absorbed by molecules in a solution. Spectrophotometers have five basic components: a **light source,** a **diffraction grating,** a **slit,** a **detector,** and a **readout** to display the output of the detector. The arrangement of these parts is shown in **figure 5.4.**

When light passes through the diffraction grating, it is split into a spectrum of colors or wavelengths, which then diverge. Sections of the projecting spectrum can be either blocked or allowed to pass through a slit, so that only one color enters other sections of the spectrophotometer.

Light that passes through the slit travels to the detector, where it creates an electric current proportional to the number of photons. If a current meter is attached to the detector, the electric current output—which represents the quantity of light striking the detector—can be measured and displayed on a meter or a digital readout. The scale is usually calibrated in two ways: **percent transmittance,** which runs on a scale from 0 to 100, and **absorbance,** which runs from 0 to 2 in most practical applications.

Before the light-absorbing properties of a solution can be measured, two and sometimes three adjustments to the spectrophotometer are necessary. First, the diffraction grating must be adjusted so that the desired color of light (measured as wavelength) passes through the slit. Rarely does a spectrophotometer have any means for calibrating this, so the settings on the dial are assumed to be true values. Second, the output of the detector must be adjusted to correct for drift in the electronic circuit. Third, a compensation must be made for dirt or contaminating colored material in the light path between the source and the detector.

To adjust for materials in the light path, a clean sample tube is filled with the same solvent (usually water) used in dissolving the dye. It is placed in the sample compartment of the spectrophotometer (see **fig. 5.5**). This tube is called a **blank** and the process is called **zeroing.** The readout should indicate 0 absorbance, or 100% transmittance, depending on which scale you are using. If it does not, an adjustment must be made. If your lab is equipped with modern digital spectrophotometers, a single control may both read the blank and zero the instrument.

If a colored solution is put in the tube in place of the pure solvent, some of the light coming from the slit will be absorbed, and some will be transmitted to the detector. The amount absorbed will be proportional to the concentration of the dye molecules.

The readout indicates the detector's output. If the transmittance scale is used, the amount of light transmitted by the solution is measured as a percentage by the spectrophotometer. This measurement is described by the following equation:

$$\text{Percent transmittance (T)} = \frac{\text{intensity of light through sample}}{\text{intensity of light through blank}} \times 100$$

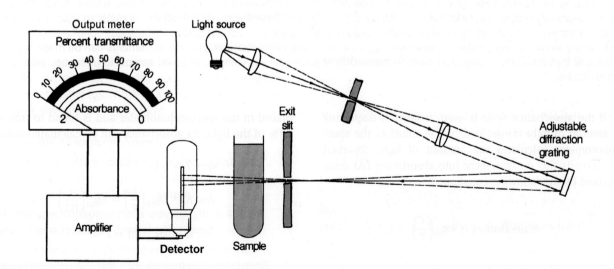

Figure 5.4 Schematic drawing of the path of light through a spectrophotometer. Newer spectrophotometers will have a digital readout rather than an output meter.

measurements could be further refined using smaller intervals, but such precision is not necessary for your work in this lab.

8. Using your data in table 5.2, refine your measurements and record the readings on the last three lines. Remember to zero the instrument at each new wavelength.

Common Errors in Spectrophotometry Experience indicates that students often get poor results when using spectrophotometers because they:

Incompletely mix solutions before placing in cuvettes;

Forget to blank the spectrophotometer each time wavelength is changed;

Hold cuvettes incorrectly so that fingerprints are made on optical surfaces;

Use dirty glassware or cuvettes.

9. Analysis Plot the measured *absorbance* as a function of wavelength on the graph paper at the end of this lab topic or on a computer using graphics software. Instructions for drawing graphs are in appendix B. Read them! What is the independent variable?_____

When all points are plotted, draw a smooth curve connecting the points. Because absorbance is not linearly related to wavelength, it would not make sense to draw a straight line through the points.

The plot you have made is called an **absorbance curve** or an **absorption spectrum.**

What is the wavelength of peak absorption for bromophenol blue? _____

What are the units? _____

Constructing a Standard Curve In this section you will construct a *standard curve,* which demonstrates the linear relationship between absorbance and concentration (fig. 5.6) described in the Lambert-Beer Law. You will use this standard curve to determine the concentration of dye in an unknown.

10. Procedure Obtain a stock solution of bromophenol blue containing 0.02 mg/ml. Prepare a series of dilutions in eight test tubes, using the proportions of dye and water listed in **table 5.3**. After adding the solutions, mix on a vortex mixer or by holding the test tube in the left hand between your thumb and forefinger. Gently strike the bottom of the tube several times with the forefinger of your right hand to create a swirling motion in the tube.

11. Set the spectrophotometer at the maximum absorption wavelength for bromophenol blue as determined in the previous experiment. Record the wavelength in the heading of table 5.3. After calibrating the spectrophotometer with water as a blank, read the *absorbance* for all eight tubes in the dilution series at this wavelength. Because you are not changing wavelength, you do not need to blank the spectrophotometer between sample readings. Record your results in the last column of table 5.3.

12. Analysis To calculate the actual concentration of bromophenol blue in each tube as called for in the fourth column of table 5.3, do the following:

1. Determine the total mg of dye added to each tube by multiplying the number of ml of added dye by the dye concentration, which was 0.02 mg/ml.

2. Then divide the value by 4 ml, the total volume of fluid present after adding water.

These calculations are summarized in the following work table.

Tube	ml Dye	Total mg Dye	Dye Concentration
1	___	___	___
2	___	___	___
3	___	___	___
4	___	___	___
5	___	___	___
6	___	___	___
7	___	___	___
8	___	___	___

Table 5.3 Concentration and Absorbance for Eight Dye Dilutions at _____ Nanometers

Tube	ml of Dye	ml of H_2O	Concentration	A
1	0	4.00	0.00 mg/ml	___
2	0.50	3.50	___	___
3	1.00	3.00	___	___
4	1.50	2.50	___	___
5	2.00	2.00	___	___
6	2.50	1.50	___	___
7	3.00	1.00	___	___
8	3.50	0.50	___	___
9	4.00	0	0.02 mg/ml	___
10	unknown	0	___	___

properties. Many organic molecules absorb radiant energy because of the nature of their chemical bonds. Light-absorbing organic molecules often have a system of single and double bonds between adjacent carbons or carbon and nitrogen. For example, proteins and nucleic acids absorb ultraviolet light in the wavelength interval 240 to 300 nanometers (nm), pigments and dyes absorb visible light (about 400 to 770 nm), and many organic compounds absorb infrared energy (above 770 nm). Our perception of color is related to the ability of pigment molecules in our eyes' cone cells to absorb light energy. If an object appears to be red, it contains molecules whose chemical bonds absorb blue or green light while reflecting or transmitting red light to the red sensitive cones of our eyes.

A **spectrophotometer** is an instrument designed to detect the amount of light energy absorbed by molecules in a solution. Spectrophotometers have five basic components: a **light source**, a **diffraction grating**, a **slit**, a **detector**, and a **readout** to display the output of the detector. The arrangement of these parts is shown in **figure 5.4**.

When light passes through the diffraction grating, it is split into a spectrum of colors or wavelengths, which then diverge. Sections of the projecting spectrum can be either blocked or allowed to pass through a slit, so that only one color enters other sections of the spectrophotometer.

Light that passes through the slit travels to the detector, where it creates an electric current proportional to the number of photons. If a current meter is attached to the detector, the electric current output—which represents the quantity of light striking the detector—can be measured and displayed on a meter or a digital readout. The scale is usually calibrated in two ways: **percent transmittance**, which runs on a scale from 0 to 100, and **absorbance**, which runs from 0 to 2 in most practical applications.

Before the light-absorbing properties of a solution can be measured, two and sometimes three adjustments to the spectrophotometer are necessary. First, the diffraction grating must be adjusted so that the desired color of light (measured as wavelength) passes through the slit. Rarely does a spectrophotometer have any means for calibrating this, so the settings on the dial are assumed to be true values. Second, the output of the detector must be adjusted to correct for drift in the electronic circuit. Third, a compensation must be made for dirt or contaminating colored material in the light path between the source and the detector.

To adjust for materials in the light path, a clean sample tube is filled with the same solvent (usually water) used in dissolving the dye. It is placed in the sample compartment of the spectrophotometer (see **fig. 5.5**). This tube is called a **blank** and the process is called **zeroing**. The readout should indicate 0 absorbance, or 100% transmittance, depending on which scale you are using. If it does not, an adjustment must be made. If your lab is equipped with modern digital spectrophotometers, a single control may both read the blank and zero the instrument.

If a colored solution is put in the tube in place of the pure solvent, some of the light coming from the slit will be absorbed, and some will be transmitted to the detector. The amount absorbed will be proportional to the concentration of the dye molecules.

The readout indicates the detector's output. If the transmittance scale is used, the amount of light transmitted by the solution is measured as a percentage by the spectrophotometer. This measurement is described by the following equation:

$$\text{Percent transmittance (T)} = \frac{\text{intensity of light through sample}}{\text{intensity of light through blank}} \times 100$$

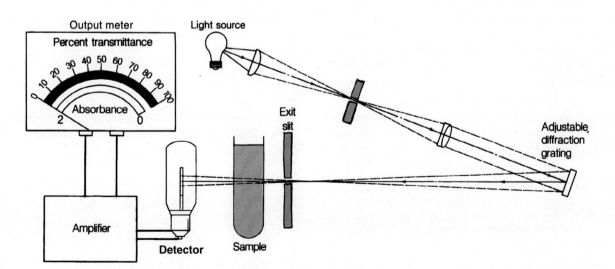

Figure 5.4 Schematic drawing of the path of light through a spectrophotometer. Newer spectrophotometers will have a digital readout rather than an output meter.

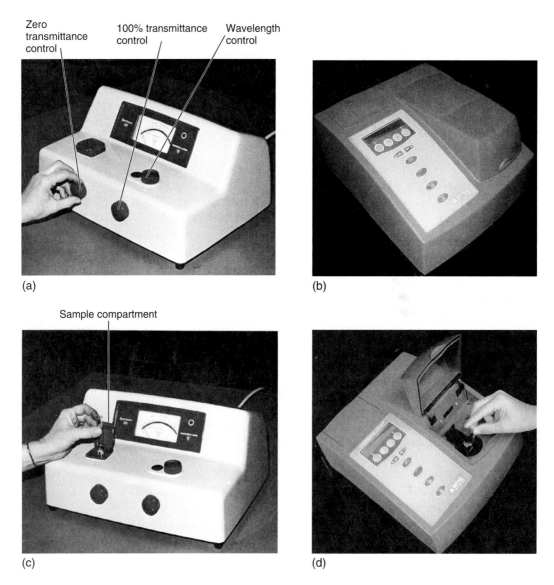

Figure 5.5 Steps for using a spectrophotometer. Two models are shown. The following steps apply to either one. (1) Turn on the instrument and wait 10 minutes for it to stabilize. (2) Adjust the wavelength to 450 nanometers. (3) Only if you are using an older machine with a meter, adjust the meter reading to zero transmittance by either rotating the zero control knob or using the zero control button. (4) Insert a tube containing only solvent into the sample compartment. Keep the index line of the tube aligned with the index line on the sample holder. (5) Close the cover and adjust the meter to read OA (100% T), by either rotating the 100% transmittance control knob or pushing the appropriate button. If the spectrophotometer is used for any length of time, recheck these readings now and then. Once standardized, measurements can be made on samples at that wavelength only. You must restandardize at each new wavelength. (6) Insert sample to be measured and record results.

If the **absorbance** scale is used, instead of displaying the amount of light transmitted by the solution, the spectrophotometer calculates the amount of light absorbed and converts this measurement into absorbance (A) units described by the following equation:

$$A(\text{absorbance}) = \log_{10}\left(\frac{1}{T}\right)$$

An example may help to clarify the relationship between transmittance and absorbance. If a dye solution is placed in the spectrophotometer and is found to transmit 10% of the light, its absorbance can be calculated thus:

Since $T = 0.10$

And $A = \log_{10}\left(\frac{1}{T}\right) = \log_{10}\left(\frac{1}{0.1}\right)$

$A = \log_{10}(10) = 1$

Scientists prefer to work with absorbance units because they are directly and linearly related to the concentration of the dye in solution (see **fig. 5.6**). This relationship is

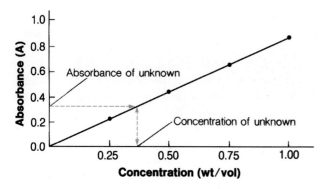

Figure 5.6 Standard curve for the absorbances of known dye concentrations. First, the absorbances of four known concentrations of dye are measured and plotted to make a graph. Then the absorbance of an unknown concentration is measured. The graph is used to find the concentration of dye that would give the measured absorbance.

described by the **Lambert-Beer Law,** which states that for a given concentration range absorbance is directly proportional to the concentration of solute molecules.

This relationship will be true regardless of the dye used, providing monochromatic (one-color) light is used. The wavelength of light used should be strongly absorbed by the dye but weakly absorbed by other molecules in solution. Often we choose the wavelength of maximum absorption to maximize sensitivity. That wavelength is usually given in the directions for any experiment or can be experimentally determined as you will do in the next section.

Spectrophotometer Experiments

8 ▶ When using the spectrophotometer, work with a lab partner. Obtain two cuvettes, a pipette, and about 25 ml of aqueous bromophenol blue (concentration 0.02 mg/ml) in a small beaker. You will use these materials for two experiments: Creating an absorbance curve where you will determine the wavelength of maximum absorption and then constructing a standard curve which will allow you to determine the concentration of dye in an unknown solution given to you by your instructor.

Familiarize yourself with the controls on your spectrophotometer, using figure 5.5 as a guide. Your instructor will answer any questions you have.

Measuring an Absorption Spectrum

1. Prepare a cuvette with distilled water, called the **blank,** and a cuvette with the bromophenol blue standard solution, called the **sample.**
2. Set the spectrophotometer's wavelength to 450 nm using the wavelength control adjustment.
3. To zero the spectrophotometer, insert the blank, close the sample compartment cover, and adjust the A to 0 (or 100% T). The display should now read 0.00 A.
4. Remove the blank and insert your sample. Remember to always wipe the outside of the cuvette before inserting it into the compartment and always insert it in with the same side forward. Record the absorbance reading in the appropriate row in **table 5.2**. *Make sure you are reading absorbance (A) and not transmittance (T).* Absorbance readings will be between 0.00 A and 2.00 A.
5. Remove the sample cuvette.
6. Change the wavelength to 475 nm and repeat steps 3–5. Continue increasing the wavelength by 25 nm intervals until you reach 650 nm and you have filled out table 5.2 except for the last three lines.
7. Look at your data in table 5.2 and determine at what wavelength the absorbance readings is highest. Because you changed wavelengths at 25 nm intervals, you do not know the wavelength of maximum light absorption very precisely. Maximum absorption could be above or below your maximum value in the intervals between readings. To refine your precision, make additional measurements of absorption on either side of the peak.

    For example, if your original data are:

475 nm	0.82 A
500 nm	0.96 A
525 nm	1.10 A
550 nm	0.88 A

    The maximum absorbance is somewhere near 525. Increase the wavelength to 530 and measure A. Let's say it gives a reading of 0.99 A.
    Now decrease the wavelength 10 nm to 520 nm and measure A. Let's say it gives 1.15 A.
    If decreasing the wavelength further to 515 nm gives 0.97 A, you are finished.

The maximum absorbance at the resolution of these hypothetical measurements is at 520 nm. The

Table 5.2	Absorbance Readings for Bromophenol Blue
Wavelength (nm)	**Absorbance Units**
450	_____
475	_____
500	_____
525	_____
550	_____
575	_____
600	_____
625	_____
650	_____
_____	_____
_____	_____
_____	_____

measurements could be further refined using smaller intervals, but such precision is not necessary for your work in this lab.

8. Using your data in table 5.2, refine your measurements and record the readings on the last three lines. Remember to zero the instrument at each new wavelength.

Common Errors in Spectrophotometry Experience indicates that students often get poor results when using spectrophotometers because they:

Incompletely mix solutions before placing in cuvettes;

Forget to blank the spectrophotometer each time wavelength is changed;

Hold cuvettes incorrectly so that fingerprints are made on optical surfaces;

Use dirty glassware or cuvettes.

9 *Analysis* Plot the measured *absorbance* as a function of wavelength on the graph paper at the end of this lab topic or on a computer using graphics software. Instructions for drawing graphs are in appendix B. Read them! What is the independent variable?_____

When all points are plotted, draw a smooth curve connecting the points. Because absorbance is not linearly related to wavelength, it would not make sense to draw a straight line through the points.

The plot you have made is called an **absorbance curve** or an **absorption spectrum**.

What is the wavelength of peak absorption for bromophenol blue? _____

What are the units? _____

Constructing a Standard Curve In this section you will construct a *standard curve*, which demonstrates the linear relationship between absorbance and concentration (fig. 5.6) described in the Lambert-Beer Law. You will use this standard curve to determine the concentration of dye in an unknown.

10 *Procedure* Obtain a stock solution of bromophenol blue containing 0.02 mg/ml. Prepare a series of dilutions in eight test tubes, using the proportions of dye and water listed in **table 5.3**. After adding the solutions, mix on a vortex mixer or by holding the test tube in the left hand between your thumb and forefinger. Gently strike the bottom of the tube several times with the forefinger of your right hand to create a swirling motion in the tube.

11 Set the spectrophotometer at the maximum absorption wavelength for bromophenol blue as determined in the previous experiment. Record the wavelength in the heading of table 5.3. After calibrating the spectrophotometer with water as a blank, read the *absorbance* for all eight tubes in the dilution series at this wavelength. Because you are not changing wavelength, you do not need to blank the spectrophotometer between sample readings. Record your results in the last column of table 5.3.

12 *Analysis* To calculate the actual concentration of bromophenol blue in each tube as called for in the fourth column of table 5.3, do the following:

1. Determine the total mg of dye added to each tube by multiplying the number of ml of added dye by the dye concentration, which was 0.02 mg/ml.

2. Then divide the value by 4 ml, the total volume of fluid present after adding water.

These calculations are summarized in the following work table.

Tube	ml Dye	Total mg Dye	Dye Concentration
1	___	___	___
2	___	___	___
3	___	___	___
4	___	___	___
5	___	___	___
6	___	___	___
7	___	___	___
8	___	___	___

Table 5.3 Concentration and Absorbance for Eight Dye Dilutions at _____ Nanometers

Tube	ml of Dye	ml of H$_2$O	Concentration	A
1	0	4.00	0.00 mg/ml	___
2	0.50	3.50	___	___
3	1.00	3.00	___	___
4	1.50	2.50	___	___
5	2.00	2.00	___	___
6	2.50	1.50	___	___
7	3.00	1.00	___	___
8	3.50	0.50	___	___
9	4.00	0	0.02 mg/ml	___
10	unknown	0	___	___

Transfer the dye concentrations in this work table to the "Concentration" column in table 5.3.

13 Use the graph paper at the end of this lab topic or your computer to plot your data with absorbance as a function of dye concentration. Instructions for making graphs are given in appendix B. Remember that the x-axis, or abscissa, is always the independent variable and the y-axis, or ordinate, is always the dependent variable. In this experiment, what is the independent variable?

Label both axes. After plotting your data points, draw a straight line that, on the average, best fits all points of the data. Do not connect the points and create bumpy curves. The best straight line technique compensates for some of the random variability in the data.

The plot of absorbance as a function of dye concentration is called a **standard curve.** By reading the graph, you can determine the absorbance of any dye concentration within the range of concentrations tested. The line on the graph may be extrapolated to predict the absorbance of concentrations beyond the highest tested, but there is always a danger that the Lambert-Beer Law does not apply at very high concentrations.

14 *Determining an Unknown Concentration* Your instructor will now give you a solution of bromophenol blue that contains an unknown amount of dye.

How can you determine the dye concentration using the spectrophotometer and the standard curve you just constructed? Briefly outline the steps below. Perform the procedure and record the result below.

Unknown concentration = _____ mg / ml

Most likely, you recorded your unknown dye concentration with three or more, decimal places. If you did, it is time to read appendix A about significant figures. In this experiment, what variable is known with the least precision? _____ How many significant figures does it have? _____ How many significant figures can you legitimately use in estimating the dye concentration in the unknown? _____

15 Your instructor will now tell you the actual concentration of dye in the unknown. Calculate your percent error.

Lab Summary

16 On a separate sheet of paper, answer the following questions as assigned by your instructor.

1. Explain the difference between precise and accurate.
2. Describe why a standard is used to calibrate a device.
3. Describe the differences between mean, range, and standard deviation.
4. Turn in your standard curve of absorbance versus dye concentration. Indicate the absorbance of your "unknown" on the graph and the dye concentration to which it corresponds.
5. Explain how you would create a standard curve to determine the amount of orange dye that had been added to a can of orange soda.

You may want to try the critical thinking questions that apply some of the knowledge you gained in doing this lab.

Learning Biology by Writing

Prepare a lab report in which you present your data on determining the dye concentration in the unknown (include your graph). Be sure you state the problem and then discuss how you solved it. In your discussion, describe in general terms some sources of experimental error in lab work and the value of working with repetitive measurements.

Critical Thinking Questions

1. Explain why the sample cuvettes (sample tubes) used with the spectrophotometer are "matched." Why were the outsides cleaned each time before placing them in the spectrophotometer? What effect does the sample volume have on readings? What effect does tube placement have on readings?

2. Describe a "normal distribution" curve. What percentage of the data is encompassed by one standard deviation on each side of the mean?

3. What is meant by "grading on the curve" or "curving" the grades?

4. A class has just finished doing the standard curve part of this exercise and determined the dye concentrations in unknown samples. All of the determinations are 25% lower than expected. Describe at least three sources of error that would give these results.

5. You have a sample of dye that contains 1 mg of dye per ml. You dilute one ml of the sample to a total volume of 5 ml. What is the dye concentration in 1 ml of the diluted sample?

Lab Topic 6

Visualizing Biological Molecules

Supplies

Resource guide available at
www.mhhe.com/labcentral

Equipment and Software

Computers with Jmol installed. Program can be downloaded from http://jmol.sourceforge.net

Download the pdb files for the following molecules by using Google or visiting the World Index of Biomolecular Visualization Resources at http://molvis.sdsc.edu/visres. (1) water; (2) glucose; (3) sucrose; (4) amylose; (5) stearic acid; (6) oleic acid; (7) glycerol; (8) triglyceride; (9) phospholipid; (10) membrane lipids; (11) alanine; (12) serine; (13) leucine; (14) phenylalanine; (15) hexapeptide; (16) peroxidase; (17) AMP; (18) DNA.

Student Prelab Preparation

Before doing this lab, read the Background and other sections of the lab topic that have been assigned by your instructor.

Use your textbook to review the definitions of the following terms:

Acid	Hydroxyl group
Amino group	Molecular mass
Base	Methyl group
Carboxyl group	Monomer
Covalent bond	Nonpolar
Ester linkage	Peptide bond
Functional group	Phosphate group
Glycosidic linkage	Phosphodiester linkage
Hydrogen bond	Polar
Hydrophilic	Polymer
Hydrophobic	Sulfhydryl group

Describe in your own words the following concepts:

Role of covalent and hydrogen bonds in three-dimensional structure of molecules

Carbon backbones in organic molecules

Role of functional groups in biochemical reactions

Condensation or dehydration synthesis reactions in polymerization and hydrolysis reactions

How shape can reflect molecular function

After finishing the prelab review, write any questions you have about terms, concepts, or techniques in the margins of this lab topic. The lab activities should help you answer these questions, or you can ask your instructor during the lab.

Objectives

1. To recognize carbon backbones and functional groups characteristic of carbohydrates, lipids, amino acids, proteins, and nucleic acids
2. To predict hydrophilic and hydrophobic interactions
3. To understand the concept of a polymer and to identify monomers of major biopolymers
4. To recognize that the shapes and sizes of biological molecules relate to their function

Background

Although virtually every chemical element can be found in living organisms, the most common are carbon, hydrogen, oxygen, nitrogen, phosphorus, and sulfur. The acronym CHONPS is a convenient way to remember these elements. Carbon is perhaps the most important element. It can chemically combine with other carbons or with H, O, N, P, S, as well as with many other elements to form thousands of chemical compounds. Carbon-containing molecules are organic molecules, and those common in living organisms are biochemicals.

Carbon forms four covalent bonds by sharing electrons with up to four adjacent atoms. For example, a carbon could form covalent bonds to four hydrogen atoms (CH_4), or it could combine with two atoms of oxygen (CO_2) where two covalent bonds link each oxygen to the carbon. Carbon can share electrons with adjacent carbons to build chains of carbons called **carbon backbones** (**fig. 6.1**). Such backbones may vary from two carbons to over 22. Other atoms can share electrons with the carbons in the backbone and form projecting "limbs" that give an organic molecule chemical and physical properties that the backbone alone does not have. Sometimes these limbs can be complex and consist of several atoms. The limbs, called **functional groups,** give unique chemical properties to each type of organic molecule.

Six types of functional groups (**fig. 6.2**) are common in biological molecules, although there certainly are others. Not all biological molecules have the same functional

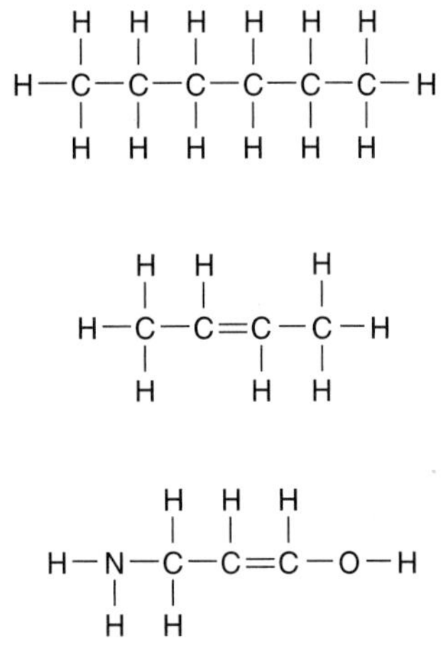

$$H-\underset{\underset{H}{|}}{\overset{\overset{H}{|}}{C}}-\underset{\underset{H}{|}}{\overset{\overset{H}{|}}{C}}-\underset{\underset{H}{|}}{\overset{\overset{H}{|}}{C}}-\underset{\underset{H}{|}}{\overset{\overset{H}{|}}{C}}-\underset{\underset{H}{|}}{\overset{\overset{H}{|}}{C}}-\underset{\underset{H}{|}}{\overset{\overset{H}{|}}{C}}-H$$

$$H-\underset{\underset{H}{|}}{\overset{\overset{H}{|}}{C}}-\underset{H}{\overset{H}{C}}=\underset{H}{C}-\underset{\underset{H}{|}}{\overset{\overset{H}{|}}{C}}-H$$

$$H-\underset{\underset{H}{|}}{N}-\underset{\underset{H}{|}}{\overset{\overset{H}{|}}{C}}-\overset{\overset{H}{|}}{C}=\underset{}{C}-O-H$$

Figure 6.1 Examples of carbon backbones in organic molecules. Each line represents a covalent chemical bond. Molecules differ in length of backbone, bonding pattern, and elements present.

molecules can have molecular masses over a wide range from approximately 100 for many amino acids to over 1,000,000 for some proteins and nucleic acids.

Biochemists talk about classes of organic molecules: carbohydrates/polysaccharides, lipids, amino acids/proteins, and nucleotides/nucleic acids. All have carbon backbones but differ from each other in the length of their backbones and in the types of functional groups that are present. There are certainly additional classes of molecules, but knowledge of the four basic classes allows one to appreciate many of life's molecular processes. In this lab topic you will explore the molecular structures characteristic of each of these classes.

The medium of life is water. Almost all biochemical reactions involve molecules that are dissolved in water. The simple structure of water (H_2O) obscures its great importance. What is so important about water? Water is a **polar molecule,** a dipole with two regions that have a tendency to be positive and a single region that tends to be negative (**fig. 6.3**). Consequently, water molecules interact with each other and with other charged and polar molecules by forming weak opposite-charge interactions called **hydrogen bonds.** Hydrogen bonds are much weaker than covalent bonds, but nonetheless are very important in biological systems. They determine what dissolves in cells and are the basis for the three-dimensional structure of proteins and nucleic acids.

Polarity explains why some molecules dissolve in water and others do not. **Figure 6.4** illustrates why table salt (NaCl) readily dissolves. Charged, Na^+ and Cl^- ions in a crystal are in constant thermal motion. As an ion starts

groups. Some contain only one type while others have several. The single line before the functional group indicates that the group will share one pair of electrons with a carbon through a covalent bond.

The molecular mass of an organic molecule is a measure of its size. For example, glucose ($C_6H_{12}O_6$) has a molecular mass of 180. This is calculated by summing the masses of all the atoms in the molecule. Biochemical

Group	Structural Formula	Ball-and-Stick Model	Chemical Properties
Hydroxyl	—OH	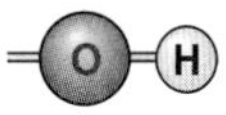	Polar; hydrophilic
Carbonyl	—C— ‖ O	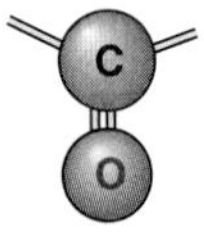	Moderately polar
Carboxyl	—C(=O)OH	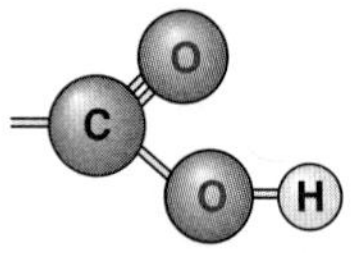	Acid; releases H^+
Amino	—N(H)(H)	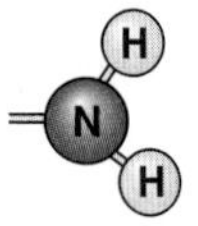	Base; accepts H^+
Sulfhydryl	—S—H	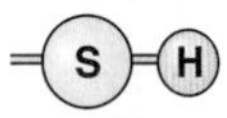	Nonpolar
Phosphate	—O—P(=O)(O⁻)—O⁻	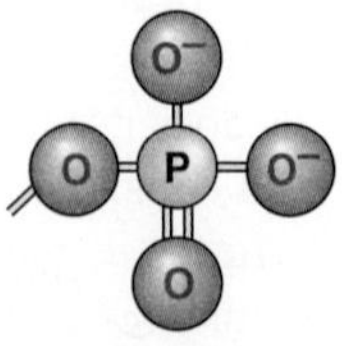	Polar; negative charge in solution

Figure 6.2 Common functional groups in biological molecules.

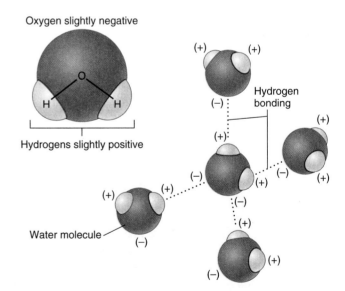

Figure 6.3 Chemical bonds in water. Two hydrogens share electrons with oxygen through covalent bonds. Oxygen attracts the electrons more strongly than do the hydrogens. Consequently, the oxygen bears a partial negative charge and the hydrogens bear slightly positive charges, creating a dipole molecule. The resulting dipoles are attracted to one another, forming hydrogen bonds that weakly hold them in position.

VISUALIZING BIOLOGICAL MOLECULES

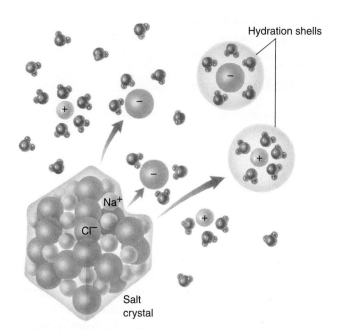

Figure 6.4 Ionized substances such as salt dissolve in water because the water dipoles are attracted to the ions, forming hydration shells around them. The hydration shells prevent the ions from binding back into the crystal. Note the difference in orientation of the water molecules surrounding the Na and Cl ions. Similar hydration shells develop around polar organic molecules when they dissolve in water.

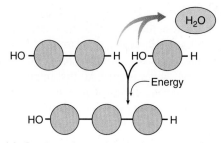

(a) Condensation or dehydration synthesis

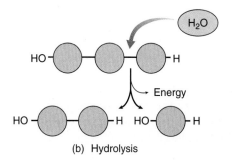

(b) Hydrolysis

Figure 6.5 (a) Condensation reactions between functional groups form biological polymers. A hydrogen from one monomer is removed along with a hydroxyl from the other to form water and a larger organic molecule. (b) The bond can be broken in a hydrolysis reaction. This is what occurs in digestion.

to escape a crystal, it is usually pulled back by ions of the opposite charge, except in aqueous solutions. In water, the escaping ion is immediately surrounded by oriented water molecules that form a **hydration shell** around it. The hydration shell prevents the escaped ion from binding back into the crystal; consequently, it remains in solution.

A molecule that readily dissolves in water is said to be **hydrophilic,** or water loving. If a biochemical molecule is relatively neutral with no polar tendency, it will not dissolve in water and is said to be **hydrophobic,** or water hating. The functional groups $-OH$, $-COOH$, $-H_2PO_4$, and $-NH_2$ are hydrophilic, whereas those with $-H$, $-CH_3$ and $-SH$ are hydrophobic. In large molecules, it is the relative number and location of different functional groups that determines solubilities, and some molecules may be partially hydrophilic and partially hydrophobic. Thus solubility properties are not black and white: there are many shades of gray. In this lab, you will predict whether a molecule is hydrophilic or hydrophobic based on its structure.

Many biological molecules are **polymers,** very large molecules made from a series of repeating subunits, called monomers, covalently linked together. Examples are proteins made from a few to several hundred amino acids and polysaccharides like starch and cellulose are polymers made from thousands of sugar molecules. Biological polymers result from **condensation** or dehydration synthesis reactions (fig. 6.5). In these reactions, the $-OH$ group of a carboxyl group combines with a hydrogen from a hydroxyl, phosphate, or amino group on another monomer to join the two together through a new covalent bond, releasing a water molecule. Because many monomers have more than one of these groups, they can combine with other monomers to build up chains often thousands of monomers in length. Such synthetic reactions usually require the expenditure of cellular energy to make the bonds. These bonds can also be broken through **hydrolysis** reactions. Digestion of starches, lipids, proteins, and nucleic acids are examples of hydrolysis reactions. In this lab topic, you will explore how biological polymers are formed.

LAB INSTRUCTIONS

This lab uses computer modeling to investigate the carbon backbones and functional groups of common biological molecules. You will look at how polymers form from monomers and at the unique shapes of several molecules.

Getting Started

Work with a partner so that you can discuss what you are seeing. On your computer, there is a folder called Molecular Modeling. In it is a program called *Jmol.jar*, which is a program for visualizing molecules. The program uses molecular data files called *pdb* (Protein Data Bank) files that have three-dimensional coordinates for the atoms found in a particular type of molecule based on x-ray diffraction data. These files should be numbered in the order that you will use them during this lab.

1. Double-click on the *Jmol.jar* program to open it. A Jmol window will appear. At the upper left of the window click on *File*, choose *Open* and navigate to the *Molecular Modeling* folder to see the molecule data files.

Open the file entitled **01-Water.pdb.**

When the file opens, you should see a ball and stick representation of water molecules on a black background.

If you put the cursor on top of a water molecule and click and hold while moving the mouse, you will rotate the molecules to change the viewing angle. You can zoom in or out by holding the shift key down, clicking and moving the cursor up or down. Smaller displays allow you to see shapes move easily.

In **table 6.1** record the colors used to represent oxygen and hydrogen. Using the atomic masses given in the table calculate the molecular mass of water and record it below.

Note the general shape of a water molecule. The bond angle between the hydrogens is about 104° and the lines from the Hs to the O represent electrons shared through covalent bonds. What the model does not show is that the mobile electrons tend to cluster more frequently around the O nucleus than around the H nuclei. Because electrons carry a negative charge, the O in the center of the angled molecule has a tendency to be negative. Clustering of the negatively charged electrons around the oxygen partially exposes the positively charged hydrogen nuclei. Thus a **dipole** is established. When biochemicals containing polar functional groups are added to water, the dipoles orient around the groups in hydration shells. Sometimes these are sufficient to keep the molecule suspended in solution. When the biological molecules are very large with hydrophilic and hydrophobic regions, the hydration shells cause hydrophilic regions to extend and hydrophobic regions to fold, altering the three-dimensional shapes of those molecules.

During this lab you will look at 17 other molecules and answer questions about their structure, shape, and solubility. This is a compare and contrast exercise to help you recognize different classes of molecules, functional groups, monomers, polymers, molecular shapes, and hydrophobic and hydrophilic characteristics.

Carbohydrates

Carbohydrates, commonly known as sugars or saccharides and by their polymer names such as starch and cellulose, are found in every living cell. They serve as sources of energy when broken down in metabolism, as chemical building units for other types of molecules, as cell to cell recognition molecules, and for structural support in cell walls of plants, fungi, and bacteria.

Sugars You will start your investigation of carbohydrates by looking at simple sugars or monosaccharides. Sugars contain C, H, and O in the ratio of one C to two Hs to one O. The number of carbons in the backbone of the molecule can vary between 3 and 7. Thus when there are five carbons, the formula is $C_5H_{10}O_5$. This would be called a pentose because it is a sugar with five carbons. How many carbons are found in a triose?_____

All sugars have multiple hydroxyl (–OH) functional groups (**fig. 6.6**) and some have phosphate and amino groups as well. Glucose is a hexose. Its structure can be represented in a number of ways as is shown in figure 6.6. Molecules of glucose in crystals exhibit the straight chain structure, but when dissolved in water convert to the ring structure. The ring structures can be portrayed in a simplified or short hand notation, where the carbons located at each angle of the ring are not drawn. As a reader of chemical structures, you have to learn to mentally fill in the carbons. Look at one of the ring structures in figure 6.6, count the number of carbons and hydrogens in the molecule.

Table 6.1		
Element	Atomic Mass	Color in Model
Hydrogen	1	
Carbon	12	
Nitrogen	14	
Oxygen	16	
Phosphorus	31	
Sulfur	32	

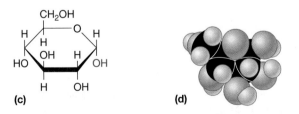

Figure 6.6 Several ways of representing the structure of glucose. (*a*) linear form found in dry crystals; (*b*) combination linear-ring; (*c*) ring form found in aqueous solution in cells; the darker lines represent 3-D aspect of structure. (*d*) space-filling model.

2 **Glucose.** Open the glucose data file. By clicking on *File* at the upper left of time window, selecting *Open*, and choosing **02-glucose.pdb**. The water file will close and a ball and stick model of glucose will be displayed. If you put the cursor on top of the molecular structure and click and hold while moving the mouse, you will be able to rotate the molecule to view it from different angles. You can grab the lower corner of the display screen with your cursor and click as you drag to increase its size. Compare the model on your screen to the ring structure depiction in fig. 6.6c as you rotate the molecule.

Count the atoms of each element found in glucose and write its chemical formula below.

Record the color of the C atoms in table 6.1.

Calculate the molecular mass of glucose by multiplying the number of each type of atom by the atomic mass. Sum these numbers to get the mass of the molecule. What is it?

From your earlier calculation, you know that water has a molecular weight of 18. How many times heavier (and larger) than a water molecule is a glucose molecule? _____

How many hydroxyl functional groups are shown in the model? _____

Hydroxyl groups are polar. The oxygen tends to be negatively charged because of electron clustering, and the hydrogens tend to be positive. Draw a straight chain glucose molecule and then indicate how water molecules would orient in a hydration shell around it. Recognize that the shell would be several water molecules thick as the molecules in the second layer orient to those in the first, and so on.

Predict whether glucose is hydrophilic or hydrophobic? Why do you say so?

3 **Sucrose.** Now open the data file for sucrose by choosing *open* from Jmol's *File* Menu and clicking on **03-sucrose.pdb**. Sucrose, also known as table sugar, is used by plants to transport photosynthetic products from leaves to roots as well as between other parts of the plant.

Resize and rotate the sucrose molecule to clearly see the two ring structures.

Sucrose is a disaccharide, meaning that it is made by a condensation reaction between two monosaccharides, glucose and fructose. What is the chemical formula for sucrose?

What is the molecular mass of sucrose?

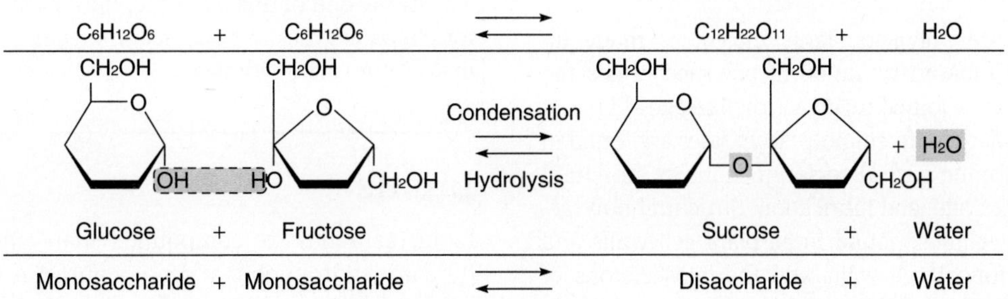

$C_6H_{12}O_6$ + $C_6H_{12}O_6$ ⇌ $C_{12}H_{22}O_{11}$ + H_2O

Glucose + Fructose ⇌ Sucrose + Water (Condensation / Hydrolysis)

Monosaccharide + Monosaccharide ⇌ Disaccharide + Water

Figure 6.7 A condensation reaction producing a glycosidic linkage between glucose and fructose forms sucrose. The linkage is broken in hydrolysis reactions. Note that most Cs and their –OH groups and –H's are left off carbons in the rings for clarity.

Using fig. 6.7 as a guide, identify which part of the molecule on your screen is derived from glucose and which from fructose.

Reactions of a hydroxyl group of one sugar with a hydroxyl group of another (**fig. 6.7**) forms a covalent bond called a **glycosidic linkage** between the two sugars. Because water is a by-product this is one example of a condensation reaction. This glycosidic linkage can be broken in a hydrolysis reaction, yielding the two starting sugars. When sucrose is made in plants, the condensation reaction occurs and when we metabolize sucrose, enzymes in our bodies perform the hydrolysis reactions. Locate the glycosidic linkage between the two sugars. Draw the atoms and bonds found in the linkage below.

Looking at the chemical structure, would you predict that sucrose is hydrophilic or hydrophobic? _____ Why?

Is this prediction consistent with your observations when you add sugar to your tea or coffee? _____

If you boil a sugar solution made with distilled water, the water evaporates as hydrogen bonds are broken by the heat and the water molecules escape as a vapor. Eventually, the sugar will recrystallize and be the same as the sugar you added. What does this experiment tell you about the relative strengths of hydrogen and covalent bonds? Which is weaker and can be broken by moderate heat?

Polysaccharides

Polysaccharides are polymers, large molecules made up of many smaller monomeric units. In polysaccharides the monomers are sugars joined by glycosidic linkages like that seen in sucrose. Many different polysaccharides are found in organisms, performing such diverse functions as structural support, energy storage, and lubrication. Structural polysaccharides include cellulose found in all plant cell walls and chitin found in fungal cell walls and the exoskeletons of many invertebrate animals. Energy storage polysaccharides include glycogen in animals and starches in plants. They are made when sugars are plentiful, then broken down when sugars are needed. Mucus is an example of a lubricating polysaccharide found in both the plant and animal kingdoms and illustrates the importance of hydration shells. The large size of polymers limits their solubilities, but water molecules nonetheless hydrogen bond to the hydroxyl groups that project from the sugar monomers. Consequently, mucus feels slippery because when such hydrated molecules are placed between two surfaces that are moving across each other, the weak hydrogen bonds are sheared and the water molecules slide by each other.

4▶ *Starch.* Open the file for **04-amylose.pdb,** sizing and positioning it for best viewing. The starch molecule you see on your screen is but a small section of a complete amylose molecule. Amylose will typically contain over 1,000 glucose units. Amylose is one form of plant starch and was chosen because its structure is a straight chain. Other starches have branches and are more complex (**fig 6.8**).

Find the glucose monomers in the starch molecule. How many are shown in this sample?

Find the glycosidic linkages between adjacent glucose units. How many are there?

Do you predict that this molecule is hydrophilic or hydrophobic?

You might think that starch would dissolve in water. It does so only partially because the molecular weight is so great. If a thousand glucose units condense to form a starch molecule, what is the approximate molecular mass of the starch?

Large molecules such as this often do not enter into true solution but are suspended as a colloidal mixture when added to water. Colloids consist of aggregations of molecules that remain suspended in a liquid.

At the end of this lab topic, find the summary table 6.2 for characteristics of biological molecules. Fill in the information for carbohydrates.

Lipids

Lipids as a class of compounds share one characteristic: they are hydrophobic and not soluble in water. Common names for different kinds of lipids include fats, glycerides, oils, phosphoglycerides, steroids, and waxes. Lipids vary widely in their functions in cells. Fats and oils store large

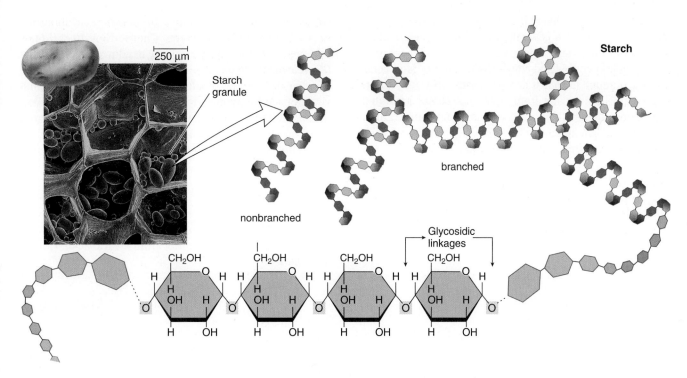

Figure 6.8 Structure of starch granules found in potato cells. Glycosidic linkages formed by condensation reactions join glucose residues into long chains. Some starches, called amylopectins, are branched; others, called amyloses, are not.

amounts of energy. When an organism is taking in excess food, fats are synthesized and stored. Later when food intake is low, the fats may be broken down so that the energy can be used. Other fats are extremely important in forming cellular structures. Phospholipids are lipids containing phosphate groups. They are the basic building units of cellular membranes, defining the cell boundary as well as forming many structures (organelles) within eukaryotic cells. Steroids are also found in membranes, and in many animals some function as homones. Waxes are waterproofing agents that prevent desiccation in terrestrial environments. In this section, you will investigate the structures of fatty acids, glycerides, and membranes.

Lipids are not polymers. They have molecular masses that are only several hundred rather than several thousand. The basic units in glycerides are glycerol and fatty acids (fig. 6.9) but they do not repeat uniformly as in polymers. Glycerol is a sugar. It has a three-carbon backbone, each carbon with a hydroxyl group. Fatty acids are long chain molecules with up to 22 carbons in their backbones. Except for a terminal carboxyl group, the only other functional group is hydrogen. Because of their composition, fatty acids are very hydrophobic and do not interact with polar water molecules to form hydration shells. Some fatty acids have double or triple bonds between some of the carbons in the backbone. They are said to be **unsaturated fatty acids**. In contrast, **saturated fatty acids** have no double bonds between adjacent carbons.

The physical state of glycerides, *i.e.*, whether they are oils or solid fats, depends on their molecular mass and state of saturation. The general rules are the longer the backbone of the fatty acid and the greater the degree of

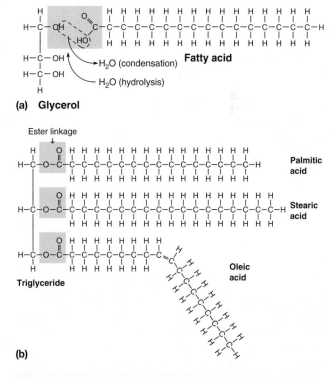

Figure 6.9 Structures of glycerol, fatty acid, and a triglyceride. (*a*) Glycerides are formed when the carboxyl of a fatty acid undergoes a condensation reaction with a hydroxyl of glycerol. (*b*) Up to three fatty acids, which may be the same or different, can condense with one glycerol to form a triglyceride. Unsaturated fatty acid such as oleic acid contain double covalent bonds, introducing a "kink" in what is normally a straight hydrocarbon chain.

saturation, the more likely it will be solid at room temperature. Plant lipids are generally oils because they contain many unsaturated fatty acids. Manufacturers of margarine have long known this and adjust the mixture of fatty acids in their product so that it spreads easily on fragile toast. Waxes are lipids made from very long chain, saturated fatty acids so that they are solids at room temperature. When you polish an apple picked directly from a tree, friction melts and spreads the natural wax.

Fatty Acids

5 *Stearic acid.* Open the data file for **05-stearic acid.pdb**.

Note the shape of its carbon backbone. What is the chemical formula for stearic acid?

Calculate the molecular mass for stearic acid. Is it less than or greater than that of glucose?

Use figure 6.2 to identify any functional group present in stearic acid and list them below.

Stearic acid is called a fatty acid because the long hydrocarbon backbone is considered typical of fats and the carboxyl/group has acid properties. It ionizes into a proton and an anion called stearate. Any chemical compound donating protons is an acid.

Do you see any regions of the molecule that could be polar?_____ How do you think hydration shells would form around this molecule? Draw a cartoon below showing them.

Based on your drawing, what do you predict should happen when you mix an oil like stearic acid with water?

6 *Oleic Acid.* Resize and move the stearic acid molecule to the top half of your screen. Choose **New** from Jmol's *File* menu to open another Jmol window, and open the **06-oleic acid.pdb** data file in the new window. Click in the title bar of the new window and drag to move it so that you can see the models for both stearic and oleic acid at the same time. If necessary, click in the lower right corner of a Jmol window and drag to adjust window size.

What is the chemical formula for oleic acid?

How does oleic acid differ from stearic acid? How is it similar?

Stearic acid is a saturated fatty acid. All carbons share only one covalent bond with each adjacent carbon. Oleic acid is an unsaturated fatty acid. Two adjacent carbons share a double bond. The double bond causes the backbone to have a kink in it.

Glycerides

Glycerides form when fatty acids condense with glycerol, forming ester linkages.

7 *Glycerol.* In the window displaying oleic acid, open the file for **07-glycerol.pdb**. Rotate and size the molecule. Note how different it is from stearic acid in the other open window and how it resembles glucose. Glycerol is actually a carbohydrate but it can react with fatty acids.

What is the chemical formula for glycerol? What is its molecular mass?

What functional groups are present?_____
Would you predict that this molecule is hydrophilic or hydrophobic? _____

Triglyceride. Carboxyl groups of fatty acids can undergo condensation reactions with one or more of the hydroxyl groups on glycerol to form **ester linkages** (fig. 6.9). Because glycerol contains three hydroxyl groups, up to three fatty acids can join with a single glycerol, forming a mono-, di-, or triglyceride. Different fatty acids can link with each hydroxyl group to form a complex molecule. A phospholipid forms when fatty acids link to two of the hydroxyls and a phosphate group links to the third.

You are now going to explore how the ester linkages form. Position the stearic acid in one window with its carboxyl group toward the hydroxyl groups on glycerol in the other window. Adjust the sizes so they are similar. Remember, molecule size is adjusted by pressing the shift key while clicking on the molecule as you move, the cursor up or down. Explain to your partner what will happen when an ester linkage forms between these two, producing a monoglyceride.

8▶ Now open the file *08-triglyceride.pdb* in the glycerol window, leaving the stearic acid window open in the background. Locate the glycerol and fatty acid components of the molecule. How many ester linkages are in a triglyceride?_____

Do you think that a triglyceride would be hydrophilic or hydrophobic? Why do you say so?

Note that the fatty acids linked to each position of the glycerol in this model are all the same. In cells, the three fatty acids could be different. Sometimes they will have different backbone lengths or different degrees of saturation. **Figure 6.10** shows how saturation affects the ability of lipids to pack closely together, leading to some being solids at room temperatures whereas others are liquids. Which of the following do you think would be an oil at room temperature—a triglyceride containing saturated fatty acids that all had backbones that were 18 to 20 carbons long or one where the fatty acids were unsaturated with backbones of 14 to 16 carbons?

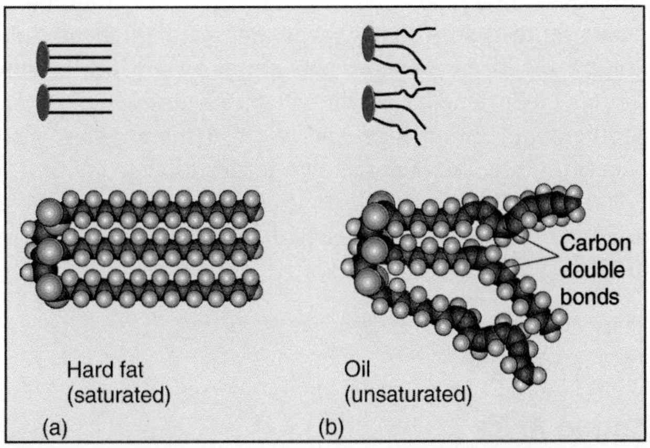

Figure 6.10 (a) Hard fats such as those found in animals (*e.g.,* bacon grease) are composed of saturated triglycerides that can pack closely together. (b) Oils such as those found in plants (*e.g.,* cooking oil) are composed of unsaturated triglycerides that cannot pack closely together. Consequently, they remain liquid at room temperatures.

Phospholipid. Phospholipids are lipids found in cell membranes. They are essentially a diglyceride with a phosphate group on the third hydroxyl group of the glycerol.

9▶ Click in the stearic acid window, and open the file *09-phospholipid.pdb.* Adjust the phospholipid and triglyceride models to be about the same size. What color is the phosphorus atom?

Compare the phospholipid to the triglyceride. What do you see that is the same and what is different about the two molecules?

The phosphate part of the molecule tends to be electronegative and may actually ionize, losing hydrogen ions from its hydroxyls. Find the glycerol-derived part of the phospholipid and lightly trace over it with your cursor.

Would you predict that a phospholipid is hydrophilic, hydrophobic, or that it has a little of both these properties.

Molecules that have both hydrophobic and hydrophilic regions are called **amphipathic** molecules. It is this property of phospholipids that allows them to form the lipid bilayers that make up cell membranes.

Membranes

When added to water, phospholipids float with the charged "heads" down into the water and the hydrophobic tails up toward the air as illustrated in **figure 6.11**. In cells, when two or more phospholipids are synthesized, they form a bilayer with the hydrophobic tails oriented toward each other, and the hydrophilic heads facing toward the water found within and around cells.

10▶ Close one of the Jmol windows. In the remaining Jmol window, open the file *10-membrane.pdb.* This shows a section of phospholipid bilayer and its associated hydration shells. For clarity, hydrogens in the phospholipid molecules are omitted and no proteins are shown

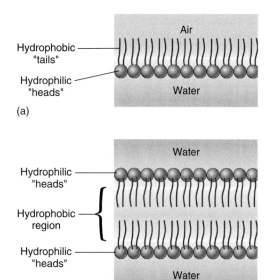

Figure 6.11 Phospholipid structural arrangements in water. (a) On water surfaces, they form films a single molecule thick with hydrophobic tails away from water. (b) Immersed in water as in cells, they form bilayers with hydrophobic tails inward and hydrophilic heads outward, facing the water.

Which parts are nonpolar and hydrophobic, and likely to be excluded from the polar environment of water?

Explain how these interactions produce the bilayer structure of the membrane.

embedded in the membrane. Rotate the membrane to see where water molecules (red and white) are located. Is water associated with the surfaces of the membrane or within its interior? _____

Now simplify the model by telling Jmol to hide the associated water. To do this, choose *Console* from Jmol's *File* menu. A new window called *Jmol Script Console* will appear. Type *hide water* after the $ prompt, then hit the enter/return key. Close the *Script Console*, and rotate the membrane model to examine its structure without the hydration shells.

What you see is an array of phospholipid molecules arranged in two layers. The glycerol head groups are marked by their associated oxygens (red) and attached phosphates (orange). Those groups are oriented toward the outside surfaces and the fatty acid tails are oriented toward the inside. The fatty acid tails contain only carbon and hydrogen; because hydrogens are not shown, you only see the carbons, colored gray. Phosphate groups in cell membranes often have another molecule, ethanolamine, linked to one of its oxygens. It includes a nitrogen atom which is the blue color you see at the surfaces.

Which parts of the phospholipid molecules are polar and hydrophilic, and likely to form weak electric bonds with water molecules?

Rotate the membrane model so you can see individual fatty acid chains near one of its corners. Some of the fatty acid backbones have bends in them. What produces such bends? _____

The kinks in the backbones of unsaturated fatty acids prevent them from packing together closely. This keeps membranes more fluid and decreases the point at which they solidify, so they stay liquid at temperatures lower than do membranes containing just saturated fatty acids. Organisms can adjust the fluidity of their membranes by changing the proportions of saturated and unsaturated fatty acids. Mammals that hibernate and allow their body temperature to drop during the winter maintain membrane fluidity by inserting more unsaturated fatty acids with shorter chain lengths as winter approaches. Cholesterol, another lipid found in eukaryotic cell membranes, also affects fluidity, but is not shown in this model.

At the end of this lab topic, find the summary table 6.2 for the characteristics of biological molecules. Fill in the information for lipids.

Proteins

Amino Acids

Amino acids are very important because they are the monomers from which proteins are assembled. There are 20 different amino acids in most organisms, and hundreds of amino acids can be linked to form large polymers called polypeptides or proteins that are important for catalysis, structure, transport, immune defense, cell adhesion, and communication in cells. The name "amino acid" recognizes two characteristic functional groups of these molecules: an amino group ($-NH_2$) and an acid (carboxyl) group ($-COOH$).

1-14 You will look at the three-dimensional structure of four amino acids: alanine, phenylalanine, serine, and leucine. To do this, open four blank Jmol windows. To open a new window, choose *New* from Jmol's *File* menu. Size the windows (click and drag the lower right corner) and position them (click and drag in the window's title bar) so you can see all four, then open the files *11-alanine.pdb, 12-serine.pdb, 13-leucine.pdb,* and *14-phenylalanine.pdb,* one in each window. To make comparisons easier, rotate the molecules so the oxygen atoms of the carboxyl groups are to the right and the amino nitrogens (blue) point down.

Start by comparing alanine and phenylalanine, the smallest and largest of the four. In solution near neutral pH, the carboxyl groups ionize, losing a proton, and the amino groups electrostatically attract an additional proton. Jmol displays this state by showing a single covalent bond for one carboxyl oxygen, and an extra bond illustrating the attraction between the amino nitrogen and a proton.

What is the chemical formula and molecular mass of alanine?

What is the chemical formula and molecular mass of phenylalanine?

Which is larger than glucose? _____
Are either larger than oleic acid? _____
List features that both alanine and phenylalanine share.

List features which differ between the two molecules.

Now look at the other two amino acids. What similarities do you see?

All of the common amino acids have one carbon, called the **alpha carbon,** which is bonded to an amino group, a carboxyl group, a hydrogen, and a fourth group that differs between amino acids. This fourth group is often called the R-group. Look at **figure 6.12** and identify the alpha carbon and the R-group for each of the amino acids shown.

As you look at figure 6.12, note that the amino acids are arranged into three categories based on R-group polarity: nonpolar, polar, and electrically charged. These properties are quite important in determining the three-dimensional structure of protein polymers.

Peptides

Amino acids can polymerize through formation of **peptide bonds,** which covalently link the amino group of one amino acid with the carboxyl group of another in a condensation reaction that also produces one molecule of water (**fig. 6.13**). One such reaction forms a dipeptide, which still has a free amino group at one end and a free carboxyl group at the other. It can then react with additional amino acids to make a tripeptide and so on. A polymer of many amino acids is termed a polypeptide or protein.

15 Using figure 6.13 as a guide, rotate two of your amino acid molecules so that the amino group of one faces the carboxyl group of another. Explain to your partner which oxygens and hydrogens react to form water, and which of the remaining atoms are covalently linked in a peptide bond.

Once you have described this, have your partner identify what would be the free amino group and free carboxyl group of the resulting dipeptide. These are available for additional peptide bond formation and polymerization with other amino acids. Then close three of your four Jmol windows.

All free amino acids have amino and carboxyl groups, which are polar in solution and promote solubility. In a polypeptide, however, the amino and carboxyl groups bonded to the alpha carbon are tied up in peptide bonds, and are no longer polar. Instead, the R-groups of these amino acids determine the structure and solubility of the polypeptide. R-groups containing only carbon and hydrogen are hydrophobic and do not electrically attract water or other polar molecules, while polar or charged R-groups attract water or oppositely charged R-groups. In solution, those features cause large polypeptides to fold into unique shapes that depend on the sequence of amino acids that they contain.

Go to the *File* menu and *Open* the file **15-hexapeptide.pdb.** Find the peptide bonds between the amino acids and note how the bond involves the carboxyl group of one amino acid and the amino group of another. How many peptide bonds are in this molecule? _____

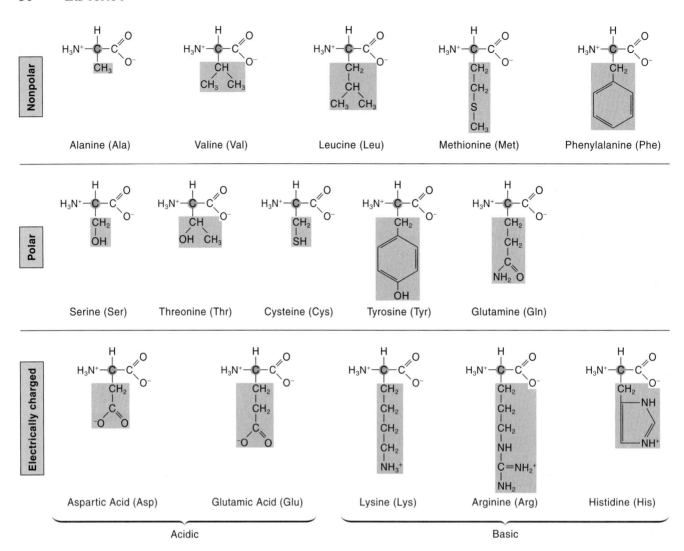

Figure 6.12 Some of the 20 common amino acids. Each has an alpha carbon (red) that is bonded to both an amino group (H_2N-) and a carboxyl group ($-COOH$). At pH 7, the prevailing form of the amino is H_3N^+ and of the carboxyl is COO^-. They are categorized according to the main properties of their R-groups (=side chains) which are shaded green. Nonpolar amino acids are hydrophobic while polar and electrically charged amino acids are hydrophilic.

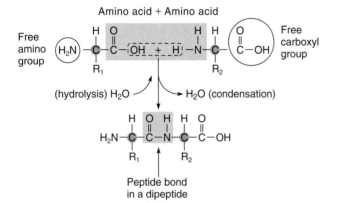

Figure 6.13 Condensation reactions between amino acids form peptide bonds. Reaction is between the amino group of one amino acid and the carboxyl group of another. Free amino or carboxyl groups on the new dipeptide can condense with other amino acids lengthening the peptide.

Identify the backbone of the polymer as it passes along the length of the hexapeptide.

Starting at the end with a free amino group, the amino terminus, trace along the backbone noting how it consists of a repeating pattern: amino nitrogen-alpha carbon-carboxyl carbon.... How many times is this repeated?_____

Note how the R-groups project from the backbone. What are the yellow atoms?_____ Using the structural formulas in fig. 6.12, identify the amino acids from the amino terminus in the hexapeptide and write the names in sequence below.

Predict which of the R-groups are potentially hydrophilic and which are hydrophobic, writing your predictions next to the sequence data above. Do you think this peptide would dissolve in water? _____

Protein Structure

Proteins are large polymers of amino acids. The hormone insulin is a small protein containing 51 amino acids. The blood pigment hemoglobin contains over 600 amino acids, and the biggest proteins can contain over 20,000. Cells contain thousands of different proteins, each of which is synthesized using information stored in the gene for that protein. This means that the types of proteins we have in our cells are inherited characteristics depending on our individual genetic legacies and mutation histories.

When describing proteins, four levels of structure are used (**fig. 6.14**). The **primary structure** is the sequence of amino acids, starting from the amino terminal end. If you had two different proteins each consisting of 100 amino acids, both would contain the same 20 types, but one might have more of one type of amino acid and less of another so that the primary sequences would be different. The English language is based on an alphabet of 26 letters and an almost infinite variety of words is possible when these are combined in different sequences. So it is with proteins, where the alphabet consists of 20 amino acids and the words are very much longer than those in English.

The primary structure of a protein is important because it dictates the overall shape of the molecule. Although all underlying governing principles are not fully understood, it is well known that proteins are rarely straight-chain molecules. Instead, the backbone is twisted and folded into a three-dimensional shape that is unique for each type of protein, resulting in a unique function. This complex subject is more understandable if one thinks about higher levels of structure in protein molecules.

At the **secondary level** of structure, local regions of the backbone may be arranged as **helices** or **pleated sheets** separated by more open random coils (fig 6.14) with all three often found along the length of a single backbone. Helices and pleated sheets are stabilized and held in position by hydrogen bonds that form among the amino acids involved. Regions containing substantial amounts of methionine, alanine, glutamate, leucine, and lysine tend to form helices. Those rich in isoleucine, phenylalanine, threonine, tryptophan, tyrosine, and valine seem to favor pleated sheet conformations. The amino acids glycine and proline are sometimes called helix breakers, marking the beginning or end of a secondary structural region.

The **tertiary level** of structure describes how the secondary structures and coils can, themselves, fold to form more complex three-dimensional shapes. Think of a protein as string of several hundred beads, with the string being the backbone. Some regions of beads along the length have been glued into fixed helical or pleated patterns, but others are flexible allowing the string to fold back on itself into loops and coils. Because cellular proteins are in aqueous solutions, the relative

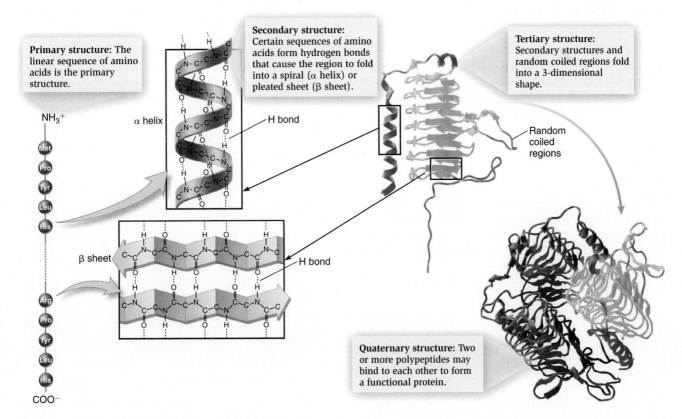

Figure 6.14 Levels of structure in a protein.

hydrophobic/hydrophilic nature of R-groups along the polypeptide chain determine which regions of the chain form the outside surface of the folded molecule and which are inside. Regions with polar and charged amino acids may extend outward while hydrophobic regions will cluster together in the interior. The loops and clusters that develop are stabilized when additional covalent, hydrogen, and ionic bonds develop between regions that come to lie near each other. Each type of protein has a unique shape because its unique sequence of amino acids determines how it folds.

Some large proteins, but none that you will look at today, also have a **quaternary level** of structure. Quaternary level of structure is the association of two or more polypeptide chains to form a complete functional protein. Hemoglobin is such a protein. It consists of four subunits, each containing approximately 150 amino acids.

16 Go to the *File* menu and *Open* **16-peroxidase.pdb.** This is a more complex molecule than those modeled so far. Peroxidase is an enzyme, and you will study its function in the next lab. It consists of over 300 amino acids. Rotate the peroxidase molecules to do a preliminary assessment of its structure. Hydrogen atoms are omitted for clarity, but you should see a globular molecule surrounded by a hydration shell.

Choose *Console* from the Jmol's *File* menu. A Jmol *Script Console* window will open. Resize and position the two windows so that you can see both model and console. Click after the $ prompt in the console window and type:

   hide water   and press return or enter

This hides the hydration shell, leaving a simplified view of the protein.

Primary Level of Structure. To identify the first and last amino acids in the primary sequence of peroxidase, we will use some of the advanced features of Jmol. In the Jmol *Script Console* window type the commands (words in Italics) shown below:

   select 1   and press return or enter
   color gold   and press return or enter

The first amino acid located at the amino terminus of the peroxidase molecule will turn gold. Click on one of the gold atoms and the console window will display information about the selected atom starting with the three-letter identification for the amino acid as capital letters in brackets followed by a number representing the position in the primary sequence from the amino terminal end. Look at fig 6.12 to get the amino acid's name and write it below.

There are 306 amino acids in peroxidase. Modify the procedure used to identify the first amino acid to identify where the last amino acid at the carboxy terminus is located. Use *magenta* as your color choice. What is the amino acid?

Between these two locations are 304 additional amino acids, arranged in a complex, folded pattern. Note how the first and last amino acid at opposite ends of the long protein are relatively near one another as a result of folding. We will look at the folding pattern in more detail to try to tease out more about the secondary and tertiary structure.

With the Jmol *Script Console* open, click on any one of the amino acids projecting from the edge of the molecule. The abbreviation of its name will appear in the *Script Console* window. Look at fig. 6.12 to get the amino acid's name, but recognize that all amino acids are not shown in the figure and you may have to choose another. Randomly sample three amino acids and write their names below indicating whether you think they are hydrophobic or hydrophilic.

A high proportion of the amino acids chosen by you and your classmates probably have hydrophilic R-groups. Discuss with your partner why such amino acids are likely to be on the outside of a globular protein and summarize your discussion as a hypothesis. Enter it below.

Jmol has features that let us further investigate the distribution of amino acid types. In the Jmol *Script Console* window, click on *Clear* to clear the window, then type:

   select polar   and press return or enter
   color lightblue   and press return or enter
   select hydrophobic   and press return or enter
   color green   and press return or enter
   display all   and press return or enter

All the atoms of polar amino acids should now be colored light blue, and all of the atoms of hydrophobic amino acids should be colored green. Rotate the molecule to identify patterns where hydrophobic and polar (hydrophilic) amino acids are located: toward the outer surface or toward the interior of the protein; intermixed or in clusters of one kind. Record your observations below and relate them to your hypothesis.

Peroxidase includes some atoms that are not part of the amino acid chain. Rotate the molecule and locate the region where carbons remained gray and nitrogens blue when you changed the amino acid colors. This is a heme group, consisting of a porphyrin ring of carbon and nitrogen surrounding an iron atom (orange). It forms part of the active site of the enzyme. Rotate the molecule again and locate two calcium ions (light green), which are not covalently bonded, but trapped in R-group shells of negative charge.

17 *Secondary Level of Structure.* Now use Jmol to look at secondary structure. Remember, secondary structures are local regions where H-bond interactions cause regular patterns of folding of the amino acid chain.

To view secondary structure, use Jmol's *Cartoon* scheme as follows

Choose *Select* and then *All* from Jmol's *Display* menu, or enter the command *select all* in the *Script Console* window.

Click on the Jmol logo in the lower right corner of the peroxidase window. If you don't see the Jmol logo, hold down the *control* key while clicking on the molecule. A pop-up menu should open.

Click on *Style*, then click on *Scheme* in the pop-up menu.

Choose *Cartoon* from the *Scheme* pop-up menu.

The *Cartoon* scheme shows α-helical regions in pink and β-pleated sheet regions in yellow. Random coils, regions of the protein backbone without regular secondary structure, are in white. Hold the cursor over any part of the structure to see a label identifying the amino acid at that spot.

What type of secondary structure is most common in peroxidase?_____

Is there just one or more than one region of each type of secondary structure? _____

In the *Script Console* window type:

select hydrophobic   and hit return or enter
color lime green   and hit return or enter

Where are most of the hydrophobic amino acids located? In helices, pleated sheets, or open coils?

18 *Tertiary Level of Structure.* Tertiary structure is how the secondary structures and random coils are arranged in space. The cartoon view shows you a map of the tertiary structure, but does not show all the bonds that contribute to that structure. When the primary structure folds, regions that are linearly far apart may now lie close together. If two cysteines come close together, a covalent **disulfide bridge** can form between them and stabilize the protein's structure. Disulfide bridges can also link peptide fragments together, as in the hormone insulin, or form closed loops in a peptide backbone.

To identify disulfide bridges, return to the ball and stick view with the original colors by reopening *16-peroxidase.pdb* from Jmol's *File* menu. Remember to remove the hydration shells by typing *hide water* in the *Script Console* window.

Rotate the molecule and locate the sulfur (yellow) containing amino acids methionine and cysteine. Only the cysteines form disulfide bridges. To see them clearly, type the following in the *Script* window:

Select cys   and press return or enter
Color white   and press return or enter
Select sulfur   and press return or enter
Color yellow   and press return or enter

All cysteines will now be white with their sulfurs yellow. The disulfide bridges between cysteines will be easy to spot. The single sulfurs, not in white amino acids, are in methionines and do not form disulfide bridges. Hold the cursor over each sulfur in a disulfide bridge to see the sequence number of the cysteine and record it in the table below. Do this for each of the eight cysteines forming four disulfide bonds in peroxidase. Increasing the window size or enlarging the molecule by holding down the shift key and dragging the mouse straight down across the molecule may help. After you have the sequence numbers, calculate the number of amino acids in each of the four loops formed by pairs of linked cysteines.

Disulfide Bridge #	First Cysteine Sequence #	2nd Cysteine Sequence #	Loop Size (# included amino acids)
1			
2			
3			
4			

Now you know the peroxidase molecule includes four loops stabilized by disulfide bridges. In the space below, draw a line with loops to show the way those loops are arranged. Label one end of your line with the name of the amino terminal amino acid and the number 1, then sketch the loops. Write the sequence numbers of the cysteines that form the bridges for each loop. Label the other end of your line with the sequence number and name of the carboxy terminal amino acid.

19 How big is the peroxidase molecule? You know how many amino acids are present. Use the average of the molecular masses you calculated for alanine and phenylalanine to estimate the molecular mass of peroxidase.

When you are finished, go to **table 6.2** and fill in the information for amino acids and proteins.

Advanced Analysis of Peroxidase Active Site

This section is best used after you have completed Lab Topic 7, "Determining the Properties of an Enzyme," because molecular modeling can provide a theoretical explanation of the experimental results.

At the end of the last section, you estimated the molecular mass of peroxidase. Its substrate, hydrogen peroxide, has a molecular mass of 34. How many times larger than hydrogen peroxide is the enzyme?

Enzymes catalyze chemical reactions by binding with the molecules on which they act, triggering the reaction and then releasing a product. The part of the enzyme involved in the catalysis is the **active site.** Jmol has some features that will allow you to perform a rather sophisticated analysis of the active site to explain the experimental results you obtain in your enzyme assay experiments.

20 Reopen *16-peroxidase.pdb* from Jmol's *File* menu. Then choose *Console* to open the *Script Console* window. At the $ prompt, type:

hide water   and then return or enter
select hem   and then return or enter
color pink   and then return or enter
select iron   and then return or enter
color orange   and then return or enter

Rotate the molecule until you have a clear view of the pink heme group with an orange iron ion in the center. Click on *Display* at the top of the screen, then *Atom,* and finally *100% van der Waals* to enlarge the iron ion for clear viewing.

Hydrogen peroxide, the substrate of peroxidase, must diffuse into the interior of the molecule down the open pathway, called the cleft, to bind at the **active site** with the iron for catalysis to occur. Anything that affects the geometry of the cleft or the heme group holding the iron atom will affect the reaction rate. The overall shape of peroxidase results from the folding pattern of the polypeptide chain, stabilized by covalent disulfide bonds, hydrogen bonds, and ionic bonds between charged R-groups.

Temperature Effects. Of the three types of stabilizing bonds, hydrogen bonds are most sensitive to temperature, especially in the helix and pleated sheet regions of the molecule. When the temperature of an enzyme is raised, random vibrations of the polypeptide chain become stronger, straining or breaking hydrogen bonds. When enough bonds break, the molecule loses its optimal shape for catalysis and the reaction slows. At very high temperatures, between 50°C and 100°C, most proteins completely lose their secondary and tertiary structure, destroying their catalytic abilities. On cooling, proteins refold but may not return to their original shapes, effectively destroying their function. This is what happens to egg white proteins when an egg is boiled or fried. The phenomenon is called **denaturation.**

Jmol allows you to view the internal hydrogen bonds in peroxidase. Type the following in the Jmol *Script Console* window after the prompt:

select all	and press return or enter
calculate hbonds	and press return or enter
hbonds on	and press return or enter
color hbonds white	and press return or enter

Dashed white lines, representing internal hydrogen bonds, now appear, usually between a carboxyl oxygen of one amino acid and the hydrogen of an amino group on another located some linear distance away along the backbone of the enzyme, but now in close proximity due to folding. Not all amino acids are involved; some form no hydrogen bonds.

To see more clearly where hydrogen bonds are localized, bring up the cartoon view of peroxidase by doing the following:

While holding down the control key, click on the molecule.

In the pop-up menu that appears, choose *Style* and then *Scheme*.

From the *Scheme* menu, choose *Cartoon*.

You now have a clear view of helices (pink), pleated sheets (gold), and random coils (white "worms") and how they are folded into the tertiary level of structure. The multiple stabilizing hydrogen bonds are clearly visible. Rotate the molecule and describe where most of the hydrogen bonds are located.

To see where the active site is located, type the following in the *Script Console* at the $ prompt:

Select hem and then press return or enter

While holding down the control key, click on the molecule.

In the pop-up menu that appears, choose *Style* and then *Scheme*.

From the *Scheme* menu choose *Ball and Stick*.

The heme group at the active site will now be added to your model as a ball and stick representation. Rotate it to see the broad pathway of the cleft into the active site. Note how hydrogen bond stabilized helices, pleated sheets, and random coils are all involved in forming the cleft. From this view alone, we cannot say which hydrogen bonds are most sensitive to heat. With a little imagination, however, you can see that the shape of the cleft should change if hydrogen bonds are broken. If a molecule like this is boiled in water, most hydrogen bonds would be broken, leading to radical change in the enzyme's three-dimensional shape, especially in the critical areas of the cleft and active site.

21 ▶ **pH Effects.** Reset the model to the starting condition by opening *File* and choosing **16-peroxidase.pdb.** Open the *Script Console* from the *File* menu and type, *hide water.*

Look at figure 6.12. The electrically charged group of amino acids are especially sensitive to changes in pH. To see their location in peroxidase, type the following in the *Script Console* window:

select charged	and press return
color lightblue	and press return

The location of the charged amino acids is now revealed. Several appear to be near the cleft leading to the active site. Among the charged amino acids, some are classified as acidic because their R-group contains a free second carboxyl group. To see their location, type the following into the *Script Console* window:

select acidic

color lime green

The basic amino acids with a free amino group remain light blue while the acidic ones are lime green. None of the acidic amino acids seem to be in the cleft, but several basic amino acids are located in the cleft leading to the active site.

The representation of the peroxidase molecule on your screen is a bit misleading. Each atom actually takes up more space than represented on the screen because of thermal vibration.

To change the display to one with more realistic proportions:

click on Jmol's *Display* menu, choose *Select,* and then choose *All* from the *Select* dropdown menu;

again open the *Display* menu, choose *Atom* this time, and then *100% van der Waals.*

In the *Script Console* window type the following:

select hem	and press return or enter
color pink	and press return or enter
select iron	and press return or enter
color orange	and press return or enter

The display will change to space-filling representations of the atoms, showing that peroxidase is really not an open molecule with lots of spaces. Rotate the molecule, looking for the cleft into the pink heme group and a view of the orange iron ion at its center. Hold your mouse over the two basic amino acids in the cleft leading to the active site to see their identity. Peroxidase's substrate, hydrogen peroxide (H-O-O-H), contains two dipoles and their interaction with these basic amino acids positions the hydrogen peroxide at the active site.

When the pH of the solution surrounding peroxidase is changed, the charges on acidic and basic amino acids will change. The initial charges are due to ionization of the additional carboxyl groups in the acidic amino acids that give up a proton or to the attraction of a proton by the additional amino groups in the basic amino acids. At low pH, fewer carboxyl groups are ionized and have a negative charge, while more amino groups are protonated and have a positive charge. At high pHs, the opposite is true. Because charges affect the hydrophilic/hydrophobic properties of amino acids as well as their attraction to their neighbors in the folded protein, pH affects shape. These changes affect the shape of the cleft, the access of hydrogen peroxide to the active site, and the shape of the active site itself.

22 *Cleft and Substrate Sizes.* To illustrate that the tight fit occurs between the enzyme and its substrate and why the shape of the cleft is so important in determining the rate of an enzyme catalyzed reaction, do the following. In the *Script Console* window, type

    select peo   and press return or enter
    color gold   and press return or enter

The data file for peroxidase contains four hydrogen peroxide molecules represented without hydrogen atoms. The peroxides are now gold colored, but scattered around the model. Rotate the model until you have a clear view of the active site and a hydrogen peroxide. Does it look like hydrogen peroxide could fit in the cleft and move to the active site? Describe how much wiggle room you see?

How about a molecule similar in size such as hydroxylamine NH_2OH which has a molecular mass of 33 compared to 36 for hydrogen peroxide?

Most enzymes have an optimum pH and optimum temperature at which the reaction proceeds the fastest. Almost all enzymes lose their ability to act as catalysts when the temperature approaches 100°C. Explain these observations using your investigation of peroxidase's structure and function.

Nucleic Acids

Just as proteins are polymers of amino acids, nucleic acids are polymers of monomers called **nucleotides.** Nucleic acids range in size from a hundred or so nucleotides in some RNAs to well over several hundred million in DNA.

Unlike the relatively simple monomers you saw in polysaccharides and proteins, a nucleotide is a complex monomer. It is made of three subcomponents: a sugar, a nitrogenous base, and a phosphate group (**fig. 6.15**). Nucleotides contain one of two types of sugars: ribose or deoxyribose. Both are five carbon sugars but deoxyribose has one less oxygen than does ribose. Nucleotides containing ribose function as temporary energy transfer

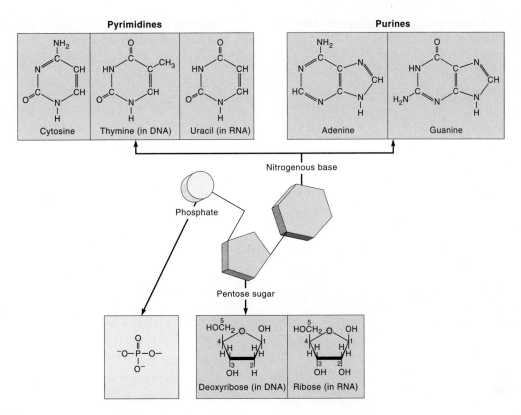

Figure 6.15 Nucleotide structure is shown in center. One of two types of 5 carbon sugar is found in a nucleotide. Nucleotides in DNA have the sugar deoxyribose and those in RNA have ribose. The nitrogenous base portion can be one of five types of bases. However, thymine is never found in combination with ribose and uracil never in combination with deoxyribose.

compounds in cell metabolism, as components of different types of RNA as well as performing other functions in metabolism. Nucleotides containing deoxyribose are found only in DNA where they make up genes. A phosphate group is covalently bonded with the sugar. It is the acid part of a nucleic acid. Also in covalent linkage with the sugar is a nitrogenous base. There are five common bases: adenine, cytosine, guanine, thymine, and uracil. Thymine is not found in ribose-containing nucleotides (RNA) and uracil is not found in deoxyribose-containing nucleotides (DNA).

23 *Nucleotides.* Open the data file *17-AMP.pdb*. Adenosine monophosphate is found in RNA.

Look carefully at the nucleotide and identify the nitrogenous base, sugar, and phosphate group in the molecule. Look again at the sugar. Is it ribose or deoxyribose? _____

Is this molecule hydrophilic or hydrophobic? Why do you say so?

Rotate the molecule for an overview of its structure. Choose *Top* from Jmol's *View* menu, and then click and drag to rotate the DNA molecule around its long axis. You should be able to see that the molecule looks like a spiral staircase, with two spiral edges separated by flattened steps. This is the basis for the description of DNA as a double helix. If you have trouble seeing the double helix, switch to a *Cartoon* view by depressing the control key while clicking on the molecule. From the pop-up menu choose *Style*, then *Scheme* and *Cartoon*. When finished, change back to the *ball and stick* scheme.

Locate the nitrogenous bases (nitrogens are blue). Where are they located in the molecule?

24 *DNA.* Open the file *18-dna.pdb,* which shows only a short section of a DNA molecule. DNA molecules in your chromosomes are hundreds of thousands times longer.

Next, locate the sugars and phosphates, and describe their location.

Adjacent nucleotides are linked by covalent phosphodiester bonds (-O-P-O-) produced by a condensation reaction. What parts of adjacent nucleotides are linked by phosphodiester bonds?_____

Now rotate the molecule until you can clearly see two nitrogenous bases extending toward the middle of the double helix. Are there any covalent bonds between those bases?_____

If there are no covalent bonds between bases, what other kinds of bonds might hold the two strands of the double helix together?_____

You can test the hypothesis that hydrogen bonds are involved. Tell Jmol to display hydrogen bonds as follows:

Choose *Console* from Jmol's *File* menu to open a *Script Console*, in the *Script Console* window after the $ prompt enter:

calculate hbonds   and press return or enter
hbonds on   and press return or enter
color hbonds yellow and press return or enter

Rotate the molecule and locate the yellow dashed lines representing hydrogen bonds. Where are they located?

Position the cursor over an atom in the nitrogenous base on one side of a set of hydrogen bonds. After a moment, a label will appear identifying that base and atom. Standard abbreviations are A for adenine, T for thymine, C for cytosine, and G for guanine. Write down the name of the base in the table below, then follow the same procedure to identify its complementary base on the other side of the hydrogen bonds. Note how many hydrogen bonds connect those two bases. Repeat with four other pairs of bases to fill in the table. You can speed this up by switching to Jmol's *Cartoon* view using the pop-up *Style* and *Scheme* menus.

Base	Complementary Base	# of H Bonds

Using your data, what can you say about the pattern in which base pairs are associated in DNA?

When you are finished, fill in the information about nucleotides and nucleic acid in the summary table 6.2 at the end of this lab topic.

25▶ Lab Summary

On a separate sheet of paper, answer the following questions as assigned by your instructor.

1. What are polar molecules? Explain how functional groups can be polar.
2. Explain the difference between covalent and hydrogen bonds.
3. What determines whether a molecule is hydrophilic or hydrophobic?
4. Explain how polar substances dissolve in water. Describe the events at the molecular level.
5. Write a critique of this statement explaining why it is wrong: "Hydrophobic substances do not dissolve in polar solvents like water. Therefore, hydrophobic substances are not important in cells."
6. If you are shown the chemical structure of biochemical, how could you tell whether it was a fatty acid, amino acid, or nucleic acid?
7. If you were shown the structure of another biochemical, how could you tell whether it was a sugar or a triglyceride?
8. What are biological polymers? Give examples of several types of polymers investigated using Jmol and describe the monomers found in each.
9. If a peptide bond forms between the amino group of one amino acid and the carboxyl group of another, how is it possible for more than two amino acids to join to each other?

10. The primary structure of a protein is the sequence of amino acids. Describe how this sequence determines the folding and shape of the protein. What is the significance of the shape of proteins?
11. Review the data you entered in table 6.2 on page 74 and answer these questions:
    (a) What are the smallest and largest molecules observed?
    (b) Which functional groups are most common in biochemicals?
    (c) What chemical elements are found in all biochemicals?
    (d) What are the names of the four types of covalent bonds found in polymers?

You may want to try the critical thinking questions that apply some of the knowledge you gained in doing this lab.

Internet Sources

The investigation of the three-dimensional structures of biological molecules is a very current area of research. Several WWW sites are repositories of data files for visualizing complex molecules such as proteins and nucleic acids. Check out at www.rcsb.org or www.ncbi.nlm.nih.gov/. After looking these sites over, how would you describe the type of work that seems to be going on? What type of a background would you need to do such work?

Learning Biology by Writing

Write a short essay describing the monomer/polymer concept as it applies to biology. Integrate into your discussion what the effects of hydrophilic and hydrophobic monomers are on polymer shape. Give examples from your lab work.

Critical Thinking Questions

1. Wood, which is composed mostly of cellulose, never completely dries and loses all of the water it contains. The wood in the walls of houses may contain 8% to 16% moisture, meaning that the indicated percentage of its weight is due to water. However, if you looked at the wood and felt it, you would say that it was dry. Explain where this water could be located and what type of chemical bonding is holding it in the wood.

2. A biochemist studying the fatty-acid composition of cell membrane phospholipids in normal and "hibernating" frogs found that the chemical composition of the membrane changed with the season. In the summer there were more saturated fatty acids present and in the winter more unsaturated. Explain how this might be adaptive and allow the frog's survival.

3. The amino acid sequence of a protein is analyzed, and one end of the amino acid chain is composed almost exclusively of hydrophobic amino acids while the opposite end has hydrophilic amino acids. Others studying this protein found that it is an antibody found on lymphocyte cell surfaces. How do these two independent observations complement one another?

4. The molecular weight of the haploid human genome is about 1.83×10^{12}. The average molecular weight of a base pair in DNA is about 610, and each base pair has a length of 0.34 nm. How many base pairs are in the human haploid genome?

5. Using the data given in question 4, how long is the DNA found in a haploid human sperm or egg? Each of the 23 chromosomes in a human sperm or egg contains a single DNA molecule. What is the average length in nanometers of DNA found in a human chromosome? Convert this to centimeters.

6. Hydrogen bonds have been called the velcro® of biological systems. By analogy, covalent bonds are the thread stitches. Explain why both types of bonds are important in the functioning of cells.

Table 6.2 Summary of Characteristics of Biological Molecules

Molecule Name	Elements Present	Estimated Molec Mass	Functional Groups Present	Hydro-philic or -phobic?	Polymer or Monomer?	If Polymer-Type of Bond
Glucose						
Sucrose						
Amylose						
Stearic acid						
Oleic acid						
Glycerol						
Triglyceride						
Phospholipid						
Alanine						
Serine						
Leucine						
Phenylalanine						
Hexapeptide						
Peroxidase						
AMP						
DNA						

Lab Topic 7

DETERMINING THE PROPERTIES OF AN ENZYME

SUPPLIES

Resource guide available at
www.mhhe.com/labcentral

Equipment

Constant temperature water baths
Spectrophotometers at 500 nm
Computers with Jmol program and peroxidase.pdb file

Materials

Blender or mortar and pestle
Fresh turnip or horseradish root
    Alt: buy peroxidase from Sigma Chemical
Markers
Spectrophotometer cuvettes
10 ml test tubes and rack
50 ml beakers
5 ml pipettes graduated in 0.1 ml units with suction devices or automatic pipetters and tips
Thermometer (alcohol)

Solutions

10 mM H_2O_2
25 mM guaiacol
2% Hydroxylamine
Citrate-phosphate buffers at pHs 3, 5, 7, and 9
Peroxidase (horseradish), 6.4 mg/L in pH 5 buffer if using Sigma Chemical source

STUDENT PRELAB PREPARATION

Before doing this lab, read the Background material and sections of the lab topic that have been assigned by your instructor.
    Use your textbook to review the definitions of:

Buffer	Peroxidase
Enzyme	Product
Inhibitor	Spectrophotometer
pH	Substrate

You should be able to describe in your own words the following concepts:
Structure of an enzyme
Effect of pH on protein structure
Effect of temperature on protein structure
Effect of competitive inhibitors on enzymes
Review appendix B

After finishing the prelab review, write any questions you have about terms, concepts, or techniques in the margins of this lab topic. The lab experiments should help you answer these questions, or you can ask your instructor during the lab.

OBJECTIVES

1. To assay the activity of an enzyme using a spectrophotometer
2. To organize data for inclusion in a lab report
3. To test the following null hypotheses:
   a. The amount of enzyme does not influence the rate of reaction
   b. The temperature does not influence the activity of an enzyme
   c. The pH does not influence the activity of an enzyme
   d. Boiling an enzyme before a reaction does not influence its activity
   e. Molecules with shapes similar to an enzyme's substrate have no effect on enzyme activity
4. To visualize the 3D structure of peroxidase with a computer and explain the pH and temperature effects seen in experimental results

BACKGROUND

The thousands of chemical reactions occurring in a cell each minute are not random events but are highly controlled by biological catalysts called **enzymes**. Like all catalysts, enzymes lower the activation energy of a reaction, the amount of energy necessary to trigger a reaction.
    Most enzymes are proteins with individual shapes determined by their unique amino acid sequences. Because these sequences are spelled out by specific genes, the chemical activities that occur in a cell are under genetic control. The shape of an enzyme, especially in a relatively

small region called the **active site,** determines its catalytic effects (**fig. 7.1**). The active site of each type of enzyme has a unique shape that allows the enzyme to bind with only certain kinds of molecules called the **substrate** of the enzyme. For example, some enzymes bind with glucose but not with ribose because the former is a six-carbon sugar while the latter has only five carbons. The binding specificity is similar to the specificity found between locks and keys. Often metallic ions, such as Fe^{+++}, Mg^{++}, Ca^{++}, or Mn^{++}, aid in the binding process, as do vitamins or other small molecules called **cofactors** or **coenzymes.**

The binding between enzyme and substrate consists of weak, noncovalent chemical bonds, forming an **enzyme-substrate complex** that exists for only a few microseconds. During this instant, the covalent bonds of the substrate either come under stress or are oriented in such a manner that they can be attacked by other molecules.

The result is a chemical change in the substrate that converts it to a new type of molecule called the **product** of the reaction. The product leaves the enzyme's active site and is used by the cell. The enzyme is unchanged by the reaction and will enter the catalytic cycle again, provided other substrate molecules are available.

Individual enzyme molecules may enter the catalytic cycle several thousand times per second; thus, a small amount of enzyme can convert large quantities of substrate to product. Eventually enzymes wear out; they break apart and lose their catalytic capacity. Cellular proteinases degrade inactive enzymes to amino acids, which are recycled by the cell to make other structural and functional proteins.

The amount of a particular enzyme found in a cell is determined by the *balance* between the processes that *degrade* the enzyme and those that *synthesize* it. When no enzyme is present, the chemical reaction catalyzed by the enzyme does not occur at an appreciable rate. Conversely, if enzyme concentration increases, the rate of the catalytic reaction associated with that enzyme will also increase.

All factors that influence binding obviously affect the rate of enzyme-catalyzed reactions. The pH or salt concentrations of a solution affect the shape of enzymes by altering the distribution of $+$ and $-$ changes in the enzyme molecules which, in turn, alters their 3D shape and substrate-binding efficiency. Temperature affects the frequency with which the enzyme and its substrates collide and, hence, also affects binding. At very high temperatures, enzyme function usually slows because such temperatures break the hydrogen bonds that maintain the shape of the enzyme. As the shape of the active site changes, the binding efficiency with the substrate is reduced. These factors will be investigated during this laboratory.

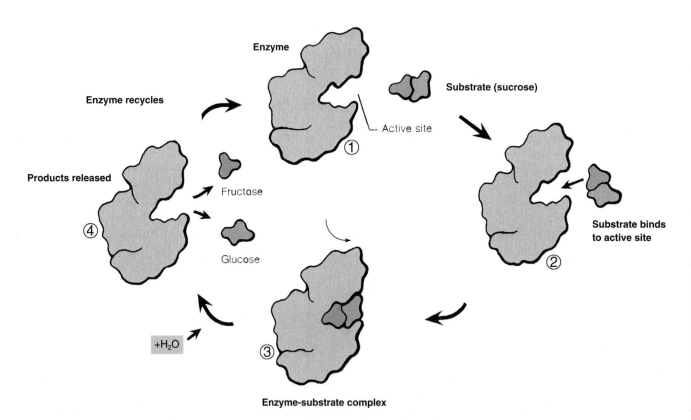

Figure 7.1 Enzymes are large proteins or ribonucleoproteins that catalyze chemical reaction. The substrates bind at the active site, then are converted into products, which are released, freeing the active site so the enzyme can again bind substrate. In this example, the enzyme, sucrase, converts sucrose into glucose and fructose. Note that enzyme and substrate sizes are not to the same scale. Enzymes are often thousands of times larger than substrates, and the active site is only a small region of the enzyme.

Peroxidase

During this lab, you will study an enzyme called **peroxidase**. It is a large protein containing just over three hundred amino acids. It has an iron ion located at its active site (**fig. 7.2**). Peroxidase makes an ideal experimental material because it is easily prepared and assayed. Turnips and horseradish roots are rich sources of this enzyme as are many animal tissues.

The normal function of peroxidase is to convert toxic hydrogen peroxide (H_2O_2), which can be produced in certain metabolic reactions, into water and another harmless compound:

$$H_2O_2 + RH_2 \xrightarrow{\text{peroxidase}} 2H_2O + R$$

In this equation, R stands for another molecule that acts as an electron donor, even though it does not bind to the enzyme active site. Some molecules that can act as electron donors change color when they are oxidized in this reaction, and the color change can be used to follow reaction progress. You will use the dye guaiacol, which turns brown when oxidized. The reaction sequence is:

$$H_2O_2 + \underset{\text{(colorless)}}{\text{guaiacol}} \xrightarrow{\text{peroxidase}} 2H_2O + \underset{\text{(brown)}}{\text{oxidized guaiacol}}$$

By measuring the oxidation of guaiacol as the development of a brown color, you can follow the rate of breakdown of hydrogen peroxide. To quantitatively measure the amount of brown color, the enzyme, substrate, and dye can be mixed in a tube and immediately placed in a spectrophotometer. As hydrogen peroxide is broken down color accumulates, causing the absorbance at 470 nm to increase. The change in absorbance can be recorded at intervals to give a measure of the rate of the enzyme catalyzed reaction. The more active the enzyme, the faster will be the change in absorbance.

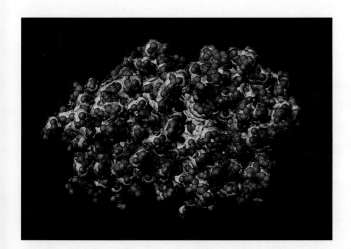

Figure 7.2 Computer-generated image of peroxidase. Green area is active site with red iron ion at its center. With a molecular weight near 31,000 peroxidase is nearly 1,000 times larger than its substrate, hydrogen peroxide.

The procedure for using a spectrophotometer was explained in lab topic 5. You should review those instructions before proceeding. (See fig. 5.5.)

LAB INSTRUCTIONS

In this lab, you will determine the effects of several factors on the activity of peroxidase. In so doing, you will be studying those factors that influence the binding efficiency between the enzyme and its substrate and the stability of the enzyme. Most of what you learn about peroxidase can be generalized to all enzymes.

Preparing Peroxidase

These steps will be done by the instructor before class to save time.

Homogenize 1 to 10 g of peeled turnip, by adding it to 100 ml of cold (4°C) 0.2 M phosphate buffer at pH 5. Grind the mixture in a cold mortar and pestle with sand or blend it for 15 seconds in a cold blender. If a blender is used, use less than a gram of tissue because the homogenization is more efficient. Filter the extract to clarify. Alternatively, a peroxidase solution can be prepared using enzyme obtained from a biochemical supply house.

Standardizing Amount of Enzyme

Extracts prepared from cells contain hundreds of different types of enzymes, including peroxidase. The activity of each enzyme will vary, depending on the size and age of the turnip; the extent of the tissue homogenization; and the age of the extract. Only peroxidase, however, will react with H_2O_2.

To determine the correct amount of extract to use in the next experiments, a trial run should be performed in which the amount of enzyme added is the independent variable.

Before starting an experiment, a null hypothesis should be stated. In this experiment, the hypothesis should relate the rate of reaction to the amount of enzyme added. An example will be given here, but you will make your own hypotheses in the other experiments done in this lab. It is reasonable to predict that more enzyme should increase the rate of reaction. A corresponding null hypothesis is:

The amount of enzyme added to the reaction will not affect the rate of reaction.

1 ▶ *Procedure* To test this hypothesis, use the following directions to set up the chemical reactions and to conduct the experiment:

1. Label four 50 ml beakers as follows: *peroxidase; buffer, pH 5; 10 mM H_2O_2(substrate);* and *25 mM guaiacol* (dye). Fill each about half full with the appropriate solution and use them as stocks for pipetting. Label four pipettes to correspond with the beakers. Alternatively, your lab instructor may have these reagents available in dispensers. If that is the case, you will be given verbal directions.

2. Number seven test tubes. The contents of each tube will be:
   1. All reactants but no enzyme; to be used as a blank in calibrating the spectrophotometer
   2. Substrate and dye
   3. Dilute concentration of peroxidase
   4. Substrate and dye (same as 2)
   5. Medium concentration of peroxidase
   6. Substrate and dye (same as 2)
   7. Concentrated peroxidase

   The exact quantities to be added to each tube are listed in **table 7.1**. Because no tube contains both enzyme and substrate, no reactions occur in the tubes as they are. By mixing pairs 2 and 3, 4 and 5, or 6 and 7 you can start the reaction when you are ready to measure it in the spectrophotometer. If you mix them before you are ready, the reaction will be over before you can make a measurement.

3. Add appropriate amounts of the stock solutions to each tube. Use of the wrong pipette or dispenser will cross contaminate your reagents and give you meaningless results.

4. Use the directions in figure 5.5 to adjust the spectrophotometer to zero absorbance at 470 nm using the contents of tube 1. This tube is used to "blank" the spectrophotometer, so that any color in the reaction mixture will not influence subsequent measurements when enzyme is present.

5. If you are working as teams of students, one person can be a timer, another a spectrophotometer reader, and another a data recorder. Note the time to the nearest second and mix the contents of tubes 2 and 3 by pouring them back and forth twice. Mixing should be completed within ten seconds.

6. Add the reaction mixture to a cuvette by pouring or using an eye dropper. Wipe the outside, and place the cuvette in the spectrophotometer. Read out loud the absorbance at 20-second intervals from the start of mixing so that the recorder can write them in **table 7.2**. If you are a little late in reading the meter, record the absorbance and change the table to show the actual time of the reading. After two minutes (six readings) remove the tube from the spectrophotometer and visually note the color change. Discard the solution as directed by your lab instructor.

7. Now mix the contents of tubes 4 and 5, transfer to a cuvette, and repeat your measurements for two minutes at 20-second intervals. Record the results in table 7.2.

8. After discarding the previous reaction mixture, mix the contents of tubes 6 and 7, transfer to a cuvette, and record the absorbances in table 7.2.

2 *Analysis* Plot the values in table 7.2 on one panel of graph paper provided at the end of this exercise or by using a computer. The abscissa (x-axis) should be the independent variable (time in seconds) and the ordinate (y-axis) the dependent variable (absorbance units). Explain why absorbance is considered the dependent variable.

Plot all three tests on the same coordinates using different plotting symbols. Appendix B discusses how to make graphs. Using a clear plastic ruler or curve fitting program on a computer, draw the single straight line that best fits the points for each of the conditions. (Curves may plateau at the end.)

In mathematical terminology, what characteristic of the graphed line is a measure of enzyme activity?

Table 7.1 Mixing Table for Trial to Determine Amount of Peroxidase to Use (values in ml)

Tube	Buffer (pH 5)*	H_2O_2	Peroxidase*	Guaiacol (Dye)	Total Volume
1 Blank	5.0	2.0	0	1.0	8
2	0	2.0	0	1.0	3
3	4.5	0	0.5	0	5
4	0	2.0	0	1.0	3
5	4.0	0	1.0	0	5
6	0	2.0	0	1.0	3
7	3.0	0	2.0	0	5

*Note how volumes are adjusted to keep the total volume constant.

Table 7.2 Results from Trial Peroxidase Amounts (record absorbance at 470 nm)

Time (Sec)	Tubes 2 and 3 0.5 ml Peroxidase	Tubes 4 and 5 1.0 ml Peroxidase	Tubes 6 and 7 2.0 ml Peroxidase
20			
40			
60			
80			
100			
120			

What are the units of activity in this experiment?

Do you accept or reject your null hypothesis regarding rate of reaction and amount of enzyme? Did the amount of peroxidase change the rate of reaction? How do you explain this?

Which amount of peroxidase gave a linear absorbance change from 0 to 1 in approximately 120 seconds? Use this amount in all subsequent experiments in this exercise unless told otherwise by your instructor. *Change the amount of peroxidase called for in mixing tables 7.3, 7.5, 7.7, and 7.9 as necessary based on your experiment.* The standardizing procedure must be repeated each time an extract is prepared. Why?

If one of the enzyme concentrations tested caused a faster reaction, why are we not using that concentration in subsequent experiments? Experience suggests that the reaction will be too fast to get accurate absorbance readings and that the spectrophotometer will be less accurate at high absorbances.

Factors Affecting Peroxidase Activity

If the scheduled laboratory period is short, your instructor may divide you into teams, each of which will investigate one or more of the following experimental variables. The results will be shared at the end of the lab period and may be included in your report.

Effect of Temperature To determine the effects of temperature on peroxidase activity, you will repeat the enzyme assay with solutions warmed or cooled to four temperatures:

1. In a refrigerator at approximately 4°C or an ice bath at 0°C
2. At room temperature (about 23°C, but should be measured)
3. At 32°C
4. At 48°C

If constant temperature baths are not available, improvise with plastic containers, adding hot and cold water to adjust the temperature.

3▶ State a null hypothesis that relates change in enzyme activity to the temperature of the reaction mixtures.

To test your hypothesis, use the following directions to set up the reactions and conduct the experiment.

Procedure Number nine test tubes. Refer to **table 7.3** for the volumes of reagents to be added to each tube. Refer to the results from the previous experiment and adjust the amount of extract and buffer added to these tubes, if needed.

Table 7.3 indicates the temperature for each pair of test tubes. Place the pairs in appropriate water baths for at least 20 minutes to be sure that they reach the target temperature. While waiting, realize that the room-temperature experiment can be performed immediately while the other tubes temperature-equilibrate. Be sure to measure the room temperature and record it in **table 7.4**.

Adjust the spectrophotometer with the contents of test tube 1. Then mix tubes 4 and 5. After mixing, measure

Table 7.3 Mixing Table for Temperature Experiment (all values in ml)

Temperature	Tube	Buffer (pH 5)*	H_2O_2	Peroxidase*	Guaiacol (Dye)	Total Volume
	1 Blank	5.0	2.0	0	1.0	8
0°C or 4°C	2	0	2.0	0	1.0	3
	3	4.0	0	1.0	0	5
Room temp	4	0	2.0	0	1.0	3
=____ °C	5	4.0	0	1.0	0	5
32°C	6	0	2.0	0	1.0	3
	7	4.0	0	1.0	0	5
48°C	8	0	2.0	0	1.0	3
	9	4.0	0	1.0	0	5

*Note volumes may change depending on results from table 7.2. Adjust accordingly.

the change in absorbance for two minutes at 20-second intervals.

After the tube pairs have been in the water baths for 20 minutes, remove tubes 2 and 3 from the 4°C water bath, mix, and measure the absorbance changes over two minutes at 20-second intervals. Record results in table 7.4.

Repeat the procedure for pair 6 and 7, leaving the last pair in its water bath.

Finally, repeat the procedure for pair 8 and 9.

The temperatures will not remain constant during the measurement, but the effects can be overlooked.

4▶Analysis Graph your results from table 7.4, using a computer or one of the graph grids at the end of this unit. Put data from all four temperatures on the same graph, but use different symbols for each temperature. Draw a best-fit straight line through the data for each temperature. What feature of this line shows the reaction rate (= **enzyme activity**) at that temperature?_____

Making a Derivative Plot From your graph of the raw data, it is not easy to see the relationship between temperature and reaction rate. To show this relationship more clearly, prepare a derivative graph during or after lab.

A **derivative** is a slope in your raw data graph; the derivatives for the lines at each temperature are reaction rates. Plotting these derivatives as a function of the independent variable temperature condenses your data into an easily understandable form. To prepare such a graph, first determine the slope of the lines for each temperature in units of absorbance change per second. Record below. Note that Δ, the Greek letter delta, is a shorthand way of saying "change in and A = absorbance."

Temperature	Enzyme activity (ΔA/sec)
4°C	
23°	
32°	
48°	

Use a new graph panel and label the x-axis *Temperature in degrees C*. Label the y-axis *Rate of reaction in ΔA/sec*. Choose scales to include all your data, and plot your data points. Does enzyme activity vary with temperature? _____ At what temperature is enzyme activity maximal? _____

Table 7.4 Temperature Effects on Peroxidase Activity (record absorbance at 470 nm)

Time (Sec)	Tubes 2 and 3 4°C	Tubes 4 and 5 Room Temp = ____ °C	Tubes 6 and 7 32°C	Tubes 8 and 9 48°C
20				
40				
60				
80				
100				
120				

Do your data suggest that a straight line would be a good description of the relationship between enzyme activity and temperature?_____ If not, draw a smooth bell-shaped curve to illustrate that relationship. You now have a graph that is easy to understand and allows you to rapidly estimate the temperature that yields the highest reaction rate, called the *temperature optimum*.

How could you improve the accuracy of your estimate of the optimum temperature?

Enzymes differ in their temperature optima. For example, enzymes of bacteria growing in hot springs often have temperature optima that are quite high, maybe 70°C, while enzymes in our own cells often have optima near our body temperature of about 37°C.

5 ▶ Effect of pH Horseradishes and turnips from which peroxidase is extracted, grow best in soils with a pH near 6. Write a hypothesis that predicts how peroxidase activity will change with the pH of the solutions used in the assay mixture.

To test your hypothesis, perform the following experiment.

Procedure Your instructor will supply buffers at pHs of 3, 5, 7, and 9. Number nine test tubes. Set up pH-effect tests by adding the reagents described in **table 7.5** to each tube. Depending on your results in table 7.2, you may have to change the volumes of extract and buffer used.

After adjusting the spectrophotometer with the contents of test tube 1, mix pairs of tubes one at a time (2 and 3, 4 and 5, 6 and 7, 8 and 9). Measure absorbance changes at 20-second intervals for two minutes for each pair before mixing the next pair. Record the results in **table 7.6**.

6 ▶ Analysis Graph your results from table 7.6, using a computer or one of the graph grids at the end of this unit. Put data from all four pHs on the same graph, but use different symbols for each pH. Draw a best-fit straight line through the data for each pH. What feature of this line shows the reaction rate (= **enzyme activity**) at that pH?_____

Making a Derivative Plot From your graph of the raw data, it is not easy to see the relationship between pH and reaction rate. To show this relationship more clearly, make a derivative graph like that described in the temperature section. To prepare such a graph, first determine the slope of the lines for each temperature in units of absorbance change per second. Record below. Note that Δ, the Greek letter delta, is shorthand way of saying "change in and A = absorbance."

pH	Enzyme Activity (ΔA/sec)
3	
5	
7	
9	

Use a new graph panel and label the *x*-axis *pH*. Label the *y*-axis *Rate of reaction in* ΔA/*sec*. Choose scales to include all your data, and plot your data points. Does

Table 7.5 Mixing Table for pH Experiment (values in ml)

pH	Tube	Buffer*	H_2O_2	Peroxidase*	Guaiacol (Dye)	Total Volume
5	1 Blank	5.0 (pH 5)	2.0	0	1.0	8
3	2	0	2.0	0	1.0	3
	3	4.0 (pH 3)	0	1.0	0	5
5	4	0	2.0	0	1.0	3
	5	4.0 (pH 5)	0	1.0	0	5
7	6	0	2.0	0	1.0	3
	7	4.0 (pH 7)	0	1.0	0	5
9	8	0	2.0	0	1.0	3
	9	4.0 (pH 9)	0	1.0	0	5

*Note volumes may change depending on results from table 7.2. Adjust accordingly.

Table 7.6 Effects of pH on Peroxidase Activity (record absorbance at 470 nm)

Time (Sec)	Tubes 2 and 3 pH 3	Tubes 4 and 5 pH 5	Tubes 6 and 7 pH 7	Tubes 8 and 9 pH 9
20				
40				
60				
80				
100				
120				

enzyme activity vary with pH?_____ What are the units for enzyme activity?_____ At what pH is enzyme activity maximal?_____

Do your data suggest that a straight line would be a good description of the relationship between enzyme activity and pH?_____ If not, draw a smooth bell-shaped curve to illustrate that relationship. You now have a graph that is easy to understand and allows you to rapidly estimate what pH yields the highest reaction rate. This is called the *pH optimum.*

How could you improve the accuracy of your estimate of the optimum temperature?

Enzymes differ in their pH optima. For example, the enzyme pepsin secreted into our stomachs along with hydrochloric acid has a pH optimum below 3 and works quite well in the acid stomach. When the stomach contents enter the small intestine, bicarbonate is added from the pancreas and liver and the pH rises to above 7. Pepsin stops working at that pH, but the pancreas secretes other enzymes with pH optima at or above 7 that take over chemical digestion.

Effect of Boiling on Peroxidase Proteins have stable 3D structures because of hydrogen bonds as well as other bond types. At high temperatures many hydrogen bonds break and the 3D structure changes to a new stable configuration that may not go back to the original shape when it cools. Such heating (usually to temperatures greater than 70°C) results in a permanent structural change in the protein called **denaturation**. This is what happens to the white of an egg when you boil it.

If heating an enzyme irreversibly alters its shape, what do you predict will happen to its ability to catalyze a reaction?

7 State a hypothesis that predicts what the activity of peroxidase will be after boiling.

To test your hypothesis, follow these directions:

Procedure Add 3 ml of extract to a test tube and place it in a boiling water bath. It must be left there for at least 15 minutes so its contents reach a temperature near 100°C. Remove the tube and let it cool to room temperature.

Table 7.7 Mixing Table for Boiled Peroxidase Experiment (values in ml)

Tube	Buffer (pH 5)*	H_2O_2	Normal Extract	Boiled Extract*	Guaiacol	Total Volume
1 Blank	5.0	2.0	0	0	1.0	8
2	0	2.0	0	0	1.0	3
3	4.0	0	1.0	0	0	5
4	0	2.0	0	0	1.0	3
5	4.0	0	0	1.0	0	5

*Note volumes may change depending on results from table 7.2. Adjust accordingly.

… might also enter the active site. Such molecules can be used as probes to learn more about the active site. For example, hydroxylamine ($HONH_2$) has a structure similar to hydrogen peroxide (HOOH). Calculate the molecular weight of hydrogen peroxide: _____. Calculate the molecular weight of hydroxylamine: _____. What do you predict will be the effect on enzyme activity of adding hydroxylamine to the assay mixture for peroxidase? State your prediction as a hypothesis.

Table 7.8 Boiled Peroxidase Results (record absorbance at 470 nm)

Time (Sec)	Tubes 2 and 3 Normal Extract	Tubes 4 and 5 Boiled Extract
20		
40		
60		
80		
100		
120		

Number five test tubes and add reagents as called for in mixing **table 7.7**.

Use the contents of tube 1 to blank the spectrophotometer. Mix the contents of tubes 2 and 3, pour the mixture into a cuvette, and read the absorbance at 20-second intervals for two minutes. Record the results in **table 7.8**. Repeat your measurements by mixing tubes 4 and 5.

8 *Analysis* Graph your results from table 7.8, using a computer or one of the graph grids at the end of this unit. Draw a best-fit straight line through the data for normal enzyme and then a second through the boiled enzyme data. Calculate both slopes and record then on the graphs.

Compare the activity of peroxidase after boiling to the activity of peroxidase kept at room temperature. How did boiling affect the activity?

Do you accept or reject the hypothesis made at the beginning of this experiment? Why?

Probing the Active Site If enzymes bind with their substrates to catalyse a reaction, it stands to reason that small molecules that look like the normal substrate

9 *Procedure* To test this hypothesis, add 1 ml of 2% hydroxylamine to 2 ml of peroxidase solution. Mix and let the mixture stand for at least 15 minutes so that the hydroxylamine has time to interact with the enzyme.

After 15 minutes, prepare five test tubes as described in **table 7.9**.

After adjusting the spectrophotometer with the contents of test tube 1, mix pairs one at a time (2 and 3, 4 and 5) and measure the changes at 20-second intervals for two minutes. Record your measurements in **table 7.10**.

10 *Analysis* Graph your data and calculate the slopes of the linear portions of the curves as a measure of enzyme activity. Record the slopes next to the lines in your graph.

Do you accept or reject the hypothesis made at the beginning of this experiment? Why?

Advanced Molecular Analysis

11 In lab topic 6, you used the molecular modeling program Jmol to manipulate 3D views of molecules. Data files for horseradish peroxidase can be downloaded from the Protein Data Bank at www.rcsb.org/pdb and viewed in this program. The directions on pages 67–70 can guide you in looking at this molecule. By looking at the 3D structure

Table 7.9 Mixing Table for Hydroxylamine Experiments (values in ml)

Tube	Buffer (pH 5)*	H_2O_2	Extract*	Hydroxylamine Treated Extract	Guaiacol	Total Volume
1 Blank	5.0	2.0	0	0	1.0	8
2	0	2.0	0	0	1.0	3
3	4.0	0	1.0	0	0	5
4	0	2.0	0	0	1.0	3
5	3.5	0	0	1.5	0	5

*Note volumes may change depending on results from table 7.2. Adjust accordingly.

Table 7.10	Hydroxylamine Results (record absorbance at 470 nm)	
Time (Sec)	Tubes 2 and 3 Normal Extract	Tubes 4 and 5 Hydroxylamine-treated Extract
20		
40		
60		
80		
100		
120		

of the molecule, you can investigate several important concepts:

1. You can view the small cleft that hydrogen peroxide must enter to gain access to the active site.
2. You can see the active site with its heme group holding an iron ion that is necessary to catalyze the breakdown of hydrogen peroxide.
3. You can view the location of diamino and dicarboxylic amino acids in peroxidase. When pH is changed, these amino acids may change their charges; those that were neutral may become + or −, and those that were charged may become neutral. This leads to changes in the shape of peroxidase which changes how easily hydrogen peroxide can enter the cleft to the active site. While the program does not illustrate the changes in shape with pH changes, this can be inferred from the location of the electrically charged amino acids.
4. You can view the helical regions of peroxidase's secondary structure. Because the helices are stabilized by hydrogen bonds, you can infer why activity changed with temperature, including explaining denaturation.

Lab Summary

12▶ You have collected quantitative information about how enzymes work, using peroxidase as an example. Your instructor will discuss the results in class and will indicate how to share the data from your experiments with the class. All data should be summarized by making graphs. Graphs are much easier to remember than are tables of numbers. You should have five raw data plots as follows:

1. Effects of amount of enzyme used
2. Effects of temperature
3. Effects of pH
4. Effects of boiling the enzyme
5. Effects of a molecule similar to normal substrate

For each of these plots, write a simple sentence that summarizes the outcome of the experiment, that is, how the rate of the reaction changed under the conditions tested.

In addition, make two derivative plots where you take the complicated graphs for the temperature data and pH data and create simple bell-shaped curves that show the temperature or pH optimum for peroxidase. Directions for making these plots were given at the end of the respective experiments.

Write a one-sentence conclusion based on the derivative plots.

In addition to summarizing the data, you should answer these questions using the concept that anything that affects enzyme-substrate binding also affects reaction rate.

1. Why does temperature affect the rate of an enzyme catalyzed reaction?
2. Why does boiling the enzyme reduce the enzyme's catalytic effectiveness?
3. Why do changes in pH of the assay mixture cause the enzyme to be more or less effective?
4. Why does a molecule like hydroxylamine alter the rate of the reaction?

You may want to try the critical thinking questions that apply some of the knowledge you gained in doing this lab.

Internet Sources

Use the WWW to locate information about the enzyme peroxidase.

Open your browser program and enter the following URL: http://www.uniprot.org/. The Universal Protein Resource (UniProt) is a collaboration between the European Bioinfomatics Institute, the Swiss Institute of Bioinfomatics, and the Protein Information Resource that maintains protein databases. Use the **"Search in"** drop-down menu to select the **Literature citations** database and type **turnip peroxidase** in the **Query** box. Then click the Search button to see published research papers on turnip peroxidase. Click on the title of the paper published in 1980 by Mazza and Welinder to see the abstract of the original description of the amino acid sequence of this protein.

LEARNING BIOLOGY BY WRITING

This exercise provides material well suited for writing a scientific report. You have tested five hypotheses concerning the properties of peroxidase. Use the general directions in . . . Appendix D to write a report on your finding.

A good lab report begins with an introduction and a statement of purpose, which summarizes the hypothesis tested in the experiments.

Cite the lab manual as your source of procedures.

In your results section, use graphs to summarize your data. (See appendix B.) Separate graphs should be made for each factor—pH, temperature, hydroxylamine and boiling. Write a descriptive legend (for example, "Peroxidase activity at different pH levels").

The results from the temperature or the pH experiments should be summarized in separate graphs in which you plot the derivative (slope) as a function of temperature (or pH).

For your discussion, read the section in your textbook about enzymes. Discuss how the experiments you performed can be interpreted in terms of how different chemical and physical treatments affect enzyme structure. What are the limitations of these experiments? Will all enzymes have the same pH and temperature optima? Will all denature at the same temperature or be inhibited by hydroxylamine? How would you do the experiments differently, if you were to repeat them? What are the sources of error?

CRITICAL THINKING QUESTIONS

1. Would performing the "temperature effects" portion of this experiment on a very hot day have any effect on the results?

2. If you used a cooked turnip as your extract source, what results would you expect? Why?

3. If you were hired to supervise a biochemical manufacturing plant that used enzymes to convert starches to sugars, how would you go about determining if the reactions were operating at the best rates?

4. A mutation occurs in the gene coding for peroxidase such that a hydrophilic amino acid is substituted for a hydrophobic one in the cleft leading to the active site. Speculate as to what might be the effect(s) on the ability of the enzyme to breakdown hydrogen peroxide.

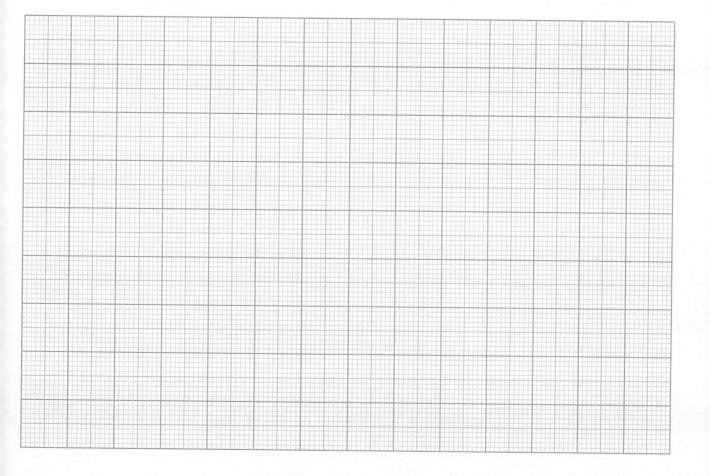

DETERMINING THE PROPERTIES OF AN ENZYME

Lab Topic 8

Measuring Cellular Respiration

Supplies

Resource guide available at
www.mhhe.com/labcentral

Equipment

Barometer
Hot plates

Materials

Distillation apparatus (see fig. 8.2)
1 L Erlenmeyer flasks
Test tubes and racks
One-hole stoppers with 1 ml × 0.01 ml pipettes inserted (see fig. 8.3)
Parafilm
Nonabsorbent cotton
Absorbent cotton
Miscellaneous beakers
Weighing pans
Pasteur pipettes
Thermometer

Solutions

Dilute corn syrup or sweetened grape juice
0.1 M $Ba(OH)_2$ (*Caution:* caustic)
I_2KI (5 g I_2:10 g KI:100 ml H_2O)
1.5 M NaOH (*Caution:* caustic)
15% KOH in dropper bottles (*Caution:* caustic)
95% ethanol in dropper bottles
Yeast packets
Germinating peas, beans, soy, or wheat

Student Prelab Preparation

Before doing this lab, read the Background material and sections of the lab topic that have been assigned by your instructor.

Use your textbook to review the definitions of the following terms:

Aerobic metabolism	Fermentation
Anaerobic metabolism	Glycolysis
Cellular respiration	Oxidation
Electron Transport System	Reduction
Ethanol	Tricarboxylic acid cycle

Describe in your own words the following concepts:

What is ATP and why it is important
How ethanol is produced by yeast
How CO_2 is produced in cellular respiration
Where O_2 is used in respiration
Review appendix B

After finishing the prelab review, write any questions you have about terms, concepts, or techniques in the margins of this lab topic. The lab experiments should help you answer these questions, or you can ask your instructor during the lab.

Objectives

1. To identify the products of aerobic and anaerobic respiration in yeast
2. To test for the effects of cold treatment on aerobic respiration of germinating seeds

Background

Cellular respiration is the oxidation of organic compounds through cellular metabolism to release energy in a form that is usable by a cell. It should not be confused with breathing, meaning lung or gill ventilation in animals. Biologists use the metabolism of glucose as a model system to study cellular respiration. All organisms, both prokaryotic and eukaryotic, have the enzyme systems, termed **metabolic pathways**, which release energy from glucose. Some metabolize glucose **anaerobically**, which requires no oxygen, others only **aerobically**, where oxygen is required, and some can do it either way, depending on the availability of oxygen in the environment.

Cellular respiration is an **oxidative** process, involving the removal of electrons from glucose. Oxygen may or may not be involved. However, these reactions are not simple electron transfers. Instead, hydrogen atoms are removed from glucose and passed to a hydrogen acceptor in the cell. Because a hydrogen atom contains an electron along with its proton, a dehydrogenation event is an oxidation (the loss of an electron) and receipt of hydrogen is a reduction (the gain of an electron), satisfying the requirement that redox reactions must occur as coupled pairs. When C to H covalent bonds are broken, energy can be released or travel with the hydrogen to enter into another reaction. When the energy is used to make ATP, then the cell has essentially transferred energy from food into a form suitable for maintaining itself, synthesizing molecules needed in growth, or performing functions such as ion transport or movement.

Anaerobic metabolism is by far the simplest metabolic pathway for harvesting energy from glucose. If we use yeast as a model for anaerobic metabolism, the overall equation can be written as:

$$C_6H_{12}O_6 \text{(glucose)} \xrightarrow{\text{yeast}} 2\ CH_3CH_2OH \text{(ethanol)} + 2\ CO_2 + \text{energy}$$

In anaerobic respiration, about 10 enzymes work sequentially on the glucose, cleaving it into halves in a metabolic pathway called **glycolysis.** After glycolysis, other enzymes remove a carbon, releasing it as carbon dioxide. The two carbon organic molecules that are left are ethanols. As the bonds that were in the original glucose were broken, energy was released. Some escaped as heat, and some was trapped in ATP molecules that the cell uses to fuel its energy needs. Overall the process is a chemical oxidation process, but no free oxygen is involved. Anaerobic metabolism is not terribly efficient and only about 2% of the energy available in glucose is usable by the cell. A small amount escapes as heat, a small amount is trapped in ATP, but most is still contained in the bonds of ethanol. However, if no oxygen is present, the cell cannot further metabolize the ethanol and it is excreted, rising in concentration in the cell's environment as more glucose is anaerobically metabolized. For yeast living in a wine barrel or in the large fermenting tanks of a gasohol plant, the ethanol excretory product eventually reaches toxic levels, killing the very yeast that produced it as the CO_2 bubbles off. For yeasts living in a rising bread dough, the CO_2 is trapped in the dough causing it to rise. Ethanol, also in the dough, is released as the sweet smell detected when the dough is baked, terminating the reaction.

Aerobic respiration is more efficient than anaerobic respiration. The overall reaction can be written as:

$$C_6H_{12}O_6 \text{(glucose)} + 6\ O_2 \rightarrow 6\ CO_2 + 6\ H_2O + \text{energy}$$

However, this simple equation conceals the fact that over 30 enzyme catalyzed steps are involved. The first 10 steps are the same glycolysis steps found in anaerobic respiration but there the similarity between the two types of respiration stops. Ethanol is not formed. Instead the product of glycolysis is sequentially acted on in over 20 enzyme catalyzed steps to ultimately produce CO_2 and H_2O. These steps are too complicated to describe here and you should consult your lecture textbook for details on the metabolic pathways known as **tricarboxylic acid cycle** and the **electron transport system.** As a result of this multistep sequential breakdown, all of the chemical bonds between carbons, hydrogens, and oxygens that were in the glucose molecule are broken. The carbons and oxygens from glucose escape as CO_2. All of the hydrogens from glucose eventually combine with oxygen to produce water. Ever wonder why we need a constant supply of oxygen in order to live? It is to supply this phase of aerobic respiration. The breaking of all the bonds in glucose releases energy. Some escapes as heat and some is trapped in ATP which the cell uses to satisfy its energy needs. The efficiency of the system is much better than anaerobic respiration, although not perfect. About 30% of the energy originally in glucose finds its way to ATP which, in turn, is used to fuel the cell's energy needs. The rest is lost as heat (entropy).

In eukaryotic cells, the reactions of aerobic respiration are compartmentalized. The glycolysis enzymes are distributed throughout the watery part of the cytoplasm. Those of the tricarboxylic acid cycle (TCA) and electron transport system (ETS) are located inside the mitochondria. Within these organelles, the ETS components are embedded in the membranes of the cristae while the TCA enzymes float in the matrix. Any cell containing mitochondria is capable of aerobically metabolizing glucose and it is the ETS system on the cristae that consumes oxygen.

Why do some organisms metabolize anaerobically if the process is so inefficient compared to aerobic respiration? Oxygen is the key factor. Those organisms that can perform anaerobic respiration can live in environments where there is no oxygen, eliminating competition with aerobic species for resources. Some organisms, such as yeasts, can take advantage of both worlds, switching from aerobic to anaerobic depending on oxygen availability. Our own skeletal muscle cells, like those of many other animals, also have this metabolic flexibility. Under nonstrenuous conditions, the blood delivers sufficient oxygen to the muscles so they metabolize aerobically. Under strenuous conditions, the muscle's need for oxygen to supply aerobic respiration surpasses the ability of the blood to deliver oxygen. The muscle cells respond by turning on anaerobic respiration. However, they do not produce ethanol after glycolysis. Instead, lactic acid is produced. The lactic acid tends to build up in the muscles and eventually reaches levels that cause muscle malfunctions that we experience as cramps, weakness, and burning sensations. The advantage of being able to do this is obvious. It allows the muscles to operate beyond their aerobic range, providing capacity for short-term burst performance, invaluable in emergencies.

Anaerobic metabolism which results in an organic molecule as a direct product, such as ethanol or lactic acid, is termed **fermentation.** Evolutionary biologists consider fermentations to be the earliest form of cellular respiration. Before the atmosphere had a significant level of free oxygen from photosynthesis, the only organisms that could live would be those that could ferment food materials. With the advent of free oxygen, the anaerobic part of fermentations (glycolysis) became the start of aerobic respiration as additional metabolic capabilities evolved.

LAB INSTRUCTIONS

You will observe the results of aerobic and anaerobic respiration in yeasts. You will then measure aerobic respiration in germinating seeds.

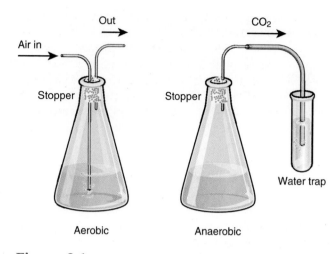

Figure 8.1 Culture setups for growing yeast. In aerobic culture, air is continously bubbled through the culture. In anaerobic culture, carbon dioxide produced in respiration forces all air (and oxygen) out of flask. Water in the trap prevents oxygen from re-entering the flask.

Respiratory Flexibility in Yeast

Yeast will metabolize glucose either aerobically or anaerobically, depending on environmental conditions. When oxygen is present, yeast uses glycolysis, the tricarboxylic acid (TCA) cycle, and electron transport to gain usable energy from glucose, producing CO_2 and H_2O as end products. When oxygen is not available, the TCA cycle and electron transport shut down. Ethanol and CO_2 become the end products of glucose metabolism as glycolysis followed by fermentation take over. In this section, you will investigate this metabolic flexibility of yeasts, demonstrating the end products of each and the relative efficiencies of the two forms of metabolism.

Culturing Yeast Before lab, your instructor added sweetened grape juice or diluted corn syrup to three flasks. Baker's yeast was added to two of the flasks but none was added to the third. It is a control, indicating if the starting solutions contain any of the products of respiration. The flasks containing yeast were set up as in **figure 8.1**. The air stream entering the aerobic flask supplied adequate oxygen to maintain aerobic respiration. The anaerobic flask allowed carbon dioxide to escape but because of the water trap no oxygen could enter. The initial oxygen present would have been consumed in the first few hours, so any yeast cells now living are metabolizing the sugars anaerobically. After about 6 to 12 hours, you should be able to see whether aerobic and anaerobic metabolism produce the same or different products.

1 ***Tests for Respiratory Products*** Your instructor will demonstrate a simple classic chemical test to determine whether the cultures are producing CO_2, a product of both anaerobic and aerobic metabolism. If the gas being emitted by a culture is bubbled through 0.1 M $Ba(OH)_2$, any CO_2 present reacts with the water in the test tube to form carbonic acid which in turn reacts with barium hydroxide to form barium carbonate, a white precipitate:

$$CO_2 + H_2O \rightarrow H_2CO_3$$
$$H_2CO_3 + Ba(OH)_2 \rightarrow BaCO_3 \downarrow + 2H_2O$$

Enter the results of the carbon dioxide tests in **table 8.1**.

2 The other products from respiration should be ethanol in the anaerobic culture and water in the aerobic. Unfortunately, you cannot test for the small amount of water produced in an aqueous culture but there is an easy test for ethanol. Ethanol can be isolated from the culture by distillation because it boils at a lower temperature (78°C) than water (100°C).

Three distillation setups should be assembled (**fig. 8.2**). About 100 ml of anaerobic culture should be added to one, 100 ml from the aerobic to the second, and 100 ml of the control to the third. Turn on the hot plates and collect about 30 ml of distillate from each still. Remember to turn off the hot plates! What you have just done is a scaled-down version of the process used in the liquor and gasohol industries.

3 Ethanol in the distillates can be detected by a procedure called the **iodoform test.** Take five test tubes and number them. First add 2 ml of I_2KI and then 1.5 ml of 1.5 M NaOH to each. If the solutions do not have an orange tinge, add more I_2KI until they do. These are the *test reagents*. Now add the following *samples* to each tube:

Tube Sample
1. 2.5 ml water
2. 1.25 ml water and 1.25 ml 95% ethanol
3. 2.5 ml of distillate from the control

Table 8.1 Results from Respiratory Studies of Yeasts

	Control—no yeast	Anaerobic	Aerobic
Producing CO_2?			
Producing ethanol?			
Describe culture density	No cells present		

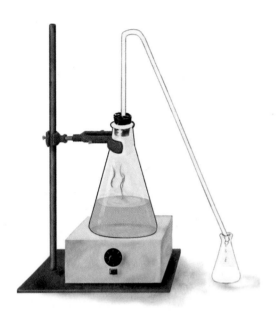

Figure 8.2 A reflux air-cooled distillation setup. The glass tube should measure about 30 cm from the flask to the curve to allow reflux action. As ethanol boils off and passes along the tube, it condenses into a liquid that will be collected and analyzed.

Tube Sample (see page 91)
4. 2.5 ml of distillate from the **anaerobic** culture
5. 2.5 ml of distillate from the **aerobic** culture

Mix all solutions and let stand for five minutes. If ethanol is present, it will react with the iodine in the presence of NaOH to form **iodoform,** which will settle out as a yellow precipitate. If no reaction occurs in tube 2 (the known sample of ethanol), add 1 ml more of NaOH to *all* tubes and mix.

Record your results in table 8.1. Use *NR* to record where there is no reaction, and + to indicate where a precipitate is found. If the amount of precipitate varies between tubes use +, ++, or +++ to show relative amounts.

Why were tubes 1, 2, and 3 included in the iodoform test procedure? What purposes do they serve?

Look at your data in table 8.1. What is the same and what is different between the products of anaerobic and aerobic respiration?

Demonstration of Respiratory Efficiency How many times have you heard that aerobic respiration is more efficient than anaerobic? No doubt several, but where is the evidence? One can count ATPs produced in the two processes in the diagrams found in a textbook, but that is rather indirect. If aerobic is superior to anaerobic then the differences should be demonstrable. Measuring ATP production by cells directly is difficult because living cells use it for growth and metabolism, and it does not accumulate. The yeast cultures provide another way to evaluate respiratory efficiency.

In the presence of oxygen, yeasts should extract more chemical potential energy from sugar. Consequently, aerobically grown yeast should divide more frequently than those in an anaerobic culture. This can be stated as a testable hypothesis. When identical cultures are grown aerobically versus anaerobically, aerobic cultures should grow faster, and have higher cell densities at any given time.

One easy way to estimate cell densities in cultures is to look at turbidity. When cells are suspended in a culture, ambient light reflects from their surfaces giving the culture a cloudy appearance, called **turbidity.** The higher the density of cells, the more turbid the culture appears.

4 Your instructor will take 5 ml samples from the anaerobic and aerobic cultures (after vigorous shaking to suspend all cells) and put the samples in clear glass test tubes of uniform diameters and wall thickness. The tubes will be sealed with parafilm, shaken to suspend yeasts and held against a dark background. You may see a difference between the two culture treatments, but if the cell densities are high, the differences can be obscured by too much turbidity.

If that is the case, a serial dilution should be used to reduce the cell density. This is done by taking 2.5 ml from each culture after shaking to suspend the cells. Mix the sample with 2.5 ml of water, reducing the density by half. If differences are still not apparent, 2.5 ml of the dilution can be added to another 2.5 ml of water, reducing the original concentration to ¼. This may be continued, depending on cell density, several more times.

If you compare the later diluted samples to earlier ones, you should be able to make a quantitative estimate of the difference. For example if the ⅛ dilution of the aerobic culture has a turbidity identical to that for the ½ dilution of the anaerobic culture, then the aerobic culture had four times as many cells.

On the last line of table 8.1, enter the dilutions of the cultures that yielded the same turbidity and estimate how many times more efficient aerobic metabolism was in producing cells. What is your estimate? _____

Look at your data in table 8.1. If the aerobic culture was more efficient, it harvested more energy from the sugars in the culture medium than the anaerobic culture did. Where did the energy go in the anaerobic culture?

oxygen and produce carbon dioxide as they metabolize sugars to produce ATP. If the CO_2 is chemically removed as it is produced, the pressure in the chamber will drop in proportion to the O_2 consumed. The simple apparatus shown in **figure 8.3** can be constructed to measure this change.

Potassium hydroxide can be used to make a CO_2 "scrubber" in the apparatus. If CO_2 passes over a wet surface, it enters the water as carbonic acid. Carbonic acid reacts with KOH in the water to form potassium carbonate. Because potassium carbonate is a solid, it has a negligible volume compared to gaseous CO_2. The reactions are:

$$H_2O + CO_2 \text{ (gas)} \rightarrow H_2CO_3 \text{ (solution)}$$
$$H_2CO_3 + 2 \text{ KOH} \rightarrow K_2CO_3 \text{ (crystalline)} + 2H_2O$$

If a small drop of dye is placed in the open end of the pipette in the apparatus in figure 8.3, it will move inward as the pressure in the chamber decreases due to oxygen consumption and move in or out due to fluctuations in room temperature and atmospheric pressure. The markings on the pipette allow direct readings of volume.

To convert measured volume to actual volume of oxygen consumed, two corrections are necessary. One correction offsets the effects of fluctuations in temperature and pressure during the experiment. The other adjusts your **measured volume** to **standard volume** at a standard pressure of 760 Torr (mm of Hg) and standard temperature of 273 Kelvin. The first correction is made by using an experimental control called a **thermobarometer** and the second by a calculation using the **combined gas laws** that you learned in your chemistry classes.

Doing Discovery Science Seedlings germinating from seeds have high aerobic metabolic rates that can be measured in the apparatus shown in figure 8.3. Let's assume that an agricultural seed company is interested in developing seeds that are frost resistant after germination begins. They believe there is a market for seeds that could be planted early in spring when there is still the possibility of frost at night that would kill the varieties that are typically planted. For initial testing, they propose that each batch of seeds be divided into two groups; one to be given a cold shock by exposure to freezing temperatures

Aerobic Respiration in Germinating Seeds

The rate of aerobic respiration can be measured by measuring the rate of oxygen consumption. If living tissues or small organisms are placed in a closed chamber, they consume

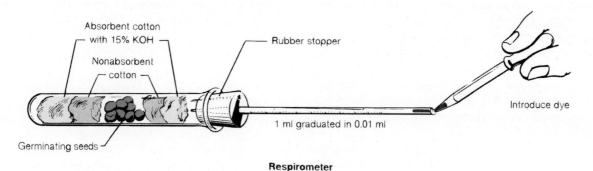

Figure 8.3 Apparatus for measuring oxygen consumption in seedlings.

and the other to be kept at room temperature. Rather than plant them and wait to see if they grow, they want to use a quick test for viability: oxygen consumption. By comparing oxygen consumption rates between cold-treated and untreated germinating seeds, they believe that they can pick the best stock for selective breeding and searching for frost tolerant genes. You and your lab partner are the lab technicians who will do the experiments, following the directions given here.

Your lab instructor will tell you what species you are to test. Peas, beans, wheat, soy or another species could be used. When the seeds come to your lab, they will have been germinated and cold-treated. You are to measure and compare the oxygen consumption of cold-treated and untreated seeds.

5 Pause for a moment and think about the experimental design. It would be faster to test just cold treated seeds. Explain why you need to test both cold-treated and untreated seeds from each strain or species.

How are the rates of metabolic processes likely to change when seeds are chilled? _____

State the temperature(s) (cold or room temperature) at which you will measure the rate of oxygen consumption for each group of seeds, and explain why you made those temperature choices.

Experimental Setup Before lab, your instructor will have cold-treated some seeds at subfreezing temperatures followed by rewarming to room temperature and kept others at room temperature. Both types will have gone through the same germination treatment.

To make the oxygen consumption measurements as comparable as possible, you will add the same number of seeds to two different tubes. However, if you look at the seeds, it's obvious that they are all not the same size. Larger seeds have more tissue and consume more oxygen than smaller ones. Therefore, the two seed samples should be weighed.

6 Label and weigh two empty weighing pans, recording the masses in the following worktable. If you are using an electronic balance, adjust it to zero (tare) with an empty weighing pan on the balance.

Table 8.2 Seed Masses

	Normal	Cold-Treated
Mass of seeds and pans	_____	_____
Mass of empty pans	_____	_____
Mass of seeds	_____	_____

Now add about 10 to 18 four-to-six-day-old normal germinating seeds to one pan and the same number of cold-treated seeds to the other. Record the masses in **table 8.2**. If you did not tare the balance, subtract the empty pan weight to obtain the tissue weight.

Transfer the seed mass values to table 8.5.

7 Take three tubes and add a small ball of *absorbent* cotton 2 cm in diameter to the bottom of each tube. Hold the tube vertically and drop in four drops of 15% KOH, so that the drops fall directly on the cotton and do not run down the sidewalls. Cover the moistened cotton with a layer of *nonabsorbent* cotton. The nonabsorbent cotton protects the seeds from contact with the corrosive KOH, the active component of many drain cleaners.

Add normal seeds to one tube, cold-treated seeds to another, and no seeds to the third. This third tube serves as the *thermobarometer*, a control for temperature and pressure changes in the room.

After the seeds are in place, add a second ball of *nonabsorbent* cotton to all three tubes and cover it with a ball of *absorbent* cotton. Moisten the absorbent cotton with two drops of 15% KOH.

Add dry stoppers with pipettes to all three tubes. The setups should look like figure 8.3. Be sure that the tip of the pipette does not poke into the cotton, which can plug it and prevent accurate readings. When each apparatus is assembled, wrap the side of the stopper and upper part of the tube with tape or Parafilm to prevent the stopper from moving outward during the experiment.

8 **Measurement** Place the tubes on the table and let them equilibrate for 10 minutes. Temperature equilibration after holding the tubes in your warm hands is absolutely essential for accurate results.

Is your body temperature above or below room temperature? What happens to the pressure exerted by a gas when temperature increases? How would a rising temperature affect your results?

Table 8.3 Raw Oxygen Consumption in Milliliters

Start time = _____

Tube	Contents	Reading Time (minutes)					
		0	3	6	9	12	15
1	Normal seeds						
2	Cold-treated seeds						
3	Thermobarometer						

Table 8.4 Thermobarometer-Corrected Data in Milliliters-Measured Volumes

Tube	Contents	Reading Time (minutes)					
		0	3	6	9	12	15
1	Normal seeds						
2	Cold-treated seeds						

Table 8.5 Derived Oxygen-Consumption Data

Tube	Contents	Slope (ml/min)	Seed Mass (g)	Mass-Specific Rate of O_2 Consumption (ml min^{-1} g^{-1})
1	Normal seeds			
2	Cold-treated seeds			

After 10 minutes, add a drop of dye into the end of each pipette using a Pasteur pipette. Try to get the front of the drop near the zero mark. When the front surface of the drop passes zero, record the time and the dye position in ml on the pipette in **table 8.3**.

Read the position of the front surface of the dye drop at two- to five-minute intervals (three minutes suggested but should be modified according to rates observed). Record the readings in table 8.3. If fluid moves out from the 0 mark toward the end of the pipette, estimate the volume change and use a minus sign to indicate readings less than zero.

Do not touch the tubes during the experiment to avoid changing the tube temperature which will affect gas pressure.

9 **Analysis** Correct the raw data by subtracting the values recorded from the thermobarometer. This tube corrects for any changes during the experiment due to changes in atmospheric pressure or room temperature. Thus, changes in the thermobarometer tube reflect environmental variation, whereas those in the other tubes reflect changes due to metabolism plus environmental variation. Subtraction of the thermobarometer value, which may be either plus or minus, corrects for the effects of environmental variation. (Remember that if you subtract a negative number, you must actually add.) Enter the corrected values in **table 8.4**.

Now plot both sets of the measured-volume-of-oxygen-consumed figures as a function of time on a computer or one piece of graph paper at the end of the exercise. Directions for making graphs are given in appendix B. Use different plotting symbols for each treatment and label all axes. Draw straight lines that best fit each data set.

10 Calculate the slope of each line and enter the slopes in **table 8.5**. Divide the slope values by the seed mass to calculate a seed mass-specific rate (a rate of oxygen consumption per gram of tissue) for each treatment.

If you are going to compare rates of oxygen consumption between the samples, why is it necessary to calculate oxygen consumption on a per-gram-of-tissue basis?

11 These values are only measured rates of consumption and must be corrected to standard conditions before they can be reported and compared to published data. By convention, gas volumes are always reported as volumes at 760 Torr (mm of Hg) pressure and 273 absolute temperature in Kelvin units, known as standard temperature and pressure (**STP**).

The gas laws, mentioned earlier, are used to make this correction, as follows:

If V_{meas} equals measured volume at P_1 (atmospheric pressure during the experiment) and at T_1 (temperature during experiment in Kelvins which are °C + 273), and if

V_{STP} equals corrected volume at standard pressure P_2 and at standard temperature T_2, then by the gas laws:

$$\frac{P_1 V_{meas}}{T_1} = \frac{P_2 V_{STP}}{T_2}$$

Rearrange to calculate the standardized volume of oxygen consumed per minute per gram of tissue,

$$V_{STP} = V_{meas} \times \frac{P_1}{P_2} \times \frac{T_2}{T_1}$$

Add the standard values:

$$V_{STP} = V_{meas} \times \frac{P_1}{(760)} \times \frac{273}{T_1}$$

Ask your lab instructor for the atmospheric pressure (in Torr = mm of Hg) and temperature readings (in °C) in the room during the experiment. If barometric pressure is obtained from a weather station, realize the reported value has been corrected to sea level to make pressures comparable over geographic regions. You will have to correct weather station value for altitude to get an actual value. Record them below.

Atmospheric pressure (P_1) = _____ Torr
Temperature (T_1) = _____ °C = _____ K

Now use the equation above to calculate the mass-specific rates of O_2 consumption at STP for both (1) untreated seeds and (2) cold treated seeds. For V_{meas} use the mass-specific rates you calculated in the last column of table 8.5.

12) What are the mass-specific rates of oxygen consumption for your experiments after adjustment to standard conditions?

Untreated seeds _____ ml g^{-1} min^{-1}
@ STP

Cold-treated seeds _____ ml g^{-1} min^{-1}
@ STP

Does cold shock inhibit respiration in your seeds? If so, calculate by how much?

Hint: % inhibition = $\left(\dfrac{Normal - treated}{Normal} \right) \times 100$

% inhibition = _____

13) Conclusion Compare your results with those of other groups that tested different seeds. Which species or strain would you recommend as the best candidate for the frost tolerant project, and why?

Lab Summary

14 On a separate sheet of paper, answer the following questions as assigned by your instructor.

1. What evidence did you collect that suggests aerobic respiration is more efficient than anaerobic?
2. Plot the data in table 8.3 on the graph paper at the end of the exercise and turn in the graph with this summary. What conclusion can you come to based on the data summarized in this graph.
3. Adjust the rate of O_2 consumption (ml min^{-1} g^{-1}) for both samples to standard temperature and pressure. Show the equations with substituted values. By what percentage did cold treatment inhibit respiration?
4. What are the advantages of aerobic respiration over anaerobic respiration? What are the advantages of anaerobic respiration over aerobic?

You may want to try the critical thinking questions that apply some of the knowledge you gained in doing this lab.

Learning Biology by Writing

The experiments in this laboratory exercise were chosen to demonstrate some of the properties of anaerobic and aerobic metabolism. The truly experimental portion of the exercise was the measurement of oxygen consumption.

Write a report that summarizes your observations and makes a recommendation to the company about what will be the effects of early planting for the seeds tested. Review the directions in appendix D for writing lab reports.

Critical Thinking Questions

1. One group of students in Denver, Colorado, and another group in New Orleans, Louisiana, performed the "aerobic respiration in peas" experiment. Are their results directly comparable? Explain.
2. Some students say that animals respire and plants photosynthesize. What evidence do you have to argue that this is a false generalization?
3. If you were looking at different types of eukaryote cells through a transmission electron microscope and found certain cells that had no mitochondria, what prediction could you make about their metabolism?
4. In very deep lakes and several inches down in the muck on the bottom of swamps the dissolved oxygen concentration is zero. Many bacteria, fungi and some other organisms live in these environments and are able to metabolize glucose. How do they do it?
5. Does the pressure exerted by a gas increase or decrease when its temperature is increased but its volume is held constant? When temperature decreases and pressure increases, what happens to the volume of a gas?

Lab Topic 9

INVESTIGATING PHOTOSYNTHESIS

SUPPLIES

Resource guide available on WWW at
http://www.mhhe.com/labcentral

Equipment

Compound microscopes

Spectrophotometers with glass, or methacrylate cuvettes; not polystyrene

Kitchen blender

Desk lamps with outdoor flood lamps

Magnetic stirrer

Photometer (optional)

Materials

Fresh *Elodea* cuttings

Slides and coverslips

Frozen spinach

Flat-sided glass tanks about 10 cm across

Ring stand and two clamps

Test tubes with one-hole stoppers

Syringe, 1 ml and 1¼" needles

1-ml pipettes with bend at the top end above 0 as in figure 9.5

Parafilm

800-ml beakers

Cheesecloth

Graduated cylinders

250-ml separatory funnel

125-ml Erlenmeyer flasks

Slides and coverslips

Solutions

Acetone

Petroleum ether

80% methanol

10% NaCl

Na_2SO_4 powder (anhydrous)

MgO powder

0.1 M $NaHCO_3$

STUDENT PRELAB PREPARATION

Before doing this lab, read the Background material and sections of the lab topic that have been assigned by your instructor.

You should use your textbook to review the definitions of the following terms:

carotenoid	spectrophotometer
chlorophyll a and b	stroma
chloroplast	thylakoid
grana	xanthophyll
photosynthesis	

Describe in your own words the following concepts:

Absorption spectrum

Calvin cycle

Light reactions

Standard temperature and pressure

After finishing the prelab review, write any questions you have about terms, concepts, or techniques in the margins of this lab topic. The lab experiments should help you answer these questions, or you can ask your instructor during the lab.

OBJECTIVES

1. To estimate number of chloroplasts per cell
2. To measure the absorption spectrum of chlorophyll
3. To determine how the rate of photosynthesis varies with light intensity.
4. To visualize starch accumulation in leaves

BACKGROUND

Photosynthesis supplies energy to virtually every ecosystem. Not only do plants, algae, bacteria (especially cyanobacteria) and some archaeans use this energy to grow and reproduce, but they also are eaten by animals or decomposed by bacteria and fungi, thus supplying these organisms' energy needs as well. In photosynthesis, light energy from the sun is captured and used to make new covalent chemical bonds in organic molecules. Carbon dioxide serves as a source of carbon, and water as a source of hydrogen in these reactions. An important by-product of photosynthesis is molecular oxygen, which enters the atmosphere, replacing that which is consumed in aerobic respiration and various other oxidations. In fact, the ancient earth's atmosphere contained no free oxygen until photosynthetic organisms evolved. In addition, oxygen in the upper atmosphere is photochemically converted to ozone. It forms a shield that reduces the transmission of mutagenic ultraviolet light to the earth's surface.

To give you an idea of the magnitude of the importance of photosynthesis, one hectare (= 2.47 acres) of corn in the Midwest results in a net annual conversion of 5,900 kg of carbon from CO_2 to carbon stored in organic molecules. Photosynthesis produces one molecule of oxygen for each carbon stored, so a hectare of corn produces enough oxygen to meet the annual respiratory needs of about 72 people. During 2008, Iowa planted 7,740,000 hectares of corn, enough to meet the annual oxygen needs of 560 million people. (The entire population of North America was about 450 million people in 2009.) Of course, much more oxygen is consumed worldwide by other people and organisms, by combustion, and by human industry and commerce.

Realize that energy, unlike carbon dioxide, oxygen, or water, does not cycle in ecosystems. Instead, energy flows in a direction toward dissipation. Plants convert light energy to chemical bond energy, storing it in the bonds of biochemical compounds. As plants use these compounds or are consumed by animals that, in turn, are consumed themselves or die and decompose, energy is lost. Loss occurs whenever the chemical bonds in biomass are broken and reformed in metabolism along food chains. As a general rule, only 10% of the energy available in ingested food is used to make new biomass; the remaining 90% dissipates as low-grade heat (entropy). To maintain food chains, constant energy input is required. If photosynthesis suddenly ceased on this planet, most life would eventually end because all energy stored in chemical bonds of biomass would be released as heat as one organism fed on another in ever-shortening food chains.

Leaves are the photosynthetic organs of most plants (**fig. 9.1**), except in those with photosynthetic stems. Leaves are well adapted to the task. Being only a few cells thick allows light to penetrate to the deepest cell layers. A covering of nonphotosynthetic epidermal tissue limits water loss and has pores called stomata that allow gas exchange. The photosynthetic machinery is represented by a few layers of cells in the center of the leaf, the mesophyll layers. Vascular tissues bring needed water and minerals from the roots and also transport photosynthetic products to fulfill the energy needs of nonphotosynthetic cells in the stem and roots.

Chlorophyll is located in organelles called **chloroplasts** in mesophyll cells (fig. 9.1). Surrounded by two membranes, a chloroplast contains stacks of **thylakoid** membranes called **grana**. Chlorophyll is embedded in the thylakoid membranes and it is only here that the light-absorption reactions of photosynthesis occur. The open space inside a chloroplast that is not filled by thylakoid stacks is the **stroma**. It contains the numerous enzymes involved in using carbon dioxide to make carbohydrates.

Chlorophyll is a remarkable molecule that absorbs light energy, transferring it to its electrons. These electrons

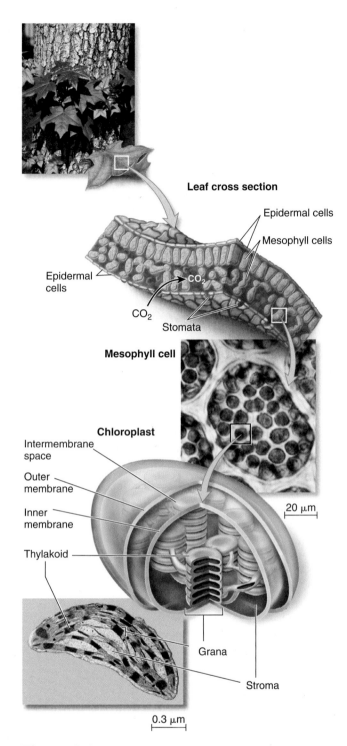

Figure 9.1 **Leaf organization.** Leaves are composed of layers of cells. The epidermal cells are on the outer surface, both top and bottom, with mesophyll cells sandwiched in the middle. The mesophyll cells in most plants are the primary sites of photosynthesis.

can then perform cellular work as they pass through the **photosynthetic electron transport chain,** a collection of compounds located in the thylakoid membranes of the chloroplasts. The energy is used to form **ATP** (adenosine triphosphate) by chemiosmosis, and the high-energy electrons may pass to the electron carrier $NADP^+$ (nicotinamide adenine dinucleotide phosphate), producing **NADPH**. Both

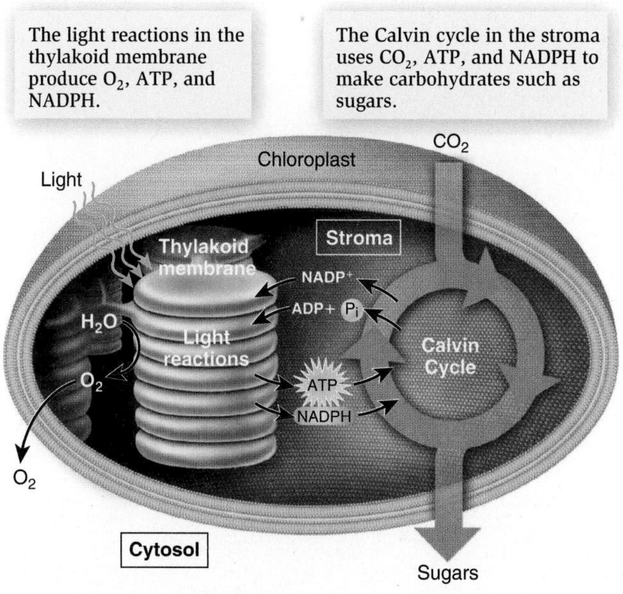

The light reactions in the thylakoid membrane produce O$_2$, ATP, and NADPH.

The Calvin cycle in the stroma uses CO$_2$, ATP, and NADPH to make carbohydrates such as sugars.

Figure 9.2 The reactions of photosynthesis occur in the chloroplast: the light reactions in the thylakoids and the Calvin cycle in the stroma.

ATP and NADPH are needed to produce new carbohydrates. These processes, collectively known as the **light reactions,** produce oxygen as a by-product. It diffuses into the atmosphere. Review the diagrams of the light reactions in your textbook and determine what is the source of oxygen released during photosynthesis. Does the oxygen come from CO$_2$ or H$_2$O?

The ATP and NADPH produced during the light reactions are used as energy and hydrogen sources in carbohydrate synthesis (**fig. 9.2**). In these reactions, CO$_2$ molecules are bonded one at a time to already existing carbohydrates, a process known as **CO$_2$ fixation.** The reactions are collectively called the **Calvin cycle,** or dark reactions. In most plants, glyceraldehyde phosphate is the product of the Calvin cycle. It can then be converted into hundreds of other compounds by other enzymes in the chloroplast or cytoplasm, resulting in the formation of sugars, lipids, amino acids, and nucleotides needed by plant cells for maintenance and growth. Because glyceraldehyde phosphate is a three-carbon compound, plants performing photosynthesis in this way are said to carry out **C$_3$ photosynthesis.**

At very high light intensities, the oxygen concentration in the chloroplasts dramatically increases as the light reactions produce massive amounts of O$_2$. When subject to high oxygen concentrations, an enzyme in the Calvin cycle responsible for carbon dioxide fixation, rubisco bisphosphate carboxylase, can catalyze a secondary reaction between O$_2$ and ribulose bisphosphate, degrading this key intermediate. Loss of the intermediate reduces carbohydrate production, wastes energy, and uses oxygen. In other words, very high light intensities reduce carbon fixation. The overall phenomenon is known as **photorespiration.** Some plants, called **C$_4$** plants, produce four carbon compounds in photosynthesis instead of glyceraldehyde phosphate as an adaptation to reduce photorespiration. Details can be found in your textbook.

LAB INSTRUCTIONS

In this lab topic, you will observe chloroplasts, determine the absorption spectrum of chlorophyll, measure how light intensity influences photosynthetic rate, and test for starch accumulation in leaves exposed to light.

Chloroplasts

Plant cells in roots, deep within stems, flowers, or fruits, or forming various outer coverings do not contain chloroplasts. Only certain types of parenchymal cells contain chloroplasts and can carry out photosynthesis. How many chloroplasts would you estimate are in a typical plant cell? One, a dozen, or hundreds?

1) To answer this question take a few moments to make a wet-mount slide of a leaf from the aquatic plant *Elodea* and look at it with your compound microscope.

Elodea leaves are only two cell layers thick and both layers are photosynthetic. Being an aquatic plant, it lacks nonphotosynthetic epidermal layers of cells found in terrestrial plants, protecting them against desiccation. Focus only on one cell layer as you study the cells. The cytoplasm is a thin layer of material inside of the prominent cell wall. It contains a large vacuole that takes up most of the space inside the walls. Because of their chlorophyll content the chloroplasts are easy to spot. Even at high power, you will not see the thylakoids because they are beyond the resolving power of your microscope.

Count the number of chloroplasts in a cell and write your estimate on the blackboard or enter it into a spreadsheet so that a class average can be calculated. What was your estimate? _____ What was the class average and standard deviation? Record below.

Average number of chloroplasts per cell = _____ +/− st. dev. _____

Put your microscope away. It will not be needed again in this lab.

Photosynthetic Pigments

Chloroplasts contain two types of chlorophyll, called *a* and *b,* differing slightly in chemical structure (**fig. 9.3***a*). Chlorophyll is a relatively large molecule with a molecular

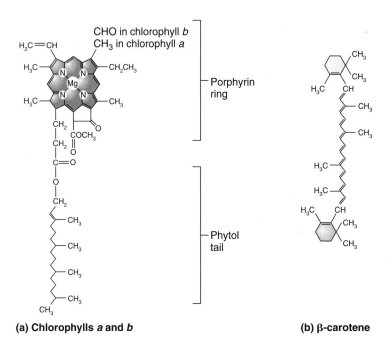

Figure 9.3 Structure of chloroplast pigment molecules. (a) Chlorophylls *a* and *b* are the same except for substitution where indicated. (b) Carotenoids and xanthophylls are similar to the B-carotene shown.

weight of approximately 900. The molecule has two parts with distinctly different functions. The complex head consists of a system of alternating single and double bonds between carbons forming a porphyrin ring with a magnesium ion at its center. It is the light-absorbing part of the molecule. The tail is derived from phytol, a long-chain alcohol. It is hydrophobic because it is primarily a long nonpolar hydrocarbon chain. The hydrophobic end anchors the whole molecule into the lipid layers of the thylakoid membranes, holding the porphyrin head close to other chlorophyll molecules and proteins that comprise the reaction centers for the light reactions of photosynthesis.

Chloroplasts also contain orange and yellow pigments called carotenoids (fig. 9.3*b*) and xanthophylls. Sometimes called accessory pigments, they absorb light energy that can be passed to the chlorophylls in the photosynthetic reaction centers, thus increasing the range of wavelengths that can drive photosynthesis. In addition, some carotenoids function in photoprotection, interacting with free radicals produced in photosynthesis which can damage the chloroplasts. The light-absorbing properties of carotenoids and xanthophylls are due to their alternating systems of single and double bonds between carbons. Obviously in a different pattern than those found in chlorophyll, these bonds absorb light energy in the yellow to orange portion of the spectrum. Because carotenoids and xanthophylls are essentially composed primarily of carbon and hydrogen, they are nonpolar and hydrophobic. Consequently, they are found embedded in the thylakoid membranes of the chloroplasts.

Extraction Procedure Photosynthetic cells, such as those found in spinach leaves, are a complex mixture of polar compounds dissolved in the water of the cytoplasm, hydrophobic nonpolar compounds dissolved in the lipids of the membranes, and nonsoluble molecules such as proteins, starches, and cellulose. The pigments found in chloroplasts are hydrophobic for the most part. The phytol tail of chlorophyll and the hydrocarbon backbones of carotenoids and xanthophylls make them more soluble in nonpolar solvents than in polar solvents. Acetone and methanol are somewhat polar solvents while petroleum ether is nonpolar. We can use these properties to extract pigments from plant cells.

If disrupted cells are treated with acetone, it mixes with any water present, penetrating structures such as chloroplasts and solubilizing the pigments. If petroleum ether is added to the acetone extract, it does not mix with the watery acetone because it is very nonpolar. The nonpolar pigments move from the somewhat polar acetone to the petroleum ether. We can then separate the different pigments by chromatography or further purify the extract so that only chlorophyll is present.

CAUTION

Petroleum ether, methanol, and acetone are highly flammable. No flames are permitted in the laboratory during this procedure. Have absorbent material handy to soak up any accidental spills. Solvents should not be poured down the drain, but should be collected for proper disposal.

2 You or your instructor will use a kitchen blender to homogenize 60 g of frozen spinach leaves in 100 ml of water, containing 0.2 g of MgO powder. The MgO protects the chlorophyll against loss of its magnesium. Pour the slurry into a 500-ml beaker, add 100 ml of acetone, cover with a watch glass or foil, and stir for 10 to 15 minutes on a magnetic stirrer. Filter the resulting slurry into another beaker. This extract is a complex mixture of fats, sugars, proteins, and pigments, including carotenoids, xanthophylls, and chlorophyll.

Add 50 ml of the filtrate to a separatory funnel, then add 50 ml of petroleum ether and 25 ml of 10% NaCl. Stopper the funnel and shake. Chlorophyll will move into the nonpolar phase as you shake the mixture. *Be careful!* Pressure will build up in the separatory funnel and must be released periodically by loosening the stopper. After vigorous shaking, place the funnel in a ring stand to allow the two solvent systems to separate. The polar aqueous phase with polar compounds will settle to the bottom, while the nonpolar petroleum ether containing chlorophyll will rise to the top (**fig 9.4**)

When separation is complete after about 15 minutes, loosen the stopper and drain the lower water-acetone layer into a chemical waste container.

Now add 30 ml of 80% methanol and 15 ml of 10% NaCl to the funnel. Shake and allow to settle. The yellow flavones will preferentially dissolve in the methanol and separate into the bottom layer. Drain it into a waste container. Add 30 ml of 10% NaCl to the chlorophyll extract and shake again to remove residual acetone and methanol. Discard the aqueous phase into a waste container. You should now have a reasonably bright green solution of chlorophyll.

Remove traces of water from the petroleum ether-chlorophyll solution by adding about 2 g of anhydrous Na_2SO_4 to the extract. Remove Na_2SO_4 by filtering. You will now measure the absorption spectrum of the clear green filtrate.

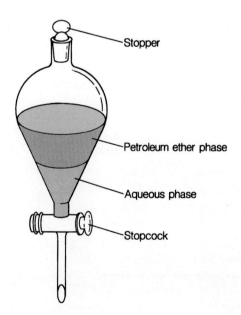

Figure 9.4 A separatory funnel, used to separate a two-phase liquid mixture. The stopper must be removed before the stopcock is opened.

3 ▶ *Measure Absorption Spectrum* In this section you will determine the light absorption properties of the chlorophyll extract using a spectrophotometer. Chlorophyll appears green in white light because its system of covalent bonds in the porphyrin ring "head" absorb photons of blue and red light, but not green light. Using this information, hypothesize what the absorption spectrum curve for chlorophyll will look like. Make a sketch below to record your hypothesis.

4 ▶ Adjust a spectrophotometer to zero absorbance using pure petroleum ether as a blank. (Review fig. 5.5 for operating instructions.)

Add the chlorophyll extract to a spectrophotometer tube and read its absorbance across the visible spectrum at 20 nm wavelength intervals. If the solution is too concentrated (absorbance greater than 2 at 400 nm), dilute it with petroleum ether. If it is too weak, blow air through the extract to evaporate some petroleum ether. This must be done in a fume hood. Remember to zero the spectrophotometer with petroleum ether at each new wavelength. Record your readings at each wavelength in **table 9.1**.

5 ▶ Observations of the color of light requested in table 9.1 can be obtained by placing a strip of white paper in the sample chamber of the spectrophotometer. If you look down into the tube with your hands cupped around the tube chamber opening, you should see the reflected light as it passes through the tube. Change the wavelength and record the colors in the table.

6 ▶ *Analysis* After lab when you are writing your report, you will graph these data. The resulting graph is called an absorption spectrum. (See appendix B for graphing instructions.) Remember the independent variable should be on the x axis. Is wavelength or absorbance the independent variable? _____ Label all axes and be sure to create a one-sentence figure legend that describes the sample.

Table 9.1 — Light Absorption Characteristics of Chlorophyll

Wavelength	Light Color	Absorbance
400		
420		
440		
460		
480		
500		
520		
540		
560		
580		
600		
620		
640		
660		
680		

Compare the actual absorption spectrum to your earlier prediction. Was your prediction close? _____

After plotting the absorption spectrum, make a hypothesis that relates rate of photosynthesis to the wavelength of light striking a plant. Would you predict that photosynthesis would be greatest or least when the wavelength of light striking a plant corresponds to the wavelength of light maximally absorbed by chlorophyll? Test your hypothesis by looking up action spectrum in your textbook and comparing your absorption spectrum to the wavelengths of light that are effective in photosynthesis.

Measuring Photosynthetic Rate

You will use the aquatic plant *Elodea* to measure the effect of light intensity on the rate of photosynthesis. Oxygen produced in chloroplasts diffuses from plant cells into the surrounding environment. Because *Elodea* is an aquatic plant, the oxygen can sometimes be seen as small bubbles forming along the margins of the leaves. If *Elodea* is placed in a container such as shown in **figure 9.5**, the enlarging bubbles force water to move into the bent, calibrated pipette, allowing the volume of oxygen to be measured.

In this experiment, you will determine how the rate of oxygen production changes as light intensity is varied by changing the distance between a flood light and *Elodea*. Light intensity is inversely related to the distance from the lamp to the specimen. Using these data, you will graphically estimate the light intensity that produces just enough oxygen to exactly meet the needs of aerobic respiration. This intensity is called the **light compensation point.** Photosynthesis does not stop at this light intensity. It is ongoing, but all of the oxygen produced is immediately consumed.

In your own words, state the purpose of this experiment.

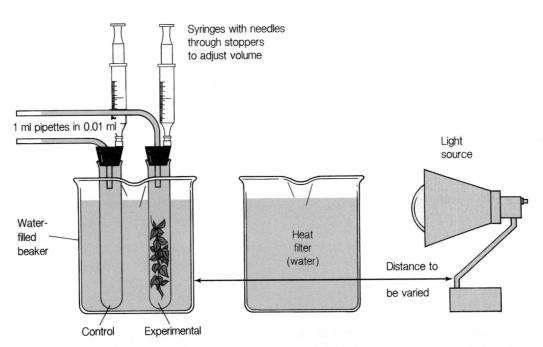

Figure 9.5 Apparatus for studying oxygen production at different light intensities.

INVESTIGATING PHOTOSYNTHESIS 105

7 ▸ Procedure Obtain a piece of *Elodea* 5 to 6 inches long. Choose a healthy specimen with an actively growing leaf bud at the tip. Make a fresh cut across the basal end and place the sprig in a test tube **with the cut end up.**

Fill the test tube with 0.1 M $NaHCO_3$ solution and add a rubber stopper containing a bent glass pipette and syringe, as shown in figure 9.5. The $NaHCO_3$ is a source of carbon dioxide for the plants. By making sure carbon dioxide is plentiful, photosynthesis will be limited only by light intensity.

The joint between the rubber stopper and the test tube must be dry to get a good seal and fluid must flow into the pipette as you push the stopper in. Wipe the stopper and inside edge of the tube with a tissue and, after tightly seating the stopper, wrap tape or Parafilm® around the tube and stopper to hold it in place. You must get a good seal. If the stopper "creeps," it will give you erroneous readings. The position of the fluid in the pipette can be adjusted by raising or lowering the plunger on the syringe once the stopper is in place.

Prepare a second tube without *Elodea* to act as a **thermobarometer**, a control for temperature and atmospheric pressure fluctuations.

Place both tubes in a beaker of water at room temperature. Set up a heat filter and a lamp as shown in figure 9.5. The lamp puts out white light, light that contains all the colors of the spectrum. Move the lamp so it is 100 cm away from the tubes with the heat filter in between. *Dim the room lights* and wait 10 minutes for equilibration.

If photometers are available in your lab, take this time to measure light intensity. Your instructor will give you directions about how to use the available equipment and the units of measurement. Enter the value in **table 9.2**. Later when you move the light forward, the intensity should be remeasured and the values entered in the appropriate tables.

If you do not have photometers, use distance between the light and the tubes as an index of light intensity.

What hypothesis are you testing in this experiment?

8 ▸ After equilibration, measure the temperature in the *Elodea* water bath and record in table 9.2. Adjust the fluids in the pipettes of both tubes so that they are at the 0.2 ml mark. Note the time and record the pipette volume readings for both *Elodea* and the thermobarometer in table 9.2.

Read the position of the fluid in the pipettes at two-minute intervals for 10 minutes. Record your data in table 9.2. If you do not get a change of about 0.15 ml in ten minutes, you should extend the reading intervals by several minutes. If you change the time interval, be sure to change the times printed in table 9.2. At higher light intensities (shorter plant-to-light distances), greater movement will be seen.

Table 9.2	Oxygen Production Readings (in ml) and Thermobarometer-Corrected Data								
		Start time = _____				Time (minutes)			
Distance (cm)					0	2	4	6	8	10
100	Intensity = _____		*Elodea* (EI)							
	Temp = _____		Thermobarometer (TB)							
			Corrected (EI − TB)							
75	Intensity = _____		*Elodea* (EI)							
	Temp = _____		Thermobarometer (TB)							
			Corrected (EI − TB)							
50	Intensity = _____		*Elodea* (EI)							
	Temp = _____		Thermobarometer (TB)							
			Corrected (EI − TB)							
25	Intensity = _____		*Elodea* (EI)							
	Temp = _____		Thermobarometer (TB)							
			Corrected (EI − TB)							

Oxygen is not very soluble in water. As *Elodea* produces oxygen in photosynthesis, the newly produced gas will force water to move into the pipette. Therefore, fluid movement in the pipette is a measure of the rate of photosynthesis.

The changes you observed in the tube containing *Elodea* are the result of two simultaneous processes (1) the production of oxygen by *Elodea*, which always results in an increase in volume, and (2) fluctuations in temperature and pressure, which may result in either positive or negative changes in volume. To obtain a "corrected" reading of oxygen production, you must subtract the thermobarometer control reading from the *Elodea* reading. Do this and enter the difference on the third line of table 9.2.

9▸ Collect three more data sets with the light source at successively closer distances (75, 50 and 25 cm) and record the results in table 9.2. Be sure to record light intensities and water bath temperatures for each distance. If you use other distances, change the numbers in table 9.2 to match.

10▸ Analysis Using a computer or a panel of graph paper at the end of this lab topic, plot the corrected cumulative volumes as a function of time. Plot the data for each light intensity (as measured by a photometer or as 1/distance between bulb and plant) on the same coordinates using different plotting symbols for each distance. Using a ruler, draw four straight lines that best fit the points for each treatment.

Now calculate the slopes for each of these lines. The slopes are rates of oxygen production with units of ml/min. Enter these values as V_{meas} in **table 9.3**. Also enter your measures of light intensity.

Recall the discussion of gas volumes in the cellular respiration experiment (see pages 95–96). Review that section before you go on. Because gas volumes change with temperature and pressure, by convention gas volumes are always reported at standard temperature and pressure (STP) according to following equation:

$$V_{STP} = V_{meas} \times \frac{273\ K}{T_{meas}} \times \frac{P_{meas}}{760\ torr}$$

11▸ In this experiment, P_{meas} is the barometric pressure in torr (= mm of Hg) during the experiment. Your instructor will provide this value. T_{meas} is the temperature in Kelvin (K = °C + 273) in the water bath surrounding the *Elodea* tubes. If you did not measure the temperature, you can use room temperature, provided that the water bath was allowed to reach temperature equilibrium before your experiment. Use the work table to correct your measured rates of oxygen production, and enter the results in the last column of table 9.3.

Use a computer or a panel of graph paper at the end of this chapter to plot your data from table 9.3, illustrating how the rate of oxygen production corrected to STP varies with light intensity.

Which variable is the independent variable and belongs on the x axis? Label both axes and include units of measurement.

What is the best way to describe the relationship between rate of photosynthesis and light intensity? Is a straight line a good description of your data?

If your data look like they fall on a straight line, draw a best-fit staight line through the points. If your data do not appear to fall on a straight line, draw a continuously changing smooth curve through the points. Once you have chosen a line or curve to describe your data, extrapolate your curve until it crosses the x axis of your graph.

12▸ At the x axis intercept point, your data predict a light intensity at which the net rate of oxygen production by *Elodea* will be zero. That is the compensation point, the light intensity at which the rate of oxygen production in photosynthesis equals the rate of oxygen consumption in aerobic respiration.

From your data, what light intensity (or distance) do you predict is the light compensation point? _____ This is a testable prediction (hypothesis).

Table 9.3	Worktable to Adjust Measured Volumes/Min to STP					
Light Intensity or 1/distance	V_{meas} (ml/min)	T_{meas} (°C)	T_{meas} (K)	273/T_{meas} (K)	P_{meas}/760 (torr)	V_{STP} (ml@ STP/min)

Should the compensation point light intensity be greater than zero, equal to zero, or less than zero? Why?

How could you experimentally determine if you predicted the correct light compensation point?

13) *Testing a Prediction* Test your prediction by running an experiment. Place the *Elodea* and control tube at the predicted light intensity for the compensation point and measure the oxygen consumption for nine minutes, entering your data in **table 9.4**, below. Subtract the thermobarometer readings values to get corrected readings.

Was your prediction correct? Must you accept or reject your hypothesis?

Now think critically about the result. If you saw no oxygen production, most would say their prediction was true: no oxygen was produced at the tested light intensity. However, this does not mean that you correctly predicted the compensation point light intensity. Why? What result would you expect if your prediction of the compensation point light intensity was too low?

Starch Accumulation in Leaves

When excess sugars are produced in photosynthesis, starch accumulates in leaves. Starch accumulation can be demonstrated by staining leaves with an I_2KI solution. Starch stains brownish purple.

Several days before the lab, a *Geranium* plant was placed under constant illumination. Some leaves were partially masked by tightly folding a 1" strip of opaque construction paper around part of the leaf and securing it with a paper clip. A fun variation of this is to cut initials or letters in part of the paper so light can get through to the leaf. A second plant with green and white variegated leaves was also placed in the light.

14) Take three leaves (unmasked, masked, and variegated) and put them in boiling water for 5 minutes. Then transfer them with forceps to hot 50% isopropyl alcohol for a few minutes to remove the carotenoids and chlorophyll. Place each leaf in a dry petri dish. Spray or soak the leaves with I_2KI solution. Describe the results.

Unmasked leaf

Masked leaf

Variegated leaf

How do you explain the differences in starch accumulation seen in these leaves?

Table 9.4	Oxygen Production Readings at Predicted Compensation Point				
	0	3	6	9	12
Elodea					
Thermobarometer					
Elodea − thermobarometer					

Lab Summary

15 On a separate sheet of paper, answer the following questions as assigned by your instructor.

1. Discuss why leaves are thin, only a few cells thick, and contain multiple chloroplasts.
2. Plot the data in table 9.1 and turn it in with this summary. Indicate with arrows on the x-axis the wavelengths of light most strongly absorbed by chlorophyll. See appendix B for directions on making graphs.
3. Plot the thermobarometer-corrected data for each light intensity in table 9.2.
4. Calculate the slopes of the four lines plotted in question 3 to obtain the rates of photosynthesis at each light intensity. Correct those values to STP. Graph the rates as a function of the light intensity. Indicate the predicted light compensation point on your graph.
5. What happened when you placed *Elodea* at the light compensation point?
6. Explain why a thermobarometer tube was included in the light intensity experiments. For what did it control?
7. Describe the results from the starch accumulation experiment. Why were these patterns of starch accumulation obtained?

You may want to try some of the critical thinking questions that apply the knowledge you gained in doing this lab.

Internet Sources

Despite the essential role of chlorophyll in photosynthesis, there are many things that we do not know about it. One type of current research has to do with how chlorophyll is synthesized including what genes are involved. Use Google to search the WWW for information about chlorophyll synthesis and review the information available on a few of the sites. Write a short paragraph describing what type of research is being performed.

Learning Biology by Writing

In this lab, you experimentally determined (1) the properties of chlorophyll and (2) the effects of light on the rate of photosynthesis. Your instructor will tell you which factors to include in your laboratory report.

To begin your report, outline the general purposes of your observations and experiments in a few sentences. Briefly describe the techniques you used. Include your results graph of the absorption spectrum of chlorophyll and/or photosynthetic rate as a function of light intensity. In the discussion, answer such questions as:

1. How does the chlorophyll absorbance spectrum relate to the expected action spectrum for photosynthesis?
2. What is the practical significance if the light level never rises above the compensation point?
3. How might you improve the experiments to gain more information?

Critical Thinking Questions

1. During the chlorophyll extraction procedure, xanthophyll and carotene pigments were extracted and discarded. What are the advantages to the plant of having these accessory pigments?
2. Deep, clear tropical ocean waters are dominated by Rhodophytes (red algae). Their color is derived from red pigments, phycobilins, that mask the chlorophyll. Why does this particular arrangement make them suited to their environment?
3. Ice and snow on lake surfaces in winter often reduces light intensity in the water below. If light intensity falls below the light compensation point, what is the implication for the ecosystem?
4. At night do green plants produce or consume oxygen? Why?

Lab Topic 10

MITOSIS AND CHROMOSOME NUMBER

SUPPLIES

Resource guide available at www.mhhe.com/labcentral

Equipment

Compound microscopes
Water bath or heating block at 60°C

Materials

Prepared slides of sectioned whitefish blastula
Prepared slide of sectioned onion root tip
Growing onions with root tips four days old. Immerse green onions from grocery in water with aeration.
Glass pestles, 4 inches × 1/8 inch; round end in flame and then file flat area on end
Small vials or tubes (about 10 ml) with caps
Small watch glass
Razor blades
Forceps
Slides and coverslips

Solutions

Fixative solution: one part glacial acetic acid to three parts 100% methanol made at start of each lab
1 M HCl (*Caution*)
45% acetic acid (*Caution*)
Fresh Feulgen stain (*Caution*)

STUDENT PRELAB PREPARATION

Before doing this lab, read the Background material and sections of the lab topic assigned by your instructor.
Use your textbook to review the definitions of the following terms:

Anaphase	Centriole
Aster	Chromatin
Cell cycle	Chromosome
Cell plate	Chromatid
Centromere	Cytokinesis
Centrosome	Interphase
Kinetochore	Prophase
Metaphase	Spindle
Mitosis	Telophase

Describe in your own words the following concepts:
How and when chromosomes replicate
Positions of chromosomes during phases of mitosis
Significance of mitosis
Review discussion of average and standard deviation in appendix C

After finishing the prelab review, write any questions you have about terms, concepts, or techniques in the margins of this lab topic. The lab activities will help you answer these questions, or you can ask your instructor during the lab.

OBJECTIVES

1. To identify stages of mitosis and cytokinesis using prepared slides of plant and animal cells
2. To use observations of the frequency of mitotic phases to determine their durations
3. To stain chromosomes in dividing plant tissues
4. To estimate the number of chromosomes in onions and to test a hypothesis by using descriptive statistics

BACKGROUND

All cells come from the division of preexisting cells: none spontaneously assemble even if all the proper ingredients are available. In prokaryotic cells, division occurs by binary fission in which the cell duplicates its DNA, then divides to produce two daughter cells. In protists, plants, fungi, and animals, cell division is more involved. It consists of a preparatory time called interphase in which parts are duplicated, a chromosome distribution process called **mitosis** or nuclear division, and finally a separation process called **cytokinesis** or cytoplasmic division. The result of these integrated processes is two identical cells where previously there had been one.

Most cells have only one nucleus. It contains multiple **chromosomes,** the carriers of the genes. The number of chromosomes varies among species. For example, potatoes have 48, donkeys 66, and mosquitoes 6. The 46 chromosomes found in humans are thought to carry between

20 to 25 thousand genes, a surprisingly small number. Each chromosome consists of a single long DNA molecule that is highly coiled and folded with the loops held in place by scaffold proteins (**fig. 10.1**). The shortest human chromosome contains about 1.7 cm of DNA and the longest about 8.5 cm.

A cell that will eventually divide must replicate all of its chromosomes so that each daughter cell receives copies of all the genes. Likewise, if it is to produce daughter cells identical to itself, it must also double all of its component cytoplasmic parts such as enzymes, carbohydrates, fats, membranes, and organelles.

If the life of a cell is viewed as a cycle from the time of formation until it divides in two, called the **cell cycle,** the period of growth is called **interphase.** It is the longest phase of the cycle. In adult humans, cells underlying the skin divide about every 24 hours to replenish skin cells that are lost. Interphase lasts about 23 of those hours and the actual division process takes only about an hour.

Events during interphase are complex. Researchers have found that DNA is not made continuously during interphase. Instead, a cell that has just been formed from a cell division undergoes a period of cytoplasmic synthesis called the G_1 subphase, lasting about 12 hours in human skin cells. Next, the DNA in all 46 chromosomes replicates during the subphase called **S** which lasts about 8 hours. Each of the 46 chromosomes now contains two long DNA molecules and associated proteins and these molecules are held together at a region in each chromosome called the **centromere.** The two duplicate parts of each chromosome are called **sister chromatids.** At this time the DNA is quite extended and the genes are active. Individual chromosomes cannot be observed and the nucleus appears to be filled with a fibrous material called **chromatin,** a collective name given to all of the extended chromosomes. Following the S phase, the G_2 subphase begins as the cell prepares for mitosis. Part of this preparation involves coiling of the extended DNA molecules to produce compact structures which will soon be visible as distinct chromosomes, each consisting of two chromatids. While compacted, the genes on the chromosomes are not active.

The G_2 phase ends when the events of mitosis start. During mitosis, the identical halves of each replicated chromosome will separate from each other and pass into the daughter cells when **cytokinesis,** the division of the cytoplasm, occurs.

You can spot a cell that is beginning mitosis because its nuclear envelope will be starting to break down, making the outline of the nucleus less distinct. At the same time, tubulin proteins in the cytoplasm will be forming microtubules, giving the cytoplasm a fibrous appearance. When these changes are visible, the cell is in **prophase** (fig 10.2a). As the nuclear envelope disappears, the microtubules arrange themselves into a diamond-shaped **spindle** (<>). Its pointed ends define the **poles** of the cell and an imaginary vertical line across the center defines the **equator.**

The spindle's function is to separate the sister chromatids from each other in an orderly manner. Some microtubules from each pole, but not all, attach to the sister chromatids at the **centromere** (fig. 10.1), the chromosomal region where the two sister chromatids are joined. As a result of tugging between the tubules from different poles, the chromosomes are dragged to the equator and align along it. When the chromosomes reach this position, the cell said to be in **metaphase** (fig. 10.2b).

Metaphase ends when the sister chromatids lose their attachment to each other at the centromere. Because one chromatid is attached to tubules going only to one pole and the other to tubules going only to the opposite pole, any tension on the tubules draws the former sister chromatids toward opposite poles. As the chromosomes begin to move, the cell is said to enter **anaphase** (fig 10.2c).

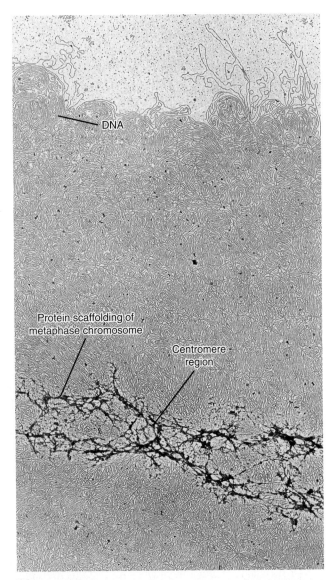

Figure 10.1 If a cell in metaphase is treated with salts and detergent, the single long DNA molecule in each chromatid unravels and spews out. It can then be photographed through an electron microscope. The protein scaffold remnants of the chromosome are also visible.

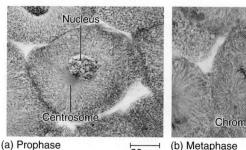
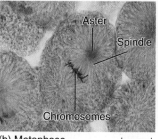
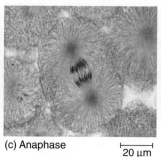
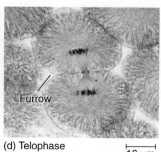

(a) Prophase 20 μm (b) Metaphase 20 μm (c) Anaphase 20 μm (d) Telophase 16 μm

Figure 10.2 Mitosis in white fish blastula cells: (a) prophase with formation of aster and chromosomes starting to condense; (b) metaphase with fully formed spindle and asters and chromosomes aligned on equator; (c) anaphase with sister chromatids separating to become independent chromosomes; (d) telophase with chromosomes starting to become diffuse and cytokinesis furrow forming. The furrow will eventually pinch the parent cell into two daughter cells.

As the chromatids, now individually called chromosomes, approach the poles, the cell is described as entering **telophase**. Once at the poles, several simultaneous events are set in motion. Nuclear envelopes form around the chromosomes at each pole, creating two genetically identical nuclei. Within the nuclei the chromosomes begin to uncoil, the genes become active, and the spindle breaks down. As all of this is happening, the cytokinesis processes that will divide the single parent cell into two equal daughter cells begin.

Cytokinesis processes differ in animals and plants. **Furrowing** in animal cells involves the cell's surface constricting inward in a way that is analogous to wrapping your hands around a soft balloon filled with water and squeezing to divide the balloon into two halves. In animal cells, this is accomplished when a ring of actin, the same protein found in your muscles, forms around the equatorial periphery of the cell. It constricts, gradually forming two daughter cells (fig. 10.2d). Each will contain half the contents of the old cell and each will have one of the new nuclei containing a complete set of chromosomes. In dividing plant cells, membrane vesicles containing cell wall components migrate to the equator between the two new nuclei. Starting at the center, these vesicles gradually fuse forming a **cell plate**. It gradually enlarges eventually partitioning the parent cell into two cells, each containing one of the new nuclei (see fig 10.3a third cell from the left).

Lab Instructions

You will identify mitotic stages in animal and plant cells. You will use selective staining techniques to stain DNA in chromosomes and determine the number of chromosomes that are characteristic for onions. It is best to start the staining procedure first and then look at the stages of mitosis in prepared slides during waiting periods in the procedure.

Mitosis in Animal Cells

Mitosis is best seen in growing tissues, which have many cells dividing at the same time. The whitefish **blastula** is an early embryonic stage in the development of a whitefish. At this stage, the embryo is essentially a disc of dividing cells on top of a globe of yolk.

Biological supply houses rapidly kill the cells of the blastula and preserve them chemically. The tissue is then embedded in wax to make it rigid and sectioned into thin slices. The sections are mounted on microscope slides and stained to make the chromosomes visible.

1▶ Obtain a prepared slide of a whitefish blastula and observe it under scanning power with your compound microscope (fig. 10.2). Note that several sections of the blastula are on the slide. Each is composed of many cells. Some contain darkly stained chromosomes that are barely visible at scanning magnification.

Center a cell containing darkly-stained chromosomes in the field of view and observe it first under medium power and then with high power. Locate the **spindle** and **chromosomes**. At the poles of the spindle, you will see "star burst" structures called **asters** which consist of microtubules radiating from a clear area called the **centrosome**. It is the center for microtubule formation. Though not visible through the light microscope, centrioles are found at the center of the asters in animal cells. Centrioles have not been found in plant cells.

Now, locate another cell in which the chromosomes are visible. Chances are that the chromosomes are not aligned in the same patterns as in the previous cell. In a blastula, cell divisions are not synchronized, so different cells are in different stages of mitosis. Furthermore, since you are looking at sectioned material, there may be cells in which neither chromosomes nor nuclei are visible because the section was taken from the edge of a cell and did not include any nuclear material. The apparent absence of nuclear material in these cells is called an artifact, a phenomenon resulting from technique and not from some biological mechanism.

Over the next 10 minutes, look at the blastula slide and identify cells in interphase (I), prophase (P), metaphase (M), anaphase (A), and telophase (T). Use figure 10.2 to help you identify the stages. Look carefully at least five cells in each phase and determine if they have a nucleus, asters, or spindle. In the work table below, put a tick in the appropriate column and row when you see the structure in a cell.

Phase	Nucleus	Aster	Spindle
I			
P			
M			
A			
T			

Can you see a furrow developing in any of the telophase cells? _____

All of the cells you see in one of the sections on your slide came from the mitotic division of a single cell, a fertilized egg. Do you think they are genetically the same or different?

Mitosis in Plant Cells

2 As time permits, obtain a slide of a longitudinal section of an onion root tip. Cells just behind the tip divide by mitosis as the root elongates. Study this area on your slide and identify cells in interphase, prophase, metaphase, anaphase, and telophase (**fig. 10.3**). Note that although the cells and chromosomes look different from what you saw in the whitefish, the patterns of chromosome movement in plants are the same as in animals. Find a cell in metaphase and sketch it below.

3 *Looking for Asters, Centrosomes, and Cell Plates*
Recall the asters with the clear centrosome area at their centers seen in whitefish cells (fig. 10.2). Can you see asters or centrosomes in any of the dividing onion cells? If so, have your instructor confirm.

Look at a cell in late telophase. Can you see a line forming across the center (equator) of the long axis of the spindle? This is the start of the **cell plate** that will separate the two daughter cells.

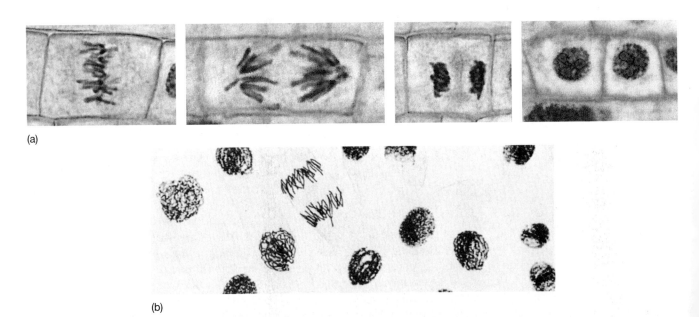

Figure 10.3 Stages of mitosis in an onion root tip as seen in two types of preparations: (*a*) root tip sectioned and stained so cells remain intact, and (*b*) root tip stained with Feulgen reagent and then squashed.

What is the genetic significance of mitosis? Are the daughter cells genetically the same, or different? Why is this important in a growing organism?

Estimating Duration of Mitotic Phases Mitosis lasts for about 90 minutes in onion root tip cells. Each of the four phases takes a different amount of time. The phase lasting the longest will be the most commonly observed. Why do you think this is so?

4) Observe 20 *dividing* onion root tip cells, and tally the frequency of occurrence of each mitotic phases in **table 10.1**. Your instructor may ask to combine your data with that of others to create a large sample size. If so, record the combined data in table 10.1.

Because frequency of occurrence is directly proportional to the length of a phase, multiply the percentage of the cells in a phase by 90 minutes, the duration of mitosis in onion cells, to obtain an estimate of the time required for that phase. Enter the time values in the last column of table 10.1.

Which phase of mitosis takes the most time?

Staining Chromosomes in Dividing Cells

Actively growing root tips, young leaves, flower buds, or other dividing plant tissues have a high percentage of cells in mitosis. These tissues can be prepared for study by a squashing technique in which the tissue is softened, stained, flattened, and observed (**fig. 10.4**).

The use of onion root tips is recommended, though it would be possible to use any other plant, bulb, or germinating seed that has vigorously growing roots.

Tissue Preparation

 1. Fixation: Pour 2 ml of *freshly* prepared methanol-acetic acid **fixative** into a small vial. Remove three, 1-inch-long roots with growing tips and place them immediately in the fixative. It rapidly kills the cells and preserves structure. Place the vial in a 60°C water bath for 15 minutes.

> **C A U T I O N**
>
> Methanol/acetic acid mixture is flammable, corrosive, and has noxious fumes. Wear goggles and protect clothing.

2. Softening: Before staining the tissue, it is necessary to **hydrolyze** it, that is, to partially digest the cells. Working in a ventilated area or fume hood, slowly pour off the fixative into a watch glass, being careful not to lose the roots. Discard the fixative in a waste container.

   Add about 2 ml of 1 M HCl to the vial containing the roots and incubate it in a water bath for 10 minutes at 60°C. *The temperature and time in this step are critical.* Too short or too long a hydrolysis may result in poor staining in the next step.

> **C A U T I O N**
>
> HCl is hydrochloric acid which, if splashed, will damage your eyes (wear goggles) or clothing. Feulgen stain appears colorless, but if it gets on your hands or clothing, it will stain them vivid pink. Household bleach will remove stains.

Table 10.1 Determining Duration of Mitotic Phases

Phase	Number Seen	% of Total	Length in Minutes
Prophase	_____	_____	_____
Metaphase	_____	_____	_____
Anaphase	_____	_____	_____
Telophase	_____	_____	_____
Total	_____		

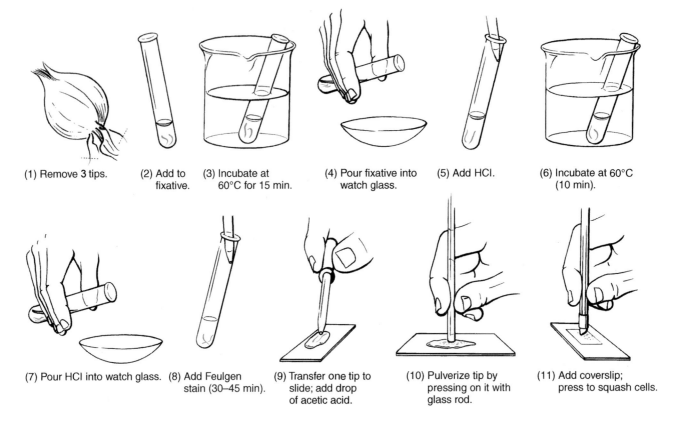

Figure 10.4 Steps for preparing tissue by the Feulgen stain/squash method.

3. Staining: After hydrolysis, pour off the acid into a watch glass, taking care not to lose the roots. Add a milliliter or so of **Feulgen stain** to the vial containing the roots. The slide may be prepared after 30 minutes but waiting longer allows the stain to intensify.

4. Making slide: Transfer one root to a very small drop of 45% acetic acid on a slide. Look at the tip carefully. It should have a prominent purple to pink band of staining which is less than a millimeter wide. This is where the dividing cells are located. Using a razor blade, cut the darkly stained tip away from the unstained lighter and older area. Discard the remaining older portion.

> **CAUTION**
> Acetic acid is corrosive with a noxious odor. Avoid contact. Flush skin with water if contacted.

5. Pulping: Use a polished glass rod to press down on the root tip on the slide. Do this 10 to 20 times to break the tip apart. The difference between making a superior and a mediocre slide lies in the pulping. Place a clean coverslip over the preparation.

6. Squashing: Lay the slide on the table and cover the slide with two thicknesses of paper towels. Put your thumb on the towel over the coverslip and press directly downward as hard as you can. You cannot press too hard! However, use care to prevent the coverslip from breaking or moving sideways. If it does move, the cells will roll on top of one another and destroy the material. If this happens, make a new slide with one of your other roots.

Common mistakes include:

- *Failure to follow staining and tissue preparation directions.*
- *Not ensuring that root tips are submerged in reaction solutions.*
- *Incompletely pulping and squashing tissue.*
- *Moving coverslip during squashing.*
- *Not scanning around slide with medium-power objective to find squashed cells.*
- *Underestimating chromosome number because not focusing up and down when counting.*

6 ▶**Observing Your Slide** Use the scanning objective on your microscope to locate any pinkish squashed material on your slide. Switch to the 10 × objective when you find a stained area on the slide. Reduce the light intensity to see the chromosomes. Switch to high power when cells are located.

Small quantities of 45% acetic acid may be added to the edge of the coverslip to prevent the preparation from drying. If the cells are not sufficiently spread, they may be squashed further by repeating the pressing procedure.

Study your slide and identify cells in anaphase and metaphase. Compare what you see to figure 10.3b. How many chromatids make up the chromosomes at:

Metaphase_____
Anaphase_____

Draw a squashed onion root tip cell in anaphase below.

Table 10.2 Number of Chromosomes Observed by You in 10 Onion Root Cells

1. _____	6. _____
2. _____	7. _____
3. _____	8. _____
4. _____	9. _____
5. _____	10. _____

moving to one or the other pole of the cell represents the diploid number for the species. Select a cell in which the chromosomes are well spread by squashing, and count the number of chromosomes at one pole. Record the count in **table 10.2**. Now count those at the other pole and record. Find another cell and repeat until you have 10 counts.

When counting, be sure to use the high-power objective and to **focus up and down** as you count so as to include chromosomes that may lie above or below the plane of focus. If the chromosomes are tangled, it may be helpful to count the visible ends of chromosomes and then to divide the count by two to estimate the chromosome number. If this method is used, always round up to the next even number when an odd number of ends is observed, because every chromosome has two ends and one was most likely missed in counting.

A facsimile of **table 10.3** will be on the blackboard or on a computer spreadsheet in the lab. Record your observations as directed. After everyone has recorded their observations, count the total number of cells seen by the group having four chromosomes, five chromosomes, and so on. Record these numbers in column B.

If the number of chromosomes is constant for a species, why were different numbers of chromosomes seen in different cells?

Testing a Hypothesis You have learned to recognize mitotic stages and to carry out a classical chromosome-staining technique. You will now apply this knowledge in a simulation of real world situations.

A technique often used by taxonomists to identify closely related species is chromosome counting, also known as **karyotyping.** These scientists often prepare samples using the same techniques you did and then statistically analyze their data. In this portion of the lab, you will estimate the diploid (2N) number of chromosomes for the species used to make your slide. The following describes what could be a real-life application of this procedure.

Imagine that you work in a biological laboratory that performs chromosome analyses for hospitals and anyone else who is willing to pay the going rate.

A major horticultural firm has come to your lab with a problem. The firm, *Seeds Are Us*, has been selling a new strain of onion seed, which has been very successful. However, another horticultural firm, *Gardeners Incorporated*, claims that *Seeds Are Us* stole the variety from them. Furthermore, *Gardeners Incorporated* wants damages because they spent years and millions of dollars developing the new variety, which is protected under a horticultural patent. Your job is to provide evidence for a court case arguing that the two varieties in question are genetically different and thus do not represent a patent infringement.

A literature search of horticultural patents yielded the basic karyotype information about the *Gardeners Incorporated* variety. It is a registered variety with a diploid number of chromosomes equal to 18. You must now determine the number of chromosomes in the *Seeds Are Us* variety.

A hypothesis to be tested by collecting data on number of chromosomes would be: The number of chromosomes in the *Seeds Are Us* variety (the onions you have in lab) is equal to 18.

As a head technician in the laboratory, you give the following directions to all other technicians:

7 *Collecting Data* To count chromosomes, look only at mid- to late-anaphase cells. The number of chromosomes

8 *Analysis* You will now statistically analyze the class data in table 10.3, using the following three steps.

1. Plot a frequency **histogram** (bar graph) of the data in columns **A** and **B** of table 10.3, using the graph paper at the end of the exercise. Appendix B contains graphing instructions. This histogram will give you a visual representation of the data variability.

2. Calculate the **average** number of chromosomes per cell from the class data using the following steps:

Table 10.3 Summary Worktable for Combined Lab Data

A # Chromosomes/Cell	B # Cells Seen	C A × B	D (A − Average)²	E B(A − Average)²
4				
5				
6				
7				
8				
9				
10				
11				
12				
13				
14				
15				
16				
17				
18				
19				
20				
	Σ _____ Total cells seen	Σ _____ Total chromosomes seen		Σ _____

(Note: Σ (Sigma) is a Greek letter meaning "sum of.")

a. In table 10.3, sum the values in column **B** and enter the value at the bottom (= ΣB).

b. For each row, multiply the value in column **A** by that in **B** and enter the product in column **C**.

c. Sum column **C** and enter value at the bottom. [= Σ(A × B)]. This equals the total number of chromosomes observed.

d. Divide the sum of column **C** by the sum of column **B** to obtain the average number of chromosomes per cell.

$$\text{Avg. \# chromosomes per cell} = \frac{\Sigma(A \times B)}{\Sigma B}$$

What is the average number of chromosomes per cell based on the class data?

Average = _____

If your calculated average is not a whole number, why does it make sense to round it to a whole number?

3. Using a calculator or computer spreadsheet, calculate the **standard deviation** of the class data by following these steps.

Standard deviation is discussed at the end of lab topic 5 and in the beginning of appendix C. Remember that:

$$\text{Standard deviation} = \pm\sqrt{\frac{\Sigma(\text{observation} - \text{average})^2}{n - 1}}$$

To calculate the expression Σ(observation − average)²:

a. For each row, subtract the average from the value in column **A**. Square this difference and enter the square in column **D**.

b. For each row, now multiply the value in column **D** by the value in column **B**. Enter the product in column **E**.

c. Sum column **E** and enter the value at the bottom.

To calculate the standard deviation, divide the sum of column **E** by one less than the sum of column **B**. Take the square root of the quotient so obtained. What is the standard deviation for your class data?

$$\text{St. dev.} = \pm\sqrt{\frac{\Sigma E}{\Sigma B - 1}}$$

Write the values for the average and standard deviation on your histogram.

9. Conclusion Based on the class data now summarized in the average, standard deviation, and histogram, what is your conclusion regarding your hypothesis? Is the number of chromosomes in your class sample (obtained from *Seeds Are Us*) the same or different from 18 (the number of chromosomes in the *Gardeners Incorporated* variety)? Must you accept or reject your hypothesis?

If your analysis showed that the two cultivars had the same number of chromosomes, does this prove they are the same cultivar?

Do all species have different numbers of chromosomes or do many have the same number?

Lab Summary

10. On a separate sheet of paper, answer the following questions as assigned by your instructor.

1. Describe the chromosomes and their positions during the following stages of mitosis:

   Prophase
   Metaphase
   Anaphase
   Telophase

2. For a species with a diploid number of chromosomes equal to four, draw a cell as it would look at metaphase.

3. Compare and contrast cell division in plant versus animal cells. Name several similarities and at least three differences based on your lab observations.

4. Submit the frequency histogram of your class data on onion chromosome counts. Indicate the average and standard deviation on the graph. State your conclusion about whether the chromosome number obtained by the class was 18.

You may want to try the critical thinking questions that apply some of the knowledge you gained in doing this lab.

INTERNET SOURCES

Much genetic research has gone into building databases that describe what traits are located on what chromosomes. More detailed studies have determined what is the nucleotide sequence in the single DNA molecule found in each chromosome. Such projects are now underway for a number of organisms, including the onion. To look for information on the chromosomes of onions, use the search engine Google at http://www.google.com. When connected, enter *Onion Genome*.

Scan the first few pages of the listing and connect to a few of the sites that are interesting to you. Answer these questions:

1. What is unusual about the onion genome?
2. Why do you think there are so many sites addressing what seems to be an obscure topic?

LEARNING BIOLOGY BY WRITING

Continuing with the situation presented under Testing a Hypothesis, imagine that you must write a report summarizing your laboratory's studies of chromosome number in onion varieties. Read Appendix D for directions about writing lab reports.

CRITICAL THINKING QUESTIONS

1. Can you think of any reasons why organisms rarely have 2N chromosome numbers greater than 100?
2. If two organisms have the same 2N number of chromosomes, are they always members of the same species? Justify your answer.

3. Why do biologists say that mitosis produces genetically identical daughter cells? How can this be true?
4. Many protists, fungi, and plants have haploid cells that divide by mitosis. How many chromosomes would you expect to find in the cells resulting from such a division?
5. If each human chromosome contains 5 cm of DNA, on average, and human cells contain 46 chromosomes, how many meters of DNA are in a nucleus in one of your cells?
6. If the human genome contains 25,000 genes, how many genes, on average, are found on one of your chromosomes? (*Hint:* use haploid number.)

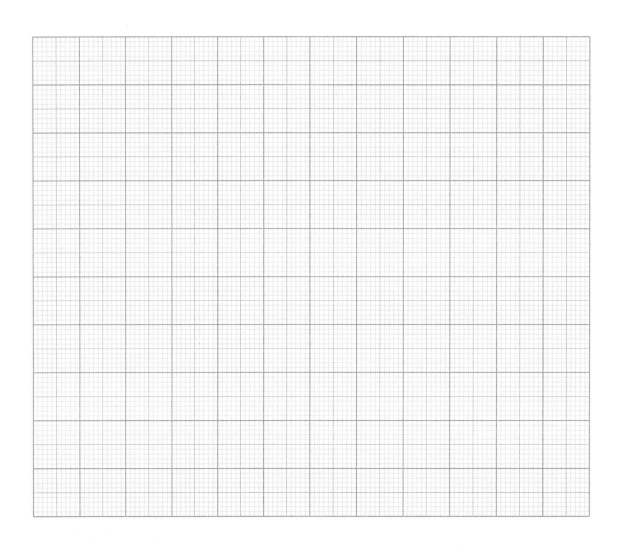

Lab Topic 11

MODELING MEIOSIS AND DETERMINING CROSS-OVER FREQUENCY

SUPPLIES

Preparation guide available at
www.mhhe.com/labcentral

Equipment
Compound microscopes

Materials
Homemade play dough or modeling clay (red and blue)
Sordaria fimicola cultures on petri plates
  (Carolina/Biological Supply Biokit) and fresh plates for subculture; wildtype and tan strains
Slides and coverslips
Toothpicks or probes

STUDENT PRELAB PREPARATION

Before doing this lab, read the Background section and other sections of the lab topic that have been assigned by your instructor.
  Use your textbook to review the definitions of the following terms:

Ascus	Ploidy
Crossing-over	Recombinant
Diploid (2N)	Recombination
Haploid (N)	Segregation
Homologous chromosomes	Spore
Independent assortment	Synapsis
Nonrecombinant	Zygote

  Describe in your own words the following concepts:
Chromosome positions during phases of meiosis I and II
Genetic significance of crossing over
How segregation occurs in meiosis I
How independent assortment occurs in meiosis I
Chromosomal mapping
Review Chi square, appendix C

  After finishing the prelab review, write any questions you have about terms, concepts, or techniques in the margins of this lab topic. The lab experiments will help you answer these questions, or you can ask your instructor during the lab.

OBJECTIVES

1. To model the phases of meiosis and fertilization with the goal of explaining the sources of hereditary variation in sexual reproduction
2. To determine experimentally the frequency of crossing-over in a living fungus
3. To perform a chi-square statistical test of a hypothesis

BACKGROUND

Meiosis is a form of nuclear division which produces genetically distinct daughter cells having half the number of chromosomes found in the parent cell. In sexually reproducing species if meiosis did not occur, the number of chromosomes would double with each fusion of egg and sperm. For example, humans have 46 chromosomes. Without meiosis, a sperm and egg would each contribute 46 chromosomes to a fertilized egg, called a **zygote,** so that it would have 92. Since mitosis always produces daughter cells with the same number of chromosomes as the parent cell, all cells in the new individual would also have 92 chromosomes. If this individual mated with another of the same generation, the third generation would have 184 chromosomes in its cells. Obviously, if this exponential progression continued, there soon would not be a cell large enough to contain the increasing number of chromosomes in each generation.

  Meiosis prevents this problem in sexually reproducing species by reducing the chromosome number, called **ploidy,** by one-half at some point in the life cycle. A **diploid** nucleus contains pairs of chromosomes; a **haploid** nucleus has only one chromosome from each pair, or half as many. In animals, meiosis always results in either haploid egg or sperm production, but this is not the case in all organisms. In fungi, for example, meiosis occurs in cells soon after fertilization, forming haploid spores from diploid zygotes. The haploid spores are then dispersed by wind. If they settle on a suitable substrate, they germinate and divide by mitosis, producing a multicellular fungus composed of haploid cells. During sexual reproduction in the next generation, cells of one fungus fuse with those of another eventually producing a diploid zygote, and the cycle repeats.

  As a result of fertilization, a zygote contains two similar sets of chromosomes: one inherited from the mother and one from the father. In humans, for example, if both parents have normal cholesterol levels both the egg and the sperm will carry a gene for normal cholesterol levels on one of their 23 chromosomes. The zygote will receive both chromosomes at fertilization and the new individual will have two normal genes. However, if one parent has elevated cholesterol (hyper-cholesterolemia) and the other normal levels, the zygote may carry one gene for elevated levels and one for normal levels on the matched chromosomes. (The resulting child would most likely have elevated levels because this gene is a physiologically dominant gene. One in 500 humans carry the gene.)

These examples illustrate that chromosomes occur in pairs in diploid cells, with one member of a pair traceable to each parent. Members of such pairs are said to be **homologous;** that is, they can be matched on the basis of the genes they carry. In the example just given, the blood cholesterol level trait in general, not the specific levels, provides the basis for the homology. Recent work in the Human Genome Project has determined which genes are found on which chromosomes and is providing a complete map of all human chromosomes. You can view the results of this research by accessing the online database for human genetics at *www.ncbi.nlm.nih.gov/mapview/* and click on *Homo sapiens* (Build).

Meiosis reduces the chromosome number by **segregating** (separating) the members of each homologous pair from a diploid cell into haploid cells. In a human with a diploid number of 46 chromosomes, each egg or sperm produced by meiosis contains one member of each pair, or a total of 23 single chromosomes. However, although a single gamete contains *only one member of each pair,* it will contain a mixture of the chromosomes derived from an individual's father and mother, a phenomenon known as **independent assortment.** There is no division mechanism that separates all maternally derived chromosomes from all paternally derived chromosomes. Therefore, meiosis produces gametes that are genetically distinct, each containing a mixture of maternal and paternal chromosomes. As you will see in today's lab, this leads to variation in future generations.

A stylized representation of meiosis for a hypothetical organism having four chromosomes is shown in **figure 11.1.** Note that there are two parts to meiosis. In **meiosis I,** the homologous chromosomes (each consisting of two chromatids) line up side by side during prophase I continuing into metaphase I, whereas in **meiosis II,** there are half the number of chromosomes (each consisting of two chromatids) and they line up singly at metaphase II. Understanding these differences in alignment is the key to understanding the differences between meiosis I and II. Why? Because once the chromosomes align as they do, the separation pattern, involving both segregation and independent assortment, is determined.

A cell that begins meiotic cell division has previously undergone a period of growth called **interphase.** As in the interphase before mitosis, this interphase includes an S subphase during which the DNA in the cell replicates. At the end of meiotic interphase, each chromosome consists of two identical chromatids joined at the centromere. As this cell enters meiotic **prophase I,** the homologous chromosomes pair off and lie side by side as a **tetrad** of

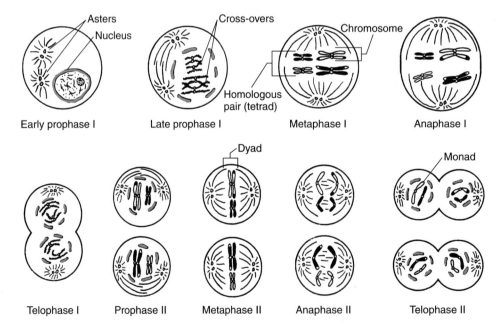

Figure 11.1 Schematic representation of meiosis. The positions of chromosomes at metaphase I and II are the key to understanding how reduction in chromosome number (segregation) occurs during meiosis because the alignment determines the pattern of separation. Meiosis is not a random halving of chromosome number; it is highly ordered. Chromosomes are color coded to represent their origins: white are maternal and black, paternal. Random alignment of maternal and paternal chromosomes at metaphase I is the basis for independent assortment. Note a double crossing-over during prophase I in the large pair of chromosomes.

four chromatids, in much the same way that one might hold four skeins of yarn clenched in one hand.

This pairing process, called **synapsis,** occurs only during prophase I and involves forming a connection between adjacent chromatids in tetrads. During this pairing process, the homologous chromosomes may reciprocally exchange parts in a process called **crossing-over,** also known as **recombination.** This means that parts of the maternally derived chromosome are actually exchanged between the paternally derived chromosome and vice versa. The net result of exchanging of pieces of chromatids is that each member of the homologous pair can become a mosaic of parts derived from both the maternal and paternal chromosomes. Consequently, new gene combinations on homologous chromosomes are created. Crossing-over has profound implications as an additional source of evolutionary variation in sexually reproducing species.

As prophase I ends, the recombined chromosomes unravel except at points of crossing-over called **chiasmata** (fig. 11.2). The homologous chromosomes remain attached at the chiasmata as they line up at the center of the cell at the phase called **metaphase I.** While these events are occurring, the nuclear membrane breaks down and a spindle forms.

At the end of metaphase I, the homologous chromosomes in each pair separate from each other as the cross-over points separate. One mosaic chromosome from each pair, consisting of two chromatids, moves to opposite poles of the cell during **anaphase I.** In most animals and many plants, the end of meiosis I is marked by the occurrence of cytokinesis, the division of cytoplasm. Each daughter cell thus produced has half as many chromosomes as the starting cell. The chromosomes are a mixture of maternally and paternally derived chromosomes due to independent assortment. Furthermore, each chromosome consists of two chromatids and each may contain a combination of maternally and paternally derived genes because of crossing-over.

In the second part of meiosis, the reduced number of chromosomes in each cell again line up at the center of the cell in **metaphase II,** and then proceed through **anaphase II** and **telophase II.** The result is the separation of the sister chromatids and the production of a total of four cells, each with the haploid number of chromosomes and each genetically different because of independent assortment and crossing-over.

LAB INSTRUCTIONS

You will model chromosome behavior in meiosis to test your understanding of the sequence of events in meiosis, especially independent assortment and crossing-over. You will then use your understanding to determine the frequency of crossing-over during meiosis. The results of your experiment will be statistically analyzed.

Testing Understanding Through Modeling

You will first model meiosis without any crossing-over, and then model it with crossing-over to see what the effects are on genetic variability. Work with a partner. The member having the most understanding of meiosis should act as a listener and the one with the least as a presenter.

1> In the supply area are bags of colored play dough or modeling clay to be used in making model chromosomes: blue and red. Get one of each.

Working together, make strands of play dough by rolling it on the table until you have four equal-length long strands (two red and two blue) and four equal short ones (two red and two blue). All should be about the thickness of a pencil. The long ones represent a chromosome that carries different genes compared to the short ones. The red ones are copies of the chromosomes that were inherited from the individual's mother and the blue ones from the father.

2> *Modeling Meiosis with No Crossing-Over* Now assume your roles. The presenter should use what is needed from your stock to create a hypothetical cell in early interphase (G_1) from the above parentage that is *diploid* with 2N = 4. The listener should double-check these directions and make sure the presentation is correct.

The next stage to be modeled is late interphase, showing the results of DNA synthesis. The presenter should show what the chromosomes will look like after the *S subphase* of interphase is over.

After pinching the identical strands together to represent *sister chromatids* joined by a *centromere,* the presenter should now arrange the chromosomes as they would appear in *metaphase I* of meiosis. Sketch your metaphase arrangement, using colors or color labels in

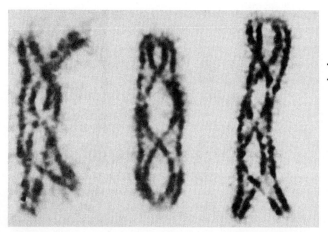

Figure 11.2 Photomicrographs of three homologous chromosome pairs from a grasshopper showing cross-over points in late prophase I. Count the number of chromatids in each of the three pairs.

the space below. The presenter should describe which chromosomes are *homologous* to each other and which are *nonhomologous*. Figure 11.1 can serve as a guide.

sketches of the chromosomes in each of the four resulting cells. Use colored pencils or label the colors of the chromosomes and any segments of a different color.

3> The presenter should move the chromosomes to positions for a cell in *telophase I,* describing the events of anaphase I. At this point, the presenter should pause and describe how *segregation* just happened.

Now think about your metaphase I arrangement that you drew above. How many other arrangements of colored chromosomes are possible? _____ Return the chromosomes to the metaphase I alignment at the equator. Explain how alternative metaphase I arrangements of chromosomes would lead to *independent assortment*. Add a description of the effect of independent assortment to your sketch of metaphase I.

The presenter should now move the chromosomes through the stages of meiosis II to *telophase II,* explaining how many cells would be formed in total and what the *ploidy* level of each cells is. A simple summary of the results of independent assortment should be given.

Modeling Meiosis with Crossing-Over Now, the presenter and the listener should switch roles.

4> In this simulation, we will assume that chromosomes have replicated with *sister chromatids* joined at their *centromeres* and that the homologous chromosomes are in *synapsis* as tetrads during prophase I. The new presenter should set the models accordingly.

The presenter should now illustrate the effects of *crossing-over* by swapping one segment between homologs in each tetrad. Then move the chromosomes into their metaphase I positions. Describe the chromatids in each pair, using the terms *recombinant, nonrecombinant,* and *recombination.*

The presenter should next move the chromosomes all the way through meiosis I and II to produce four cells from the initial metaphase I alignment. The presenter should describe each meiotic phase as the chromosome models are moved. In the following space, make diagrammatic

5> As before, the results from one metaphase alignment do not demonstrate the effects of *independent assortment.* The presenter should return the chromosomes to the metaphase I alignment, placing them in the alternative alignment and demonstrate the final effect of independent assortment happening in conjunction with crossing-over. Sketch the final chromosomes in the four daughter cells resulting from the new metaphase I alignment in the space below. Use colored pencils or label the colors of the chromosomes and any segments of a different color.

6> ***Meiosis Modeling Analysis*** Look at your sketches which are data from your simulation. Describe what the effect of crossing-over alone was and then add the effect of independent assortment to the description.

Recognize that there are often several cross-overs that occur between synapsed homologs, not just one as you modeled. Discuss what each of your homologous pairs would look like if four cross-overs occurred as shown in figure 11.2. Describe the results of your discussion below.

7 ▶ Modeling Fertilization Take one of your resulting "cells" from the last simulation and think of it as a sperm. Turn to another group in the lab and ask them to choose one of their "cells" to act as an egg. Lay your chromosomes next to theirs to simulate fertilization. Discuss with the group whether the resulting *zygote* has the same combination of colored chromosomes that any of the parents had. What is your conclusion regarding the effects of crossing-over and independent assortment on hereditary variation among offspring?

Choose a second "cell" to act as a sperm to be combined with a second, new egg. Perform the fertilization and look at the chromosomes. Are they the same as in the first simulation? _____

What are the implications of this for genetic similarities and differences between siblings?

8 ▶ Analysis If you spent the time to randomly combine all of your sperm with all of the other group's eggs, how many random combinations would be possible? (*Hint:* a 4 × 4 combination table with four sperm types across the top and four egg types down the side will help you figure this out.)

In textbooks you will often find statements to the effect that, segregation, independent assortment, multiple crossing-over, and biparental inheritance all contribute to genetic variability among offspring in sexually reproducing species. What *evidence* do you have from your simulation of meiosis and fertilization that this is true?

One source of variation in sexually reproducing species that was not illustrated in the simulation is mutation. When it happens, it usually occurs only on one of the chromatids in a homologous pair and meiosis proceeds as you demonstrated. How would that affect the likelihood of a such mutation being passed on to offspring?

Separate your play dough strands into their respective colors and put them back into the bags for others to use in later labs. Any short segments that are mixed colors should be discarded so as not to introduce confusion.

In the next lab activity, you will observe the effects of crossing-over in a living organism.

Measuring the Frequency of Crossing-Over

Background on Experimental Organism The effects of crossing-over can be demonstrated in *Sordaria fimicola*, an ascomycete fungus. The biology of ascomycete fungi is discussed in more detail in lab topic 19. The life cycle of *Sordaria* is shown in **figure 11.3a**. In nature, the fungus grows in the dung piles of herbivorous animals. When mature, a reproductive structure called a **perithecium** develops. It contains **ascospores**, small single haploid cells surrounded by a tough outer wall. The ascospores are forcibly ejected and stick to the leaves of grasses or other low-growing plants. Grazing herbivores ingest the spores. They pass through their digestive systems and are defecated. The ascospores germinate in the new dung to produce thin haploid filaments called **hyphae**. The hyphae grow by mitosis, eventually producing a network of filaments called a **mycelium** (fig. 11.3b).

Under appropriate conditions, an **ascogonium** having several nuclei is formed by the fusion of two haploid hyphae from different mating types. Special hyphae, called ascogenous hyphae with two nuclei per cell, are produced by mitosis from this structure. The ascogenous hyphae eventually produce **asci,** saclike structures in which ascospores are produced. This process starts with the two nuclei in the cell fusing, the equivalent of fertilization. The diploid zygote nucleus then undergoes meiosis with crossing-over, forming four genetically distinct haploid nuclei. Each of these nuclei then undergoes mitosis, giving rise to eight ascospores in the ascus (fig. 11.3a).

About a hundred asci will develop in the area of hyphal fusion. As they develop, the asci are surrounded by a growing mass of other hyphae, forming the perithecium (**fig. 11.4**) with an opening at its apical end. As individual asci mature, they eject their ascospores through the opening of the perithecium. When ingested by a herbivore, the cycle repeats.

The color of the ascospores is genetically controlled. Normal or wild strains of *Sordaria* produce black ascospores, while mutant forms of *Sordaria* produce tan ascospores. Consequently, in a mixed population of *Sordaria*, three types of fusions (matings) are possible: tan and tan, black and black, and tan and black. A conceptual model of the chromosome behavior in these matings would appear something like **figure 11.5**.

This figure is based on two experimentally verified conclusions: that the cells in a single ascus result from meiosis followed by a mitosis in each of the meiotic daughter nuclei; and that the eight ascospores in an ascus are arranged in linear order, reflecting the order of the steps in the meiotic and mitotic divisions that occur in the ascus. When the diploid stage is **homozygous** (both

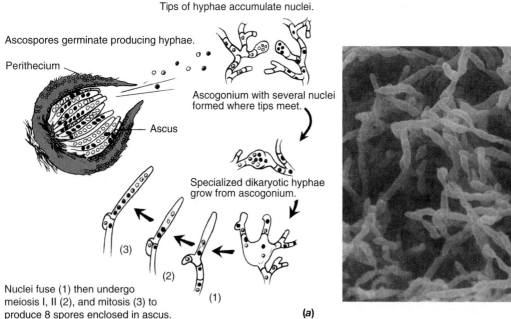
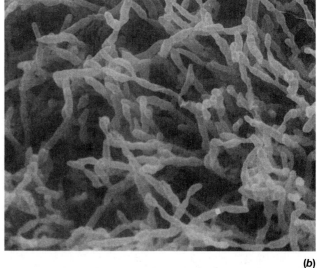

Figure 11.3 Stages in the life cycle of the fungus *Sordaria*: (*a*) diagram of sexual phases of life cycle; (*b*) scanning electron microscope photo of *Sordaria* mycelium.

Open perithecium   Many asci each containing 8 ascospores

Figure 11.4 Perithecium of *Sordaria* has burst, releasing more than 50 asci, each containing eight ascospores.

be present, corresponding to the three matings previously described.

Use a dissecting microscope with a white background to look at the surface of the mycelium. With a toothpick or dissecting needle scrape a few perithecia from the surface and make a wet mount on a microscope slide. Do not dig into the agar growth medium; just scrape off the perithecia! Add a coverslip and, using a pencil or probe, very gently press on the coverslip to squash the perithecia. When a perithecium bursts, it will release many asci, each containing eight ascospores (fig. 11.4).

Common mistakes include:

- *Rupturing fragile perithecia by digging into agar when taking a sample from culture.*
- *Pressing too hard to release asci; consequently both perithecium and asci break open If this happens, make a new slide.*
- *Pressing too lightly so asci are not adequately released from perithecium or the asci are not adequately spread for viewing. Can press a little harder on such slides, but may press too hard. Need to experiment— it is an art not a science!*
- *Identifying immature wild (black) spores as mutant (tan) spores. Immature black spores appear light in color but are also granular. Record such as black not tan.*

fusion members are genetically the same—for example, black and black), all haploid ascospores are the same. (See fig. 11.5a and b.) When the diploid stage is **heterozygous** (for example, black and tan), six possible results can occur. Two of these results are obtained only when crossing-over fails to occur and are called **nonrecombinants**. The two different chromosomes duplicate, segregate, and undergo an additional mitosis forming the ascospore pattern shown in figure 11.5c. However, when crossing-over does occur, it produces the **recombinant** ascospore patterns shown in figure 11.5d. Because the ascospores cannot pass by each other in the ascus, they remain aligned in the order that they were produced. Thus, only the process of crossing-over can account for the patterns shown in figure 11.5d.

Though all of this may seem complex, it is really a simple biological system. The orderly patterns of colored ascospores in an ascus result from meiotic divisions in which crossing-over did or did not occur.

If you accept the chromosome model shown in figure 11.5 for fusion and ascus development in *Sordaria*, you can determine how often crossing-over takes place by counting the number of asci having recombinant or nonrecombinant arrangements of tan and black spores.

9▶ ***Procedure*** Obtain a petri dish in which a cross was set up between two *Sordaria* strains about a week before your laboratory. Note the "fuzzy" growth in the plate, consisting of thousands of intertwined branching hyphae in a mycelial mat. On the surface of the mycelial mat will be small black objects, the perithecia that each contain many asci. Each ascus, in turn, contains eight spores. One parental strain produced only black spores and the other only tan. In the dish, three types of perithecia will

10▶ Scan the slide with a microscope on scanning power to find the asci, and then switch to medium power while adjusting the light intensity to clearly see the spores. If the perithecia contain only asci with all black or all tan spores, make a new slide using material from a different region of the culture dish. Continue making slides until you find perithecia having asci with both types of spores. These are hybrid asci resulting from black by tan crosses. Use a marker to draw a circle on the underside of the plate where you found recombinant asci so it will be easier for others to get a sample.

The possible spore arrangements in the ascus are shown in figure 11.5.

11▶ Count all asci of types (c) or (d) that you can see on your slide. You should count at least 50 asci for statistical accuracy. Record the counts as tic marks in the space below and then transfer the totals to **table 11.1**.

Count tallies:

(c) Nonrecombinant

(d) Recombinant

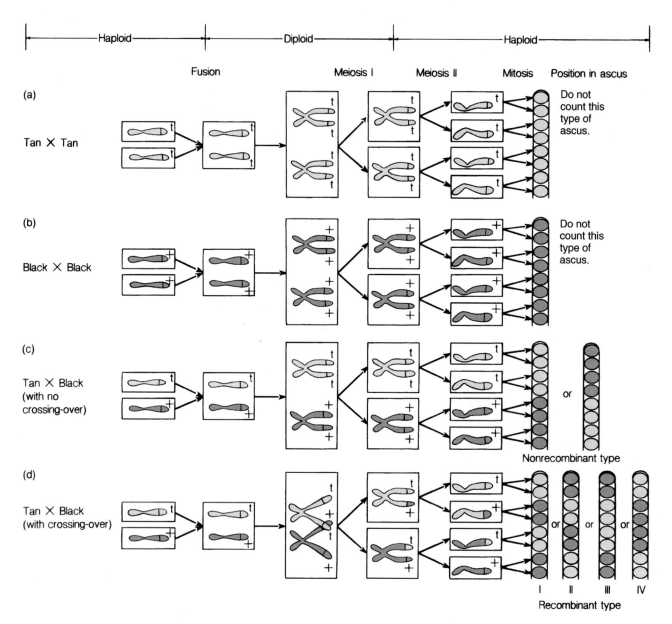

Figure 11.5 Alternative models of chromosome behavior during *Sordaria's* sexual reproduction. Models (a) and (b) will be common. You must find areas where hyphae from tan and black strains have fused to form perithecia before you will observe asci of types (c) and (d).

Analysis The relative frequency of recombinant asci equals the number of recombinant hybrid asci observed divided by the total number of hybrid asci observed, or:

$$\% \text{ recombinants} = \frac{\# \text{ recomb.}}{\# \text{ recomb.} + \# \text{ nonrecomb.}} \times 100$$

12 Calculate the % recombinants found among the asci you observed.

Table 11.1 Results from *Sordaria* Ascospore Counts

Type	Cross	Ascospore Pattern	Number Observed
(c) Nonrecombinant	Tan × black	4:4	_____
(d) Recombinant	Tan × black	2:2:2:2 or 2:4:2	_____

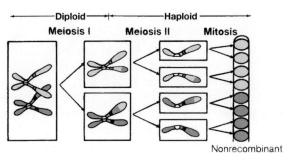

Situation A—Gene locus close to centromere; no cross-over in region between gene and centromere

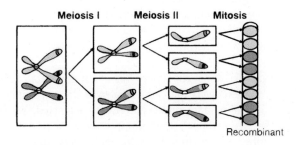

Situation B—Gene locus far from centromere; cross-over occurs between gene and centromere

Figure 11.6 Abbreviated "what if" models showing that cross-over must be in the chromosomal region between the spore color locus and the centromere to give a recombinant-type ascus.

13 Go to the chalkboard and write the % recombinants you observed. When all students have listed their values, calculate a class average. What is the average percent of recombinants observed by the class? What is the standard deviation of the class data? (Refer to Appendix C for method of calculation.)

The relative distance in chromosomal map units of the spore color gene from the centromere of its chromosome can be calculated from the percentage of recombinants (fig. 11.6). Geneticists apply a logical relationship to calculate map distances—the greater the distance between a gene locus and the centromere, the greater will be the frequency of crossing-over. When the gene locus is at the centromere, no crossing-over will occur.

Why do you think there will be little or no crossing-over when a gene is located very close to the centromere or when two genes are very close together?

In your own words, explain what the number you calculated describes?

Map units are obtained by multiplying the percentage of recombinants by one-half because each recombinant ascus counted represents a tetrad of chromatids in which there has been an exchange involving only two of the four chromatids. So, if you find that 30% of the asci are the recombinant type, the gene for spore color would be said to be 15 units from the centromere.

14 Calculate the distance of the spore color gene locus from the centromere using the class average for percentage of recombinants.

Advanced Statistical Analysis When students do this experiment, they often want to know what is the "right" answer. How many map units should they have? Researchers who have done this experiment agree that 28 map units separate the gene for spore color from the centromere, but this may vary with culture age and strains of fungi used.

If analysis of your class data does not yield this value, you are faced with a dilemma. Did your class perform the experiment in the wrong way or are you simply seeing random variation in results? Your class should discuss how the experiment was done and how they collected data to rule out any errors.

If all data are judged to have been collected by acceptable means, random variation may be the answer. To see if your class answer to map distance is within an acceptable limit of random variation, a statistical test called the **chi-square** (χ^2) test can be employed. This test compares actual results to expected results and will tell you with a certain probability whether differences may be due to chance deviations. You should read Appendix C, which describes this test in more detail.

Before performing a chi-square test, you must formulate a **null hypothesis (H_o)** and an **alternative hypothesis (H_a)**. The H_o is that there is no difference between the observed results and the expected results other than due to chance. You found _____ map units between the centromere and the spore color gene using the class data and expected that number to be 28. The H_a is that there is a significant difference, which may be due to error or to biological changes in the fungus. You would not be able to determine why the difference was obtained without doing more experiments.

15 ▶ Procedure To do a chi-square test on your data, first convert the expected map units back to the percentage of recombinants and nonrecombinants expected, 56% and 44%, respectively. Now multiply these % times the total number of asci counted by the entire class to get the expected frequency of asci of each type. Enter these values in the following table on the second line.

On the first line of the table, enter the number of recombinant and non-recombinant asci observed by the class. Subtract the second line from the first line to get the difference between the expected and observed.

	Recombinant	Nonrecombinant
Number Observed	_____	_____
Number Expected	_____	_____
Difference	_____	_____
(Difference)²/Expected	_____	_____

You are now ready to calculate a χ^2 statistic that summarizes the variation in your data. χ^2 is defined as:

$$\chi^2 = \Sigma \frac{(\text{difference})^2}{\text{expected}}$$

Square the difference for recombinants in the table and divide that number by the expected, entering the value on the last line. Repeat these calculations but use the values in the nonrecombinant column. Sum both these answers to calculate a χ^2 for your class data.

What is the χ^2 value for the class data?

If you think about how χ^2 was calculated, its meaning should be obvious. When a calculated χ^2 is a large number, there is a lack of agreement between the observed and expected (the null hypothesis must be rejected); when χ^2 is small, the agreement is good (the null hypothesis would not be rejected). The problem you now face is to decide whether the χ^2 you calculated is large or small.

Statisticians use a standard table to aid in these decisions. This table provides χ^2 values for various experimental designs and levels of confidence where deviations are due to chance. Your experimental design has two categories: recombinant or nonrecombinant. You should use the value listed in the standard table under one degree of freedom (d.f.) because **degrees of freedom** are defined as being equal to one less than the number of date *categories*. Next you must decide on a probability level. Biologists usually use 95% confidence level or a 5% chance that rejecting the null hypothesis will be a wrong decision. (An alternative way of expressing this is that you have 95% confidence that a decision to reject the null hypothesis is correct.)

16 ▶ Conclusion If you look at table C.4 in appendix C, a critical value of 3.8 will be found at $\chi^2_{.95}$ and d.f. = 1. If your calculated χ^2 is less than or equal to this critical value, you should not reject the null hypothesis (i.e., your answer generally agrees with the expected value of 56% recombination). If your calculated χ^2 is greater than 3.8, you should reject the null hypothesis and infer the alternative hypothesis is true.

What is your conclusion? Do you accept or reject the null hypothesis?

---

17 ▶ Lab Summary

On a separate sheet of paper, answer the following questions assigned by your instructor.

1. A hypothetical mammal has a diploid (2N) number of 6. Draw a picture of metaphase I and metaphase II as they would appear in this species. Add a third panel to your drawing in which you sketch metaphase of mitosis as it would appear in this species. Write a short paragraph that highlights the differences among these three pictures.

2. In your own words describe how segregation, independent assortment, and crossing-over result in daughter cells that are genetically distinct.

3. Without describing the details of meiosis and crossing-over, write a description of sexual reproduction in *Sordaria*.

4. You observed the results of crossing-over in crosses between black and tan *Sordaria*. Describe in your

own words how crossing-over occurs and results in the ascospore patterns seen.
5. What frequency of crossing-over did you measure for *Sordaria*? Is this statistically the same as the value published in the literature? How did you reach this conclusion?

You may want to try some of the critical thinking questions that apply the knowledge you gained in doing this lab.

Internet Sources

Several research laboratories work on the genetics of recombination using *Sordaria*. Use the Internet to locate three different labs that are doing such work. Open your browser program and use Google to search for the word *Sordaria*. You will get a long list back.

Look over the list and click on those entries that are research reports. Make a list of three labs that are doing research on *Sordaria* and briefly describe the research projects.

Learning Biology by Writing

In a lab report entitled "Determining the Frequency of Crossing-Over in a Fungus," briefly describe *Sordaria*, indicating why it is an ideal organism in which to study crossing-over. Describe how you measured the frequency of recombination and present the class results as a histogram. Use the chi-square analysis to see if your class got the "expected" answer. Discuss how crossover frequency can be converted to map units and indicate the value your class obtained. Discuss the effect crossing-over has on genetic variability from generation to generation. Finally, indicate possible sources of error in the experiment.

Critical Thinking Questions

1. Why would you expect the diploid (2N) number of chromosomes always to be an even, not an odd, number in every species?
2. In Mendelian genetics, the principles of segregation and independent assortment are used to predict outcomes from genetic crosses. Describe the basis for these principles in meiosis.
3. Crossing-over creates new combinations of alleles on chromosomes that are passed on to the next generation. Explain the evolutionary significance of such genetic variability. What are other sources of genetic variation in sexually reproducing species?
4. Can meiosis occur in a haploid cell? Why? Can mitosis occur in a haploid cell?
5. Why do scientists use critical value tables in statistical analysis? Shouldn't each researcher decide what is acceptable variation?

Lab Topic 12

ANALYZING FRUIT FLY PHENOTYPES AND GENOTYPES

SUPPLIES

Resource guide available on WWW at
www.mhhe.com/labcentral

Equipment

Compound microscopes
Dissecting microscopes

Materials

Prepared slide of giant chromosomes from fruit fly larva as demonstration
Drosophila melanogaster cultures (Carolina Biological Supply)
   Wild type (sexes separated)
   Single chromosomal mutants (sexes separated)
   White eyes
   Apterous wings
   Unknowns constructed by crossing wild type with any or all of above mutants (sexes separated) to get F_1
Drosophila virilis cultures at third instar larva stage
Widemouthed vials for culture chambers
Cotton or sponge stoppers
White cards
Camel's hair brushes
Petri plate
Fly Nap® (Carolina Biological Supply) or carbon dioxide generator from bicarbonate or dry ice
Slides and coverslips
Dissecting needles and forceps
Dead flies for practice phenotyping
Alcohol lamp

Solutions

Instant *Drosophila* medium
   Pour dry mixture directly into culture chamber.
   Add water. Each chamber should be filled to a depth of 3/4 to 1 inch. Sprinkle a few grains of dry yeast on the surface of the medium; the yeast will reproduce and act as a food source. Stopper the chamber with foam plugs.
Jar of ethanol to kill and hold flies
Acetocarmine stain
0.8% saline

STUDENT PRELAB PREPARATION

Read the Background material and sections of the lab topic that have been assigned by the instructor.

Use your textbook to review the definitions of the following terms:

allele	homologous chromosome
dominant	homozygous
F_1 generation	locus
F_2 generation	mutant
gene	phenotype
genotype	recessive
heterozygous	wild-type

Describe in your own words the following concepts:

Mendelian dominant and recessive
Mendelian independent assortment
Mendelian segregation
Use of a Punnett square

After finishing the prelab review, write any questions you may have regarding terms, concepts, or techniques in the margins of this lab topic. The lab experiments should help you answer these questions or you can ask your instructor during the lab.

OBJECTIVES

1. To identify the life stages of the fruit fly, *Drosophila melanogaster*
2. To observe gene loci on giant chromosomes found in a species of *Drosophila*
3. To identify dominant and recessive alleles from crosses of pure-bred stocks
4. To observe the effects of meiotic segregation when parents are heterozygous
5. To determine if parental flies with unknown genotypes are homozygous or heterozygous at loci studied

BACKGROUND

The discovery of the basic laws of heredity was made by Gregor Mendel, a Czech (then part of Austrian empire) monk and scientist, who in 1866 worked out a conceptual model of how seven characteristics are inherited in peas. However, the significance of Mendel's theories was not recognized by the scientific community until the early 1900s, about 20 years after his death. Researchers reviewed his published papers and realized that this far-sighted scientist had discovered the principles of heredity that governed inheritance in not only plants, but animals, fungi and protists as well.

Why was Mendel's work unappreciated when it was first published? At that time, the processes of cell division and fertilization were not well understood. The prevailing theories of inheritance suggested that liquids, not particles, were responsible for the inheritance of characteristics. In addition, Mendel's arguments were based on statistics from experimental crosses rather than from direct observation of biological structure. For all these reasons, no one paid much attention to the curious writings of the little-known monk-scientist.

The theories suggested by Mendel have been verified repeatedly. Furthermore, Mendel's methods are an excellent role model for beginning scientists: Ask simple questions, collect quantitative data, derive your explanations from data analysis, and publish your results.

What were Mendel's contributions? They included:

- The factors responsible for inheritance were particles not fluids;
- The particles occurred in pairs in the peas;
- One particle could dominate over the other when both were present;
- The particles segregated when pollen and eggs were produced (Principle of Segregation);
- The particles governing an inheritable characteristic segregated independently of the particles governing other characteristics (Principle of Independence).

In the years since the rediscovery of Mendel's work, additional principles of heredity have been discovered. For example, Walter Sutton demonstrated that the biological mechanism for Mendel's principle of segregation was the separation of homologous chromosomes during anaphase I of meiosis. Sutton also showed that Mendel's principle of independent assortment was controlled by the random alignment on the spindle of maternally and paternally derived chromosomes during metaphase I of meiosis. Sutton's work provided evidence that Mendel lacked: the biological mechanisms for the statistical results Mendel observed. As a result of the work of Sutton and others, the science of inheritance became a field of intense investigation.

Since chromosomes, meiosis, and the operation of Mendel's principles of inheritance seemed universal in all eukaryotic organisms, it followed that mechanisms found in one eukaryotic organism were most likely applicable to all other eukaryotic organisms. Biologists then began looking for the best experimental organism to use in studying more complicated inheritance patterns.

For a number of reasons Thomas Hunt Morgan at Columbia University and others in the early 1900s decided that the common fruit fly (*Drosophila melanogaster*) was an ideal experimental animal. Fruit flies are small and feed on yeast, so that large populations can be kept in a small space and on a small budget. *Drosophila* lay hundreds of fertile eggs after mating and produce a new generation in 10 to 14 days, so results are quickly and easily obtained. And finally, the diploid number of chromosomes in the fruit fly is only eight, which makes it possible to determine on which chromosome a gene for a particular trait is located.

In his early studies of fruit flies as a model genetic system, Morgan was the first scientist to provide experimental evidence that a specific gene is part of a specific chromosome. He found that a gene determining eye color was located on a chromosome called the **X** chromosome, also called the sex chromosome. In fruit flies as in mammals, females have two **X** chromosomes (*i.e.*, a homologous pair), but males have only one. In place of the second **X** chromosome, males have another chromosome called the **Y** chromosome and it carries no gene determining eye color. When females produce haploid eggs by meiosis, every egg will contain an **X** chromosome. When males produce sperm by meiosis, half of the sperm contain an **X** chromosome and half contain a **Y**. Genes on the **X** chromosome, and the traits they cause, are said to be **sex-linked genes** or **traits**. The inheritance of a sex-linked trait can modify the Mendelian ratios expected from genetic crosses.

LAB INSTRUCTIONS

In this lab, you will first learn the life cycle of fruit flies and to identify males and females. From a demonstration cross, you will determine which traits are dominant and which recessive. You will set up a mating pair of flies with dominant phenotypes, but unknown genotypes. By analyzing the progeny from this cross, you will determine the parent's genotypes by applying Mendel's principles.

Fruit Fly Life Cycle

1 When male and female fruit flies mate, about 1,000 sperm are transferred in a packet, which is stored in the female's reproductive tract. As eggs descend the female's reproductive tract, sperm are released and fertilize them. A female may lay up to 800 eggs during her reproductive period. Mating occurs about six to eight hours after the female emerges from the pupa, and eggs are first deposited on the second day after copulation. Each egg looks like a small grain of rice and is about 0.5 mm long with two thin filaments projecting from its anterior end (**fig. 12.1**).

After about a day at 25°C, a larva emerges from the egg. The larva is eyeless and legless but has black mouthparts and conspicuous spiracles (air pores) through which

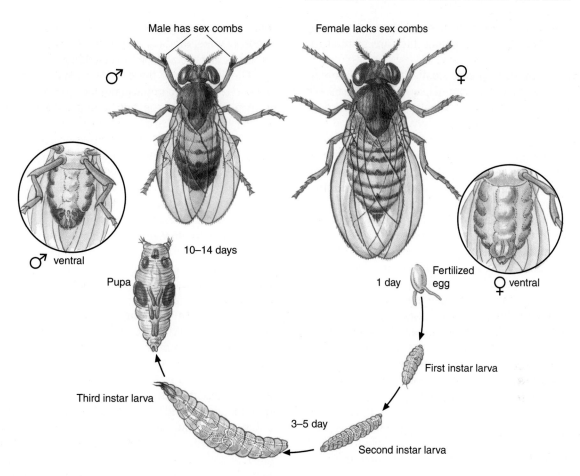

Figure 12.1 Life cycle of the fruit fly, including differences between adult sexes. Stages are not drawn to scale.

it breathes. In nature, the larva pushes its way through rotting fruit, eating fungi and fruit tissue. As it grows, the larva goes through three stages called **instars.** At the end of the first and second instars, the larva molts and sheds its exoskeleton and mouthparts, allowing the organism to increase in size. When the third instar ends, a **pupa** is formed. Before transforming into the pupal stage, the larva stops feeding and crawls onto a dry surface.

Inside the pupa, major reorganizational events occur. The larval structures and tissues break down and are reabsorbed, and adult tissues begin to grow from imaginal discs (clumps of adult tissue producing cells) that have been dormant in the larva. This tissue reorganization is completed in only four days. At the end of this period, the adult fly emerges from the pupal casing.

When the fly emerges, it is elongated and light in color, and the wings are folded. During the next hour, the body becomes shorter and more rounded, the wings expand to their normal shape, and the body color darkens. Adult fruit flies live an average of three weeks but do not grow any larger.

About five to eight hours after emergence, courting behavior begins. The purpose of this behavior seems to be sex identification. A male initiates courtship with every fly, male or female, it contacts. Courtship gestures include tapping the other fly with his foreleg and circling closely around the other fly while rapidly vibrating the wing closest to it to stimulate mating behavior in a receptive female. If the female is not receptive or if the other fly is a male, a sharp kick ends the encounter. A receptive female, however, turns her abdomen and allows the male to mate.

Fruit Fly Anatomy

2 Before doing any genetic crosses you need to be familiar with the general anatomy of fruit flies. Obtain several fruit flies that have been killed or anesthetized. Place the flies on a white index card or fly pad and observe them with a dissecting microscope using reflected light.

Identify the following structures on several of the flies.

Head
    antennae
    proboscis (mouthpart)
    ommatidia (compound eyes)

Thorax
    wings
    halteres (reduced second wing pair)

Abdomen
    Count segments and note pigmentation color and pattern.

3▶ Use a camel's hair brush to separate your flies into two piles by sex. Male and female flies are pictured in figure 12.1.

Males are generally a little smaller than females with five, rather than seven, abdominal segments. The male's abdomen also tends to be more rounded and blunt with darker markings at the tip, especially on the dorsal surface. These dark markings do not fully develop until a few hours after emergence. Therefore, newly emerged males can easily be confused with females. The front legs of the male have dark hairlike projections called **sex combs,** which the female lacks. At the posterior end of the abdomen on the under surface, the male is darkly pigmented and has two **claspers,** which the female lacks.

After you have sorted your flies by sex, have your instructor check your identifications. Dispose of these flies as directed.

4▶ Now, take a jar in which there are larvae. Pick a few larvae out of the culture, and look at them under the dissecting microscope. Can you see the mouthparts and the outline of the digestive system?

Look for pupae on the sides of the container or on the dry paper wick above the medium. Larvae seek dry places in which to pupate so that the emerging adult will not become mired in sticky food materials. If you place a pupa on a slide in a drop of water and observe it with your compound microscope, you may be able to see the developing adult inside. Use low magnification and high light intensity.

Giant Chromosomes (Optional)

When the third instar larvae of one species of fruit fly, *Drosophila virilis,* pupate, they secrete protein materials from their salivary glands that "glue" the pupa to a surface. In these larvae, the salivary gland cells contain giant chromosomes that are actively involved in RNA synthesis.

The large size of these chromosomes is due to three factors: (1) these chromosomes are partially condensed as in prophase; (2) the two homologous chromosomes lie side by side in permanent synapsis; (3) most important, each chromosome has been replicated a number of times (up to ten), and the copies are joined in register along their length. Though such giant chromosomes are not usual in most animals, their occurrence in fruit flies greatly aids genetic analysis, because it is possible to see chromosomal regions that nearly correspond to genes.

In order to view these chromosomes, you must dissect out the salivary glands, stain them, and observe them through your compound microscope (**fig. 12.2**). Alternatively, a prepared slide can be viewed on a demonstration microscope.

5▶ Locate a third instar larva crawling up the jar wall or the largest ones in the growth medium. Transfer one to a drop of saline on a microscope slide and observe under a dissecting microscope.

Push a dissecting needle through the head just behind the black mouthparts and push a second needle through the midbody or hold with forceps. Pull the needles apart, and, with a little luck, the head will separate from the body, trailing the salivary glands. The glands will be two elongated, grapelike, transparent clusters of tissue as shown in figure 12.2. Be careful not to confuse the salivary glands with the whitish glistening fat bodies that stick to them.

Clean the fat body off the glands with your needles and move the glands to a clean area of the slide that is devoid of saline. Cover the glands with a drop of acetocarmine stain and let sit for five minutes.

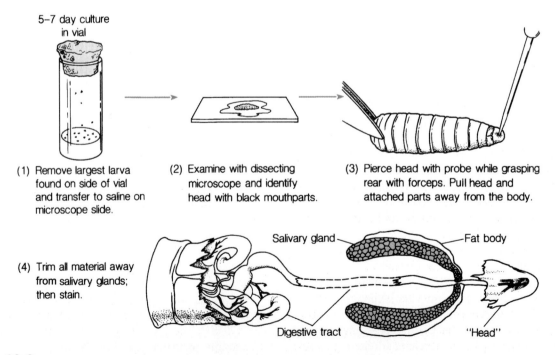

Figure 12.2 Instructions for removing salivary glands from third instar larvae.

When ready to observe, add a coverslip to the preparation, cover it with a piece of tissue, and press firmly and directly down, to squash the cells and nuclei. Do not let the coverslip move sideways, or a poor slide will result. Observe with your compound microscope using the medium power to find the tissue and the high power to observe chromosomes. If nuclei are not broken, remove the slide from the microscope and press again on the coverslip.

How many chromosomes do you see? _____

Can you see banding in the chromosomes? _____

Sketch part of a chromosome below.

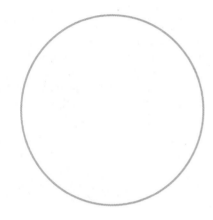

Theoretical Background

We will now review some theoretical background before performing experimental genetic crosses with fruit flies.

Genes are located on chromosomes and chromosomes are found as homologous pairs in diploid organisms. The location on a chromosome where a gene is found is the **locus** for that gene. Thousands of gene loci are found on each chromosome. If we could look at a single locus in a population of fruit flies, ignoring all other loci, we might find that some flies have exactly the same gene on both chromosomes in the homologous pair. These flies would be described as being **homozygous** for that gene. We could write this down in the following way.

Let "**A**" be the symbol for a specific gene. A homozygous diploid individual would have two "**A**" genes present and this would be written as **AA**. When you write down the genes that are found in an individual, you are specifying the **genotype**. However, as we studied a population of fruit flies, we might find some that had a variation of the gene, one that was not exactly the same. This would be called an **allele** of gene "**A**" and could be written as "**a**." Hypothetically, some flies could be homozygous for this allele and would have a genotype that was written **aa**. Some, however, could have the "**A**" form of the gene on one chromosome and the "**a**" form of the gene on the other chromosome in the homologous pair. The genotype of this individual would be written as **Aa,** and it would be described as being **heterozygous** at this chromosomal locus.

While the preceding discussion of genes was instructive, it is not possible to look at genes through a microscope and tell what kind are on a chromosome. The only way to determine what genes are present—short of doing DNA analysis, which is not easy—is to do genetic crosses and then to reason what genes must have been present to give the results. When you do genetic crosses, you deal with **phenotypes**—what an organism looks like, not with genotypes—what genes are present. Phenotypes are the result of gene expression (along with environmental influences) so the occurrence of a phenotype is usually a good indicator that a gene is present. What the phenotype does not tell you is the heterozygous or homozygous nature of the genotype for the trait. Therefore, one of the initial steps in a genetic study is determining the dominant or recessive nature of that gene.

Fruit flies have eight chromosomes: three pairs called **autosomes** and two called **sex chromosomes** that are designated **X** and **Y**. Females have two X sex chromosomes, so they have four homologous pairs. Males have one X and one Y sex chromosomes, so they have three homologous pairs plus an X and Y.

Because of these chromosomal differences between the sexes, one would expect to see differences in the inheritance of traits, depending on whether the genes for the traits were carried on autosomes or sex chromosomes.

Alleles on Autosomes'

If a gene locus for a trait with two different alleles is on one of the three pairs of autosomes, then a cross between two homozygotes for contrasting traits (alleles *A* and *a*) can be diagrammed as follows:

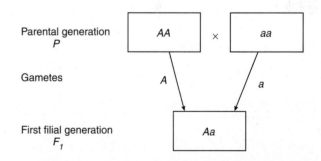

Obviously, one parent must be male and one female, but it makes no difference which parent is homozygous for one allele (*A*) and which is homozygous for the other (*a*). The result will always be the same: progeny will always have two different alleles (heterozygous) regardless of whether they are male or female.

When Mendel did experiments like this, he invented the concepts of **dominant** and **recessive** to explain his results. Because the parents are homozygous for different alleles, they look different; *i.e.,* they have different phenotypes. Hypothetically, we could say AA is blue because the "A" allele causes blueness and aa is white because the "a" alleles produce no color development. If a blue is crossed with a white, what color will the offspring be? The only way to tell is to do an actual genetic cross. When Mendel did similar experiments, he found all progeny were the equivalent of blue. He explained it by saying the allele for blue **dominates** over white which is **recessive** whenever both are present in the same individual.

If the heterozygous offspring from the first filial generation (F_1) are allowed to mate, the results demonstrate that the recessive allele is present in the heterozygote because the recessive allele segregates from the dominant allele during meiosis and reappears in the next generation when the gametes combine. This can be demonstrated by using a **Punnett square,** a paper and pencil method of keeping track of the kinds of gametes that can be produced and the combination of gametes at fertilization as follows:

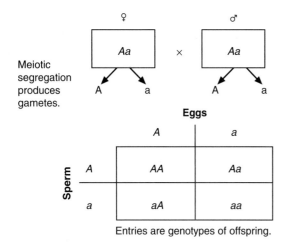

The genotypic results from a cross between two heterozygotes would be 1 *AA*: 2 *Aa*: 1 *aa*, or a phenotypic ratio of 3 dominant (blue) to 1 recessive (white) if *A* were dominant to *a*. Because the parents have the same genotype, it makes no difference which one is male or which is female.

Performing a Genetic Cross

Observing Phenotypes You will perform a genetic cross with flies having contrasting phenotypes, some may have wings, others lack them. Some may have light, dark, or other lighter body colors. You instructor may provide other contrasting phenotypes as well. Each contrasting pair of traits is determined by a pair of alleles from one locus; one member of the pair dominant over the other.

8 In the lab are dead flies (or photos of flies) exhibiting the contrasting phenotypes with which you will work. They came from true breeding stocks, indicating that they were homozygous for the alleles. The first step in this lab is to become familiar with the phenotypes.

Obtain the flies in matched pairs: winged/wingless; dark/light; or whatever phenotypic pairs your instructor is providing. Your job is to determine the sex of the flies and to become familiar with the contrasting phenotypes. In **table 12.1**, indicate the trait determined by each locus and the contrasting expressions of the alleles at that locus.

9 ***Determining Dominant and Recessive*** Stocks of these true breeding flies that you just analyzed were used as parents to produce a new generation, called the F_1 generation. All members of the F_1 generation will be heterozygous at the loci included in this study. Why?

Table 12.1	F_1 Phenotypes from Crosses Between True Breeding Stocks			
		Number of		
		Males	Females	Conclusion
Locus 1 determines _____				
First allele expressed as _____				
Second as _____				
Locus 2 determines _____				
First allele expressed as _____				
Second as _____				
Locus 3 determines _____				
First allele expressed as _____				
Second as _____				
Locus 4 determines _____				
First allele expressed as _____				
Second expressed as _____				

Because each heterozygote has both alleles, the allele that is expressed in the offspring will by definition be the dominant allele. By observing the phenotypes seen in the F_1, you can determine which allele in each pair is dominant and which is recessive.

Look at the F_1 progeny and count the males and females having one or the other of the contrasting traits at each locus. Record the numbers of males and females expressing the alternative phenotypes in table 12.1. From this data you should be able to reach a conclusion about which alternative expression of the gene is dominant and which is recessive at each gene locus. Record your conclusions in the last column of table 12.1.

10. Determining Genotypes

In the lab are several stocks of fruit flies that all look alike because they are expressing the dominant allele at the loci you are studying. However, some may be homozygous at one or more of the loci we are working with and others could be heterozygous at one or more loci. Your job will be to determine, to the extent that you can, the genotypes of a pair of unknown flies given to you by your instructor.

Before you receive the flies, you need to do a little background work to develop a hypothesis that will be tested by your results.

For each trait, you know which allele is dominant and which is recessive from your analysis of the F_1 generation, recorded in table 12.1. Because the unknown flies are expressing the dominant phenotype, you know that they are either homozygous or heterozygous but you cannot look at a fly and say which is which. Depending on whether one or both of the flies are homozygous or heterozygous, three matings are possible. Using the notation U^+ to represent any unknown dominant allele and u to represent any unknown recessive, diagram the three possible matings in the blank Punnet squares below.

First alternative; cross between _____ × _____

Second alternative: cross between _____ × _____

Third alternative: cross between _____ × _____

Look at the genotypic outcomes from all three possible crosses. Are they all the same or different? Is there a critical outcome that will allow you to say that both parents are heterozygous? What is it?

Now think about phenotypes that come from the expression of the alleles in the genotypes. Because the flies you will be given are all expressing the dominant phenotype, you do not know whether they are homozygous or heterozygous at one or more than one locus. Therefore, you have to look for the critical outcome. What phenotypes, if they appeared among the offspring from your flies, would tell you that both of your parental flies were heterozygous?

Can you use these theoretical genotypic and phenotypic outcomes to formulate a hypothesis that could be answered by looking at the offspring from the two flies you will be given? Write your hypothesis below.

What is the critical outcome you are looking for?

11. Setting Crosses

Get a culture vial from the supply area. Following the directions given by your instructor, add flakes of growth medium, distilled water, and yeast to the vial. Let the medium hydrate for about 5 minutes. Label the vial with your name and then add the code number you are given to identify the unknowns.

When male and female flies are separated within four hours of emergence from their pupal casing, they are virgins and can be used in controlled mating experiments. Female fruit flies mate once and store sperm in a seminal receptacle. They will lay eggs for several days after mating. Each time eggs are produced some sperm are released from the seminal receptacle and fertilize the eggs.

Fruit flies can be anesthetized by treatment with carbon dioxide. A small amount of dry ice put in an Erlenmeyer flask with a one-hole stopper is a safe carbon dioxide generator. A small rubber tube from the stopper can be inserted past the stopper into a vial to anesthetize the flies.

Your instructor will anaethetize several stocks of flies that represent the unknown genotypes. You will be given a pair, one male and one female, and the code number for the unknown stock. Write the code on the vial and in **table 12.2**.

Once the anaesthetized flies are in your vial, add a sponge stopper, but do not turn the vial upright. Doing so will cause the flies to fall on the wet growth medium and they will stick there. The vials can be stood upright when the anesthesia wears off and the flies become active. After the flies have recovered, they may be incubated at room temperature or in a 27°C incubator. New flies will start appearing in the culture vials in about 10 days. You will harvest and count them in two weeks. Some care will be necessary in the interim.

12 *Tending the Flies* After a week, you should do the following:

1. Remove the F$_1$ parents, the flies you mated. None of the offspring will have metamorphosed into adults yet. Determine the parents' phenotypes at each of the loci being studied and record them in table 12.2.

2. Examine the growth medium. If it seems dry and shows flakes, add a several drops of distilled water from a squeeze bottle. If, on the other hand, it seems wet, take a small piece of tissue and push this into the medium to wick away moisture. Leave the tissue in the vial. Return the vials to the incubator.

13 *Counting Offspring* Two weeks after setting the cross, you will be able to count the offspring. Before doing that, write what each locus governs in column 2 of table 12.2. Review what is dominant and what is recessive for each locus in table 12.1.

Following instructions from your instructor, over-anesthetize your flies so that you kill them. Pour them out of the vial onto a white index card, which can be set on a block of dry ice for several minutes to be sure the flies are dead. Examine them, using a dissecting microscope. Use a fine brush to separate them into piles with different phenotypes for the loci being studied. When all are separated, record the number showing the dominant or recessive phenotype for each locus in table 12.2.

Table 12.2	Phenotypes Among Offspring from Unknown Pair # _____

Parental phenotypes were _____ × _____

Locus #	Expressed as	# Dominant	# Recessive
1			
2			
3			
4			

14 *Analysis* Review the hypothesis you made for the theoretical outcomes when you set up the cross two weeks ago. At what loci do you see the critical result?

At what loci were your flies heterozygous?

If they were heterozygous, Mendelian principles tell us that you should get a ratio of 3:1 expressing the dominant versus the recessive phenotype. What were your ratios at the loci you think are heterozygous?

Did the ratios match the expected results? Why?

Can you say with certainty that both of your flies were homozygous at the other loci? Why or why not?

Your instructor may ask you to turn in your conclusions along with the code number for the pair of flies you were given.

Lab Summary

15 On a separate sheet of paper, answer the following questions assigned by your instructor.

1. Describe the life cycle of *Drosophila* in detail and describe how to distinguish between male and female flies.
2. Discuss how Mendel's principle of segregation can be used to explain your results.
3. When, during meiosis, does segregation occur? Draw a chromosomal model with eight chromosomes that explains the results of your experiment.
4. If two or more of the loci being studied were on the same chromosome instead of being on different chromosomes, how would that have affected your results?
5. If one of the loci were on the X chromosome, how would that have affected your results?

You may want to try the critical thinking questions that apply some of the knowledge you gained in doing this lab.

INTERNET SOURCES

FlyBase is a large World Wide Web database that compiles information on the genetics of *Drosophila*. It is located at http://flybase.org/ You can use this database to find additional information about the gene loci you studied in this lab. When connected to FlyBase, enter the name of one of the phenotypes you studied. Click on *Search*. When the page for that gene appears, click on *Summary Information* in the list at the bottom of the page.

Look for information to answer the following questions:

a. On what chromosome is the locus located?
b. How many alleles have been described at this locus?
c. What does the gene do?
d. Describe how this gene's function is related to the phenotypic effect of the mutation.

LEARNING BIOLOGY BY WRITING

Instead of simply turning in your results, you may be asked by your instructor to write a lab report describing how you determined the genotype of the unknown. Do this by first clearly stating the problem you are trying to solve. Discuss the alternative (what-if) scenarios very briefly. Your results should be a facsimile of tables 12.1 and 12.2. Your discussion should then explain how you determined the genotype of the unknown.

CRITICAL THINKING QUESTIONS

1. In humans, brown eyes are dominant to blue and are autosomal. A child who is brown-eyed wonders whether she is homozygous or heterozygous. She has a sister who is blue-eyed. What is the "unknown's" genotype?
2. In humans, the ability to metabolize the sugar galactose is genetically determined. The gene locus is on an autosome. The allele for normal metabolism is dominant, and the recessive allele in the homozygous condition cause galactosemia—an uncomfortable but not deadly disease. A woman who is galactosemic marries a normal man and they have one child who is normal. Is this convincing proof that the man is homozygous for the normal gene?
3. Read the classic scientific papers written by Mendel and Sutton. Mendels is available on the World Wide Web at: http://www.mendelweb.org and Sutton's at http://www.esp.org/foundations/genetics/classical/wss-02.pdf.
4. If you were studying the inheritance of both eye color and wings in the same flies, how would you use Mendel's principle of independent assortment to predict the results?

Interchapter

An Outline of Sterile Technique

Sterile or aseptic technique minimizes the chances that foreign bacteria and fungi will be introduced into a pure culture. Bacterial and fungal spores are virtually everywhere. When working with bacterial cultures, you want to avoid unintentionally letting spores enter the growth media that you are using. Spores are single cells surrounded by heavy cell walls that make them resistant to drying, heat, and often chemicals. In medical labs, food and water quality labs, and research labs, sterile technique is used to prevent getting false readings. The unintended introduction of bacteria into a culture is called **contamination**. Although you will not work with disease-causing bacteria, those who do such work use sterile techniques like these to protect themselves.

Sterile technique starts with sterile materials. The most common way to sterilize these is to heat them in a pressure cooker-like device called an **autoclave**. It has a large chamber that can be loaded with materials to be sterilized. The chamber is sealed and steam introduced. The pressure builds to 15 lb/in^2, which yields a temperature of 120°C. This is held for 15 minutes. Any living organisms or their spores are killed by this treatment. This sterility is forever, unless the materials get contaminated.

To maintain sterility requires constant diligence and common sense. There are six things to consider to prevent contamination.

1. Work in an area that is free from blowing air. Sometimes a clear plastic hood is used to minimize drafts. Disinfect your work area with a wipe of 10% bleach. Lay out what you will need: source culture, packaged sterile transfer tools, new growth medium, *etc*. Inspect any new growth medium to see if it is cloudy and contaminated. Mentally go through the steps of the procedure before doing it, so that you will not have long pauses as you try to recall what to do.

2. Use only sterile tools to transfer materials. Micropipetters, and inoculating loops, are the primary tools for transferring bacteria in labs. Loops may be reusable metal loops or disposable plastic ones. To reduce likelihood of contamination, open sterile packs just prior to use. If a metal loop is being used, it should be flamed until it just starts to glow red and then allowed to cool before use.

3. In this manual, you will be asked to make transfers from liquid to agar, from agar to liquid, or from agar to agar. The specific techniques are described below and will be demonstrated by your instructor.

4. *Liquid-to-agar transfer:* Loosen caps and mentally review your procedure before starting. Avoid drafts and control your breathing to minimize air currents. Transfers can be made with a loop as described in **figure I.1** or with a pipette as follows.

   a. Select an appropriate pipetter for the volume to be delivered, and set the volume. Open the sterile pipetter tip box by tilting the lid to one side. Insert the pipetter in a sterile tip and exert pressure to secure the tip to the pipetter. Remove the tip from the box and reclose immediately.

   b. Hold the pipetter so that your thumb can operate the plunger and your little finger of the pipetting hand, is free for grasping the cap of a tube.

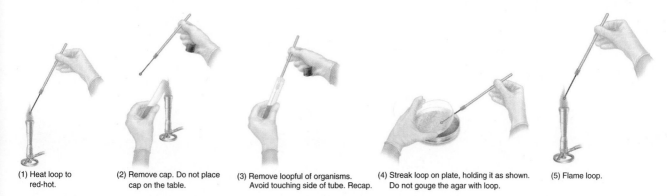

(1) Heat loop to red-hot. (2) Remove cap. Do not place cap on the table. (3) Remove loopful of organisms. Avoid touching side of tube. Recap. (4) Streak loop on plate, holding it as shown. Do not gouge the agar with loop. (5) Flame loop.

Figure I.1 Procedure for inoculating a petri plate agar culture from a liquid culture.

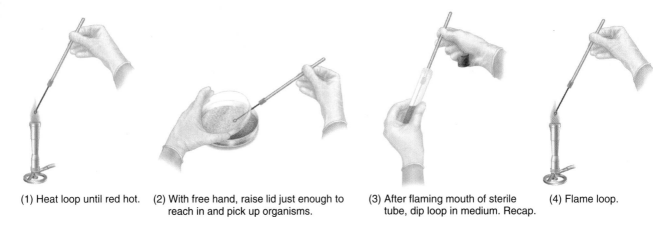

(1) Heat loop until red hot. (2) With free hand, raise lid just enough to reach in and pick up organisms. (3) After flaming mouth of sterile tube, dip loop in medium. Recap. (4) Flame loop.

Figure I.2 Procedure for inoculating a liquid culture from an agar plate.

    c. Pick up specimen tube with other hand, grip cap with little finger of pipette hand and twist off. Do not lay the cap down.

    d. Insert just the tip of the pipette into the fluid in the tube and withdraw the required volume.

    e. While avoiding air currents, open petri dish lid just enough to insert the pipette tip and expel fluid in a drop. Close dish and dispose of the pipette tip as directed.

    f. Take a sterile loop and again lift the plate lid just enough to insert tool. Spread the drop of fluid evenly on the agar. Turn plate 90° and repeat spreading. See figure 13.4.

    g. Clean area as directed and dispose of materials in biohazard bags for autoclaving.

5. ***Agar-to-liquid or agar-to-agar transfer:*** Mentally review procedure. Hold a sterile loop like a pencil and never touch anything nonsterile with tip. The following procedures are diagrammed in **figure I.2**.

    a. Touch loop to a colony and lift it off without getting agar. Remember that you need only the smallest amount on the loop.

    b. If transferring a sample to liquid, open a tube containing the growth medium. Insert the loop and swish it in the medium. Replace the cap.

    c. If transferring a sample to another agar plate, open the petri plate lid just far enough to insert the loop and streak the plate in a zig-zag pattern. Turn the plate 90° and repeat this zig-zag pattern.

    d. Clean area as directed and dispose of contaminated materials in biohazard bags for autoclaving.

6. Always wash your hands after you have finished working with bacteria. Also, wipe down the work area with 10% bleach. All tools used in transfers and all old cultures should be autoclaved or treated with bleach before disposal.

Lab Topic 13

Isolating DNA and Transformation with Plasmids

Supplies

Resource guide available on WWW at www.mhhe.com/labcentral

Equipment

Water baths at 42°C and 60°C
Incubator at 37°C
Refrigerator
Spectrophotometer

Materials

E. coli transformation kit from BioRad (pGLO) includes
    E. coli
    Plasmids with ampicillin resistance and green fluorescent protein genes (GFP)
    Luria broth and Luria broth agar
    Ampicillin
    Arabinose
    Petri dishes
    Bacteriological inoculation loops
    2.0 ml plastic microcentrifuge tubes
    UV pen light
Micropipetter with sterile tips
Test tubes
Hot plate and 800-ml beaker with boiling water
Beakers, 30 ml
Marker pens
Vortex mixer
Clean glass rods
Test tubes
Cuvettes
Aluminum foil
Autoclave bags for waste
Medium-sized onion
Ice

Solutions

10% household bleach
Woolite/enzyme/salt solution
95% ethanol (stored in freezer)
DNA dissolved in 4% NaCl
4% sodium chloride
Diphenylamine solution
50 mM $CaCl_2$

Student Prelab Preparation

Before doing this lab, read the Background section and other sections of the lab topic assigned by your instructor. Use your textbook to review the definitions of the following terms:

competent cells	selection medium
genomic DNA	spectrophotometer
plasmid	transformation
reporter gene	

Describe in your own words the following concepts:

Solubility of DNA in polar and nonpolar solvents

How a spectrophotometer is used to construct a standard curve (lab topic 5)

The difference between genomic and plasmid genes

How you could "genetically engineer" a bacterium to produce a protein normally found only in humans

What is involved in sterile technique after reading the Interchapter on pages 143 and 144

After finishing the prelab review, write any questions you have about terms, concepts, or techniques in the margins of this lab topic. The lab experiments should help you answer these questions or you can ask your instructor during the lab.

Objectives

1. To isolate genomic DNA from an onion
2. To measure the amount of DNA in a sample
3. To conduct a plasmid transformation for ampicillin resistance and green fluorescent protein in E. coli.
4. To estimate the efficiency of transformation.

Background

For almost 100 years it has been possible to isolate DNA from cells, but we have known its function for only about 70 years. In the 1940s, Oswald Avery, improving on earlier experiments of Fred Griffith, demonstrated that if DNA was isolated from a pneumonia-causing strain of bacteria and added to cultures of nonvirulent bacteria, some recipient cells were **transformed** into virulent types that caused pneumonia. This showed that genes could be artificially transferred in the laboratory, aided the identification of DNA as the genetic material and provided an important tool of genetic engineering.

Genetic engineering involves the isolation of specific genes from one organism and their insertion into a second organism of the same, or a different species. The results can have a positive and/or negative impact. Government and private industry spend millions of research dollars each year in the genetic engineering field. Results include the production of human hormones such as insulin for diabetics; various antigens/antibodies to fight diseases; agricultural seeds with higher resistance to pests, greater nutrient content, or higher protein quality, and many other useful proteins. The creation of **genetically modified organisms** (GMOs) generated either by nature or geneticists has raised concerns about their potential to do harm to our environment or health. Such concerns are active areas of scientific research and public debate.

Modern genetic engineering techniques depend upon four critical factors: (1) techniques that allow scientists to isolate single, whole genes, (2) a suitable host organism in which to insert the foreign gene, (3) a vector (transmission agent) to carry foreign genes into the host, and (4) a means of isolating the host cells that take up the foreign gene.

Modified strains of the bacterium *Escherichia coli* are widely used in genetic engineering as host organisms in which foreign genes can readily be inserted for research or protein production. *E. coli* has a single large circular DNA molecule, called **genomic DNA**, containing about two and a half billion base pairs (fig. 13.1). We now know how these base pairs are arranged into the approximately 4,300 genes in *E. coli*.

A **vector** in molecular biology is a mechanism used to insert genes from one organism into another. It may be a virus; a naked DNA molecule; or a process involving particle bombardment, heat, or electrical shock.

A **plasmid** is a small circular DNA molecule containing 1,000 to 200,000 base pairs that naturally exists in the cytoplasm of many bacteria (fig. 13.2). Plasmids replicate

Figure 13.2 Bacterial cell with circular plasmid DNA molecules surrounding it. Plasmid DNA is small in relation to the host cell: see size of genomic DNA in figure 13.1.

independently of the genomic DNA as the cell grows and pass on to each of the daughter cells during cell division. Plasmids that carry genetic information beneficial to the host cell are favored by natural selection and increase in frequency in the host cell population. For example, plasmids that confer antibiotic resistance provide obvious benefits to some bacteria as they compete with fungi and other bacteria that produce those antibiotics.

Some *E. coli* cells can take up plasmid DNA from their environment and express the plasmid genes. Cells that can take up DNA are said to be **competent cells.** Competency can be induced in the lab by treating cells with divalent cations like calcium ions, followed by a heat shock or electrophoresis. This makes the cell porous, increasing the likelihood that a plasmid can cross the cell membrane. Once in a cell, a plasmid may be replicated by the cell's normal DNA replication enzymes and passed to the daughter cells during cell division. Though only a small fraction of cells take up DNA, and are genetically transformed when exposed to this process, *E. coli* grows very rapidly, dividing every 20 minutes, so a single transformed cell can produce a billion transformed descendants in a matter of 10 hours.

Transformed *E. coli* can be easily isolated if the plasmid included a gene that changes the bacteria's metabolic capabilities. For example, if the plasmid carries a gene for resistance to the antibiotic ampicillin, only those cells that have taken in the plasmid will grow when placed in a **selection medium** containing ampicillin. A selection medium is a growth medium in which bacteria with the targeted gene (from the plasmid) can grow, but bacteria without that gene cannot grow. An alternative is to include a **reporter gene**, which makes transformed bacteria visible, *i.e.* reports the presence of the plasmid. For example, a gene from the jellyfish *Aequoria victoria* codes for **green fluorescent protein (GFP)** which fluoresces brilliant green when illuminated with ultraviolet light, and is often used as a reporter gene in genetic engineering studies.

Genetic engineers often use plasmids to insert foreign genes into *E. coli* and other bacteria. The basic strategy is

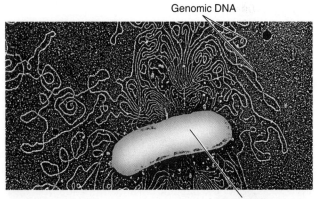

Figure 13.1 Electron micrograph of a ruptured bacterial cell showing the genomic DNA spewing from cell.

as follows. A plasmid is modified by inserting a specific foreign gene, perhaps the gene for human insulin, along with a gene for resistance to an antibiotic like ampicillin into the plasmid DNA. The modified plasmid is then mixed with competent *E. coli* which are then exposed to a heat shock. Those *E. coli* are then transferred to a selection medium containing the antibiotic. Only those cells that acquired the plasmid can grow. They also are the cells that carry the gene (in their new plasmid) for insulin. These cells can then synthesize insulin, which can be isolated from the bacterial cultures and used for insulin therapy of diabetics. Essentially any protein for which we can isolate a gene can be manufactured using this technique.

LAB INSTRUCTIONS

You will explore some basic techniques in molecular biology. You will isolate genomic DNA from an onion and estimate the amount of DNA obtained. You will carry out a plasmid transformation of *E. coli* to insert genes for ampicillin resistance and GFP, and use both a selection medium and the GFP reporter gene to identify transformed bacteria.

Isolating Genomic DNA

For most of us, DNA is an abstract substance, one that we have never seen or handled. In this portion of the exercise, you will isolate DNA from an onion and observe it.

The isolation procedure is fairly direct. You will rupture onion cells, releasing their cell contents: proteins, DNA, RNA, lipids, organelles, and various small molecules. You will then precipitate the DNA from the suspension by treatment with ethanol.

Rupturing Onion Cells The following recipe should make enough DNA for six to eight student groups.

1. Chop a medium-sized onion into 10 or so pieces. Add the pieces to a blender along with 100 ml of a detergent/enzyme/salt solution made up by diluting the commercial product Woolite in 1.5% sodium chloride. This solution contains enzymes that degrade proteins, a detergent that emulsifies fats, and salt that creates a polar environment to disassociate chromosomal proteins and dissolve the DNA.
2. Blend the onion until it is completely homogenized. If the homogenate is too foamy, add another 50 ml of the detergent/enzyme/salt solution.
3. Pour the homogenate into a beaker and incubate in a 60°C water bath for exactly 15 minutes. Time and temperature are critical in this step. Higher temperatures or longer incubations will denature the DNA. A DNA double helix is composed of two strands held together by hydrogen bonds. Heat can break hydrogen bonds. What do you think happens to the double helix when DNA is heated?

4. Cool the mixture in an ice bath for five minutes and then filter the cool homogenate through cheesecloth to remove pieces of the onion.
5. Each student group should take a few milliliters of the filtrate in a test tube. The onion DNA is in this filtrate and somehow must be isolated. How can this be done?

DNA Precipitation DNA solubility decreases in less-polar solutions and as temperature drops. We will use these properties to precipitate DNA at the boundary between our salt solution and very cold ethanol (**fig. 13.3**). If we isolate the precipitate, we have purified the DNA, leaving behind most proteins and other molecules that were extracted from the onion cells.

6. The next step is a bit tricky and requires a steady hand. The technique is outlined in figure 13.3. Very slowly add a few milliliters of extremely cold ethanol (stored in freezer) from a beaker or pipette so that it flows down the side of the tilted test tube. If done correctly, the less-dense ethanol will float as a layer on top of the aqueous solution of DNA.
7. Let the tube sit undisturbed for two to three minutes and you will see a whitish, stringy substance form at the interface of the two liquids. What do you think this is?

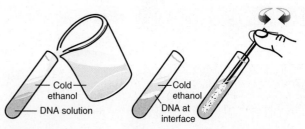

Carefully pour cold ethanol, from the beaker, down the side of the test tube. DNA precipitates as fibers. Slowly rotate glass rod back and forth between fingers. Fibers will adhere to rod.

Figure 13.3 Technique for spooling DNA.

Ethanol is a less-polar solvent than a 1.5% saline solution. DNA is a polar molecule. As the ethanol diffuses into the saline at the interface, DNA comes out of solution as a gelatinous precipitate.

8. Take a clean glass rod or a wooden stick and extend it into the tube past the interface of the two layers. Do not stir, but roll the rod or stick between your fingers. The DNA molecules will adhere and wrap around it. The technique is called **spooling.**

9. Remove the rod or stick and press it gently against the side of the tube above the liquids. This will expel excess ethanol from the spooled DNA. The DNA can be redissolved by stirring the rod with the spooled DNA in another tube containing 4% NaCl. This solution of purified DNA can now be tested by the following colorimetric technique to quantify the amount present.

Determining the Amount of DNA in Solution

DNA chemically reacts with diphenylamine to give a blue reaction product in proportion to the amount of DNA present. The amount of blue product can be quantified by reading the absorbance at 600 nm in a spectrophotometer.

To determine the amount of DNA that you have isolated, you will first prepare a standard curve by measuring the absorbance of known concentrations of DNA. Then you will measure the absorbance of your sample and compare it to the standard curve.

1. Number six test tubes. Add the solutions called for in mixing **table 13.1**.

2. Cap all six tubes with aluminum foil and place them in a boiling water bath for 5 to 10 minutes. Remove and place the tubes in a beaker of tap water to cool.

3. Calculate the number of micrograms of DNA added to tubes 2 through 5 (*Hint:* quantity = concentration of known stock multiplied by the volume used) and record in **table 13.2**.

4. Review the directions for using a spectrophotometer (see fig. 5.5). Set it to 600 nm and use tube 1 to adjust the absorbance to zero, so that any color due to contaminants is blanked. Now read the absorbance of the tubes 2 through 6 and record in table 13.2.

5. Plot absorbance as a function of concentration for the known solutions of DNA on the graph paper at the end of this lab topic. Draw a straight line that best fits the plotted points. Find the absorbance for your unknown on the y axis and read the amount of DNA present by dropping down to the x axis and reading the amount of DNA that gives that absorbance. This represents the yield from the DNA isolation procedure for the volume of DNA solution you started with. How much DNA was in your sample? Record the amount in table 13.2.

Table 13.2 Absorbance of DNA Solutions at 600 nm

Tube	Amount of DNA (μg/Tube)	Absorbance
1	0	
2		
3		
4		
5		
6	Unknown	

What is the concentration of DNA in your unknown? _____

Transformation by Plasmids

In this experiment you start with *E. coli* bacterial cells that (1) cannot grow in the presence of the antibiotic ampicillin (it kills them) and (2) do not have a gene for a protein that causes green fluorescence (GFP). You will try to get them to take up an artificially constructed plasmid containing genes for both ampicillin resistance and GFP. The plasmid is an example of recombinant DNA, containing genes from both another bacterium and jellyfish. The GFP gene in the plasmid was inserted in the arabinose operon, a set of bacterial genes that are turned on only when the sugar, arabinose, is present in the culture medium.

Table 13.1 Mixing Table for DNA Standard Curve (all values in milliliters)

Tube	Diphenylamine	4% NaCl	Known DNA 500 μg/ml	Unknown DNA from Onion
1	3.0	3.0	0.0	0.0
2	3.0	2.8	0.2	0.0
3	3.0	2.5	0.5	0.0
4	3.0	1.5	1.5	0.0
5	3.0	0.0	3.0	0.0
6	3.0	0.0	0.0	3.0

4 How can you test the hypothesis that some bacteria you expose to the plasmid acquire the ampicillin resistance gene? Write your answer in the space below:

How can you test the hypotheses that those bacteria also acquired the GFP gene? Write your answer in the space below.

Any transformation experiment should include a negative control, designed to make certain that the outcome measurement shows acquisition of a new trait, not one that was originally present. How can you make certain that the original bacteria, before exposure to the plasmid, did not have the ampicillin resistance or GFP genes? Write your answer in the space below.

Note: All the following procedures require using sterile technique, solutions, and labware. Read the Interchapter, "An Outline of Sterile Technique" on pages 143 to 144 before doing the experiment. Labware should be sterilized, and disposable pipettes, loops, and tubes should be collected and autoclaved before disposal. Disposable gloves should be worn throughout the process. Why?

Before starting this procedure and after its completion wash the table where you will work with 10% household bleach. Why?

Making Competent Cells

5 1. Label one closed sterile microcentrifuge tube +plasmid (or +P) and another −plasmid (−P). Label both with your group's name.

2. Uncap one tube and use a micropipette with a sterile tip to add 250 μl of cold 50 mM $CaCl_2$. Recap the tube and push it into crushed ice. Repeat with the other tube.

3. Obtain a petri plate containing Luria agar on which *E. coli* lacking genes for ampicillin resistance and GFP production have been growing for about 12–24 hours at 37°C.

4. Use a sterile loop to remove a single colony from the *E. coli* plate (**fig. 13.4**). Be careful not to pick up any agar. Select the −plasmid tube and immerse the loop into the $CaCl_2$ at the bottom of the tube. Twirl the loop between your index finger and thumb until the entire colony is dispersed in the solution. Cap the tube, vortex if available, and return it to the ice. Pack ice around the tube.

5. Using a new sterile loop, add bacteria in the same way to the +plasmid tube. Then use a micropipetter to transfer 10 μl of plasmid DNA containing the

(a) Tilt lid, insert loop, touch to agar surface to cool, then scrape colony off agar.

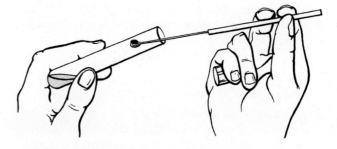

(b) Open tube as shown and swirl loop in medium.

Figure 13.4 Technique for transferring bacterial colonies from agar to liquid media.

ampicillin-resistance and GFP genes to the +plasmid cell suspension. Cap the tube, vortex the contents, and return it to the ice. Dispose of used loops as directed by your instructor. DO NOT add plasmid to the − *plasmid* tube. Why?

6. Incubate the tubes in direct contact with ice for at least 10 minutes. You can use this time to label petri plates as described in step 9.

6▸ Transformation Procedure To be successful in getting *E. coli* to take up the plasmid DNA, follow the temperature and timing in these steps *exactly*. Make certain that the part of the tube containing the bacteria makes good contact with crushed ice or the warm water bath.

7. After 10 minutes, remove both tubes from the ice bath, and **immediately** place them in a 42°C water bath for 50 seconds to heat shock the cells. After 50 seconds, return both tubes **immediately** to the ice for at least 2 minutes (longer is better).

8. Remove tubes from the ice. Open one tube and use a micropipette with a sterile tip to add 250 μl of Luria broth. Recap the tube and "gently" mix the contents. Repeat the procedure with a new sterile tip for the other tube. Then incubate both tubes at room temperature for 10 minutes.

7▸ Selective Growth To assess the results of your transformation experiment, you will grow both +plasmid and − plasmid bacteria on three different kinds of culture media:

Luria agar + arabinose	(ara)
Luria agar + ampicillin	(amp)
Luria agar + ampicillin + arabinose	(ara + amp)

9. While your tubes are incubating, obtain two plates of arabinose agar, and label the bottom (the agar side) of both plates with ara, your initials, and date. Label one plate +P and label the other − P. Repeat for the other two kinds of agar labeling them amp or ara + amp. You will have six labeled petri plates.

10. Follow these directions to inoculate the − plasmid plates with the − plasmid bacteria.
    a. Gently mix the contents of your − plasmid tube again, then using a micropipette with a sterile tip, pipette 100 μL of the bacterial suspension onto each of your three − plasmid plates. Recap the tube.
    b. Use a sterile loop to spread the bacterial suspension evenly across the surface of each plate by skating the flat surface of the loop across the agar. Rub back and forth across the whole surface plate, then repeat at a right angle to your previous direction. Use a new sterile loop for each plate. Be careful not to dig into the agar.

11. Now inoculate the three +plasmid plates with your +plasmid bacterial suspension, following the procedure in step 10.

12. Stack your +plasmid plates and tape them together. Likewise, stack and tape the −plasmid plates. Place the plates lid side down in a 37°C incubator until the ampicillin/+plasmid plate has visible colonies, about 12–24 hours. Then refrigerate until plates can be read.

Serial Dilution All bacteria will grow on the arabinose plate, but only transformed bacteria will show green fluorescence. If one wants to determine the fraction of bacteria that are transformed, then the number of fluorescing colonies on the +plasmid/arabinose plate has to be divided by the total number of colonies on that plate. However, when too many bacteria grow on a plate, the colonies merge and cannot be accurately counted. You can solve this problem by performing serial dilutions of your +plasmid solution to reduce the number of bacteria plated. Read page 180 to 181 in lab topic 15 for a description of the serial dilution technique.

8▸ Use the following directions to do your serial dilutions. Obtain five more arabinose plates. Label them ara 10 + P, ara 100 + P, 1,000 ara + P, ara 10,000 + P, and 100,000 + P. Add your initials to each plate. In the following steps, be careful not to reuse dirty pipette tips—always use a new sterile tip for each pipetting operation.

13. Make five serial dilutions of your +plasmid suspension from step 8 by using the amounts indicated in the following mixing table. Be certain to agitate each tube to suspend the bacteria uniformly before drawing fluid from it. Obtain five sterile microcentrifuge tubes and label them 1, 2, 3, 4, and 5. Then prepare dilutions according to the table below:

Tube #	Add 50 μl	Add 450 μl	Mix to Produce
1	Suspension from step #8	Luria broth	10 × dilution
2	Tube 1 suspension	Luria broth	100 × dilution
3	Tube 2 suspension	Luria broth	1,000 × dilution
4	Tube 3 suspension	Luria broth	10,000 × dilution
5	Tube 4 suspension	Luria broth	100,000 × dilution

Table 13.3 Results from Transformation Experiment

	LB agar + arabinose	LB agar + ampicillin	LB agar + ampicillin & arabinose
−plasmid			
+plasmid			

14. Using a separate sterile pipette tip for each dilution, pipette 100 µl of each of your five serial dilutions onto the corresponding agar plate. Then use a new sterile loop for each plate to spread the bacterial suspension across the plate, as described in step 10.

Tape the five plates together and place them lid side down in a 37°C incubator. Remove and refrigerate these dilution plates when you remove your other samples from the incubator.

Discard all disposable materials according to the directions from your lab instructor. Wash the bench area where you worked with 10% household bleach.

Analysis When visible colonies appear, look at the bacterial growth on the surface of the agar. Three results are possible: *no growth*, isolated *colony growth*, and bacterial *"lawn" growth*. Colonies are produced when a single cell grows and divides to produce a small mound of whitish cells. Lawn growth occurs when there are so many bacteria growing and dividing that colonies grow into their neighbors, making a continuous "lawn" on the surface of the agar.

9 Look at your plates and score them using the above classification. On any plate showing isolated colony growth, observe the colonies through the bottom of the culture plate and count the number of colonies present. Use a marker to put a dot over each colony as it is counted to keep from counting the same colony twice. Record the number of colonies in **table 13.3**.

For each plate showing bacterial growth, test for the presence of GFP by looking at the colonies under ultraviolet light. Determine whether none, all, or some of the colonies on a plate fluoresce green. For plates covered with lawn growth, individual colonies are tiny, so you may have to use a dissecting microscope to detect bacteria expressing GFP. Record the result in table 13.3. Sit down with your lab partner and systematically discuss the purpose of each plate and cell suspension combination in this experiment.

After you have finished analyzing your plates, either autoclave or flood the surface of the agar with 10% household bleach solution and place the plates in a bucket for disposal. Why?

10 Answer the following questions:

Which plate demonstrates that the experimental procedure did not affect the ability of *E. coli* to grow on the Luria culture medium?

Which plate shows that the plasmid treatment did not kill the cells?

Which plates show that the plasmid DNA was taken up and expressed by the cells?

Which pair of plates shows that the ampicillin prevents growth of untransformed *E. coli*?

Which plates show that transformation is a rare event?

Which plates represent selection media that allow only transformed bacteria to grow?

Which pair of plates show that GFP expression requires arabinose in the culture medium?

Which plate or plates show that both GFP and ampicillin resistance are transferred together?

11 *Transformation Efficiency* Genetic engineers often speak of transformation efficiency and express it as the number of transformed bacteria per µg of plasmid DNA. How do you determine how many bacteria were transformed per microgram (µg) of plasmid DNA used?

Each colony on the two +plasmid plates containing ampicillin grew from a single transformed bacterium, so the total number of colonies on those two plates is equal to the number of transformed bacteria present in the amount of +plasmid solution you added to those plates. Therefore

$$\text{Transformation efficiency} = \frac{\text{\# colonies on amp plates}}{\mu\text{g plasmid DNA in bacterial suspension on plates}}$$

How much plasmid DNA was in the bacterial solution you spread across those two plates?

$$\mu\text{g plasmid DNA} = \frac{\mu\text{g DNA}}{\mu\text{l plasmid solution}} \times \frac{10\,\mu\text{l plasmid sol.}}{510\,\mu\text{l in step 10 tube}}$$
$$\times \frac{100\,\mu\text{l from step 8 tube}}{\text{plate}} \times 2\text{ plates}$$

Get the concentration of plasmid DNA from your instructor, and calculate your transformation efficiency in the space below:

12 **% Bacteria Transformed** To determine what percent of the bacteria were transformed, you need to divide the number of fluorescing bacteria on the ara +P plate by the total number of bacteria on that plate. On the ara +P plate so many bacteria are most likely present that the plate is covered with a lawn of merged colonies, yielding **lawn growth**. You cannot distinguish individual colonies and because no one colony is very large, UV light does not reveal big spots of GFP. Therefore, we will now turn to the serial dilution plates you made up to get the numbers we need.

Choose a dilution plate on which you can count individual *GFP colonies,* ignoring the non-GFP colonies. Multiply the number of GFP colonies on the plate by the dilution factor for that plate to calculate the number of transformed bacteria in 50 µl of the original bacterial suspension.

Number of GFP colonies on plate × dilution factor for that plate =

Now choose a dilution plate on which you can count the *total* number of colonies, including *both GFP and non-GFP colonies,* and use that count and the dilution factor for that plate to calculate the total number of bacteria in 50 µl of the original suspension. All of the other serial dilution plates can be ignored. They were "just in case dilutions" chosen to provide a range of dilutions should you have needed a greater dilution.

Total colonies on plate × dilution factor for that plate =

What fraction of the bacteria were transformed?

Calculate the percent of the bacteria transformed and write it below.

Lab Summary

13 On a separate sheet of paper, answer the following questions as assigned by your instructor.

1. Explain the difference between genomic and plasmid DNA in bacteria.

2. Why does DNA precipitate at the interface between salty water and cold ethanol? Why is the precipitate "stringy"?

3. Explain how you used a standard curve to determine the amount of DNA in a sample. Do not restate the steps of the technique. Present the reasoning behind the procedure.

4. Why were the cells placed in $CaCl_2$ and heat shocked during the transformation experiment?

5. What is selection medium? A reporter gene? How are they involved in this experiment?

6. Discuss the experimental design of the transformation experiment, describing the purpose of each petri dish that was inoculated.

7. What evidence do you have that the bacteria were transformed in this experiment, and that plasmids were the transforming element?

8. Why do you think the fraction of bacteria that were transformed was so small?

9. What are some possible sources of error in your transformation experiment?

You may want to try the critical thinking questions that apply some of the knowledge you gained in doing this lab.

Internet Sources

The transformation of competent *E. coli* cells is something that is done routinely in research laboratories. For this reason, many researchers are concerned with optimizing this procedure. Use a browser to search the Internet for information on the optimization of transformation.

In your Internet browser, choose a search engine such as Google, and search on *efficiency of E. coli transformation*. You will get a long list back. Read a few of the articles. What are the authors suggesting to improve the efficiency of transformation?

Learning Biology by Writing

If you did the transformation experiment with the plasmid, you have a well-controlled experiment for a lab report. State the hypotheses that the experiment tests. Indicate the technique used. Give the data obtained and draw conclusions from the data.

Critical Thinking Questions

1. Three critical factors required for genetic engineering techniques are described in this exercise. How was each of these factors met in your transformation experiment?

2. Design a procedure that theoretically would allow you to take the gene for insulin and insert it into the cells of a person with diabetes. What would be the major problems encountered in doing such a procedure?

3. Bacteria containing plasmids die in nature. When they do, bits and pieces of their DNA, including plasmids, are released into the environment. Given your observations in this lab, explain how horizontal gene transfer can occur between species of bacteria.

4. Why is transformation important in the rapid evolution of multidrug-resistant bacteria in settings like hospitals where antibiotics are frequently used?

By understanding how these factors change gene frequencies in populations, we can understand how evolution occurs. Brief discussions of the effects of natural selection, population size, mutation, migration, and nonrandom mating follow:

Natural selection, as a cause of evolution, acts on the total phenotype of an organism, not directly on its genes or genotype. In essence, it is a test by the environment of the organism's hereditary phenotype. In sexual reproduction, no two offspring are genetically alike, because of biparental inheritance, crossing-over, independent assortment, and mutations. This can lead to subtle differences in how well some offspring function. Some function better than others under certain environmental conditions and produce more offspring. Because offspring tend to resemble their parents more than other members of a population, the next generation in this sequence will be better adapted to its environment. This selection process repeated over generations increases the population's fitness as certain alleles, and the genotypes in which the alleles occur, accumulate. Identifying natural selection as a directing force was Darwin's most important contribution to our current theory of evolution.

Not all changes in allele frequencies are directed by natural selection. Some are random changes, especially in small populations. Biologists call this concept **genetic drift.** This can be explained using statistical sampling theory. Let's consider a classical example, coin tossing. The probability of getting a heads in a coin toss is one half: likewise for tails. Does this mean that when tossing a coin twice, you will get one heads followed by one tails? It does not. There are equal chances that you will get a tails first followed by a heads, or two heads, or two tails. However, if you were to toss the coin 100 times, there is a better chance that you will get very close to having heads one half of the time and tails one half of the time. The probability of getting close to the theoretical result increases even more if you were to toss the coin 1,000 times. Statistical sampling theory tells us that the bigger the sample size, the more likely we are to get the predicted result. When Hardy and Weinberg said that populations must be large in order for their theorem to apply, they were addressing the sample size concern. However, all populations are not large. Many endangered species consist of only a few breeding individuals. Consequently, the Hardy-Weinberg law would not apply and we would expect to see random changes in gene frequencies in these populations due to statistical sampling error.

Mutation is a never ending process that occurs when random mistakes happen in DNA replication before cell division. Many mutations are spontaneous without a directly observable cause while others can be induced by certain chemicals or radioactive materials. When mutations occur in cells that will become eggs or sperm, the gametes carry the mutation. If an egg or sperm carrying a mutation participates in a fertilization, then the individual thus created has the mutation as part of its genotype even though the parents did not. Modern molecular biology has shown that many mutations are neutral and have no discernable effect on an organism's phenotype. Other mutations may be harmful but, because they are also recessive, most offspring do not express the mutation and only act as carriers. In later generations if two carriers mate, then the mutation can be expressed as a phenotype in any homozygous recessive offspring and natural selection will act on it. Some mutations may be beneficial and improve the organism's phenotype when tested against the forces of natural selection in a particular environment.

Students often ask, Where do alleles come from? The answer is that mutation creates them. As mutation is always happening at a low rate, alleles are always being created during DNA replication.

Other factors causing evolutionary change are migrations into or out of the population, and nonrandom mating. **Migration** can add alleles to a gene pool or it can remove them, depending on the genotypes of those that enter or leave. **Nonrandom mating** occurs when a species practices mate selection or consists of subpopulations that breed only among themselves. You can see from the preceding discussion, populations in nature rarely meet all five conditions for the Hardy-Weinberg equilibrium model. Therefore, they are always evolving.

LAB INSTRUCTIONS

In this lab, you will investigate several aspects of the microevolution model. You will start by playing a game that simulates the impact of natural selection on allele frequency in a population. You will then use a computer simulation to investigate the long-term effects of natural selection and population size on allele frequencies in randomly mating populations. The lab ends with an experiment investigating how lethal mutations can be induced in bacteria.

Simulating Impact of Natural Selection on Phenotypic Frequency

Based on ideas from the late Charlie Drewes

Darwin's concept of evolution is summarized in his own words as "descent with modification through variation and natural selection." Since his time, modern genetics has informed us of the sources of variation, and consensus has been reached on the definition of natural selection and its consequences. Today, biologists define **natural selection** as a process acting on populations living in environments where some individuals because of hereditary variations are less likely to survive and reproduce while others tend to flourish and reproduce. "Nature" is the agent of selection. When this process occurs over many generations, it leads to **adaptation** in future generations with members that are better suited to their native environment than were their ancestors. While some would argue that

Internet Sources

The transformation of competent *E. coli* cells is something that is done routinely in research laboratories. For this reason, many researchers are concerned with optimizing this procedure. Use a browser to search the Internet for information on the optimization of transformation.

In your Internet browser, choose a search engine such as Google, and search on *efficiency of E. coli transformation*. You will get a long list back. Read a few of the articles. What are the authors suggesting to improve the efficiency of transformation?

Learning Biology by Writing

If you did the transformation experiment with the plasmid, you have a well-controlled experiment for a lab report. State the hypotheses that the experiment tests. Indicate the technique used. Give the data obtained and draw conclusions from the data.

Critical Thinking Questions

1. Three critical factors required for genetic engineering techniques are described in this exercise. How was each of these factors met in your transformation experiment?

2. Design a procedure that theoretically would allow you to take the gene for insulin and insert it into the cells of a person with diabetes. What would be the major problems encountered in doing such a procedure?

3. Bacteria containing plasmids die in nature. When they do, bits and pieces of their DNA, including plasmids, are released into the environment. Given your observations in this lab, explain how horizontal gene transfer can occur between species of bacteria.

4. Why is transformation important in the rapid evolution of multidrug-resistant bacteria in settings like hospitals where antibiotics are frequently used?

Lab Topic 14

Modeling Processes in Evolution and Mutagenesis

Supplies

Resource guide available on WWW at www.mhhe.com/labcentral

Equipment

Incubator at 37°C

Ultraviolet light box (see fig. 14.1)

Desktop computers with Hardy-Weinberg simulation program (EVOLVE in BioQuest Collection at http://bioquest.org recommended)

Materials

Printed fabrics from local fabric store representing different environmental habitats (busier is better); about 2 × 6 ft/type. Cut 1-cm-diameter discs from five colors of craft foam with cork borer; 150 discs per color.

Clock or stop watch

Large plastic cups

Sterile
    Petri dishes with Luria's agar or Nutrient Agar
    Automatic pipetters and tips
    0.85% saline, 9.9 ml in capped tubes
    Glass rod bent into "hockey stick"

Culture of *E. coli* K12 in Luria broth

Wax pencils

Alcohol in covered beakers

Alcohol burners

Household bleach

Ultraviolet-shielding safety glasses

Rubber gloves

Student Prelab Preparation

Before doing this lab, you should read the Background material and other sections of the lab topic that have been assigned by your instructor.
    Use your textbook to review the definitions of the following terms:

allele	homozygote
dominant	phenotype
genotype	population
heterozygote	recessive

Describe in your own words the following concepts:

Allele frequency
Gene pool
Genetic drift
Hardy-Weinberg equilibrium
Mutation
Natural selection
Random mating

After finishing the prelab review, write any questions you have about terms, concepts, or techniques in the margins of this lab topic. The lab experiments should help you answer these questions, or you can ask your instructor during the lab.

Objectives

1. To simulate the conditions of the Hardy-Weinberg equilibrium
2. To test hypotheses regarding the effects of selection and genetic drift on gene frequencies in populations
3. To test a hypothesis that lethal mutations can be induced by ultraviolet light

Background

Populations, not individuals, evolve as a result of gradual changes in the relative frequencies of alleles in their gene pools over many generations. These changes result from the combined effects of natural selection, migration, genetic drift, and mutation. How these factors act, singly and in concert, are well understood. For example, selection has been used for centuries in agriculture to increase the quality and productivity of crops and livestock by choosing the best of the current generation to be the parents of future generations, repeating the process over and over again. In this lab topic, you will explore the effects of these agents of evolutionary change to develop an understanding of modern evolutionary thought.
    Models of evolution have been developed since 1858 when Charles Darwin published his version of the theory of evolution. His work was monumental in that it established the role of natural selection in evolution. Darwin worked 50 years before Mendel's genetic discoveries; thus, Darwin was unaware of the basic concepts that you know about genes, inheritance, and DNA. Researchers in

the last 100 years have established the hereditary mechanisms that support Darwin's theory.

Modern biologists view a population of organisms as a **gene pool**, which theoretically is composed of all the copies of every allele in the population at a given moment. In diploid organisms, the genes in the gene pool occur as pairs in each individual, and individuals may be homozygous or heterozygous for the alleles governing a particular trait. The genes found in an individual comprise its **genome**, and it is all of the genomes in a population that comprise the gene pool. Population genetics deals with the frequency (percentage) of alleles and genotypes in a population's gene pool and attempts to quantify the influences of mutation, selection, and other factors causing evolution.

In the early 1900s, when biologists first started to think about the genetics of populations, a common misconception was that a dominant allele would drive a recessive allele out of a population after several generations. People holding this view observed that the recently rediscovered Mendelian genetics indicated that every time two heterozygotes mated, 75% of the offspring expressed the dominant trait. Furthermore, every time a homozygous dominant individual mated with a homozygous recessive individual, all offspring had the dominant trait. They reasoned that the dominant allele, regardless of its phenotype, would, over several generations, become the only allele in the population.

In 1908, two mathematicians, G.H. Hardy and G. Weinberg, independently considered this concept and proved it wrong. They showed that no matter how many generations elapsed, sexual reproduction in itself does not change the frequency of alleles in a gene pool. Changes, if they occur, must be due to the action of other factors—the real causes of evolution.

To understand the insights of Hardy and Weinberg, imagine a hypothetical population of plants that have red flowers and white flowers. Red flowers result from a dominant allele, and white flowers are found only in individuals homozygous for the recessive allele. A field study of 10,000 plants of this species indicated that 84% of the plants had red flowers and 16% had white. Laboratory analysis of the red-flowered plants indicated 36% of the total plants were homozygous for the red allele and 48% of the total plants were heterozygous. Thus, the genotypic frequencies in the population were:

    36% AA where A = red allele, dominant
    48% Aa a = white allele, recessive
    16% aa
    ―――――
    100%

The number of alleles in the population's gene pool can be obtained by a series of simple multiplications from this information:

Every individual with the genotype AA contributes two A alleles to the gene pool and every individual with the genotype Aa contributes one A allele. Given 10,000 individuals, the number of A alleles in the gene pool equals:

$$2 \times 36\% \times 10{,}000 + 48\% \times 10{,}000$$
$$= 12{,}000 \; A \text{ alleles}$$

In the same population, the number of a alleles in the gene pool can be calculated by similar reasoning (aa individuals contribute two a; and Aa, one a). Therefore,

$$2 \times 16\% \times 10{,}000 + 48\% \times 10{,}000$$
$$= 8000 \; a \text{ alleles}$$

On a frequency (= percentage) basis in the gene pool:

$$\left(\frac{12{,}000}{12{,}000 + 8000}\right) \text{ or } 60\% \text{ are } A$$

and

$$\left(\frac{8000}{12{,}000 + 8000}\right) \text{ or } 40\% \text{ are } a.$$

This information describes the gene pool at the time of the study. What happens to allele frequencies when this population reproduces sexually?

The answer to this question can be determined by creating a theoretical model as did both Hardy and Weinberg. They proposed that random mating occurred in a population; i.e., every individual of one sex could mate with any individual of the opposite sex. This means that the alleles of the gene pool can randomly combine with each other. If no selection or mutation occurs in this large, isolated population and every pollen grain has an equal opportunity to contribute sperm to every egg, the frequencies of all genotypes in the next generation can be predicted by common sense reasoning.

To do this, we create a modified Punnett square for the population to show all the possible combinations of gametes in producing the next generation:

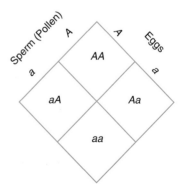

To this calculation, we can add the concept of probability. If 60% of the gene pool is the A allele, then 60% of the eggs and sperm will carry the A allele and 40% will carry the a allele. The *multiplication law of probability* is then used to predict the frequency of each type of fertilization and thus the frequency of genotypes of the next generation. This law states that *the probability (as a decimal equivalent of %) of two independent events occurring*

together is equal to the arithmetic product of their individual relative frequencies. To apply this law to populations, you need only to recognize that egg and sperm production are independent events and that when egg and sperm occur together, fertilization results.

If we now add probability information to the modified Punnett square we see how often specific genotypes result.

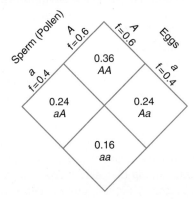

Summarizing the table, we are predicting that the genotypes in the new generation will be 36% *AA*, 48% *Aa*, and 16% *aa*, or 84% red flowers and 16% white. Compared to the parent generation, the genotype and allele frequencies are the same. In fact, the frequencies of alleles and genotypes would remain the same if we repeated these calculations over hundreds of generations. The dominant allele does not drive the recessive allele out of the gene pool as early critics suggested. The two reside in the gene pool in equilibrium, called **Hardy-Weinberg equilibrium**—as long as no selection, mutation, or other agents of evolution are acting on the population.

The hypothetical situation just described can be generalized. If we say that the frequency of the dominant allele is not a specific percentage but **p** and that of the recessive allele is **q**, then in any population then there are only two alleles at one gene locus:

$$p + q = 1 \quad \text{and} \quad p = 1 - q \quad \text{or} \quad q = 1 - p$$

Because of these relationships, we need only measure the frequency of one allele and we can calculate the frequency of the other.

Once the frequency of alleles in a gene pool is known, the frequency of genotypes can be predicted by:

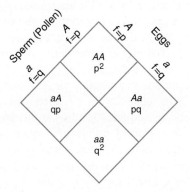

This model indicates that for a population in Hardy-Weinberg equilibrium, the homozygous dominant will be found **p²**% of the time, the heterozygote **2pq**% of the time, and the homozygous recessive **q²**% of the time. This mathematical representation is called the Hardy-Weinberg law. In the earlier hypothetical flower population, the red flowers would have had a frequency of

$$p^2 + 2pq = 84\%$$

and the white flowers of

$$q^2 = 16\%$$

If we had known this relationship then we could have quickly determined allele frequencies without counting. Because q^2 is the frequency of the homozygous recessive which has a distinct phenotype (white), we need only determine the percent of white flowers in the population and express it as a decimal equivalent. Then using the Hardy-Weinberg relationship:

$$q = \sqrt{q^2} = \sqrt{0.16} = 0.40$$

Knowing the value for q we can get the value for p from the relationship

$$p + q = 1$$

which yields

$$p = 1 - 0.40 = 0.60$$

The Hardy-Weinberg equilibrium model is very useful in studies of evolution. It is used as a null hypothesis—a baseline against which to measure populations. The model only applies when the underlying assumptions are true for a population being studied. These are:

No natural selection occurs.

No mutation occurs.

The population is large.

Mating is random.

No migration in or out occurs.

If a population is mating randomly and no other factors change the frequencies of alleles, the frequencies of genotypes in the population should be those predicted by the Hardy-Weinberg equation. If the frequencies are quite different, then this is taken as strong evidence that an agent of change is influencing the population. It is then up to the biologist to seek what is causing the change. This is done by asking such questions as:

Is natural selection occurring against a genotype?

Is mutation occurring?

Is the population small?

Is mating not random?

Are individuals migrating into or out of the population?

By understanding how these factors change gene frequencies in populations, we can understand how evolution occurs. Brief discussions of the effects of natural selection, population size, mutation, migration, and nonrandom mating follow:

Natural selection, as a cause of evolution, acts on the total phenotype of an organism, not directly on its genes or genotype. In essence, it is a test by the environment of the organism's hereditary phenotype. In sexual reproduction, no two offspring are genetically alike, because of biparental inheritance, crossing-over, independent assortment, and mutations. This can lead to subtle differences in how well some offspring function. Some function better than others under certain environmental conditions and produce more offspring. Because offspring tend to resemble their parents more than other members of a population, the next generation in this sequence will be better adapted to its environment. This selection process repeated over generations increases the population's fitness as certain alleles, and the genotypes in which the alleles occur, accumulate. Identifying natural selection as a directing force was Darwin's most important contribution to our current theory of evolution.

Not all changes in allele frequencies are directed by natural selection. Some are random changes, especially in small populations. Biologists call this concept **genetic drift.** This can be explained using statistical sampling theory. Let's consider a classical example, coin tossing. The probability of getting a heads in a coin toss is one half: likewise for tails. Does this mean that when tossing a coin twice, you will get one heads followed by one tails? It does not. There are equal chances that you will get a tails first followed by a heads, or two heads, or two tails. However, if you were to toss the coin 100 times, there is a better chance that you will get very close to having heads one half of the time and tails one half of the time. The probability of getting close to the theoretical result increases even more if you were to toss the coin 1,000 times. Statistical sampling theory tells us that the bigger the sample size, the more likely we are to get the predicted result. When Hardy and Weinberg said that populations must be large in order for their theorem to apply, they were addressing the sample size concern. However, all populations are not large. Many endangered species consist of only a few breeding individuals. Consequently, the Hardy-Weinberg law would not apply and we would expect to see random changes in gene frequencies in these populations due to statistical sampling error.

Mutation is a never ending process that occurs when random mistakes happen in DNA replication before cell division. Many mutations are spontaneous without a directly observable cause while others can be induced by certain chemicals or radioactive materials. When mutations occur in cells that will become eggs or sperm, the gametes carry the mutation. If an egg or sperm carrying a mutation participates in a fertilization, then the individual thus created has the mutation as part of its genotype even though the parents did not. Modern molecular biology has shown that many mutations are neutral and have no discernable effect on an organism's phenotype. Other mutations may be harmful but, because they are also recessive, most offspring do not express the mutation and only act as carriers. In later generations if two carriers mate, then the mutation can be expressed as a phenotype in any homozygous recessive offspring and natural selection will act on it. Some mutations may be beneficial and improve the organism's phenotype when tested against the forces of natural selection in a particular environment.

Students often ask, Where do alleles come from? The answer is that mutation creates them. As mutation is always happening at a low rate, alleles are always being created during DNA replication.

Other factors causing evolutionary change are migrations into or out of the population, and nonrandom mating. **Migration** can add alleles to a gene pool or it can remove them, depending on the genotypes of those that enter or leave. **Nonrandom mating** occurs when a species practices mate selection or consists of subpopulations that breed only among themselves. You can see from the preceding discussion, populations in nature rarely meet all five conditions for the Hardy-Weinberg equilibrium model. Therefore, they are always evolving.

LAB INSTRUCTIONS

In this lab, you will investigate several aspects of the microevolution model. You will start by playing a game that simulates the impact of natural selection on allele frequency in a population. You will then use a computer simulation to investigate the long-term effects of natural selection and population size on allele frequencies in randomly mating populations. The lab ends with an experiment investigating how lethal mutations can be induced in bacteria.

Simulating Impact of Natural Selection on Phenotypic Frequency

Based on ideas from the late Charlie Drewes

Darwin's concept of evolution is summarized in his own words as "descent with modification through variation and natural selection." Since his time, modern genetics has informed us of the sources of variation, and consensus has been reached on the definition of natural selection and its consequences. Today, biologists define **natural selection** as a process acting on populations living in environments where some individuals because of hereditary variations are less likely to survive and reproduce while others tend to flourish and reproduce. "Nature" is the agent of selection. When this process occurs over many generations, it leads to **adaptation** in future generations with members that are better suited to their native environment than were their ancestors. While some would argue that a perfect fit would eventually develop between an organism and its environment, this viewpoint does not recognize that environments are multifaceted and change over time so that what is adaptive today in one location is not adaptive at another location or in the future as environments change. Human impacts, climate change, changing sea levels, continental drift, and extinction of predators and prey are but a few examples of such changes. Consequently, evolution is an ongoing process.

To illustrate Darwin's concept of how natural selection can lead to adaptation, you will do a simulation in which students will act as predators and small foam discs are prey. Prey will be different colors to simulate different hereditary phenotypes. Members of the prey species, regardless of color, will be placed in several different habitats which will be large pieces of cloth printed with different patterns and colors, allowing you to explore if there is a perfect adaptive phenotype. Sometimes the prey will blend with the habitat and be camouflaged, but other times they will stand out. You, as the predator, will have to gather as many discs as you can in 20 seconds from an assigned habitat and place them in a cup to simulate eating them. The simulation will be repeated over three generations to determine if there are any adaptive trends.

In this simulation you are working only with phenotypes. There is no attempt to look at allele frequencies in gene pools or the frequency of genotypes in the hypothetical populations. You need only know that the color of the discs (phenotype) is hereditary, but it is not necessary to know anything about the underlying genetics. In essence, you will look at the populations in your experiments as Darwin did. In the final analysis, natural selection acts on an organism's phenotype, not its genes.

Exploring the Habitat Working in groups of four, two will serve as predators, one will be a timer, and the fourth will be a supervisor to assure rules are followed and data recorded. Read all of the instructions before beginning. Select different teams to work in the different habitats using the same color sets of prey for each habitat. This will allow you to compare whether natural selection can vary by habitat.

1) Each group should choose a habitat for their experiment. Get five bags of different-colored discs and a plastic cup to serve as the predator's stomach.

Record the name of the habitat in which you will be hunting in **table 14.1** and the names of your group members.

Count out 25 disks of each of the five color phenotypes and place them in a large cup, 125 in all. Enter the colors of the phenotypes in column 1 of table 14.1 and the number of discs of each color in column 2. Note that each phenotype has a starting frequency of 20% in the population, but you need not record this.

Have the predators turn away from the habitat as the other group members spread a habitat cloth on the table and scatter the discs randomly, more or less, over the habitat. Each predator should have a cup (stomach) in one hand ready to digest its prey while the other hand is used to feed.

Rules of the Hunt The predators will only have a limited amount of time to hunt (20 seconds). When the "timer" says **GO**, the predators turn toward the habitat and quickly hunt and "devour" prey. A predator may pick up ONLY one disc at a time and must individually place the disc in its stomach (the cup) before picking up another disc. The supervisor should watch that the predators do not pick up more than one disc at a time. The predators should work quickly to capture as many prey as they can in a feeding frenzy. After 20 seconds, the "timer" says **STOP** to end the predation period.

Start the first hunt!

2) ***Analyze Impact of Predation*** After the hunt, both predators empty their cups on the lab table and everyone sorts the captured prey into corresponding colored groups, combining the results of the feeding pair.

Count how many prey of each color were eaten and record in table 14.1, **First Generation**. Return these discs to the original bags.

Subtract the number of each color eaten from the starting number to get the number surviving. Enter the values in column 4. Sum the total number of survivors from all five phenotypes and record the total at the bottom of the column.

Calculate what percent of the total surviving population each phenotype (color) represents and record in the last column. Do this by dividing the number of survivors for each color by the total number of survivors and multiply by 100.

Compare the frequencies of the phenotypes after the first generation of predation with the starting frequencies (each phenotype was 20% of the population to start). What changes do you see? What predictions would you make about the effects of natural selection on different phenotypes in your habitat if the experiment were repeated?

3) ***Impact of Natural Selection over Generations*** The generation that survived the predation period will now mate. Assume that the prey are long-lived and meet all of the requirements for the Hardy-Weinberg equilibrium, except that selection is occurring. Assuming equal numbers of both sexes, each couple produces two offspring, doubling the number of individuals with each phenotype in the population. If the number surviving is an odd not an even

Table 14.1 Data from Natural Selection Simulations

Habitat Name _____ **Group Members** _____

First Generation

Phenotypes	# at Start	# Eaten	# Surviving	% of Total Surviving
			$\Sigma =$	

Second Generation

Phenotypes	# at Start	# Eaten	# Surviving	% of Total Surviving
			$\Sigma =$	

Third Generation

Phenotypes	# at Start	# Eaten	# Surviving	% of Total Surviving
			$\Sigma =$	

$\Sigma =$ Sum of # surviving

number, round up to the next even number. From your stock bags, add the required number discs of each color to a large cup. This new generation plus the survivors dispersed in the habitat represent the *starting population* for the second generation of the simulation. Record the starting numbers for each phenotype in table 14.1, Second Generation.

Mix the new batch of phenotypes so colors are dispersed in the cup. Have the predators turn away from the habitat and disperse the second generation in the habitat to join the survivors from the first generation. It may be a good idea to also change the positions of the survivors to prevent a "wiley" predator from remembering where certain discs had been left. Start the hunt again. Record the new data as the second generation in table 14.1 and perform the necessary calculations.

What trends do you see developing over the two generations?

Do the results support your predictions? _____

4 Repeat the simulation for a third generation, by adding a new generation based on the survivors from the second generation Disperse them, do the hunt, and perform the necessary calculations as you did previously. Record the results in table 14.1, Third Generation.

When you are finished with the third generation, collect all the disks from the habitat and cups, sort by colors, and return to the corresponding bags. Fold the habitat and return materials to the supply area.

Analysis of Natural Selection

5 What overall trends do you see across the generations of your experiment?

Was a particular body-color phenotype an advantage or disadvantage in your habitat? Why?

How would you characterize the color and pattern of your habitat?

If this experiment were continued for several more generations which phenotype would be most adaptive to the conditions of your environment? What do you predict would be the effect on the frequency of phenotypes if the predation cycle went on for hundreds of generations?

Share your results with others in your lab section. Were the same phenotypes adaptive in all habitats? Why do you think this is the case?

As in your simulation, many natural species live in more than one habitat. For example, snakes in Ohio live in habitats with light backgrounds on the limestone islands in Lake Erie but also in habitats with dark soils in the middle of the state. Snakes are prey for gulls, hawks, and crows that hunt by sight from the air. Northern snake populations tend to have light-colored phenotypes, while those from southern populations are dark-colored due to differential predation in the two geographic areas. Because individuals migrate between the two populations, neither has become completely light nor completely dark but maintains a variable phenotypic composition. For the total population of these Ohio snakes, there is not a perfect phenotype that matches their environment. The same is true for many other species that are widely dispersed.

6 If you analyze the experiment just performed by looking at the class data and not just you own, you studied a predator–prey interaction where the prey lived in a number of different habitats. If you redefine the prey population to include all of the lab groups and theoretically repeated the experiment hundreds of times with random migration between habitats, do you think that one phenotype would emerge as "best" or would a number of phenotypes be adaptive for the species a whole?

Hardy-Weinberg Equilibrium Simulation

Based on BioQuest Curriculum Consortium's EVOLVE program

While the simulation of natural selection was instructive for short-term changes in phenotypic frequency due to survival, it did not address several of the ideas underlying the Hardy-Weinberg equilibrium. The activity was more of a Darwinian approach to thinking about evolutionary change and did not incorporate a genetic interpretation. In this section, you will use EVOLVE, a computer program that allows you to change reproductive and mortality rates, population size, and migration rates into and out of the population while tracking frequencies of alleles in the gene pool and genotypes in the population. The program is based on the Hardy-Weinberg equilibrium principle and essentially performs the calculations outlined in the Background section of this lab topic. What makes it very useful is that it repeats these calculations over 200 generations for the conditions you specify, and the results are displayed as graphs, so trends are easily observed.

The program models a hypothetical sexually reproducing species having a single genetic locus where only two alleles can occur. The alleles are represented as **A** and **B** and not by the usual notation where alleles at the same locus would be indicated A, a, and maybe *a*. This is not a problem as long as you remember **A** and **B** are alternative forms of one gene at one locus. The genotypes in a population using this notation would be written as **AA** (homozygous), **AB** (heterozygous), and **BB** (homozygous). You are never told what the phenotypes are other than if A is dominant then both the homozygous dominant and the heterozygous genotypes will exhibit the trait and the homozygous recessive will have the recessive phenotype. So, unlike the natural selection activity, this computer simulation is about allele and genotype frequencies and how they do or do not change under the conditions you set up.

Verifying the Hardy-Weinberg Equilibrium Model

The first set of conditions to explore are the underlying assumptions of the Hardy-Weinberg equilibrium. Recall that a population in equilibrium will

1. Be large
2. Mate randomly
3. Experience no migration of individuals in or out
4. Experience no mutation
5. Experience no selection

7 Open the EVOLVE program by clicking on its icon. ~~Double-click on Default Problem~~ to open the *Parameters* window. To set up a population meeting the Hardy-Weinberg assumptions, enter the words and numbers below that are in italics into the appropriate boxes on the *Parameters* screen.

Title: *Equilibrium Conditions*
Generations: *200*
Genotype Number Total = *8,000*
Allele Frequencies: A = *0.5*, B = *0.5*
Genetic Drift: Max Pop: *8,000* and Post-Crash Pop *8,000*
Natural Selection—make no changes
Gene Flow—make no changes

You have told the program that you want a randomly mating population of a steady 8,000 individuals reproducing over 200 generations with the starting allele frequencies in the gene pool being 50% A and 50% B, with no selection, no migration, and no mutation. The program has used your entries to adjust the frequency of each genotype in the starting generation using the relationship

p = freq of allele A = 0.5
q = freq of allele B = 0.5
p^2 = freq of AA. What is the frequency (not number) of AA? _____
$2pq$ = freq of AB. What is the frequency (not number) of AB? _____
q^2 = freq of BB. What is the frequency (not number) of BB? _____

Before running the simulation, use your understanding of the Hardy-Weinberg equilibrium to predict whether allele and genotype frequencies will change significantly under these conditions. Write your prediction below.

Now click on *Update* and *Done* at the bottom right of the window. In the new window that appears, click on the *Freq* tab on the lower-right side. A row of numbers appears that tells you the starting conditions. They are:

- *FreqA* = the frequency of the A allele in the gene pool of 8,000
- *FreqB* = the frequency of the B allele in the gene pool of 8,000
- *FreqAA* = the frequency of the genotype AA in the population of 8,000
- *FreqAB* = the frequency of the genotype AB in the population of 8,000
- *FreqBB* = the frequency of the genotype BB in the population of 8,000

Now click on the tab for *Graph Controller* at the upper left. This is going to let you specify what you want to graph. In the menu that pops out, click on *Freq* and a new list appears. Put a check in front of the first five (frequencies of A, B, AA, AB, BB). Also check *Grid*. You have specified that you want to see a graphical output over 200 generations for the frequencies of the A and B alleles in the gene pool, and for the frequencies of the AA, AB, and BB genotypes in the population.

When done, click on *Go* at the lower right of the main window. The EVOLVE program now takes the starting population and calculates what the allele and genotypic frequencies will be in the second generation. Using the frequencies from each generation, the calculations are repeated for 200 generations. The results are plotted and the actual numbers recorded to the table below the graph.

Sketch the graphical output from this simulation below, labeling each line with A, B, AA, AB, or BB.

Describe your results for the frequency of A and B in the gene pool. Do the frequencies of these alleles change significantly over 200 generations or do they remain similar? Confirm by scanning the number table. What is your conclusion?

Describe the results for the frequency of the genotypes in the population. Do they change significantly over 200 generations or do they remain more or less the same?

To check on the repeatability of the results, click on *New Experiment* at the bottom of the window. The *Parameters* window reappears. Click *Done* and you will return to the graph window where your first graph is now shown in gray. Click *Go* and a second set of graphs appears in color on top of the first experiment in grey. Did you get essentially the same results as the first time?

From this brief sampling, what would you say happens to allele frequencies in the gene pool and genotypic frequencies in the population when a population is large, randomly mating, and experiencing no migration, mutation, or selection?

Quit

Reopen

Now that we have established a baseline, we can start to change the starting conditions systematically to determine the effects of the changes on allele and genotypic frequencies.

8 *Effect of Strong Selection for a Dominant Allele in a Large Population* You will first explore the effect of natural selection on allele and genotypic frequencies. The program will select for (favor) the dominant allele that is expressed both in homozygous dominant (AA) and heterozygous (AB) genotypes. Selection for a dominant allele is really the same as selection against a recessive allele, which is only expressed as a phenotype in the homozygous recessive.

Click on *File* at the top of the screen. Double-click on **Strong Selection for Dominant Allele** to open the *Parameters* window. Enter the values given in italics in the appropriate boxes:

Title: *Selection for Dominant Allele, Large Population*

Generations: *200*

Genotype Number: Total = *8,000*

Allele Frequencies: A = *0.05*, B = *0.95*; note that B is recessive and set high to start

Genetic Drift: Max pop: *8,000* and Post-Crash Pop *8,000*

Look at, but do not change the natural selection values. Note that the genotypes having the dominant phenotype tend to have a better survival rate and produce more offspring in each generation compared to the recessive genotype.

Predict what you think will happen to (1) the allele frequencies in the gene pool and (2) to the genotype frequencies in the population over 200 generations under these conditions. Write your predictions below.

Click on *Update, Done,* and open the *Graph Controller* at the upper left. Select frequency and check A, B, AA, AB and BB. Now click. *Go* to run the simulation under the specified conditions. Sketch the graphical output from this simulation below, labeling all lines. How and why are these results different from those in your first simulations?

Go to Graph Controller and select freq and check freq A, B, AB

From the graph, it looks as if the recessive allele may drop to zero in the gene pool. Scan down the number table to see if this is true. If not, why does the recessive allele persist in the population after 200 generations of selection against it?

This explains why recessive alleles with negative effects persist in populations. Because of heterozygous carriers, there is a probability that homozygous recessive offspring will be produced. The frequency of many human fatal hereditary diseases that appear in newborns can be explained in this way.

9 After you run the initial simulation, you might consider changing the fraction surviving or the number of offspring produced to see how the results are affected. Click on *New Experiment* to return to the *Parameter* window and make your changes. Predict how these new values will change the results obtained before. Write your changed starting conditions add prediction below.

Do the results confirm or contradict your prediction? Why?

Effect of Small Population Size All computer simulations so far have been run with large population sizes, one of the assumptions underlying the Hardy-Weinberg equilibrium. What would happen if the population was small?

10 To investigate the effect of small population size, go to *File* at the top of the screen and choose *New Problem* from the drop-down menu. Double-click on *Genetic Drift, Pop 80–100.*

Set the following parameters as shown in italics:

Title: *Small Population*

Generations: *200*

Genotype Number of Total: *80*

Allele Frequencies: A = *0.5*, B = *0.5*

Genetic Drift: Max Pop: *100* and Post-Crash Pop *80*

Natural selection and gene flow are blanked out and are set to have no influence on the outcomes.

The specifications that you have set for the model are the same as those used in your first experiment, except the population is 100 times smaller. Comparison of the results from the first simulation with the results from this one will allow you to see how population size alone affects allele frequencies in gene pools and genotypic frequencies in populations.

What do you predict will be the outcomes? The same as your first experiment or different in some way? How?

Click on *Update, Done,* and *Go* to run the simulation.

Repeat the simulation by clicking *New Experiment, Done,* and *Go* and then repeat the sequence to display a third set of results.

Did you get the same result for each trial or were they different? Think about statistical sampling theory to explain your results. An analogy might help. If you flip a coin twice, will you get H 50% of the time and T the other 50%? If you flip a coin 2,000 times, will you get very close to H 50% of the time and T 50%? How does this idea apply to randomly mating populations?

Change the parameters according to your experimental design and record them above, next to the parameter names.

Run the simulation and sketch the results, labeling all lines and axes.

11 *Testing Combined Effect of Changes in Assumptions* For this simulation, go to *File* at the top of the screen and choose *New Problem* from the drop-down menu. Double-click on *Gene Flow* and *Selection*.

In the *Parameters* window that appears, note that you can change:

Number of generations	_____
Allele frequency	_____
Population size	_____
Survival rates	_____
Reproductive rates	_____
Adults entering population	_____
Adults leaving population	_____

Discuss the possible effects of each with your lab partner and devise an experiment that will allow you to determine the combined effects of migration and natural selection on a small population.

State the question you hope to answer in this simulation.

Did the simulation answer your question? What did you learn from it?

12 Having tested the underlying assumptions of the Hardy-Weinberg equilibrium, can you now see how a null model that describes the conditions under which evolution does not occur can be used to explore how evolution does, in fact, occur? _____

Inducing Mutations

In nature, mutation is the process that creates new alleles. **Mutations** are mistakes that occur during DNA replication before cell division, or abnormalities that develop in chromosomes during cell division. Whatever the source, mutations contribute to genetic variability in populations. However, their effect is usually much less than that of genetic recombination that occurs as a result of crossing-over, independent assortment, and biparental inheritance. Mutations can be neutral, having no effect, detrimental, or beneficial. Mutations often create recessive alleles but some may be dominant, such as achondroplasia dwarfism in humans. When recessive, the trait should not be expressed until it appears in a homozygous individual, a process that can take several generations.

Natural selection is the agent in nature that determines whether a mutation is "good" or "bad." A detrimental mutation (allele) would be selected against; *i.e.*, organisms with the mutation would not function as well and would produce fewer offspring. Over time, the frequency of genotypes carrying a detrimental mutation should decrease in the population, but as you saw in the computer simulation will not disappear. The opposite would be true for a beneficial one.

In this part of the lab topic, you will experimentally induce mutations in a bacterium, *E. coli*. Bacteria are good organisms to use in mutation studies because they are haploid. If a mutation occurs, it is expressed because there is not a second allele present to mask it. You will study mutations affecting viability. Such mutations are called lethal mutations because the mutant gene fails to produce a needed product and the cell dies.

Mutations can be induced by several means. Chemicals called mutagens can change an organism's DNA, causing changes in hereditary information. Ultraviolet radiation has similar effects and will be used in your experiment. Because mutations caused by UV exposures are random, many different mutations will be induced.

Create a hypothesis to test in an experiment that relates the length of exposure to UV light to the amount of mutation expected.

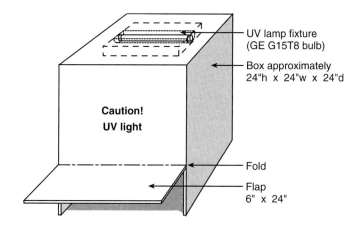

Figure 14.1 Ultraviolet light irradiation box. Insert petri dish into box and remove lid for appropriate time (table 14.2). Replace lid and remove from box. When working at the box, protect your eyes by wearing glasses made of ultraviolet-absorbing glass. Protect your skin by wearing a long-sleeved shirt and rubber gloves.

CAUTION

Never look at a UV lamp, because it can damage cells in your cornea (surface of eye).

The lamp should be in a box, and anyone near the box should wear safety glasses that will filter out ultraviolet light. Anyone reaching into the box should wear a rubber glove and a long-sleeved shirt to protect their skin from the UV light. Why?

13▶ Procedure About 12 hours before the lab, *E. coli* was transferred from a stock culture into 100 ml of nutrient broth and cultured at room temperature. You or your instructor should take this culture, and make a sterile, serial dilution.

Add 0.1 ml of the culture to 9.9 ml of 0.85% sterile saline. Cap and shake the container well.

Take 0.1 ml of this dilution and add to a second 9.9 ml of sterile saline; cap and shake it well.

Take 0.1 ml of this second dilution and add it to a third 9.9 ml of sterile saline. Shake it well.

The cell suspension in the third tube represents a millionfold dilution of the original culture. The first and second dilutions will not be used. Add 10% bleach to them and let sit before disposing.

About 30 minutes before you are going to use it, a germicidal UV lamp enclosed in a box should be turned on and allowed to stabilize (**fig. 14.1**).

14▶ Take seven petri plates containing a growth agar and number the bottoms 1 through 7. Write your initials on each one.

Review the Interchapter, "An Outline of Sterile Technique" on pages 143 to 144 for making a sterile transfer.

Shake the third dilution tube well to distribute the cells. Refer to **table 14.2** to determine the volume of the bacterial dilution to be added to each plate. Note that the plates to be irradiated longer receive a larger volume of cells to compensate for the lethal effects of UV exposure. Use a sterile micropipetter to add the required amount. Then use a sterile inoculation loop or sterile bent glass rod ("hockey stick") to spread the culture evenly across the surface of the agar (**fig. 14.2**). The "hockey stick" can be sterilized between the times you spread bacteria on each plate by putting it in a beaker of 70% ethanol. Before using it again pass it quickly through a flame to burn off the ethanol. Let the rod cool for 15 to 20 seconds before spreading the bacteria on a new plate. Alternatively, use a fresh hockey stick for each plate.

Table 14.2 Results from UV Irradiation of Bacteria

Plate	Sample Vol. (μl)	Cumulative UV Exposure (sec)	Total No. Colonies	Corrected No. Colonies	% Surviving
1	25	0 (control)	___ × 4*	___	___
2	50	10	___ × 2*	___	___
3	50	20	___ × 2*	___	___
4	100	30	___	___	___
5	100	40	___	___	___
6	100	60	___	___	___
7	200	80	___ × 0.5*	___	___

*Note: Number of colonies should be adjusted to compensate for differences in sample size plated.

Plate 1 is a control and nothing more should be done to it. Plates 2 through 7 will be irradiated with increasing amounts of UV light.

15 Refer to table 14.2 and determine the exposure time for each plate. Take one plate at a time and slide it into the UV light box. Reach in with a gloved hand and *remove the top* for the time indicated. This is done because glass and many plastics absorb UV light and little UV would reach the bacteria if the lids were left on during the irradiation period. Replace the top and remove the plate. Repeat the procedure for the other plates for the appropriate times.

When all plates have been irradiated they should be incubated overnight at 37°C. When colonies are clearly visible, the plates should be placed in a refrigerator and colonies counted in the next lab meeting.

16 *Analysis* When you examine the plates, count the total number of colonies on each plate. A colony is a clone of a single cell that was spread on the plate. The number of colonies represents the number of viable cells in each sample you spread on the agar surface. Record the results in table 14.2. Note that the plates receiving short UV exposures received lesser volumes of the cell suspension, and the numbers need to be corrected by multiplying the colony counts times a volume correction factor. This calculation makes all counts for all plates directly comparable. Enter the corrected number of colonies for each treatment in table 14.2.

Calculate the percent surviving by dividing total colony count from plate #1 into colony counts for all other plates and multiplying by 100. On the panel of graph paper at the end of the exercise, plot the percent surviving (from total counts) as a function of irradiation time.

Why do you think there are fewer total colonies formed in those samples that were irradiated longer?

Return to the hypothesis you made and come to a conclusion about accepting or rejecting your hypothesis.

Lab Summary

17 On a separate sheet of paper, answer the following questions assigned by your instructor.

1. Define natural selection. Describe what happens to the frequency of a dominant allele in a population when selection is against the dominant phenotype?

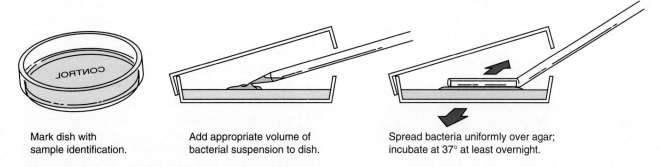

Mark dish with sample identification.

Add appropriate volume of bacterial suspension to dish.

Spread bacteria uniformly over agar; incubate at 37° at least overnight.

Figure 14.2 Inoculation of petri plates. Label the bottom of the petri dish with a marker. Open the dish from one side and add culture liquid containing bacteria. Use a sterile glass rod bent like a hockey stick to spread bacteria over agar surface. Turn plate 90° and repeat to ensure even spreading.

2. What happens to the frequency of a recessive allele when selection is against the recessive phenotype? Explain why this result is different from that in question 1.

3. How can you determine the frequency of a recessive allele in a population? Of a dominant allele at the same locus, assuming there are only two alleles at the locus in a population?

4. PKU is a hereditary disease in humans. It is caused by a recessive allele whose expression in a homozygous recessive results in an inability to metabolize the amino acid phenylalanine. About 1 in 10,000 people are homozygous recessives. Use the Hardy-Weinberg equations to determine the frequency of the allele in the human gene pool. Also determine the frequency of the dominant gene.

5. When natural selection causes changes in allele frequencies in a gene pool, biologists say adaptation is occurring. Why are not all species totally adapted to their environments?

6. Discuss the results of the experiments investigating the impact of natural selection on phenotypes found in populations. What summarizing principle can be drawn from these experiments?

7. What are the five underlying assumptions of the Hardy-Weinberg Equilibrium?

8. What did your computer simulations reveal when selecion is against the dominant allele? The recessive?

9. What is genetic drift, and why is the concept important in understanding evolution? Based on your computer simulation work, give comparative examples of what happens to allele frequencies over many generations in small populations compared to large ones.

10. Plot the data from table 14.2 as directed in the **Analysis** section. What is it about these plots that suggests ultraviolet light is mutagenic? Explain.

11. Hand in your graphs of the effects of UV irradiation on bacterial survival.

Internet Sources

Use Google to find information on the topic of human mutation databases. Search one or more of the databases for mutations that are dominant. List three such mutations, describing the phenotype and its frequency in humans. Repeat the search for three, recessive mutations and provide the same information. If the effects of the mutations are detrimental, why do they persist in the population?

Learning Biology by Writing

In this lab you did two experiments. One was a computer simulation to test hypotheses related to selection and population size, and the other was a classical "wet lab" experiment demonstrating how mutations can be induced with UV light. Although related, these two experiments are not directly comparable. Consequently, your instructor may ask you to write a lab report on only one of the activities.

For either lab report, follow the directions in appendix D. When biologists sit down to write a scientific paper, they often start by writing an abstract of the work they are reporting. This helps them organize their thoughts. You might try to do the same.

Critical Thinking Questions

1. Huntington disease is a fatal nervous system disorder causing degeneration of the brain. It is caused by a defective gene on chromosome #4. The homozygous condition is lethal to the fetus. This is an autosomal dominant trait affecting 1 in 20,000 people who would all be heterozygous. However, it is a delayed action gene that is not expressed until the affected individual is in his or her late 30s or 40s. Using the Hardy-Weinberg equation, determine (1) the frequency of carriers in the population, (2) the frequency of fetuses affected by the lethal homozygosity, and (3) the frequency of unaffected individuals.

2. In the Lake Maracaibo region of Venezuela, there is a family of about 3,000 people who are descendants of a German sailor who had Huntington disease. Would sampling this population reflect the whole population of Venezuela? What mechanism of evolution is occurring? Explain.

3. Present prenatal screening tests (amniocentesis and chorionic villi sampling) allow parents to screen babies for possible genetic defects. As well, reproductive technologies using surrogacy and sperm donation are becoming more common with couples seeking sperm donors or seeking surrogate mothers with "desirable" traits such as high IQ, tallness, and so on. Speculate on the implications these procedures have on the population as a whole with regard to evolution.

4. Natural selection can be viewed as the sum total effects of the environment on the organism and includes both abiotic (physical) and biotic (other organisms) components. Does natural selection have its effects by acting directly on the genes, genotypes, or phenotypes? Explain your answer.

5. Explain how genetic drift, and natural selection might lead to allopatric speciation if a single population was divided into two isolated populations by a natural event.

6. Ultraviolet rays in sunlight are responsible for tanning of human skin. Many see this as a healthy activity. Are you aware of any evidence that is contrary to this opinion?

7. Ultraviolet lamps are sometimes called germicidal lamps. Why?

8. You are given a culture of bacteria that grow well on an agar containing glucose, but will not grow on one containing lactose. If you split the culture in half and exposed one-half to UV light, but not the other, and plated both types of bacteria on agar with glucose or agar with lactose, four plates in all, and saw growth of the UV-treated bacteria on the lactose agar, how would you interpret the results?

Lab Topic 15

INVESTIGATING BACTERIAL DIVERSITY

SUPPLIES

Procedural guide available on WWW at www.mhhe.com/labcentral

Equipment

Compound microscopes
Dissecting microscopes
Oil immersion objectives and oil (optional)
Incubators or water baths at 37°C and 42°C
pH meter or pH tape
Balance, 0.1g sensitivity

Materials

Alcohol lamps
Grease pencil
Bacterial loops
Spring-type clothes pins
Sterile pipettes and suction bulbs
Microscope slides and coverslips
Whole milk
Living cultures
    Bacillus megaterium or *Bacillus subtilis*
    Pseudomonas fluorescens
Mixed cyanobacteria cultures containing any of:
    Anabaena
    Gloecapsa
    Merismopodia
    Oscillatoria
Soil samples
Prepared slides for demonstration
    Composite of bacterial types (cocci, bacilli, and spirilla)
Autoclave bag for disposal of cultures

Solutions

Gram stain kit with Gram iodine, crystal violet, safranin, and counterstain
Nutrient agar made with 0% and 6% NaCl in petri dishes
Sterile water
Sterile 0.85% saline, packaged 9.9 ml per capped tube
India ink
70% ethanol
95% ethanol
Test substances, such as antibiotics, metal salts, or pesticides; filter disks, 1 cm diameter

STUDENT PRELAB PREPARATION

Before doing this lab, read the Background material and sections of the lab topic assigned by your instructor.

Use your textbook to review the definitions of the following terms:

bacillus	domain
bacteria	eukaryote
coccus	Gram stain
colony	prokaryote
cyanobacteria	spirillum

Describe in your own words the following concepts:

Agar and growth media
Bacterial diversity
Differential growth
Sterile technique (read Interchapter on pages 143 to 144)

After finishing the prelab review, write any questions you have about terms, concepts, or techniques in the margins of this lab topic. The lab experiments should help you answer these questions, or you can ask your instructor during the lab.

OBJECTIVES

1. To recognize the diversity of prokaryotic cell types among bacteria
2. To perform a Gram stain on unknowns
3. To observe differential growth in bacteria extracted from soils
4. To demonstrate that several natural foods are produced and preserved by lactic acid fermentation
5. To measure bacterial contamination in milk samples, by using sterile technique and serial dilution

Background

About half the earth's biomass is composed of prokaryotic cells, living in just about every conceivable environment. Fossil remains date back 3.5 million years and indicate that these were the first type of cells on earth. From these ancient prokaryotic stocks evolved the bacteria, archaeans, and eukaryotes living today. When photosynthesis evolved among early prokaryotes, the earth's environment profoundly changed as oxygen appeared for the first time in our atmosphere. This lab will focus on bacteria as representative of the prokaryotes. Although Archaea are also prokaryotes, they are not as easy to work with and are not included here. Subsequent chapters in this lab manual will take a detailed look at the eukaryotes.

Bacterial diversity is staggering. Although only 7,500 species have been described, the total number of different types might be 1,000-fold more. Studying them is difficult because they are small, hard to see, and we usually do not know how to culture them for further study. To solve these problems, microbiologists are applying gene-sequencing technology to study bacterial diversity in soils and water, describing new species by nucleotide sequences rather than by how they look or behave. Hundreds of new species are now described each year, most of which have never been cultured or investigated in any detail. We know these organisms only as distinctive DNA sequences with little, if any, knowledge of their physiology or ecological roles. The gene-sequencing data suggest that 30 or so phyla exist.

All bacteria cells share a number of features. Being prokaryotes, they have no nuclei and lack membranous cytoplasmic organelles, such as mitochondria, endoplasmic reticulum, and vacuoles. Bacterial genomic DNA is a circular molecule (see fig. 13.1) and not a linear one as is found in the chromosomes of eukaryotes. Bacteria are haploid and normally have only a single circular DNA molecule per cell. When mutations occur they are immediately expressed. Some bacteria contain small auxilliary molecules of DNA called **plasmids** (see fig. 13.2).

The structurally simple prokaryotic cell is surrounded by a cell wall composed of **peptidoglycans,** a type of polymer consisting of sugar polymers cross-linked by small peptides. No other organisms have cell walls composed of these substances. The cell wall prevents the cell from osmotically bursting and protects it from attacks by viruses and predatory bacteria. Some bacteria have two cell membranes, one located beneath the cell wall next to the cytoplasm and another, called the outer envelope, outside the cell wall. Frequently, a mucilaginous **capsule** surrounds the cells. Composed of polysaccharides and proteins, it protects the bacterial cell from drying, binds bacterial cells into aggregates, and may protect the cells from attack by other cells. Dental plaque represents capsule material that binds bacteria to our teeth. Often bacterial cell membranes are folded inward, creating internal structures in the cytoplasm. In cyanobacteria and other photosynthetic bacteria, these folded membranes contain chlorophyll necessary to capture light energy. In one group of bacteria, the folded membranes contain a magnetite crystal of iron, allowing the bacteria to orient in the earth's magnetic field. Many bacteria are motile because they contain gas vesicles that can fill or empty, allowing them to float or sink in aquatic environments. Other bacteria have **flagella** that rotate like a propeller, allowing them to swim. Some bacteria have **pili,** threadlike surface extensions, that allow them to glide over surfaces.

Bacteria reproduce asexually by **binary fission.** This involves a doubling of cellular components including DNA and then dividing in half. The daughter cells may separate into unicells or remain attached to form filaments or grape-like clusters. Under good conditions, growth and division might take less than an hour, creating the potential for explosive population growth. Under adverse conditions many bacteria produce spores, called **endospores.** These contain a quiescent cell that can live for years in suspended animation. When conditions improve, the endospore germinates to start a new population. Endospores recovered from fossil bees encased in amber thought to be 25 million years old grew in the lab!

Bacteria do not sexually reproduce, but genetic materials are exchanged between individuals by three methods not directly coordinated with reproduction. Some bacteria **conjugate** with others of their own species by forming cytoplasmic tubes that link two cells together so that DNA can be exchanged, allowing recombination. In a second type of exchange called **transformation,** bacteria can absorb genetic material released into the surrounding environment by the death and fragmentation of other bacteria. Because this is not species specific, bacteria can incorporate genes from other species into their own DNA. It is this ability to pick up DNA from the environment and to incorporate genes that makes bacteria excellent tools in biotechnology. You studied how this occurs in lab topic 13. The third type of genetic exchange that occurs is **transduction.** It happens when a virus invades one bacterium and replicates, picking up some of the host's DNA. When the host cell dies, the virus is released and may enter another cell, carrying genes from one bacterium to the next. Transformation and transduction are examples of how **horizontal gene transfer** occurs between species. It is estimated that 17% of the genes found in *E. coli,* (a common bacterium in vertebrate intestines) came from other bacteria by horizontal gene transfer.

As a group, bacteria are more metabolically diverse than all other types of organisms. Heterotrophic bacteria consume preexisting organic molecules to gain energy to fuel cellular processes. Some heterotrophic bacteria are obligate aerobes that must have oxygen to live. Others are obligate anaerobes and will die if exposed to oxygen. Still others will use oxygen when it is present, but switch to anaerobic metabolism when it is not. What is truly amazing, however, is their metabolic flexibility as a group. There are probably no naturally occurring or synthetic organic compounds that bacteria cannot metabolize. The U.S. Environmental Protection Agency funds research on the use of bacteria in **bioremediation,** the restoration of contaminated sites to near natural conditions by the action of living organisms. Their hope is that someday we will identify a mixture of bacteria that could be applied to an oil spill site or an old industrial dump containing toxic organic compounds. The bacteria would degrade the toxin to harmless carbon dioxide, water, and maybe ammonia, rendering the site suitable for other "clean" uses. As we discover new species of bacteria and understand their metabolic capabilities, we take a step toward achieving such a goal.

Autotrophic bacteria are able to make organic compounds using sunlight or inorganic chemicals as energy sources. The photosynthetic cyanobacteria (= blue-green algae) are important producer organisms. They probably contribute more oxygen to the atmosphere worldwide than do rainforests, not to mention their carbohydrate contribution to food chains. Some bacteria are chemoautotrophic and can derive energy from oxidation of inorganic molecules such as NH_3, H_2S, H_2, and S to make carbohydrates from carbon dioxide. Purple bacteria use CO_2 and H_2S to make carbohydrates with sulfur as a by-product.

Our lives are intertwined with those of bacteria. By far, most bacteria are beneficial as important members of ecosystems where they function as decomposers, photosynthetic producers, and important basal components of food chains. Without bacteria (and fungi) to recycle carbon from dead organic matter as carbon dioxide, plant photosynthesis would be greatly reduced. Of course, as they decompose materials, bacteria grow and reproduce becoming the basis for many food chains. Their decomposition capabilities also recycle a number of other materials such as nitrogen, phosphorus, and sulfur. Their ecological roles cannot be overstated. Symbiotic bacteria, living in the roots of plants classified as legumes, convert atmospheric nitrogen into ammonia. The plants provide carbohydrates to the bacteria and the bacteria provide the plant with ammonia that is used to make plant amino acids and nucleotides. Without nitrogen fixing bacteria, plant life and consequently fungal and animal life would be drastically limited by the availability of organic nitrogen. In the intestines of vertebrates, symbiotic bacteria make vitamins that the host cannot. About half of our antibiotics (*e.g.,* streptomycin, neomycin, and tetracycline) come from bacterial sources. In nature, the soil bacteria producing these use them to fend off competitors.

The bacteria that attract most of our attention are disease-causing organisms. Although the list of such pathogenic bacteria is small relative to the total number, their impact can be significant. The bubonic plague of the 1300s was caused by bacteria. Cholera, Lyme disease, syphilis, and tuberculosis are other examples of diseases caused by parasitic bacteria. Some bacteria produce toxins. Botulism, a type of food poisoning, occurs when the bacterium *Clostridium botulinum* grows anaerobically on improperly canned foods. As little as 1 microgram (one millionth of a gram) of the exotoxin produced by this organism can kill a human.

Because bacterial cells are very small they are difficult to observe even with the finest microscopes. Before DNA-sequencing technology became common, a number of factors were taken into consideration when trying to identify bacteria. These include: colony characteristics, cell shape, staining characteristics of cells, conditions required for growth, and ecological niche. In this lab, you will explore some of these factors.

LAB INSTRUCTIONS

In this lab you will explore the use of several techniques that are typically used by microbiologists to study bacteria.

Finding Bacteria

Agar plates are used to grow bacteria in the lab and are especially useful in studying colonies of bacteria. The overlapping lid of a petri plate makes a loose seal with the base, preventing airborne bacteria from entering while allowing carbon dioxide and oxygen to pass. The **agar** in the plate is a thickening agent made from a polysaccharide extracted from the cell walls of marine red algae. When added to hot water at a concentration of 1% to 2%, the multiple hydroxyl groups on the surface of the agar molecules hydrogen bond to and immobilize the water molecules, resulting in the formation of a gel. In the lab, if nutrients such as sugars, amino acids, and vitamins are first dissolved in the water, a nutrient agar is produced. At temperatures around 85°C, nutrient agars are liquids that can be poured into tubes or petri dishes. As they cool to around 32°C to 40°C, the agars solidify, allowing a person to easily move culture dishes without spilling the contents. When bacteria are introduced onto the surface of a sterile agar containing the required nutrients, the bacteria will grow and divide every few hours. Usually after about a day, a single cell will produce a clump of cells called a **colony** on the agar surface. All cells arising from a single cell represent a **clone** of the original cell.

1 ▸ *Making a Culture* Your instructor will provide you with a sterile petri plate containing a nutrient agar and some sterile cotton swabs. Dampen a swab with one drop of sterile water and use it to wipe a surface that you think is "clean." For example, a desk top, a soda machine, or a pencil you use. You could coordinate with another group, combining your plates into an experiment. Swab one person's hands and use the swab to inoculate a plate. Have the person wash his/her hands and take a second swab for inoculating a new plate.

After collecting the sample, open the dish by lifting one edge of the lid (**fig. 15.1**). Do not remove the lid

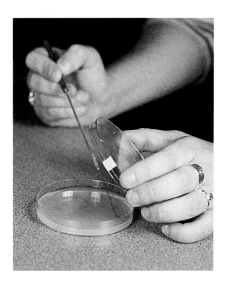

Figure 15.1 When opening petri dish, do it by raising one edge so that airbourne bacteria do not fall on plate. Do not allow hands to pass over plate as falling bacteria will contaminate culture. Control breathing so you do not exhale into an open plate.

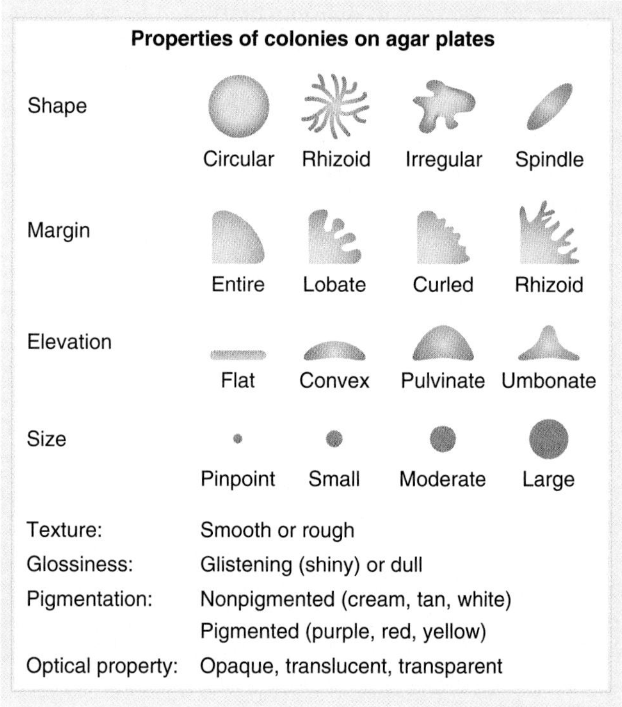

Figure 15.2 Descriptive terms used to characterize bacterial colonies.

entirely. Rub the swab in large arcs over the entire surface of the nutrient agar. Incubate the plates overnight and look at them during the next lab.

2 ▸ Analysis Do not open the plates. Look at the several types of colonies growing on the plates with your naked eye and dissecting microscope. Large circular colonies that appear to be "fuzzy" along the edges are not bacterial colonies. They are fungal hyphae that have grown from spores and should not be included in your analysis. Describe the bacterial colonies that you see according to the characteristics given in **figure 15.2**. Record your observations for at least five colony types in **table 15.1**.

Dispose of your plates as directed by your instructor. They should be flooded with bleach or autoclaved to kill the bacteria.

Bacterial Morphology

In addition to colony morphology, the shape of bacterial cells is used as a key characteristic in identifying bacteria.

Three basic cell shapes occur (**fig 15.3**). Depending on the species, bacterial cells are **cocci,** small spheres, while others are rod-shaped **bacilli** or helix-shaped **spirilla** or **spirochaetes.** Most bacteria are **unicellular,** although some species typically aggregate into **colonies.** These may be transient and break apart or may be more or less permanent and characteristic for a species.

3 ▸ Cell Shapes Obtain a prepared microscope slide of mixed bacterial cell types. Look at it first with your 10× objective, switching to 40× for a better view. If your microscope is equipped with an oil immersion objective, review the directions in Lab Topic 2 for using this. Because such objectives provide magnifications up to 1000×, you should be able to see the cell shapes more clearly. However, because such objectives are also expensive, your instructor may have this as a demonstration slide. Sketch the basic shapes of bacteria below. Use your ocular

Table 15.1	Characteristics of Bacterial Colonies.						
Sample taken from _____							
Colony #	Color	Shape	Margin	Elevation	Texture	Appearance	Optical Property
1							
2							
3							
4							
5							

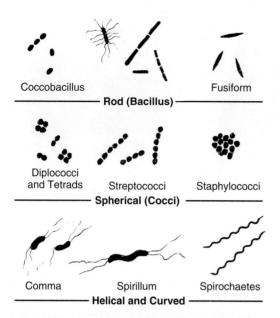

Figure 15.3 Shapes of cells are used in first step of identifying bacteria. Within the three basic shapes (rods-spheres-helices), there may be unicellular or colonial forms.

micrometer to estimate each size of a cell and add the dimensions to your drawings.

Cell shape and whether the cells occur singly or in colonies are two characteristics that are used to narrow down the possibilities when identifying bacterial species.

Cyanobacteria Cyanobacteria are easy to distinguish from other bacteria because they contain chlorophyll and are greenish in color. Cell shapes vary from those you have observed and cells are frequently aggregated into filaments or colonies (**fig. 15.4**). Like many bacteria, they produce spores that are resistant to drying, allowing cyanobacteria to survive unfavorable seasonal conditions. They may also be blown to different places by wind, a major factor in dispersal of the species. Cyanobacteria live in freshwater, marine, and wet terrestrial environments. Their gelatinous capsules and the toxins they produce make them poor food for many animals. They can be a nuisance in water supplies where they cause taste and odor problems at low concentrations and can be poisonous at higher concentrations.

4 Go to the supply area and make a wet mount slide from the mixed cyanobacteria culture. Before adding a coverslip, dip a dissecting needle in India ink and then touch the tip to the drop of culture on the slide. This will transfer a small amount of carbon particles to the drop and will allow you to see the mucilaginous sheaths that surround the cells.

After adding a coverslip, look at the preparation with the 10× objectives, switching to the 40× objective when you find something of interest. Because this is a mixture, you should see more than one species (fig. 15.4). Read about the different genera in the following text. Make a simple sketch of each species you see, adding measurements made with your ocular micrometer and draw an example in the circle below based on your observations.

Anabaena sp. have colonies of cells arranged as filaments. Filamentous organization may be a defensive mechanism making it difficult for microscopic animal grazers to ingest the entire colony. You should see some differentiation along the filament. Interspersed with numerous vegetative cells are elongated cells called **akinetes**. They are spores and will survive desiccation. Other cells are more spherical and are called **heterocysts**. They have enzymes that allow them to convert atmospheric nitrogen gas into compounds that can be used to make amino acids and nucleic acids. Some species of *Anabaena* produce neurotoxins and others produce toxins that interfere with liver function.

Gloeocapsa sp. The 40 species in this genus form colonies. Cells form new mucilaginous sheaths as they divide, so that sheaths develop within sheaths. Unlike *Merismopodia* that has cells arranged in regular arrays, the cells are arranged in a more random fashion. This species is commonly found growing on moist rocks and flower pots in greenhouses. Often the dark stains seen on buildings that are intermittently damp are due to dried cells of this species. Some species of *Gloeocapsa* are found associated with fungi in lichens as the photosynthetic partner.

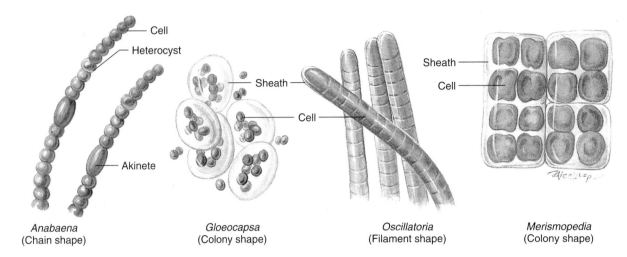

Anabaena
(Chain shape)

Gloeocapsa
(Colony shape)

Oscillatoria
(Filament shape)

Merismopedia
(Colony shape)

Figure 15.4 A few common cyanobacteria (blue-green algae).

Merismopodia sp. has colonies of single cells in a regular array within the gelatinous sheath. Usually you can see a unit that is 4 cells wide by 4 cells tall within a colony and this may be repeated several times.

Oscillatoria sp. also has a filamentous organization but there are no real cell specializations as in *Anabaena*. If you scan along the length, you will see where there are a few dead cells in the filament. Typically, the filament will fragment at this point, producing two filaments that will continue to grow. This genus is well known for its ability to move. If you do not touch your slide, you may be able to see the filaments move back and forth.

As you look at these different species of cyanobacteria, answer these questions:

- Do you see nuclei in any of the cells?

- Is their green color due to a concentration of chlorophyll in any region of the cell, or is the chlorophyll generally dispersed throughout the cell?

- Would you expect to find any of these cells dividing by mitosis or meiosis? Why?

Gram Staining

Bacteria are also identified by their staining response to certain dyes. One such staining protocol is known as the **Gram stain,** named for Christian Gram who developed the test in 1884. He found that crystal violet binds irreversibly to cell wall components of only some bacteria. Today, this observation is used as a lab test to identify bacteria as belonging to one of two groups. Bacteria that retain Gram stain when washed with alcohol are said to be **Gram positive;** those that are decolorized by the alcohol wash are said to be **Gram negative.** Cell walls from Gram positive bacteria are rich in **peptidoglycan.** Gram negative bacteria have less peptidoglycan, and a second membrane, the **outer envelope,** covers the cell wall, blocking the stain's access (**fig. 15.5**). The Gram staining reaction is an additional tool in bacterial identification. As a child, you may have had a throat culture when you were sick. Among other things, the doctor wanted to determine if you were

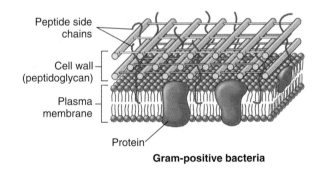

Gram-positive bacteria

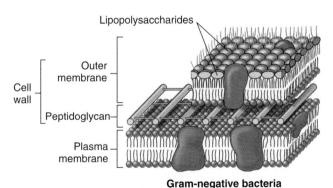

Gram-negative bacteria

Figure 15.5 Cell wall structure in Gram+ and Gram− bacteria. The additional outer second membrane and small amount of peptidoglycan in Gram− bacteria greatly reduce crystal violet stain binding.

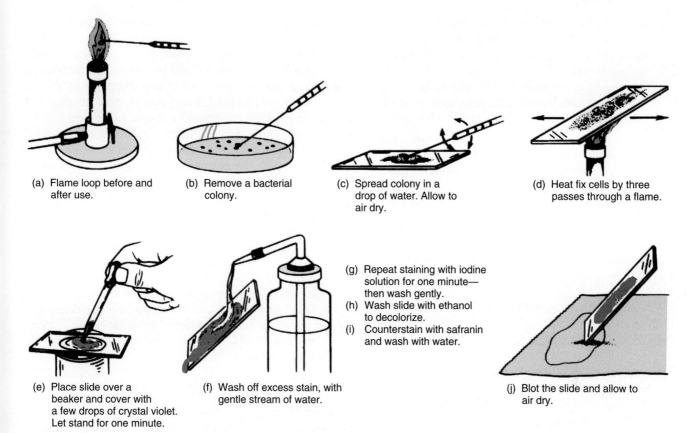

Figure 15.6 Directions for Gram staining.

infected with Gram + or − bacteria, because certain antibiotics are effective only against Gram + bacteria.

5 ▶ **Procedure** You will perform the Gram stain on two 18- to 24-hour cultures of bacteria available in the laboratory. One culture contains *Bacillus megaterium,* and the other *Pseudomonas fluorescens.* Both species are relatively large, but one is Gram positive and the other Gram negative. You will determine which bacterium is Gram+ by using the following procedures (**fig. 15.6**).

1. Wash two microscope slides with soap and water to remove oils. Dip the slides in a beaker of alcohol and let them air dry. Use a diamond pencil or wax pencil to put a "B" on one slide where you will put *Bacillus* and a "P" on the other for *Pseudomonas*.

2. If you start with colonies from petri plates, put two tiny drops of distilled water on the slide. If you start from a liquid culture, water is not needed.

   Flame a bacterial loop and let it cool. Scoop a small amount, not a whole colony, off the agar plate surface. Dip the loop in the water drop and then spread the bacteria evenly on one-third of the slide.

   Repeat the procedure for the second species on the other slide. Gently heat the slides by passing them back and forth through a low flame until the water evaporates and the cells adhere to the slide, a process known as **heat fixing.** Do not hold the slide in the flame so that the drop boils. If it does, start over!

   Alternative step #2. In 2009, students at St. Mary's University of MN under the direction of J. M. Minnerath researched an alternative method for fixation that improved subsequent sticking of the cells to the slide and their stainability. Their work is accepted for publication in a journal. Although not a standard practice, the directions are repeated here for those who would like to use the students' technique.

   First, extinguish all flames in the lab. Before bacteria are added to the slide, draw a 2-cm circle on it with a wax pencil. Add 20 μl of a bacterial culture to the circle's center. Flood the drop with 200 μl of 100% methanol and let stand for two minutes. Pour off excess methanol into a waste container. After the slide has air dried, continue with step 3 to stain and view cells.

3. Put the slides on top of a beaker and flood the surface with crystal violet stain. After one minute, use a spring-type clothespin or test tube tongs to carefully pick up the slides one at a time and wash off the stain with a gentle flow of water. Wearing rubber gloves will protect your skin from the stain.

BE CAREFUL!

This stain is very difficult to remove from your skin and clothing.

4. Now the slides should be flooded with Gram's iodine. The crystal violet stain is positively charged and sticks to the negatively charged peptidoglycan in the cell walls. The iodine forms complexes with the stain so that it is not readily washed away in the next step. It does not stain the cells itself. After one minute, gently wash the slides again with water.

5. Take a squeeze bottle containing 95% ethanol and *gently* squirt it on the surface of each slide until the solvent running off the slide is colorless. This will take about 20 to 30 seconds and should not be overdone. If you squirt too hard, the bacteria will be washed off.

   Gently rinse the slides with a stream of water.

> **BE CAREFUL!**
> Ethanol is flammable and should not be used near an open flame.

6. Flood the surface of both slides with safranin, a **counterstain** that helps you see Gram-negative cells by staining them light pink. After 30 seconds, wash off the counterstain with a stream of water.

7. Blot the excess water at the edge of the slides onto paper toweling and allow them to air dry.

8. Examine with a high-power objective, or an oil immersion objective if available.

6 *Analysis* Which species of bacterium is Gram positive? Which is negative? Sketch a few of the cells below and describe in words how they look.

Differential Growth

Colony characteristics, cell shape, and reaction to Gram stain are used to assign unknown bacteria to groups. More precise identification involves other staining characteristics or investigation of the ability to grow under different conditions and on different growth media. When available, nucleotide sequencing of DNA has proven to be an excellent tool. You will also perform serial dilutions so that you can count the number of bacteria in a sample.

To illustrate that different species grow better under different culture conditions, a simple experiment can be performed. You or your instructor should take a gram of rich organic soil and add it to 10 ml of sterile 0.85% saline. Agitate the mixture to suspend the bacteria while allowing the mineral soil particles to settle. Undoubtedly, there are many species of bacteria suspended, some few in number and others quite plentiful. A study of soil bacterial diversity in 2005 estimated that one gram of soil might contain a billion bacterial cells representing nearly a million species!

You will investigate whether some of the soil bacteria preferentially grow better in a normal nutrient agar or one that has had 6% NaCl added to it. You will also determine if some of the species preferentially grow better at room temperature or at 42°C. This is a classic 2 × 2 experimental design as follows:

	23°C	42°C
No NaCl	Lo temp, no salt	Hi temp, no salt
6% NaCl	Lo temp, hi salt	Hi temp, hi salt

Procedure To do this experiment, you have to inoculate four petri plates, two containing normal nutrient agar and two containing nutrient agar plus 6% salt. Each of these plates will be incubated for a day at the appropriate temperature and then stored in the refrigerator to slow growth until next week's lab when you will analyze them.

7 A **dilution streaking technique** will be used to inoculate the plates. First, get a sterile nutrient agar petri plate. While keeping it closed, turn it over and use a marker to draw four quadrants on the bottom of the plate. Now take the 10 ml soil suspension and shake it. Open the top and dip a sterile transfer loop into the fluid. The loop will hold about 10 microliters of fluid. Open one of the petri dishes from the edge (fig. 15.1), and streak the wet loop in large arcs across about 1/4 of the plate near the edge. See **figure 15.7** for the streaking pattern. Close the dish and flame the loop to sterilize it.

Rotate the dish 90° and open it again, inserting the now sterile loop, dragging it across the previous streaks to spread some (but not all) of the bacteria in a new direction (fig. 15.7b). Close the dish, flame the loop, rotate the plate 90°. Open and streak a third time. Close the plate, flame the loop, rotate, open and streak for a fourth and last time (fig. 15.7d). You have now spread the original sample across much of the plate with fewer and fewer bacteria in each of the subsequent streaks. This lowers the density of bacteria and will allow some to grow without competition from others. It is actually possible to thin the density to a level where only one bacterial cell is in one region of the streak. After a few days of growth, the

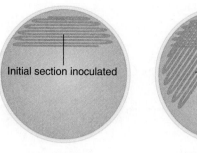
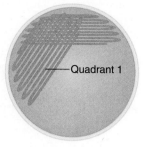

(a) Orient your plate as depicted here.

(b) Lift the lid, and use a sterile loop to make lines, or streaks, across the agar as shown in quadrant 1. Close the lid.

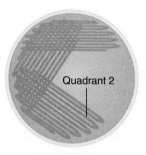
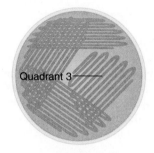

(c) Flame the loop, lift the lid, and make streaks as shown in quadrant 2. Close the lid.

(d) Flame the loop, lift the lid, and make streaks as shown in quadrant 3. Close the lid. Incubate the plate as directed.

Figure 15.7 Streaking pattern used in spreading sample on surface of agar.

results will look something like **figure 15.8**. The small circular colonies seen in the lower-right quadrant are colonies that have grown from a single cell. They represent a clone. Methods like this allow microbiologists to separate a single species from complex mixtures.

You must repeat the procedure described for the first plate to inoculate three other plates, so that all four have the same amount of material on them and have been streaked in a similar pattern.

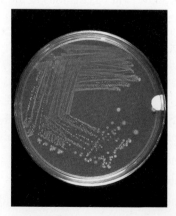

Figure 15.8 Results from streak culture after incubating. Note the dense first streak at top and how the number of bacteria gradually falls until colony growth is seen at bottom right.

Write your name on all four plates. Label one no salt plate 23°C and the other 42°C. Do the same for the 6% salt plates. Put the plates in the appropriate incubator.

Analysis At the start of next week's lab, you will analyze the plates after the bacteria have had a chance to grow. To prevent overgrowth, your instructor will put the plates in a refrigerator after 24 hours of incubation.

8) When you analyze the plates, use the terms in figure 15.2 to characterize the colonies. Record your observations below:

Plates grown at 23°C

   No NaCl

   6% NaCl

Plates grown at 45°C

   No NaCl

   6% NaCl

Do all colonies look the same regardless of the growth conditions? Explain.

Do all plates have the same number of colonies on them? Explain.

What do these results tell you about the bacterial diversity in the soil sample? Do all bacteria have the same growth requirements? Can all species tolerate the same conditions?

Dispose of your plates as told by your instructor. They must be flooded with bleach or autoclaved to kill the bacteria.

This basic experimental design could be adapted to investigate many other properties of growth. For example, dilute petroleum samples could be added instead of NaCl to see if certain bacteria grew well in a medium containing that material. If so, one might want to try to isolate them and develop dense cultures that could be spread on petroleum-contaminated soil to eliminate the contamination. Presumably they grew well because they could metabolize the petroleum and gained some benefits from it. One could also determine in a similar way the effects of pesticides or other potential soil contaminants.

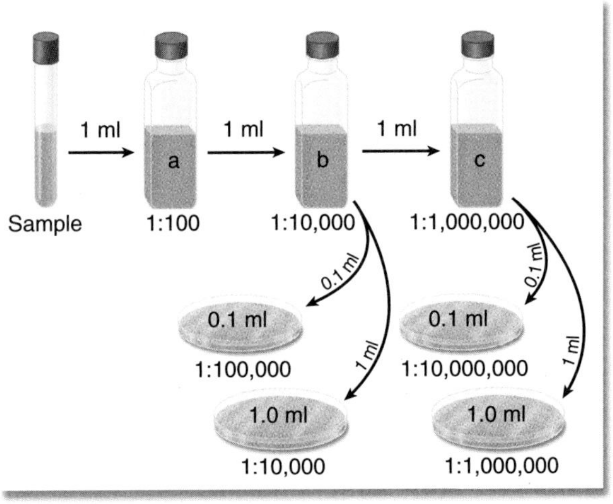

Bacterial Population Counts

Biologists frequently need to estimate the number of individual bacteria in complex materials, such as in soil, food, or body fluids.

This is done by taking a measured sample of the material and diluting it with sterile saline. A small amount of that dilution is spread on the surface of nutrient agar in a petri dish. For example, if 1 ml of milk is mixed with 99 ml of sterile saline, and 0.1 ml of the mixture is placed on the agar, then 1/1000 of the milk sample has been "**plated**." Any bacteria in the sample will be spread on the agar surface. They will draw nutrients from the agar medium and each single bacterium will repeatedly divide to form visible colonies. The colonies can be counted, and by multiplying by 1,000, the number of bacteria in the original ml of milk is obtained.

If the bacteria are plentiful in the sample, a 1/1,000 dilution may be insufficient; when a petri plate is covered by thousands of colonies they can grow on top of each other; counting is tedious often subject to error. To avoid this, **serial dilutions** are often made (fig. 15.9). In this technique, a 1 ml sample is suspended first in 99 ml of saline, and then 1 ml of that dilution is added to a second 99 ml of saline to further dilute it: in this case to 1/10,000. One milliliter of this suspension may then be diluted in a third 99 ml to give a 1/1,000,000 dilution and so on. Describe how you would make 1/10,000,000 serial dilution below.

Figure 15.9 Serial dilution technique. Each bottle has 99 ml of sterile water in it. By transferring 1 ml of sample to the first bottle, a 1:100 dilution is achieved. If 1 ml of that dilution is added to a fresh bottle, a 1:10,000 dilution is achieved. A third transfer of 1 ml to 99 ml results in a 1:1,000,000 dilution. By varying the amount of that dilution added to a culture plate, further dilutions can be achieved.

milk and "old" milk that has been opened, half-emptied, and left on the counter for at least 12 hours will be used. Half of the students will analyze one sample and the half the other. The goal is to determine the number of bacteria in each sample and then to compare the bacterial counts.

Commercial milk is heated to kill bacteria, a process called pasteurization. The standard for bacterial content in milk varies by state, but the pasteurization process usually reduces it to around 500 bacteria per ml. In Iowa, the regulatory limit is 20,000 per ml. When milk is not refrigerated and warms, the residual bacterial population can grow, souring the milk.

9▶ Here is a question for you: If you have a fresh container of milk with 1,000 bacteria per ml in it and these bacteria can divide every half hour at room temperature, how many hours will it take to reach a population density of 20,000/ml? Show your calculations in the space below.

By plating different (but convenient) amounts of the dilutions, the number of bacteria, hence colonies, per plate can be further reduced. The goal is to achieve a number of colonies between 30 and 300 on a plate. The number of bacteria in the original sample is obtained by multiplying the number of colonies by the dilution/plating factor. This is called the **Standard Plate Count** technique.

You will use the Standard Plate Count technique to determine the number of bacteria in samples of milk. Fresh

10▶ *Procedure* To measure the density of bacteria in your assigned milk sample, you will need the following:

Milk sample, a few ml

3 sterile tubes containing 9.9 ml of saline

A micropipette with at least 6 sterile tips to deliver 0.1 ml

3 petri plates with nutrient agar

Transfer hood

Sterile glass "hockey sticks"

Table 15.2 — Bacterial Counts from Milk Samples

Sample	# Colonies	Dilution	Plate Factor	# Bacteria per ml of Milk
Fresh				
Old				

Class data statistics:
    Fresh milk   Average = _____   st. dev. = _____
    Old milk   Average = _____   st. dev. = _____

To reduce the volumes of reagents needed, all sample dilutions are going to be based on using 0.1 ml of sample and diluting it into 9.9 ml of saline. If this were done once, what would be the dilution?

If a second serial dilution was made with the same volumes, what would be the dilution?

11 The first step is to prepare the dilutions. Take 0.1 ml of sample and add to a tube with 9.9 ml of saline. Cap and shake it well or mix on a vortex mixer. Label the tube 1:100. Using a fresh pipette tip, take 0.1 ml of this dilution and add it to 9.9 ml of sterile saline. Cap and mix as before and label 1:10,000. Using a fresh pipette tip, take 0.1 ml of this sample and add it to the third tube of sterile saline. Mix well as before and label this tube 1:1,000,000.

12 Take your three petri plates and label them with the same dilution numbers that are on the tubes. Add your initials and the words *Fresh* or *Old* to describe the sample.

You will now inoculate each plate with the corresponding dilution. Do this by mixing the dilution and withdrawing 0.1 ml with a fresh pipette tip. Open the petri plate by tilting the lid, and dispense the sample. Use a sterile glass hockey stick to spread the sample evenly on the plate. (See figure 14.2 for technique.)

When all plates have been inoculated, give them to your instructor for incubation. After 24 hours, they will be stored in the refrigerator until next week when you will analyze them.

13 *Analysis* When you come to lab next week, get your plates from the refrigerator and look at them. Some will have too many colonies to count and some may have no colonies. However, there should be at least one that has a countable number of colonies, preferably in the range 30 to 300 per plate.

Each colony you see represents a single bacterium that was not visible to the naked eye in the sample that you spread on the agar. While the plate was incubated, those single cells grew and divided to make "little piles" of cells that now can be seen as colonies. If you want to know how many cells were in 0.1 ml of your dilution, simply count the colonies on the plate that has at least 30 colonies! Enter your results in the first column of table 15.2. You can disregard the other two plates. They were "just in case" plates to cover if there were more or less bacteria present.

Dispose of your plates as directed by your instructor. They must be flooded with bleach or autoclaved to kill the bacteria.

To determine the number of bacteria in a milliliter of milk before dilution, you must multiply by the dilution factor and by the plating factor—10 in your case because you added only a 0.1 ml of your final dilution. Make the appropriate entries in table 15.2.

Go to the blackboard in the lab and record your bacterial counts in one of two columns: *Fresh milk* or *Old-milk*. When everyone has recorded their data, calculate a class average and standard deviation for the bacterial content per ml of fresh and old milk. Enter these results on the line below table 15.2.

Answer this question: Do you ever want to drink milk left out on the countertop again? _____

Lab Summary

14 On a separate sheet of paper, answer the following questions assigned by your instructor.

1. List the characteristics that separate organisms in the Domain Bacteria from those in Domain Eukarya. Which ones were you able to see in this laboratory?

2. What evidence do you have from this laboratory that bacteria are diverse? Cite specific examples as you explain this diversity.

3. Describe how to do a Gram stain procedure. Why is this important?

4. Explain how the number of colonies seen on an agar surface in a petri plate can be related to the number of bacteria in a sample.

5. What is a serial dilution and why is the technique used?

6. What evidence do you have from your experiments that bacteria are metabolically different?

You may want to try the critical thinking questions that apply some of the knowledge you gained in doing this lab.

Internet Sources

Use the Google search engine (http://www.google.com) to locate additional information on cyanobacteria. When connected, type in the phrase, *cyanobacteria in drinking water*. Scan the list that is returned and answer critical thinking question 4.

Learning Biology by Writing

Write a procedure that would allow you to determine the number of bacteria found in potato salad collected from a salad bar in a restaurant. Include how you would use sterile techniques to avoid false readings.

Critical Thinking Questions

1. "To make strawberry jam, mash 4 cups of berries with 7 cups of sugar (sucrose). Bring to a boil and continue boiling for twenty to thirty minutes. Ladle into sterilized jars and cover with melted paraffin wax or sterilized lids."

    Relate jam making to the growth requirements of different bacteria. Would these steps be adequate for preserving (canning) vegetables? Meats? What other methods could be employed to preserve foods?

2. Should a plant nursery that plans to grow pea and bean seedlings use a sterilized potting soil mix?

3. It is well known that many pathogenic bacteria are developing resistance to antibiotics. Explain why this is happening.

4. Why should you be concerned if cyanobacteria are found in your drinking water? Use Internet Sources to locate the information that you need.

5. Why are bacteria, rather than say human cells, used as experimental organisms in molecular biology studies?

6. Bioremediation is a process that uses bacteria to remove pollutants from soil or water. What characteristic(s) of bacteria make them excellent organisms for doing this?

7. Do you think of bacteria as being "bad" pathogens or do you think of them as being "good," essential components of all ecosystems. Justify your position.

8. Devise a procedure based on serial dilution that would allow you to assay the number of bacteria in a sample from a salad bar.

Lab Topic 16

Evolution of Eukarya and Diverse Protists

Supplies

Resource guide available on WWW at
www.mhhe.com/labcentral

Equipment

Compound microscopes
Dissecting microscopes

Materials

Cultures
  Volvocales mixed algae culture
  Mixed ciliate culture
  Mixed amoeba culture
  Euglena
  Physarum (slime mold) culture kit
Cultures or preserved specimens of red algae:
  Polysiphonia, Bangia, or *Callithamnion, Porphyra, Rhodymenia, Gigartina* or *Corallina*
Herbarium sheets with examples of red, and green algae
Prepared slides
  Trichomonas vaginalis
  Oedogonium
  Trypanosoma sp.
  Foraminiferans
Diatomaceous earth
Slides and coverslips
Sample food labels listing carrageenan or agar additives

Solutions

0.85% saline
Protoslo or methyl cellulose
1% water agar for slime molds in petri dish
Unprocessed (old-fashion) oatmeal

Student Prelab Preparation

Before doing this lab, read the Background material and other sections of the lab topic that have been assigned by your instructor.

Use your textbook to review the definitions of the following terms:

algae	isogamy
autotrophic	kingdom
cilia	multicellularity
clade	oogamy
colonial	phylum
flagella	phylogeny
genus	protist
heterogamy	protozoa
heterotrophic	species

Describe in your own words the following concepts:

Differences between bacteria and protists

Diversity among the protists

How ancestral protists might have evolved into multicellular organisms

After finishing the prelab review, write any questions you have about terms, concepts, or techniques in the margins of this lab topic. The lab experiments should help you answer these questions, or you can ask your instructor during the lab.

Objectives

1. To observe the diversity among several clades of protists
2. To learn the life cycles of biologically important protists
3. To recognize evolutionary relationships among the protists, plants, fungi, and animals

Background

The group of organisms commonly called protists has been studied for hundreds of years, since the invention of microscopes. You would think that biologists would have described all the species and agreed upon their evolutionary relationships, but this is far from the case. What is agreed upon is that they are the earliest eukaryotes, appearing early in the evolution of life, perhaps as much as 2 billion years ago. In 1969, the kingdom Protista was proposed and defined as including more or less any eukaryotic organism that was not a plant, fungus, or animal. Some 30 phyla, containing on the order of a quarter million species, were thus lumped together, but Protista was a proposal for classification convenience rather than one that represented evolutionary relationships that could be tested.

For categories in a classification to represent evolutionary relationships, called phylogenies, the categories in the classification must contain organisms that are derived from common ancestors, *i.e.* the group should be monophyletic. With the advent of new technologies in the past decade or so, it has been possible to study nucleotide sequences in the total DNA of organisms to define the genes they carry and to compare amino acid sequences found in proteins that two or more species share. Using computers with powerful processors and statistical tests, it is now possible to group organisms based on molecular similarities. The reasoning that is applied is that if two species are similar at the genetic level, then they must be related to each other in the same way that your DNA is more similar to your sibling's than it is to your lab partner's. In the new approach to phylogenies, organisms that show similarities are assigned to a category called a **clade**. It is not a kingdom, phylum, or other taxonomic category. It is grouping of genetically related organisms. A clade represents a hypothesis that can be tested as new data become available and as biologists determine the nucleotide sequences in more and more organisms. Use your Web browser to connect to http://www.ncbi.nlm.nih.gov/ and then click on *Taxonomy* and then *TaxBrowser* to see the extent of the database to date.

DNA sequence and protein comparisons have led to the proposal that all eukaryotic organisms can be divided into "supergroups," an informal category that is more inclusive than kingdom, but less inclusive than domain in the formal categories of taxonomy. As of 2009, scientific papers in such prestigious journals as the *Proceedings of the National Academy of Sciences* (USA) recognized five supergroups of all eukaryotes (**table 16.1**) based on massive molecular comparisons across 1000s of genes and proteins in 100s of organisms.

Within the supergroups are numerous clades of eukaryotic organisms that share molecular similarities, and presumably evolutionary lineages. Table 16.1 lists some of the larger clades but by no means all of them. The next several labs afford you an opportunity to examine in some detail the listed clades found within these supergroupings. This lab topic covers those organisms collectively called the protists and later lab topics explore the relationships within the plants, fungi, and animals.

The most interesting revelations in this table are that data spanning the entire domain Eukarya indicate that (1) organisms once considered to be in a single protist kingdom fall into five different supergroups, *i.e.* they are not monophyletic; and (2) protists in certain supergroups and clades share more molecular similarities with plants, fungi, or animals than they do with other organisms that are called protists.

Protistan members of the supergroups appear in the fossil record before we see fossils of plants, fungi, or animals, indicating that protists were the first to have a eukaryotic cell plan. Eukaryotes have a true nucleus surrounded by membranes, DNA organized into chromosomes that condense during mitotic or meiotic cell division, formation of a spindle from microtubules during cell division, and membranous organelles that organize the cytoplasm into compartments where specific functions occur. Some biologists speculate that the first eukaryote was formed when a bacterium and an ancestral archaean fused to form a symbiotic relationship. Mitochondria developed when this ancestral eukaryote formed a second endosymbiotic relationship with an aerobic bacterium. Chloroplasts may have first appeared when this second-level ancestor formed a symbiotic relationship with a cyanobacterium.

Most protists are unicellular, although many "algal" species are colonial or truly multicellular. By studying the colonial and multicellular forms, we can gain information that allows us to speculate how multicellularity might have developed in fungi, plants, and animals. Although protists are often described as being simple organisms, their cellular organization and metabolism are every bit as complex as those found in the so-called higher organisms. In fact, higher organisms are often much simpler at the cellular level because their many cells are specialized into tissues that perform particular functions while single protistan cells perform all functions necessary for life as independent organisms.

Protists live everywhere there is water: as plankton in the oceans and fresh water, in puddles, in damp soils, and, as symbionts in the body fluids and cells of multicellular hosts. Many are autotrophic, making their own food materials through photosynthesis, while others are heterotrophic, absorbing organic molecules or ingesting larger food particles. They are ecologically very important in food chains, especially the algae (phytoplankton) that are the energy base for most aquatic ecosystems.

Life cycles are varied. Some species reproduce only asexually by mitosis. Others have sexual phases to the life

Table 16.1 Supergroups Within the Domain Eukarya

Supergroup	Clade	Specimens to be Studied
Excavata	Parabasalids	*Trichomonas*
	Euglenozoans	*Euglena*
	Kinetoplastids	*Trypanosoma*
Archaeplastida	Rhodophytes	Various red sea weeds
	Chlorophytes	Volvocene series, *Oedogonium*
	Plants	Covered in topics 17 and 18
Chromalveolata	Alveolata	Mixed ciliates
	Stramenopila	Diatom strew
Rhizaria	Foraminiferans	Mixed examples
Unikonta	Amoebozoa	Mixed amoebae, slime mold
	Opisthokonta	Fungi and animals covered in topics 19, 20, 21, 22, & 23

cycle so that there are distinct diploid and haploid stages as are found in fungi, plants, and animals. A cyst stage is found in the life cycle of many protists, which allows a species to lie dormant during temporarily harsh conditions.

LAB INSTRUCTIONS

During this lab you will look at 11 different protists, representing the supergroups given in table 16.1. The intent is to illustrate protistan diversity and to present some of their interesting biology.

Supergroup 1: Excavata

This monophyletic group of unicellular organisms evolved early in the eukaryotic lineage. It is named for the "feeding groove" present on the cell surface of many, but not all, members of the group. The groove allows these organisms to ingest particulate food by phagocytosis and to digest it intracellularly. It is divided into two subgroupings. In some species, their mitochondria are highly modified and may have tubular, discoid, or laminar cristae. Most have two or more flagella. Excavates are free-living and symbiotic and some are parasites of animals, including humans. The clades that are included are placed together because of molecular similarities.

Parabasalid Clade Unicellular, anaerobic, flagellated protists with highly modified mitochondria; the nuclear envelope persists during mitosis; group characterized by a modified Golgi apparatus located at the base of the flagella called a parabasal apparatus. Some members have very large genomes coding for nearly 60,000 proteins, one of the largest coding capacities among all eukaryotes.

① Obtain a prepared slide of *Trichomonas vaginalis* from the supply area. Transmitted during sexual intercourse, this parasite resides in the lower genital tract of an infected woman or in a man's urethra. Commonly known as "trich," it causes irritation, and in severe cases purulent discharges from the vagina or inflammation of the male's urethra and prostate. It is estimated that infection rates may be as high as 20%. The parasite multiplies by dividing by mitosis and symptoms develop within a month of exposure. It is treatable with antiprotozoal drugs, although resistant strains are appearing. It does not form any cyst stages and can live only briefly outside the body. *Trichomonas* feeds by pinocytosis and phagocytosis as well as by absorbing nutrients from the host's body fluids across its membrane. Although they have highly modified mitochondria, they obtain energy by anaerobic metabolism.

Observe the slide first with the 10× objective, switching to high power when examples are found. Compare what you see to **figure 16.1**.

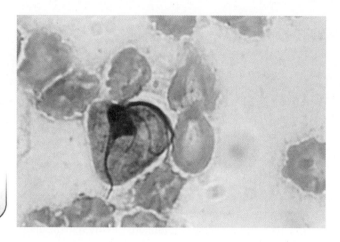

Figure 16.1 *Trichomonas vaginalis:* a parabasalid human pathogen.

Note the prominent nucleus. Identify the flagella. How many can you count? _____ Near the nucleus you should be able to see the parabasalar body, a modified Golgi apparatus, which characterizes this group. Note the undulating membrane, which appears as a fold of membrane-like material along the cell body. Sketch one of these below.

Euglenozoan Clade The 1,000 or so free-living species in this group have one long flagellum used in locomotion and a second shorter one. A flexible proteinaceous **pellicle** forms an outer covering. Beneath the plasma membrane are ribbon-like protein strips that give a striated appearance to the organism. *Euglena* is a well-known genus. Although euglenids are typically thought of as photosynthetic, they are descended from heterotrophic ancestors that had a feeding pocket adjacent to the flagellar opening. Many species remain heterotrophic or are "mixotrophic," *i.e.,* depending on environmental conditions they may be heterotrophic or autotrophic. Reproduction is asexual by modified mitosis, with the spindle forming inside the nucleus that then divides to produce two nuclei. The cytoplasm divides by longitudinal fission.

② Make a wet-mount slide of *Euglena* from the stock culture and observe through your compound microscope. If the organisms are swimming too fast to be studied, make a new slide but add methyl cellulose, a thickening

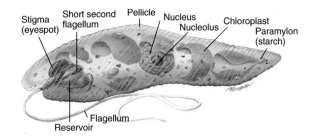

Figure 16.2 Anatomy of *Euglena* sp.

agent, or shred a small piece of lens paper or cotton into the drop so as to trap the organisms. Reducing the light intensity will help you see internal structures.

Euglena is usually oblong with the anterior somewhat blunt and bearing a long flagellum. A second flagellum is present but is too short to be seen. If you reduce the light intensity and carefully focus, you should be able to see the whipping of the flagellum. Does it push or pull the organism through the water? _____

Watch the organism closely. What evidence is there that the surrounding pellicle is flexible?

As you study *Euglena*, find as many of the structures shown in **figure 16.2** as you can.

What color is the **pigment spot,** also called the stigma, near the base of the flagellum? _____.
Although this is light sensitive allowing the animal to orient to light, it is not image forming.

Is the chlorophyll of *Euglena* localized in structures, or distributed throughout the cell as in cyanobacteria? _____

Euglena's chloroplasts differ from those in green algae and plants because they are surrounded by three membranes suggesting that they originated from green algae that were ingested and became symbionts in the ancestors of the species you are observing.

Excess sugars produced during photosynthesis are converted into **paramylon,** a unique form of storage starch. Can you see any paramylon storage structures in the cytoplasm? _____

Kinetoplastid Clade Members of this clade are unicellular and characterized by the presence of a unique organelle, the **kinetoplast,** located at the base of the flagellum. Electron microscope studies show that it is a modified mitochondrion containing a large amount of DNA. Reproduction is asexual by modified mitosis. Organisms in this group are free-living or parasitic. Three cause serious human disease.

African sleeping sickness and Chagas disease are called trypanosomiasis, a reflection of the genus name for the causative agent. Chagas disease affects up to 18 million people in the rural tropical Americas. It is transmitted from one infected individual to another by "kissing bugs," biting insects that take blood meals, by blood transfusions, and also can pass transplacentally from mother to fetus. The protozoan is in the feces of the bug and is transmitted when a person rubs the fecal material into the wound from the insect's bite or into the eyes. Most people who are infected do not seek medical attention and the infection runs its first course in 4 to 8 weeks. Symptoms include fatigue, fever, rash, and enlarged liver/spleen. Often 10 to 20 years later, people then develop the most serious symptoms which include cardiac problems and have difficulty in swallowing, leading to reduced life span.

African sleeping sickness, caused by a related species, is a serious disease and strikes almost a quarter of a million people annually in tropical Africa and can affect cattle as well. The blood feeding tsetse fly transmits it from an infected individual to others. Initial symptoms are fever, headaches, rash, and edema progressing to meningoencephalitis and death within several weeks. It is treatable with drugs.

Trypanosomes have attracted the interest of molecular biologists. Studies show they survive in the blood stream by tricking a mammal's immune system. Normally when a foreign cell enters the blood, an antibody to it is made that kills it. For the immune system to do this, it must "recognize" the chemicals found on the foreign cell's surface. Biologists have found that trypanosomes are able to change the types of molecules found on their surfaces every few days and thus always stay one step ahead of the body's natural defenses.

3 Obtain a prepared microscope slide of a blood smear from an individual with trypanosomiasis (African sleeping sickness or Chagas disease). Look at it first with your 10× objective, and when a specimen is located, with the 40× objective. You should see round red blood cells, irregular shaped white blood cells with large nuclei, and long, thin cells which are the trypanosomes (**fig. 16.3**). These organisms live and reproduce in the host's blood stream.

Draw a few of the organisms below. Can you see the flagellum? The kinetoplast?

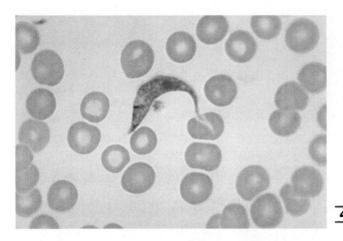

Figure 16.3 A blood smear showing trypanosome parasite against a background of red blood cells.

Before starting the next section, fill in the information for the three specimens observed in this supergroup in table 16.2 on page 198. Think comparatively about the last three specimens you studied. If you were shown a slide of one of them, how could you distinguish one from the others?

Supergroup 2: Chromalveolata

Most members of this group are unicellular, but it also includes large multicellular organisms. Based on DNA nucleotide sequence and protein data, some authors divide this group into two clades: Alveolata and Stramenopila.

Alveolata Clade This clade includes the ciliates, apicomplexans, and dinoflagellates. Some groups have small indentations in their surface membranes, called alveoli, beneath the material covering their outer surface.

Ciliophora (Ciliates) About 8,000 species of ciliates live in oceans, fresh water, and wet soils. All are unicellular. Most are free-living heterotrophs, although a few are parasitic. The phylum is characterized by using **cilia** for locomotion and feeding. They may cover the surface or be arranged in tufts and rows, even fused together to make appendage-like structures. They also have a surface covering, called a **pellicle**. A complex feeding apparatus with an oral groove leads to a fixed area where food is ingested into **food vacuoles**. Freshwater species have a **contractile vacuole**. Most ciliates have two types of nuclei. A single, large **macronucleus** regulates the normal physiological functioning of the cell. In addition, one or more **micronuclei** are involved with sexual reproduction, a process called **conjugation**. During conjugation, two ciliates, temporarily join along their oral areas, and exchange micronuclei. Then the cells separate, reorganize their nuclear material, replace their macronuclei with the newly organized genetic material, and subsequently divide. Asexual reproduction is more common. It occurs by mitosis followed by **longitudinal fission**.

4) Make a wet-mount slide of the mixed ciliate culture found in the supply area. Look at it first with 10× objective and switch to the 40× when you find something of interest. Compare it to **figure 16.4**. Continue to switch back and forth as you scan the slide. How many different types of organisms can you see? _____ Note: the cultures may contain organisms from other clades that are food for the ciliates. You may find it useful to adjust the condenser height and diaphragm opening on your microscope to improve your views.

Observe the different locomotory movements: forward, rotating, and swerving. What evidence can you see that all have cilia?

Now, make a second slide, but this time add a drop of methyl cellulose to the culture drop before you add a coverslip. This chemical is very viscous and slows the ciliates so that you can see details of cellular structure. There will be several different species of ciliates on the slide (fig. 16.4). Check them off as you are able to identify them. These may include the following:

Blepharisma sp.—These medium-size ciliates have a rose color (but not in strong light) that makes them easy to spot. *Blepharisma* feeds by sweeping small bacteria into the cytosotome using a thin membrane (not visible) and fused cilia called cirri that you should be able to see beating in waves. At the base of the cytostome, the bacteria are ingested as food vacuoles. A contractile vacuole may also be seen. It controls the volume of water in the cell.

Colpidium sp.—These are probably the smallest ciliates on your slide. Along with other ciliates, they are important in food chains in aquatic ecosystems. They eat bacteria which are probably living on dead and decaying materials. In turn, the ciliates are consumed by small animals and their energy passes along to top predators such as fish or birds.

Paramecium sp.—Several different species in this genus will be on the slide. They are medium-sized

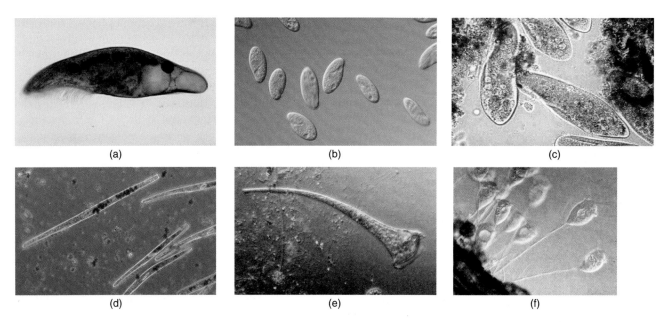

Figure 16.4 Types of ciliates likely to be found in mixed culture: (a) *Blepharisma* sp.; (b) *Colpidium* sp.; (c) *Paramecium* sp.; (d) *Spirostomum* sp.; (e) *Stentor* sp.; (f) *Vorticella* sp.

cells. Some will be green because of symbiotic algae. Others will be translucent. Known as slipper animals because of the body shape, they feed on bacteria. The stiff outer surface of the cell is composed of several layers of membrane, which along with other structures, make up a pellicle. You should be able to see the contractile vacuole(s). How often do they contract? _____

Sketch a *Paramecium* below, as well as other representatives of the group. On the sketch indicate the macronucleus, micronuclei, oral groove, gullet, contractile vacuoles, and cilia. Describe the major similarities of *Paramecium* and *Blepharisma*. What are the differences?

Spirostomum sp.—These will be very large, long, tubular cells with a macronucleus that extends the length of the cell. It looks almost like a string of beads. *Spirostomum* has a remarkable characteristic. It can contract to 1/4 of its length in 6–8 ms. Contraction can sometimes be induced by touching the coverslip. *Spirostomum* feeds on bacteria which are swept into the cytostome with a row of specialized fused cilia.

Stentor sp.—These trumpet-shaped ciliates are large, up to 2mm long! They feed on bacteria that are swept into a gullet by a crown of cilia. When free-floating, the cilia on the surface give *Stentor* mobility, but it can also attach to substrates such as dead leaves of aquatic plants. Tap on the coverslip and watch it change from a trumpet shape to a blob of cytoplasm. Some species have symbiotic green algae that use the *Stentor*'s waste products in photosynthesis and supply the host with carbohydrates. You may be able to see the macronucleus in the cytoplasm. A contractile vacuole should also be visible, and if you spend some time observing a single specimen, you should see it cycle.

Vorticella sp.—These ciliates create water currents with cilia that bring food particles to the organism. Individual *Vorticella* are capable of responding to various stimuli in their environment by rapidly coiling up their stalk and retracting toward the substrate they are attached to. If you're observing *Vorticella*, trying tapping on the table top next to your microscope while watching the organism. How does it respond?

Before starting the next section, fill in the information for this clade in the summary table 16.2.

If you were given an unknown specimen, how could you determine if it was a member of clade Euglenophyta or of clade Ciliophora?

Figure 16.5 Diatoms exist in an exquisite variety of geometrical shapes. Here an artist has arranged several species into a beautiful composition on a microscope slide using a single human hair to drag each one into position.

Stramenopila Clade This clade include diatoms, brown and golden algae, and water molds. They all have unique flagella with hairlike projections along the length of the shaft. Photosynthetic diatoms are found in huge numbers in aquatic ecosystems where they form the basis for many food chains.

Bacillariophyceae (Diatoms) The 100,000 species of diatoms share a common characteristic: a cell wall consisting of two valves made of silica. They are often yellow in color because of large amount of the pigment, fucoxanthin, which masks the green of the chlorophylls. Diatom chloroplasts are surrounded by four membranes rather than two as found in plants and algae. They contain chlorophyll types a and c, rather then a and b as in plants and green algae. These properties suggest that diatom chloroplasts were derived by forming a symbiotic relationship with red algae. In life, diatoms are the major basis for productivity in many aquatic ecosystems. Their life cycles include an asexual phase where the cell divides by mitosis and a sexual phase with meiosis and gamete fusion. When the diatom cell dies, the siliceous valves do not disintegrate but accumulate as sediments (**fig. 16.5**). In California, deposits of diatoms 300-feet deep are mined to produce diatomaceous earth, a fine powdery material. It is used as a filtering material in swimming pools and as a fine abrasive in silver polishes and toothpastes. Why do they work well for these purposes?

The delicate ornamentation of the valves is often used as a test of the resolution of microscopes. In a poor microscope, only the outline of the valve will be visible, while in very good microscopes the fine indentations and perforations will be apparent. Sketch a few different types of diatom valves below. Indicate the location of the **girdle,** the region of overlap between the two valves, and note how the valves fit together like the base and lid of a petri dish.

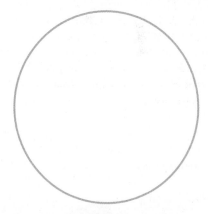

5) Make a wet-mount slide from the diatomaceous earth available in the laboratory. Place a drop of water on the slide and then add a very small amount of diatomaceous earth to the drop. Stir well before adding a coverslip and viewing.

You should easily find the diatom valves, the skeletal remains of cells that lived tens of thousands of years ago. In a top-down view, some valves will be round, others triangular, ovoid, and irregular. When viewed from the side, these same valves appear rectangular or ovoid.

When diatoms reproduce asexually by mitosis, the valves separate along the girdle line and each daughter cell receives one valve and must make the second. The cell receiving the larger lid secretes a smaller base to fit it. The daughter receiving the smaller base also uses is it as a lid and makes a smaller base to fit it. Thus, two unequal-sized daughters result. If this continues over several generations as in an algae bloom, the average size of the cells in the diatom population decreases and could approach a non-viable volume.

Size is restored through sexual reproduction. Diatoms are among the few protists that have gametic life cycles. All cells in the life cycle except gametes are diploid. When diatoms sexually reproduce, the partners bind together in a mucilaginous ball and both undergo meiosis, each producing a gamete. They fuse to produce a zygote. The zygote undergoes mitosis, producing two naked cells that each secrete both valves, restoring the diatom population to its normal size. Fill in the information required for this group in table 16.2.

Supergroup 3: Archaeplastida

This polyphyletic grouping of eukaryotes includes red and green algae as well as land plants. Their plastids are surrounded by two membranes, suggesting one was derived from an endosymbiotic cyanobacteria and the other from the enveloping membrane of the ancestral eukaryote that engulfed it. Their cells typically lack centrioles and have a cellulose cell wall. As you will see, some are unicellular, others colonial, and many multicellular.

Rhodophyte (Red Algae) Clade Over 5,000 described species of red algae are multicellular and macroscopic, living in marine environments. All lack flagella and centrioles, use floridean starch for energy storage, and have unstacked thylakoids in their chloroplasts. Most species are marine, growing as encrustations or as miniature "treelike" organisms. About 200 species are found in fresh water. The red color comes from phycoerythrin (a phycobilin-type pigment). However, not all species appear red. Life cycles include an alternation of generations between sporophyte and gametophyte stages. Cell walls consists of cellulose embedded in a matrix of other polysaccharides which include agar and carrageenan, that are commercially used as sources of thickening agents in food. Some, called coralline algae, secrete $CaCO_3$ around their cell walls, which can contribute to coral reef formation.

Most red algae grow at great depths in the oceans where the red accessory pigment phycoerythrin aids in photosynthesis. It is associated with the external membranes of their chloroplasts and absorbs blue light while reflecting red. Because blue light penetrates deeper in water than red, red algae can live at greater depths than most other algae. Some species have been found at depths greater than 600 feet. Because they are not exposed to wave actions, their body forms can be delicate (**fig. 16.6**).

6 In the lab, look at the herbarium sheets with dried red algae and at the preserved or living specimens available. Depending on the collection at hand, you may be able to observe species from the following genera:

Polysiphonia sp., *Callithamnion* sp., or *Bangia* sp. are highly branched and feathery.

Porphyra sp. or *Rhodymenia* sp. are flattened sheets, only a cell or two thick. *Porphyra* is used as the wrapping for sushi.

Based on general body size and shape, would you say that these are unicellular or multicellular? Do you see any examples of cell specialization?

Describe any morphological evidence for a division of labor among the cells in this organism.

Look at the example of *Corallina* sp. Species in this genus are coralline red algae whose cell walls are impregnated with secreted calcium carbonate. These algae are important reef-building organisms and, together with corals, are responsible for forming what we generally refer to as coral reefs, such as the Great Barrier Reef in Australia.

In the lab will be labels from various recognizable food products such as salad dressings, ice cream, *etc.* Look at the labels to see if they contain agar or carrageenan, thickening agents commercially isolated from the cell walls of red algae.

Figure 16.6 Growth forms of red algae. (a) *Botryocladia*. (b) *Gigartina* (c) *Gelidium*.

Before starting the next section, fill in the information for this clade in the summary table 16.2 at the end of this lab topic.

Chlorophyte (Green Algae) Clade

There are 7,000 or so species of green algae with about 90% found in fresh water. Some species are unicellular, while other species are colonial or multicellular. Their bright green chloroplasts have the same type of plastids and chlorophyll as those found in plant cells. Life cycles include asexual and sexual phases with flagellated gametes. Flagella are always present in multiples of two. Some species have a true alternation of generations as found in plant life cycles. Several species form symbiotic relationships with fungi to form lichens and others do so with sponges and cnidarians.

Students often ask, what is the difference between colonial and multicellular organisms? After all, both consist of many cells. Why use different words to describe similar organizational plans? The differences are subtle, but significant. Colonial organisms consist of the same types of cells that aggregate to produce the organism. All of the cells perform the same functions. There is no specialization of cells to perform unique functions as in our bodies. Multicellular organisms also consist of many cells. However, the cells are specialized to perform specific functions. Consequently, there is a division of labor among the cells, and the shape and functioning of the total organism is due to the sum of the individual contributions of each cell type. Any account of the evolutionary development of plants, fungi, and animals must offer hypotheses as to how multicellularity might have developed. As we look at diversity within this group, some hypotheses related to the origin of plants from ancestral green algae will be discussed.

Sexual reproduction in colonial and multicellular green algae varies from **isogamy** where both genders produce flagellated gametes of equal size that cannot be distinguished from each other, to **heterogamy** (=anisogamy) where males and females produce different-sized but otherwise similar gametes. Generally this takes the form of **oogamy** where the female produces a much larger, nonmotile egg and the male produces smaller flagellated sperm. Plants are all oogamous, leading some to suggest that it was heterogamous algae that gave rise to kingdom Plantae.

7 ▶ **Colonial Series and Heterogamy** In the lab there is a mixed culture of green algae called a volvocales culture. It contains several species ranging from unicells to colonies of varying shapes and sizes (**fig. 16.7**). First make a wet-mount slide of the culture without a coverslip to see the colonies. Later, add a coverslip to observe more detail. This single slide will illustrate an evolutionary trend from simple single cells to more complex colonies. The culture may have any of the following species:

Chlamydomonas sp.—This genus contains about 500 species of unicellular photosynthetic algae found worldwide in soils, fresh water, oceans, and even in snow on mountaintops. The cells have a single cup-shaped chloroplast and two flagella that pull it through the water in a breast-stroke fashion. The cells can divide asexually by mitosis or can reproduce sexually (**fig. 16.8**). The gametes look alike (isogamous) and are haploid with + and − mating types. Gametes fuse to form a diploid zygospore with an outer wall that protects it from adverse environmental conditions. When conditions improve, the diploid zygote undergoes meiosis and releases four haploid cells that resume the vegetative life cycle.

Eudorina sp.—Organisms in this genus have a colonial organization of 16 to 64 cells (mostly 32) covered by a gelatinous envelope in which the cells are quite distinct and separated from each other. The cells individually resemble *Chlamydomonas*. The two flagella found in each cell project to the outside of the colony's envelope and provide mobility. Sexual reproduction is heterogamous, but we will not go into details.

Gonium sp.—Organisms in this genus consist of nearly spherical cells, usually 8 to 16, arranged in a flat plate in a gelatinous matrix. Each cell is similar in appearance to *Chlamydomonas*. Because the cells are flagellated, the colony is able to swim.

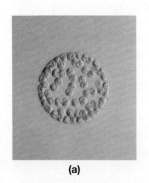
(a)

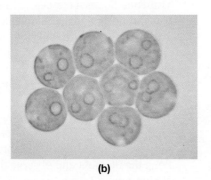
(b)

(c)

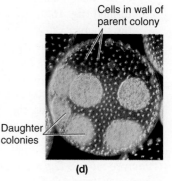
(d)

Figure 16.7 Several types of colonial green algae likely to be found in mixed culture: (*a*) *Eudorina* sp.; (*b*) *Gonium* sp.; (*c*) *Pandorina* sp.; (*d*) *Volvox* sp.

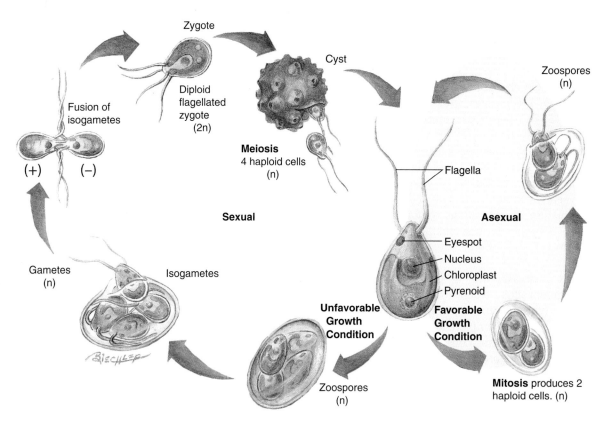

Figure 16.8 Life cycle of *Chlamydomonas,* a unicellular green alga.

Pandorina sp.—Usually 4 to 32 closely pressed cells are found in the gelatinous envelope of the spherical colonies of these algae. Again the cells resemble the base structure of *Chlamydomonas*. New colonies can be formed by asexual reproduction. Sexual reproduction is heterogamous.

Volvox sp.—This genus cannot be confused with any of the others. The spherical colonies, consisting of 500 or more cells, are the largest of this grouping and can be seen by the naked eye. Again the cells are similar to those of *Chlamydomonas* and the flagella give the colony mobility. Most of the cells are vegetative cells, but some are specialized for reproduction. Sexual reproduction in *Volvox* is oogamous; nonmotile eggs and motile sperm are formed (**fig. 16.9**) from single cells that differ from the others.

As you observe these colonies, you may see daughter colonies inside a parent colony. During asexual reproduction, some cells of a *Volvox* divide, bulge inward, and yield new colonies. When the parent colony dies, the daughter colonies are released.

In the following blue circle, draw examples of *Chlamydo-monas, Eudorina,* and *Volvox*. Discuss how the progression from simple to more complex could be considered a model for the development of multicellularity.

Filamentous Algae and Terminal Growth Some colonial green algae have their cells arranged as linear filaments. Transverse cell divisions add new cells to the growing tip of a single filament, in much the same way a plant grows by adding cells to the stem tip. Furthermore, some filamentous species are not only heterogamous they also produce gametes in specially differentiated regions called gametangia. Plants also produce gametes in gametangia.

8 You will look at one species of a filamentous alga that has gametangia. Obtain a prepared slide of *Oedogonium* and look at it through your compound microscope. Scan along the filament and observe the vegetative cells. Switch to high power to see the net-shaped chloroplast that surrounds the cytoplasm and nucleus.

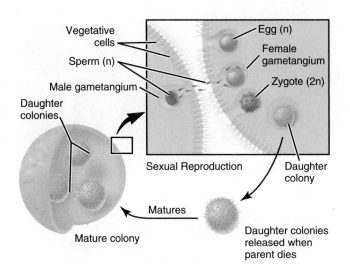

Figure 16.9 Sexual life cycle of *Volvox*, a multicellular, spherical green algae.

You should be able to see that some of the cells in the filament are dark colored and swollen (**fig. 16.10**). These are female gametangia called **oogonia** and contain a single egg. In fact, the genus name *Oedogonium* means enlarged egg cell. You should also be able to find other cells in the filament that are short and disc-shaped. These are male gametangia called **antheridia** and each produces motile sperm. Mature sperm have a crown of flagella on the anterior end. When released, sperm swim to and enter the oogonium where fertilization occurs. Is *Oedogonium* isogamous or heterogamous? _____

The development of gametangia, specialized regions for gamete development, was a necessary adaptation for algae to colonize terrestrial environments. In it, the gametes develop in a closed container which on land would protect them from drying.

Zygotes are identifiable by a thick cell wall that surrounds the oogonia. If you think you see one on your slide, ask your instructor to confirm and then share the view with others. Retention of the fertilized egg and further development of the zygote within the female is similar to the kinds of events that are also seen in plant life cycles, suggesting again that this structure may have been a second preadaptation that allowed some algae to colonize land.

Zygotes eventually divide by meiosis to produce four haploid **microzoospores.** When they escape from the oogonium, they grow to form new filaments which consist of haploid cells. When *Oedogonium* produces eggs and sperm, which type of cell division is involved? Mitosis or meiosis? _____

Asexual reproduction occurs when vegetative cells differentiate into large **macrozoospores,** which leave the filament and give rise to new filaments.

In the filamentous green algae, you have observed four characteristics that these algae share with plants. What were they?

1. _____
2. _____
3. _____
4. _____

Plant Clade Plants are also photosynthetic organisms, but unlike algae, most species of plants are terrestrial. Plants have numerous adaptations that allow them to live on land. Plant diversity will be studied in topics 17 and 18. Brief diagnostic characteristics include multicellular with distinct tissues; life cycle always involving an alternation of generations; eggs always fertilized inside a protective organ, the gametangium, where the young embryo also

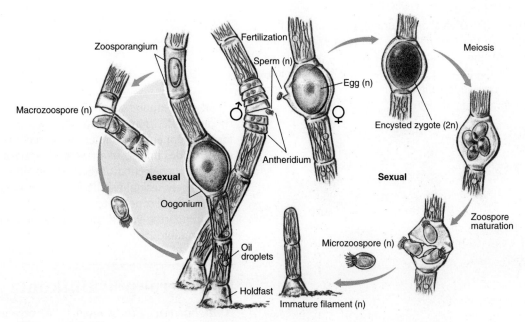

Figure 16.10 Life cycle of *Oedogonium*, a multicellular, filamentous green alga.

develops. Summarize your observations by adding the required information to table 16.2.

Supergroup 4: Rhizaria

A unicellular group that includes many of the amoebae with thin, hairlike extensions of their cytoplasm that are used in feeding. Many produce shells or external skeletons. Includes radiolarians and foramiferans.

Foraminiferan Clade Foraminiferans are a highly diverse group of protists. At least 35,000 species have been described. The name of this group means "hole bearers" and refers to the tiny holes in their shells (also called "tests") through which their pseudopodia protrude. The shell can be composed of organic material, cemented mineral particles, or calcium carbonate secreted by the organism. Most species have shells consisting of multiple chambers (fig 16.11). In some multichambered species sizes of up to a centimeter in diameter can be produced—quite remarkable for a single-celled organism. Some species use their pseudopodia to capture diatoms or bacteria as their food source. Other species of foraminiferans contain endosymbiotic photosynthetic organisms such as diatoms, green algae, or dinoflagellates, allowing them to function as photoautotrophs. Life cycles are complex and involve both asexual and sexual phases.

Because foraminiferans produce shells, they are exceptionally well represented in the fossil record. In fact, 28,000 of the 35,000 known species are found only as fossils. In addition, the species of foraminferans vary through geologic time, allowing scientists to date rock deposits. The oil industry makes use of fossil foraminferans to identify strata that are likely to contain oil deposits. Foraminiferan fossils are also used by scientists to learn about ancient global climates.

9 Examine a prepared slide of foraminiferans. Look first at the slide with the 10× objective and then examine the specimens with the 40× objective. What you are looking at is the shell of a foraminiferan—the living cell is no longer present, so pseudopodia will not be visible. Make a sketch of a foraminiferan below.

Figure 16.11 Diversity of calcareous shells (=tests) of different "forams." As the single-celled organism grows, chambers are added to the test. Thin, branched pseudopodia can be extended through pores in the test, creating a web that traps food particles. Most species live in benthic marine environments, although a few species are planktonic or occur in freshwater. Accumulation of tests forms chalky geologic deposits whose species composition correlates with past temperatures and atmospheric carbon dioxide levels.

What similarities and differences do you observe between foraminiferans and diatoms? Between foraminiferans and ciliates?

If a computer connected to the Internet is readily available, navigate to this website: http://www.foraminifera.eu/foraminifera-fossil-record.html. This website provides images of a few fossil species of foraminiferans. View a selection of the images. Which fossil species most closely resembles the foraminferan that you viewed in your microscope?_____

Summarize your observations of foraminiferans by filling in table 16.2.

Supergroup 5: Unikonta

Amoebozoan Clade Similar to organisms in clade Rhizaria, these organisms produce pseudopodia; however, the pseudopodia are considerably wider than those produced by Rhizarians. Nutrition is by phagocytosis of small particles and intracellular digestion. Most species produce cysts, although the slime molds produce spores. Organisms with such names as naked amoebas, shelled amoebas, and slime molds are included.

Gymnamoebas Commonly known as amoebas, these unicellular organisms live in fresh and salt water as well as moist soils. Some species are parasitic, such as *Entamoeba histolytica,* which causes amoebic dysentery. Other species, such as *Amoeba proteus,* are free-living. Other groups of amoebae ("shelled" or testate amoebas) include species that secrete a proteinaceous material that binds fine sand fragments together to build a protective "shell" or test. Reproduction is always asexual, involving mitotic-type cell divisions. Amoebas are common in soils, fresh water, and marine habitats where they feed on detritus or nonmotile bacteria and algae.

10 Go to the supply area and locate the culture of mixed amoebae. Unlike the other organisms you have observed today, the amoebae do not swim; they crawl on the bottom. Use a pipette to pick up some of the "scum" from the bottom and put a small drop on your slide. Add a coverslip and observe first with the 10× objective to find the plane of focus and then with the 40× objective as you spot something of interest. In viewing this slide, it is best to be patient. When amoebae are disturbed, they contract into a blob or into their shells. When they are not disturbed, they begin to spread out and to move about. When this happens, the viewing can be quite interesting.

Several different amoebae will be found in this mixture (fig. 16.12). These include:

- *Amoeba* sp.—This is a large "typical" amoeba that you may have seen in a previous biology class. There are numerous species in the genus and they live in fresh or salt water. Species in this group produce lobose cytoplasmic extensions called **pseudopodia** which are used in locomotion and to engulf prey. There usually is a single nucleus.
- *Arcella* sp.—These freshwater amoebae use fine sand grains which they bind together to build a **test.** The amoeba hangs in the chamber inside the test. An aperature at the bottom allows it to extend pseudopodia for feeding and locomotion. Tests can be yellowish to brown. Some host symbiotic algae and are green. These amoebae form cysts in adverse conditions.
- *Centropyxis* sp.—These are also testate amoebae but often the test has spines. The test surface is rough, and can be encrusted with diatom valves. The pseudopods frequently branch. These amoebae live in water drops that accumulate in damp areas or in freshwater streams and lakes.
- *Difflugia* sp.—Another testate amoeba, members of this group build a test containing both large and small particles and parts of diatoms that are cemented together by organic material that is laid down in rings. They live in wet bottom sands and marshes.

Myxogastridians (Slime Molds) Plasmodial slime molds are common in moist leaf litter and on decaying wood. There are about 500 known species. During the growth phase of its life cycle, it has a body form called a **plasmodium,** a nonwalled mass of cytoplasm containing hundreds of nuclei. Plasmodia may be bright orange or yellow in color (fig. 16.13). The cytoplasm in a plasmodium streams back and forth and allows the organism to locomote, moving several meters until it "finds" food. If you can imagine a large cytoplasm several inches long "crawling" about, you recognize where the common name slime mold originates. If food or moisture are reduced, plasmodial slime molds differentiate to form **sporangia** which produce spores that are haploid. The spores are resistant to harsh conditions but will germinate in a moist environment with food. Cells emerging from spores may fuse to start a new plasmodium.

11 Your instructor will have *Physarum poylcephalum* growing in the lab on 1% water agar with a few flakes of oatmeal for food. Several hours ago a piece of the slime mold was transferred to a fresh petri dish and was set up under a demonstration microscope so that you can see the cytoplasmic streaming. Watch for a few minutes and note how it changes direction. What purpose might this

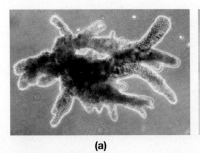 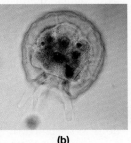 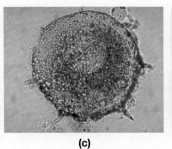 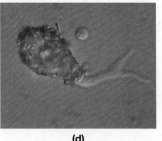

(a)   (b)   (c)   (d)

Figure 16.12   Types of amoeba likely to be seen in mixed culture: (*a*) *Amoeba* sp.; (*b*) *Arcella* sp.; (*c*) *Centropyxix* sp.; (*d*) *Difflugia* sp.

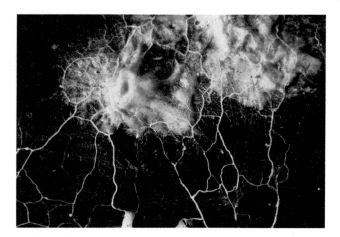

Figure 16.13 North American plasmodial slime mold (*Physarum polycephalum*) growing on a log.

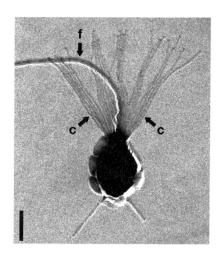

Figure 16.14 Shadow cast transmission electron micrograph of the choanoflagellate *Monosiga ovata*. Note the single, central flagellum (f) surrounded by microvilli (c), cytoplasmic extensions which form the characteristic collar. Bacteria suspended in water drawn through the collar by the beating of the flagellum are trapped on the microvilli and consumed. About 244 species of choanoflagellates are found worldwide in fresh and salt water habitats where they live usually attached to a substrate by a cytoplasmic thread. Genetic analysis indicates an evolutionary relationship to animals. Side bar = 2 μm.
Source: Proceedings of the *National Academy of Sciences* (USA), vol. 105, no. 43, 16,641–16,646.

serve in a large organism that feeds by secreting digestive enzymes and absorbing the products of digestion?

12. Your instructor may provide you with a petri dish containing water agar and a few flakes of oatmeal. Use a clean spatula to transfer a small amount of large slime mold in the lab to your petri dish. This is an inoculum. Take it back to your room for the next week. Each night look at it and record your observations. Bring this "journal" to lab next time.

Before putting your microscope and slides away, fill in the information for this clade in the summary table 16.2 on page 198.

Opisthokontan Clade Members of this clade are characterized by producing a single posterior flagellum in those cells that are flagellated. The flagellum propels the cell forward, as in a human sperm, rather than drawing it forward as seen in *Euglena*. This clade includes both kingdom Fungi and kingdom Animalia, as well as protists.

Choanoflagellates The ancestors of this small monophyletic group of unicellular and colonial organisms are most likely also the ancestors of the animal kingdom. They are characterized by a funnel-shaped collar of closely packed cytoplasmic filaments (microvilli) that surround a single flagellum on one end of an ovoid cell (**fig. 16.14**). This is exactly like the collar cells seen in sponges The choanoflagellate cell body is surrounded by an external covering called a periplast that may be polysaccharide, proteinaceous, or a silicate. A 2008 study found that several genes that code for cell adhesion and extracellular matrix proteins only occurred in choanoflagellates and animals, strongly suggesting shared origins. Most choanoflagellates are marine, but 50 freshwater species have been decribed. About 244 species are known and 16 are available in culture from the American Type Culture Collection. You will not study them in today's lab.

Fungi You will study this clade in lab topic 19. Brief diagnostic characteristics include heterotrophic feeding by absorptive means; body form as long hyphal filaments; cell walls containing chitin; usually without flagellated stages in life cycle; asexual and sexual reproduction by hyphal fusion.

Animals (Metazoans) You will study this clade in lab topics 20 through 23. Brief diagnostic characters include multicellular with cells held together by specialized cell junctions; no cell walls; production of proteinaceous extracellular matrices containing collagen; sexual reproduction is the norm with large eggs and smaller sperm bearing one flagella.

Lab Summary

13. On a separate sheet of paper, answer the following questions as assigned by your lab instructor. Your entries is Summary table 16.2 will help you answer the questions.

1. Name several characteristics that separate protists from bacteria.

2. Distinguish between unicellular, colonial, and multicellular and give an example of each from your lab work.
3. Name the five supergroups of eukaryotes that include "protists" and describe a unique characteristic of each.
4. Which clades have flagellated cells?
5. Which clades are photosynthetic?
6. Which clades contain parasitic species?
7. Name several human diseases caused by protists.
8. Describe three ways that protists propel themselves.
9. Describe how asexual reproduction occurs in protists.
10. Give an example of a protist that reproduces by sexual reproduction and discuss its life cycle.
11. What is the role of cysts in the life cycle of protists? How does a cyst differ from a test?
12. Give an example of a parasitic protist that has two hosts in its life cycle.
13. Plants are most closely related to protists in what clade?
14. What is meant by the phrase alternation of generations? Protists that exhibit this kind of life cycle are found in what clade?
15. Animals and fungi are most closely related to protists in what clades?
16. Give several examples of where protists are important members of ecosystems?

You may want to try the critical thinking questions that apply the knowledge gained in doing this lab.

Internet Sources

On the World Wide Web there are several databases that list information about diseases caused by protozoa. The U.S. Center for Disease Control (http://www.cdc.gov) is one and the World Health Organization (http://www.who.int/tdr) is another. The WHO lists 10 major tropical diseases. Of these, four are caused by protists that are transmitted by biting insects. Two of these diseases, malaria and leishmaniasis were not covered in this lab. Connect to one of these sites, use the search function, and read about how the diseases are transmitted, what the symptoms are, how they are treated, and where the diseases are common.

Learning Biology by Writing

In this lab you looked at protist representatives of the five supergroups of eukaryotes. Outline and write an essay about how eukaryotes might have originated from prokaryotic ancestors, giving rise to the incredible diversity of eukaryotes seen today. Explain the rationale behind the modern approach to phylogeny based on gene sequencing and protein analysis and how this has changed our views about how eukaryotic organisms are related to each other.

Critical Thinking Questions

1. Sexual reproduction has never been observed in some protists. What are the genetic implications of this observation?
2. Diatomaceous earth is classified as an "insecticide" by gardeners and is used to control such pests as ants, snails, slugs, and caterpillars. Explain its pesticidal properties.
3. Speculate on how colonial aggregates of cells may have led to multicellularity.
4. Because the classification of protists is a work in progress, compare the classification used here with that in your textbook. How are they similar? How do they differ?
5. Plants and green algae are in the supergroup Archaeplastida. Speculate on the evolutionary steps that led from algae to plants.

Table 16.2 Summary Table Protistan Diversity

Organism Observed	Supergroup	Clade	Uni, Colonial or Multicellular	Flagella/Cilia	Sexual/Asexual	Autotrophic/ Heterotrophic	Distinguishing Characteristics
Trichomonas							
Euglena							
Trypanosoma							
Paramecium and related							
Diatom strew							
Red algae							
Volvox and related							
Oedogonium							
Foraminiferans							
Amoebas							
Slime mold							

Lab Topic 17

ANCESTRAL AND DERIVED CHARACTERISTICS OF SEEDLESS PLANTS

SUPPLIES

Resource guide available on WWW at www.mhhe.com/labcentral

Equipment

Compound microscopes
Dissecting microscopes

Materials

Living or preserved plants
   Nitella
   Marchantia with gemmae
   Moss gametophytes with sporophytes (*Polytrichum*)
   Growing protonemata
   Mature ferns
   Where possible, examples of lower seedless plants
   Psilotum or herbarium specimen

Prepared slides
   Nitella w.m. with reproductive organs
   Marchantia antheridia, longitudinal section
   Marchantia archegonia, longitudinal section
   Marchantia sporophyte, longitudinal section
   Mature moss capsule, longitudinal section
   Mature moss antheridia and archegonia, longitudinal sections
   Psilotum, stem cross section
   Fern antheridial and archegonial gametophyte
   Fern sori, w.m.
   Fern sporophyte growing from archegonium

Slides, coverslips

STUDENT PRELAB PREPARATION

Before doing this lab, read the Background material and sections of the lab topic that have been assigned by the instructor.
   Use your textbook and lab manual to review the definitions of the following terms:

antheridium	archegonium
antheridiophore	Bryophyta
archegoniophore	diploid
gametangium	ploidy level
gamete	protonema
gametophyte	Pterophyta
haploid	sporangium
Hepatophyta	spore
Lycophyta	sporophyte
oogamy	vascular plant

   Be able to describe in your own words the following concepts:

Alternation of generations
Vascular tissue (xylem and phloem)
Reproduction in seedless plants
Adaptations to land

   After finishing the prelab review, write any questions you have about terms, concepts, or techniques in the margins of this lab topic. The lab experiments should help you answer these questions, or you can ask your instructor during the lab.

OBJECTIVES

1. To recognize that green algae share ancestral characteristics with plants
2. To observe the body plans of liverworts, mosses, and ferns
3. To learn the life cycle stages in the alternation of generations found in liverworts, mosses, and ferns
4. To collect evidence that tests the following hypothesis: Seedless plants can be arranged in a sequence demonstrating increasing adaptation to terrestrial environments

BACKGROUND

Plants are a diverse group of multicellular, photosynthetic organisms ranging from simple mosses to complex flowering plants. All plants, and some green algae, share certain characteristics, which include: (1) chloroplasts with thylakoid membranes stacked as grana and containing chlorophyll, (2) starch as a storage polysaccharide in the chloroplasts, (3) cellulose in cell walls, (4) cytoplasmic division by cell plate formation, (5) complex life cycles involving an alternation of generations in which a diploid sporophyte stage alternates with a haploid gametophyte stage; and (6) heterogamy with distinct male and female gametes produced in organs called gametangia.

What distinguishes plants from green algae is their mostly terrestrial life style and retention of the zygote as it develops into an embryo, thus protecting it from drying and allowing the "mother" to provide nutrients to the developing embryo. For these reasons, plants are called embryophytes.

Phylogeny is the study of evolutionary relationships among groups of organisms. The evolutionary story of plants is linked to the colonization of the terrestrial environment, an event that started about 430 million years ago (**fig. 17.1**). Most plant biologists agree that plants arose from an ancestral green algae. For the first 4 billion years of earth's existence, no life existed on the land, but at the end of the Silurian period, the first land plants appeared and rapidly diversified. Ten phyla of plants are recognized today (**table 17.1**). The appearance of producer organisms in the terrestrial environment, other than cyanobacteria, established a strong basis for terrestrial food chains. Terrestrial plants were then rapidly exploited by the evolution of terrestrial fungi and animals from aquatic ancestors.

As do many algae, all plants show an **alternation of generations** in their life cycles. In plants, there are two alternating multicellular stages, a diploid **sporophyte**

Table 17.1	Plant Phyla	
Phylum	Common Name	Estimated Living Species
Nonvascular Plants		
Seedless plants		
Bryophyta	Mosses	15,000
Hepatophyta	Liverworts	9,000
Anthocerophyta	Hornworts	100
Vascular plants (Tracheophyta)		
Lycophyta	Club mosses	1,200
Pterophyta	Ferns and Horsetails	12,000
Seed plants		
Gymnosperms		
Coniferophyta	Conifers	600
Cycadophyta	Cycads	170
Ginkgophyta	Ginkgo	1
Gnetophyta	Gnetae	75
Angiosperms		
Anthophyta	Flowering plants	250,000

generation and a haploid **gametophyte** generation. These two stages are not always obvious to the casual observer because one of them is often microscopic or seasonal. One goal of this laboratory topic is to give you a chance to observe both stages for liverworts, mosses, and ferns.

Figure 17.2 illustrates the stages in an abstract form. The gametophyte stage develops from a haploid plant spore when it divides by mitosis to produce a haploid multicellular gametophyte. In the adult gametophyte, haploid egg and sperm (gametes) are produced by mitosis. When the gametes fuse at fertilization, the resulting diploid zygote is the first cell of a new sporophyte generation. The zygote develops by mitosis into a diploid, multicellular sporophyte stage. In specialized structures called sporangia, cells called sporocytes divide by meiosis, producing four haploid cells called spores. The spores develop into gametophytes, completing the life cycle. All plants have both stages in their life cycles, but one must look carefully, often using a microscope, to see both. *For now, remember gametophytes are multicellular, haploid, and produce gametes by mitosis. Sporophytes are multicellular, diploid, and produce spores by meiosis.*

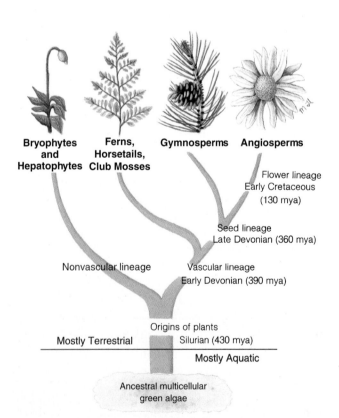

Figure 17.1 A summary of plant evolution. Note the early origin from the green algae (mya = millions of years ago).

Plant Adaptations to Land

The critical step for the ancestors of plants in making the transition from aquatic to terrestrial environments was the development of adaptations to prevent desiccation. In the earliest evolving groups of land plants, gametes were produced inside a protective covering, the **gametangium**. It is a jacket of nonreproductive cells that prevents the

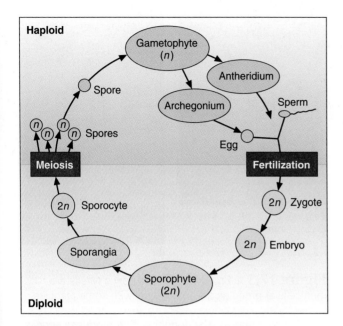

Figure 17.2 All plants have an alternation of generations in their life cycles. They produce spores that develop into gametophytes which produce gametes by mitosis. The zygote is the first cell of the sporophyte generation. Sporophytes produce spores by meiosis.

drying of gametes and following fertilization, the drying of embryos while providing nutrients to them. Furthermore, some land plants developed basal organs (roots) that absorbed water and minerals from the substrate to replace water lost by evaporation. The aerial body parts developed a waxy **cuticle** that reduced water loss. Hereditary variations of this type arising in ancient shoreline algae would have allowed the ancestors of plants to enter an environment that, though harsh in its physical characteristics, was free from competitors for resources. Such an environment would have allowed the colonizers to establish themselves and further diversify.

Once the transition to land occurred, two major lines of evolution developed among the land plants. The direction taken by the nonvascular plants (table 17.1) saw the gametophyte generation become the predominant photosynthetic stage in the life cycle. Following fertilization, the gametophyte retained the embryo as it developed into a multicellular sporophyte that was dependent on the gametophyte for nutrition and was typically short-lived. The sporophyte produced spores with a waterproofing agent, sporopollenin, in their cell walls that made them resistant to desiccation. Spores became the primary means of dispersal.

The other evolutionary line gave rise to vascular plants (table 17.1). In these plants, cells located in the stems, roots, and leaves became elongated and specialized for transport and support. In some members of this group a vascular cambium developed, which allowed secondary growth, increasing the girth of the stem and root. These tissues allowed vascular plants to grow to large sizes, which, in turn, favored development of more efficient anchorage and absorptive systems. In vascular plants, the sporophyte stage of the life cycle became more prominent than the gametophyte stage that dominates the nonvascular plant's life cycle. At the same time, reproductive adaptations occurred, giving rise to the seed plants, which you will study in lab topic 18.

In seed plants, such as conifers and flowering plants, the gametophyte phase of the life cycle became reduced to a microscopic stage dependent on the sporophyte stage. Male gametophytes became desiccation-resistant airborne **pollen**, which produced sperm that no longer needed water to swim to the eggs. Eggs were produced by a female gametophyte that was retained in a protective organ. When fertilized, the zygote and surrounding tissues developed into a **seed** containing an embryo and food reserves. In this group, seeds, rather than spores, became the effective dispersal mechanisms.

Adaptations to land over eons of time allowed plants to invade many different environmental niches. Relatively simple changes allowed the colonization of environments where water was readily available, while more complex adaptations were necessary for life in desert environments. As you study the trends outlined in this lab topic, remember that contemporary plants in different phyla have not descended from one another. They have similarities because they share common ancestors. Present species are related as cousins or more distant relatives, not as you are to your parents or siblings.

LAB INSTRUCTIONS

You will study first a group of green algae that share some common features with plants, suggesting what the ancestors of plants might have looked like. Then you will look at the terrestrial adaptations of liverworts, mosses, and ferns. You will learn the distinctions between the sporophyte and gametophyte stages in the alternation of generations that characterizes all plants.

Algal Preadaptations to Land

We start our investigation of plant diversity by looking at a group of aquatic green algae (Clade Chlorophyta) commonly called stoneworts. The "stone" part of the name comes from the fact that they secrete $CaCO_3$ on their surfaces and "wort" means herb or small plant. This small group of about 300 species is distributed worldwide in freshwater streams and lakes. What makes them interesting is the fact that they have certain structures considered necessary for life on land as well as many other shared characteristics with plants. Were organisms like these the progenitors of land plants? When you are finished looking at them, we think you will be asking why these are not considered plants. Why do we say this?

Stoneworts (in the taxonomic order Charales) have several plantlike characteristics. These include:

- **Multicellular,** consisting of many specialized cells that perform only certain functions
- **Gametangia,** reproductive organs surrounded by nonreproductive cells that protect the gametes as they develop
- **Oogamy,** the production of two different types of gametes, one smaller and motile and other larger and immotile
- **Sporopollenin,** a waterproofing compound found in the walls of plant spores and pollen
- **Plasmodesmata,** small strands of cytoplasm connecting adjacent cells
- Rosette-shaped cellulose-synthesizing complexes involved in cell wall formation
- Similar mechanisms in mitosis and cytokinesis involving microtubules
- DNA sequence similarities
- Terminal growth from an apical cell

While these shared characteristics argue persuasively for stoneworts being closely related to plants, two essential characteristics are missing. Stoneworts do not have an multicellular embryo developing from the zygote and consequently do not show an alternation of generations with distinct multicellular gametophyte and sporophyte stages.

Stonewort Anatomy and Life Cycle

Our representative green alga will be the stonewort *Nitella* sp. **fig. 17.3a**. It is found worldwide with 30 species in the United States growing in acid bogs and lakes.

1) Body Plan Look at the whole specimens available as demonstrations in the lab. This is a multicellular, filamentous green alga and can grow to be 30 cm tall. Note the "stem" with nodes that produce whorls of what look like "leaves." These are neither stems nor leaves as your microscopic observations will prove. The base of the central stalk has branched rhizoids, long thick cells that anchor it. If the specimen is in reproductive condition, some nodes will have **gametangia,** organs where the gametes develop. It is these structures which interests us the most because similar structures are found in the gametophyte stages of plants. In the dry land environment, gamete production must occur inside protective organs, so its occurrence in aquatic algae preadapts them for a transition to land.

2) Obtain a slide of a whole mount of *Nitella* sp. with gametangia and look at it with your scanning objective. Switch to 10× to observe structures in more detail.

Note that the long internodal areas consists of a single, large **internodal cell.** At the nodes are several small **nodal cells.** Division of nodal cells gives rise to the whorls

(a)

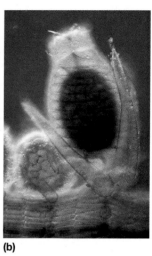
(b)

Figure 17.3 Stonewonts in the genus *Nitella* are multicellular green alga (Chlorophyta) with some plantlike characteristics. (a) whole organism; (b) enlarged view showing gametangia. The oogonium that produces eggs is the large green structure and the antheridium that produces sperm is the small orange-colored structure.

of leaflike **laterals** at each node. Note that they have the same cellular construction as the main axis of the plant and are not leaves. The stage you are looking at could be considered a gametophyte stage in a life cycle because it is haploid and bears specialized reproductive structures called **gametangia.**

Draw a stonewort in the circle below. Label all structures printed in boldface type in the Body Plan Section.

Sexual Reproduction Look at several nodes to locate gametangia (fig. 17.3b). Male gametangia, called **antheridia** (sing. antheridium), are round and orange in color at maturity but the staining may obscure color. Inside a surrounding jacket of nonreproductive cells, other cells will divide by mitosis to produce biflagellated sperm cells. If the sperm cells are produced by mitosis, what must be the ploidy level of the sperm? _____

Gametangia that are oval, with a spiral pattern of ridges, are female gametangia called **oogonia** (sing. oogonium). Each contains a single egg produced by mitosis. What is its ploidy level? _____.

Sperm released from the antheridia swim to the oogonium. Openings at the tip of the oogonium allow sperm to enter and fertilize the egg. The resulting zygote is diploid. It is retained in the oogonium. During maturation it synthesizes a thick wall containing sporopollenin around itself. Internal fertilization of an immobile egg followed by zygote retention is a characteristic of plants. *Nitella* differs from plants in that the zygote does not develop into a multicellular embryo. Instead, the oogonium releases the mature zygote. It divides by meiosis to produce four haploid cells. Three degenerate, leaving a fourth that divides by mitosis to produce a filament of cells, becoming a new individual. Note that as a result of biparental inheritance and the crossing-over and independent assortment during meiosis, each individual is genetically distinct. What is its ploidy level? _____

Earlier it was stated that stoneworts do not have an alternation of generations life cycle. What is missing is a multicellular diploid sporophyte stage. The only diploid stage of the life cycle is the single-celled zygote. The diploid phase ends with the meiotic division of the zygote. It is, in part, this omission in the life cycle that keeps us from calling stoneworts plants. The multicellular haploid stage that follows, however, is a fine candidate for a gametophyte.

Biologists think that the alternation of generations type of life cycle may have started when an ancestral stonewort retained its zygotes. They then divided by mitosis to produce a multicellular diploid stage. Some cells in this stage then divided by meiosis to produce haploid cells that were encased in tough walls, forming what we call spores. These, in turn, developed by mitosis into a multicellular haploid stage.

3▶ Summarize Observations Provide the information required in the first column of **table 17.2** at the end of this lab topic.

Choose a lab partner and describe to one another the life cycle of a stonewort.

In the next two sections, you will study liverworts, mosses, and ferns that clearly show not only haploid multicellular gametophyte stages but also diploid multicellular sporophyte stages.

Bryophytes

The term "bryophyte" has been used traditionally to describe about 25,000 species of small plants belonging to three phyla: Hepatophyta—the liverworts, Anthocerophyta—the hornworts, and Bryophyta—the mosses. The bryophytes illustrate a clear alternation of generations in which a multicellular gametophyte is the predominant stage of the life cycle. It is the plant that you would look at and say, "It's a moss." The multicellular sporophyte stage of the life cycle is reduced in size and function, is typically short-lived, and is nutritionally dependent on the gametophyte for its survival. Reproduction occurs sexually by gametes and spores and asexually by fragmentation. Although land plants, bryophytes are not completely adapted to a terrestrial way of life. Their motile sperm still must swim through water films to travel from the male gametangium to the eggs held in the female gametangium for fertilization to occur. Most bryophytes are small, less than 10 cm tall, due to their lack of vascular tissues.

You will look at two of the three phyla of plants called bryophytes.

Phylum Hepatophyta (Liverworts)

The name of this phylum is derived from the fanciful resemblance of the shape of some of these plants to the shape of the human liver and from the word *wort* meaning herb.

Gametophyte Stage

4▶ Body Plan Examine living specimens of *Marchantia*. It is commonly found on the surfaces of damp rocks and soil. Individual plants consist of a body called a **thallus** (fig. 17.4). It is the haploid gametophyte stage of the life cycle. The surface of the thallus has a diamond-shaped pattern, which corresponds to air chambers located beneath the upper epidermis. Note the numerous multicellular scales and single-celled, hairlike **rhizoids** on the lower surface. These are an adaptation for anchorage and their surface area allows water and mineral absorption, although they lack the specialized conducting cells found in true roots.

On the upper surface of the thallus may be tiny **gemmae cups.** These produce asexual **gemmae,** small, multicellular discs of green tissue. When dispersed to suitable locations, they develop into new gametophytes. If the cells of gemmae develop from thallose tissue by mitosis, what is their ploidy level? _____

Sexual Reproduction Sexual reproduction in *Marchantia* involves separate male plants that bear gametangia (sex

Figure 17.4 Leaf-like thalli of the liverwort *Marchantia* with sexual reproductive structures growing at ends of stalks.

organs) called **antheridia** (sing.-antheridium) and female plants that bear gametangia called **archegonia** (sing.-archegonium). Note the significant similarity here with the green alga *Nitella*. Gamete formation occurs inside a protective jacket of nonreproductive cells forming a gametangium. What could be the advantage of this arrangement for a plant living on land?

In the proper season, a new body part appears in the gametophyte. Small, stalked male **antheridiophores** or female **archegoniophores** grow from the notches in the tips of thalli (**fig. 17.5**). The gametangia develop on the caps of these stalks. Antheridiophores look somewhat like an inverted, flat cone on a stalk and will have antheridia on their upper surface. Archegoniophores bear archegonia on their lower surface. They look like fleshy minature palm trees.

5 Obtain a slide of a longitudinal section of an antheridiophore and look at it with the scanning objective. Note the general shape and the multicellular nature of the structure. On the upper surface, find the several ovoid **antheridia.** Sperm develop by mitosis inside the sterile jacket and are released through an opening on the upper surface (fig. 17.5). Several hundred sperm are produced by mitosis in each antheridium. Are sperm haploid or diploid? _____

6 Now look at a slide of a longitudinal section of an archegoniophore, the female reproductive structure. In

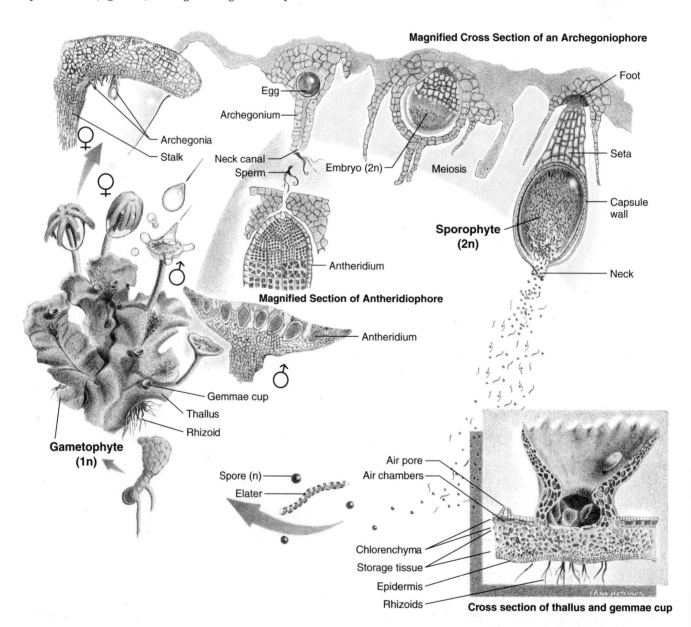

Figure 17.5 Stages in the life cycle of the liverwort *Marchantia*. Yellow shading highlights sporophyte generation.

rows on the underside of the cap are flask-shaped **archegonia** (fig. 17.5). At the base of each archegonium is a swollen area containing a single **egg.** Eggs are produced by mitosis, as are sperm. Are eggs haploid or diploid? _____ How do you know?

Extending downward from the egg in the archegonium is a **neck canal.** Sperm must swim through this canal to fertilize the egg. Fertilization requires water. Water droplets from rain or heavy dew falling on the antheridiophores splash sperm randomly onto adjacent archegoniophores. Using their flagella, the sperm swim through a water film to the egg. The resulting zygote (2N) develops into a multicellular sporophyte stage by mitosis while it is embedded in the female gametophyte's tissue.

Sporophyte Stage

7▶ Body Plan Obtain a slide of a maturing sporophyte. Following fertilization, the zygote divides repeatedly by mitosis, forming a microscopic but multicellular sporophyte stage within the tissue of the female gametophyte (fig. 17.5). There it is protected and receives nutrients from the female gametophyte. Find the sporophyte stage on the slide and look at it using scanning power.

Most of the cells in the sporophyte form a **capsule,** which surrounds other cells that become **sporocytes** and **elaters,** the latter being slender, elongate cells (fig. 17.5). Together these cells make up the contents of the **sporangium.** Thus, the sporophyte generation of the thalloid liverwort is only a microscopic sporangium attached to the lower surface of the archegoniophore growing from the gametophyte thallus.

Sexual Reproduction (continued) Inside the sporangium, each sporocyte divides by meiosis, forming four haploid cells. They mature into spores with thick cell walls. The walls contain sporopollenin which prevents drying.

When the sporangium matures, it ruptures, exposing a cottony mass of spores and elaters. The elaters, intertwined in this mass, are hygroscopic. Spiral thickenings in their cell walls cause the elaters to twist and coil as they dry or take up moisture, dislodging spores from the mass. The spores fall out of the sporangium and are disseminated by the wind. If they land in a moist area, they germinate and divide by mitosis to form a new gametophyte generation. What will be the ploidy level of the gametophyte's cells? _____

Note: Individuals developing from different spores are genetically unique because of two events in meiosis: crossing-over and independent assortment.

8▶ Summarize Observations When you finish this section, quiz your lab partner about the life cycle of *Marchantia.* Be sure that both of you understand what the gametophyte and sporophyte stages look like. Know where mitosis and meiosis occur in the life cycle. Fill in the information required in table 17.2 at the end of this lab topic.

Return all materials to the supply area.

Phylum Bryophyta (Mosses)

Many organisms commonly called "mosses" are not mosses. Reindeer moss is a lichen, Irish moss is an alga, and Spanish moss is a vascular plant. True mosses are members of the phylum Bryophyta and have the anatomical and reproductive features which you will study in this section.

Your instructor may have a display of living mosses for you to look at before you start your microscope work.

Gametophyte Stage

Body Plan Haploid moss spores germinate and divide by mitosis to produce haploid **protonema,** algalike filaments of cells, which will develop into the familiar moss plant with gametangia. What are gametangia and why are they important in terrestrial plants?

9▶ Remove some protonemata from the stock culture tube, make a wet-mount slide of it, and observe through your compound microscope. Alternatively, you may use a prepared slide of a protonemata and a young gametophyte. Each protonema is a branching thread of single cells joined end to end, similar in general appearance to a filamentous algae.

Sketch what you see on your slide. What is the ploidy level of the cells in a protonema?

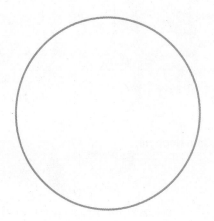

Leafy, green **gametophytes** will develop from **buds** on the protonemata (**fig. 17.6**). The gametophyte is the readily recognized stage in the life cycle of a moss. The central shaft is not really a stem. It has no vascular tissue. The leaves are not true leaves. They are only a single cell thick, except at the center, and lack stomata.

As the gametophyte matures, male and female gametangia differentiate at the tips of the stalks, but you cannot see them with the naked eye.

10 *Sexual Reproduction* To see the detail of the female gametangia, examine a prepared slide of a longitudinal section of a female gametophyte. Look at it first with your scanning objective, increasing magnification where necessary to see detail. At the tip look for the bowling-pin-shaped structures (fig. 17.6), called **archegonia**. Note that each archegonium has a jacket of cells that protects the egg during development. A single egg should be located in the swollen base, but will not be seen in every section. The egg is produced by mitosis. Will it be haploid or diploid? _____

In the narrow upper region of the archegonium, you may be able to see the **neck canal**. Sperm must swim down this canal to fertilize the egg.

11 Now obtain a prepared slide of a longitudinally sectioned tip of a male gametophyte. At the tip, find the oval-shaped **antheridia** (fig. 17.6). The antheridia have a thin layer of jacket cells that protect the sperm during development. The tightly packed angular cells inside the jacket are sperm in various stages of development. Are sperm formed by mitosis or meiosis? _____

If a water film develops on the tip of a sexually mature male gametophyte, the sterile jacket of cells splits open releasing sperm into the water film. After a few minutes they actively swim, using two flagella. Moss sperm require water films to get from the antheridium to the

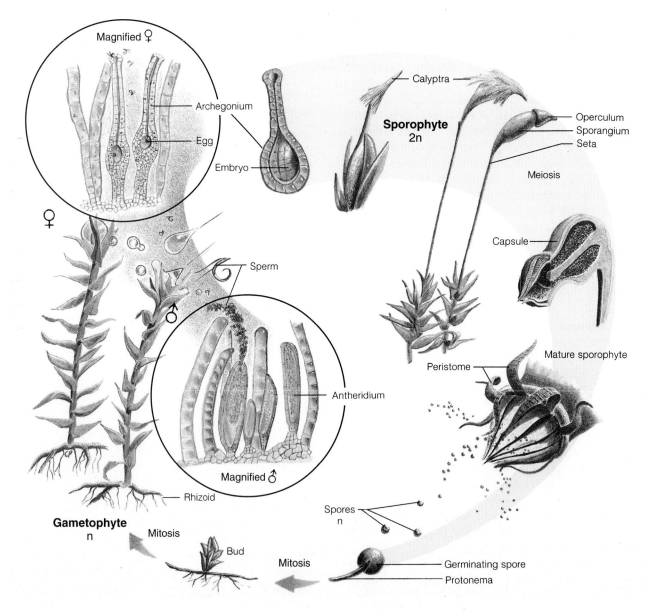

Figure 17.6 Stages in life cycle of a moss. Yellow highlights sporophyte generation.

eggs in the archegonia. Water droplets from rain or heavy dew will randomly splash sperm from the antheridium on one gametophyte to the archegonium on another.

Note the significant similarity here with the green alga *Nitella*. Gamete formation is in a sex organ (gametangium) consisting of a protective layer of nonreproductive cells. What could be the advantage of this structural development to a land plant?

Sporophyte Stage

Body plan Zygotes develop into multicellular, diploid embryos by mitosis while surrounded by the protective jacket of archegonial cells. The embryo marks the start of the sporophyte generation. The retention of the zygote and its growth into a multicellular diploid sporophyte stage differs from the life cycle seen in the green alga *Nitella* but is similar to what you saw in liverworts.

The embryo grows by mitotic division, becoming a long stalk of sporophyte tissue, the **seta** (fig. 17.6). The end of the seta differentiates into a **capsule**. The top of the capsule is closed by a lid, or **operculum**. Inside the capsule, meiosis occurs, producing haploid spores. Spores remain viable for several years and are the primary agents of dispersal for the species.

12 If mature capsules are available, observe them under a dissecting microscope, removing the calyptra and operculum to release the spores. Alternatively, observe a longitudinal section of a mature capsule on a prepared microscope slide.

When the capsule matures, the operculum drops off. Some species have a **peristome,** a ring of toothlike units that flex with changes in humidity releasing the capsule's spores during periods of dryness. The spores will germinate to produce protonemata, completing the life cycle. As a result of biparental inheritance followed by crossing-over and independent assortment during meiosis, no two protonemata will be genetically identical.

Mosses have been called the amphibians of the plants. In what ways are mosses adapted to land? In what ways are they still dependent on water?

3 **Summarize Observations** A major difference between *Nitella* and bryophytes is that the embryo developing from the zygote is retained in the archegonium as it develops into a mature sporophyte. How is this an advantage to a plant that lives on land?

Pair off with someone who has finished observing mosses. One person should describe the gametophyte generation, including how and where gametes are formed. The other should describe fertilization and the sporophyte, generation, including how and where spores are formed.

Return all materials to the supply area. Provide the information requested in the third column of table 17.2 at the end of this lab topic.

Vascular Plants

Vascular plants (also known as tracheophytes) have a number of adaptations to living in terrestrial environments not seen in the bryophytes. There is a general trend across the group wherein the plant body is differentiated into a subsoil root system for the absorption of water and minerals, and an aboveground stem that bears photosynthetic leaves. This regional specialization of the body and an increase in body size required the development of new types of plant tissues, the vascular tissues. The cylindrical cell walls of **xylem** function in carrying absorbed water and minerals to the aboveground parts of the plant, while the elongated **phloem** cells conduct the products of photosynthesis from the leaves to all other parts of the plant. The cell walls of the xylem are reinforced with lignin, allowing xylem to assume the additional function of being a support tissue. Consequently, most vascular plants can grow taller than the nonvascular plants.

The vascular plants include both seedless and seed-bearing plants. The seedless vascular plants include extinct seed ferns and the extant club mosses, horsetails, and ferns (see **fig. 17.7**). In lab topic 18 you will study the vascular plants that produce seed.

Aspects of Body Plan *Psilotum,* a tropical whisk fern, is unique among the vascular plants because it lacks roots and leaves. Root function is performed by a stem that grows underground and symbiotic endomycorrhizal fungi (see lab topic 19) that absorb water and minerals. Leaf function is carried out by a photosynthetic stem.

14 Obtain a slide of a cross section of a whisk fern (*Psilotum*) from the supply area and look at it through your compound microscope using the scanning objective.

We are using this specimen because it clearly illustrates vascular tissue organization in its stems. Note the obvious organization of the tissues in the stem. Look at the group of thick-walled cells at the center of section. These are vascular tissues, more specifically **xylem** cells, that function to carry water from the below-ground parts

Figure 17.7 Sporophyte stages of the fern allies: (a) *Lycopodium*, a club moss sometimes called a ground pine or cedar, has a below-ground stem that produces roots and an aboveground stem that bears leaves and sporangia; (b) *Psilotum*, a whisk fern, has an aboveground and belowground stem; (c) *Equisetum*, a horsetail, has jointed aboveground stems that bear whorls of leaves and below-ground true vascular roots. Some stems bear a terminal reproductive structure called a strobilus (foreground) which produces spores that give rise to the gametophyte generation.

to the above ground parts of the plant. To the outside of these cells are **phloem,** cells that transport photosynthetic products from sites of photosynthesis to non-photosynthetic cells. Beyond the phloem are thin-walled parenchymal cells that function in storage. The surface is covered by epidermal cells with periodic openings between them, permitting gas exchange. Beneath the epidermis is a three- to-four-cell thick photosynthetic layer. Draw a 90° pie-wedge segment of the stem below, showing the general organization of the stem. You need not sketch cellular detail, although you might want to add notes on cell wall thickness.

Phylum Pterophyta (Ferns and Horsetails)

In this section, you will study the life cycle stages of ferns as an example of seedless, vascular plants. The sporophyte generation is the most conspicuous stage in the life cycle of ferns, but they also have an independent gametophyte stage.

15 *Gametophyte Stage* Get two prepared slides of male and female gametophytes, also known as prothalli, from the supply area. Alternatively, your instructor may have grown gametophytes from spores, using a C-fern (*Ceratopteris*) culture kit available from suppliers. If that is the case, work with live material making wet-mount slides of the gametophytes with the lower surface oriented upward.

16 *Body Plan* Look at one of the slides using the scanning objective of your compound microscope. Rarely seen in nature because they are small, about the size of your little fingernail, and short-lived, fern gametophytes have heart-shaped, leaflike bodies, called **thalli (sing. thallus)**, that are typically one cell thick (**fig. 17.8**). They grow by mitosis from germinating haploid fern spores that are air blown to suitable locations. Thalli are photosynthetic, harvesting sunlight energy to fuel their needs, and have rootlike rhizoids on the undersurface that absorb minerals and water.

Gametangia develop on the undersurface of the thallus. Male gametangia, called **antheridia,** produce flagellated sperm and female gametangia, called **archegonia,** that produce nonmotile eggs. While many textbooks show that a single thallus will have both antheridia and archegonia, this is typically not the case in nature. When several thalli grow close together, the earliest developing ones produce eggs and the slower ones sperm. A hormone

cally determined as individuals can be male or female, depending on the environment they develop in. When a gametophyte develops alone with no neighbors, it can produce both antheridia and archegonia with resultant self-fertilization. A mating system such as this promotes cross fertilization while assuring survival. What is the advantage of cross fertilization?

Sexual Reproduction On the male slide look for darkly stained antheridia scattered out from the centerline as bumps on the lower surface of the thallus. Use higher magnification and **figure 17.9** to identify the ring and cap cell. Inside the sperm will be densely packed. In mature antheridia, water will cause the cap cell to pop off and the sperm will swim through the water film using flagella. Splashing rain or dew drops will randomly transfer sperm from one thallus to another.

17) Now look at the female slide for larger, darkly stained archegonia. With higher magnification, find the canal that leads to an egg located at the base of the archegonium. Water films allow the sperm to swim to and enter into the canal, fertilizing the egg.

- What is the ploidy level of the gametophyte? _____
- What type of cell division produces gametes in the gametangia? _____
- The zygote that is formed is the first cell of the sporophyte stage of the life cycle. It will divide by mitosis to produce a multicellular adult sporophyte. What is the ploidy level of its cells? _____

18) ***Sporophyte Stage*** Obtain a whole mount slide of a young sporophyte growing from an archegonium of a gametophyte. Look at it with the low power of your compound microscope and compare it to figure 17.9.

Body Plan Initially, the sporophyte grows out of the surface of the gametophyte, but as it enlarges and tissues begin to form, the gametophyte withers and dies, leaving a young, independent sporophyte. Find the developing vascular tissues in sporophytes and note how they extend from the base with developing roots into the young leaves. It is this system of conducting cells that will allow the fern sporophyte to achieve a larger body size than you saw in any of the previously studied plants. After several cell divisions, the sporophyte will begin to look like the ferns that you see growing in nature.

The stems of most ferns are horizontal just beneath the soil surface and are called **rhizomes.** However,

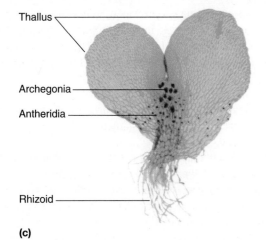

Figure 17.8 Ferns: (*a*) fern fronds are the sporophyte stage of life cycle; (*b*) sori on undersurface of the frond contain sporangia, which produce spores; (*c*) the gametophyte stage of fern life cycle develops from a spore. Rarely are both antheridia and archegonia on the same thallus.

produced by the fast-developing gametophytes influences gametangia development in slower-growing ones causing them to become exclusively male. Sex is not geneti-

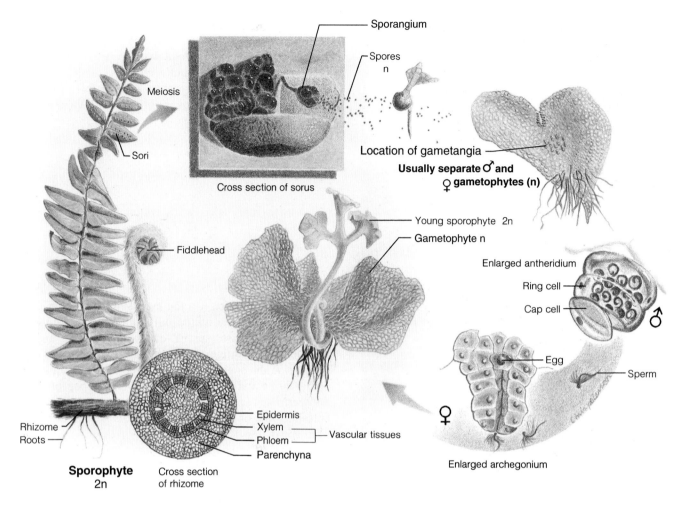

Figure 17.9 Stages in the life cycle of a fern. Yellow highlights sporophyte generation.

these structures are quite different from the rhizoids found in the fern gametophyte stage or in the mosses. Rhizoids are single, elongated cells joined end to end, which do not connect with vascular tissue. Rhizomes, on the other hand, are multicellular and contain vascular tissue.

Leaf buds along the rhizome produce leaves. These leaves break through the soil surface in a coiled position called a **fiddlehead.** It gradually unrolls to produce the fern **frond,** a single compound leaf. Roots with root hairs also develop from the rhizome and provide the sporophyte stage with water, minerals, and anchorage. Since the cells making up these structures arise from the zygote by mitosis, are they haploid or diploid?_____

At maturity, spore-producing sporangia will develop on the undersides of a fern's leaves (fig. 17.8). Often in clusters, called **sori,** sporangia produce haploid spores by meiosis. When released the desiccation-resistant spores allow for the dispersal of ferns to new locations. Spores are often sold as fern "seed" but you will see in topic 18 that spores are quite different from seeds.

19 Obtain a prepared slide of sori on the underside of a leaf from the supply area. Look at it and compare it to **figure 17.10.** Note the many sporangia in each sorus. Off to the side of the leaf you may be able to see scattered individual sporangia, some open, releasing spores, and others closed. If spores land in suitable environments, they germinate to produce prothalli (gametophytes) by mitosis. What is the ploidy level of the gametophyte's cells?_____

As a result of biparental inheritance followed by crossing-over and independent assortment during meiosis, no two spores or subsequent gametophytes will be genetically identical.

20 If ferns with mature sporangia are available, try this experiment. Remove a leaflet from a leaf and place it on a piece of white paper. Shine a bright, hot light on it while observing through a dissecting microscope. You may see explosive movements as the sporangia open.

21 *Summarize Observations* Return all slides and materials to the supply area. The life cycle of a fern is summarized in figure 17.9. Recall the stages you have observed as you study this illustration. The adaptations of the fern to the terrestrial environment include the desiccation-resistant, dispersible spores; the well-developed vascular system; the root system; and the erect, desiccation-resistant photosynthetic fronds. Most of these adaptations are in the sporophyte stage, which, in contrast to the mosses, is the dominant stage in the life cycle.

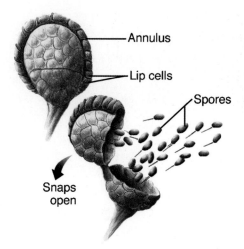

Figure 17.10 Spores are forcibly released by snapping back of the reinforced annulus on one side of the sporangium as humidity changes.

Pair off with a partner and provide the information requested in table 17.2 on the next page.

One person should explain to the other the sporophyte phase of the life cycle. Switching roles, the other person should discuss the gametophyte phase.

Now here is a question for both to discuss. What characteristics of ferns allow biologists to say that ferns are more adapted to the terrestrial environment than *Nitella* or bryophytes?

Lab Summary

22 On a separate sheet of paper, answer the following questions as assigned by your instructor.

1. Using table 17.2 as a guide, make a list of at least six ancestral characteristics that you observed which are shared by stoneworts and the three phyla of plants studied in this lab topic.
2. Using table 17.2 as a guide, list the derived characteristics that you observed which separate the three plant phyla studied in this lab topic.
3. Did any of the organisms studied in this lab topic lack a gametophyte stage? If they did, they could still complete their life cycles? Could any of the studied plants skip the sporophyte stage and still complete their life cycles?
4. List five characteristics of the plants studied in this lab topic that are adaptations to living in a terrestrial environment. Briefly explain how each enhances the chances of the plant surviving.
5. How do spores differ from seeds?
6. What event starts the sporophyte stage of the life cycle? When and in what structure does it occur in liverworts? Mosses? Ferns?
7. Answer the same questions stated in 6 for the gametophyte stage.

You may want to try the critical thinking questions that apply some of the knowledge you gained in doing this lab.

Internet Sources

Read more about an exciting new experimental system for investigating the biology of the life cycle of ferns at www.c-fern.org

You might want to use a search engine to look for other research work using ferns as experimental organisms. What are the advantages of ferns as experimental organisms in the study of developmental genetics?

Learning Biology by Writing

Write a one or two page essay that describes the alternation of generations in plants, using examples from your lab work. Describe what a sporophyte is and how it produces spores. Describe why spores are genetically dissimilar and their role in dispersal and promoting genetic outcrossing. Describe how a gametophyte develops and what gametangia are. Discuss the genetic consequences of producing gametes by mitosis. Finally, discuss how a sporophyte develops and how genetic diversity develops among sibling sporophytes.

Critical Thinking Questions

1. Were plants or animals first to invade terrestrial environments? Why do you say so?
2. Two roommates with little botanical knowledge thought they could make a quick profit by growing ferns for the commercial florist market. They thought it would be very easy. All they had to do was collect the thousands of spores produced by a fern on their desktop and plant these like tiny seeds in individual pots. If they placed the pots in the sun, fertilized and watered the spores, and waited about a month they would have beautiful plants that they could sell for several dollars apiece. Why are they doomed to failure?
3. A person interested in studying crossing-over during meiosis in ferns prepared microscope slides of an archegonium and antheridium with sperm and eggs at different stages of development. Why will this study fail?
4. A species of fern has a diploid number of chromosomes equal to 18. How many chromosomes would you expect to find in a spore? In a zygote? In a cell from a leaf? What would you expect if the species was a moss?
5. Explain how vascular tissues are an advantage in terrestrial plants.

Table 17.2 Summary: Seedless Plant Observations

	Stoneworts	Hepatophyta	Bryophyta	Pterophyta
Common name				
Habitat				
Produces spores (Y/N)				
Distinct gametophyte (Y/N)				
Distinct sporophyte (Y/N)				
Gametophyte develops by				
Gametophyte ploidy level				
Gametes produced by Mit/Mei				
Structure where egg produced				
Egg ploidy				
Structure where sperm produced				
Sperm ploidy				
Where zygote formed				
Zygote ploidy				
Sporophyte develops by				
Sporophyte ploidy				
Structure where spores produced				
Spores produced by Mit/Mei				
Spore ploidy				

Lab Topic 18

DERIVED CHARACTERISTICS OF SEED PLANTS

SUPPLIES

Resource guide available on WWW at
www.mhhe.com/labcentral

Equipment

Compound microscopes
Dissecting microscopes

Materials

Living plants
- *Gladiolus* flowers
- Dicots in flower
- Where possible, examples of gymnosperms other than pines
- Various angiosperms
- Various grocery store fruits and vegetables—to include apples, peppers, beans, melons and peas; specimens cut in longitudinal and cross sections

Plant specimens
- Needles of various conifers
- Staminate and ovulate cones, various species
- Various flowers
- Oat inflorescence
- Corn tassel
- Corn ovulate flower
- Ear of corn
- Pine pollen

Prepared slides
- Pine stem, cross section
- Pine pollen cone, longitudinal section
- Pine pollen, w. m.
- Pine ovulate cone, longitudinal section
- Pine embryo, longitudinal section
- Lily ovary, cross section, various stages
- Germinating pollen, whole mount
- Lilium anther, cross section, various stages
- Lily stigma, longitudinal section

Slides, coverslips, razor blades

Solutions

12 % raw sugar solution (do not use white or brown sugar) made up in 0.01% boric acid and 0.02% $CaCl_2$

STUDENT PRELAB PREPARATION

Before doing this lab, read the Background material and sections of the lab topic that have been assigned by your instructor.

Use your textbook to review the definitions of the following terms:

angiosperms	megasporangium
anther	megasporocyte
Anthophyta	microsporangium
carpel	microsporocyte
conifer	monocot
endosperm	ovary
eudicot	ovule
gametophyte	pistil
gymnosperms	pollen
heterosporous	seed
integument	sporophyte

Describe in your own words the following concepts:
Alternation of generations
Double fertilization
Germination
How seeds are formed
How seeds differ from spores
Plant adaptations to land
Pollination versus fertilization

After finishing the prelab review, write any questions you have about terms, concepts, or techniques in the margins of this lab topic. The lab observations should help you answer these questions, or you can ask your instructor during the lab.

OBJECTIVES

1. To observe the alternation of generations found in the life cycles of gymnosperms and angiosperms
2. To appreciate how pollen, ovarian development, seeds, and fruits are reproductive adaptations to the terrestrial environment
3. To continue to collect evidence (started in previous lab) that tests the following hypothesis: Plants can be arranged in a phylogenetic sequence that demonstrates increasing adaptation to a terrestrial environment.

BACKGROUND

In the previous lab topic, you saw how bryophytes could have evolved from green algal ancestors that were pre-adapted for life on land. Of great significance was the development of the gametangium that protected gametes from drying during maturation. Furthermore, because the egg was retained in the gametangium and fertilization occurred internally, the zygote could develop into an embryo in a protective environment. This paved the way for the development of an alternation of generations life cycle.

The stage of the life cycle with gametangia is called the gametophyte. Following fertilization, zygotes are retained in the female gametangium and develop by mitosis using the gametophyte's resources, producing a multicellular sporophyte stage in the life cycle. The sporophyte develops a sporangium at its tip that produces haploid spores by meiosis. The microscopic spores, surrounded by a desiccation-resistant wall made of sporopollenin, are dispersal agents that are air-carried without drying. Male and female gametophytes develop from the spores and will sexually produce a new sporophyte generation as the cycle repeats.

In ferns, the gametophyte and sporophyte became independent stages but they remained linked in an obligatory sequence where one stage produced the other. Starting with haploid spores that divided by mitosis to produce gametophytes bearing gametangia, the fern life cycle resembles that of a bryophyte. In both groups, sperm move through water films to fertilize eggs held in female gametangia where subsequent zygotes develop by mitosis into mature sporophytes. However, in ferns the gametophyte stage dies, leaving the sporophyte stage to live and mature on its own before developing sporangia that produce spores to continue the alternation of generations.

In this lab you will continue your study of plant life cycle evolution and adaptations to land by studying seed plants. While seed plants retain the basic life cycle strategy of alternating generations, the independence of the gametophyte and sporophyte stages are reversed from what occurs in seedless plants. Moreover, a new factor appears: the seed.

A **seed** contains a diploid multicellular sporophyte embryo and food reserves surrounded by a protective coat. Seeds seem to serve two purposes: (1) allowing the species to survive harsh conditions that can kill the parents, and (2) allowing for dispersal of the species. Two monophyletic clades of plants are seed plants: the gymnosperms, which we will represent with conifers, and angiosperms, which are the flowering plants.

You are familiar with these two groups of plants from everyday life, but common experience does not support the idea that seed plants have two life cycle stages. Where in the life cycle are gametophytes and sporophytes? A pine tree, juniper bush, marigold, or oak tree are all sporophyte stages. When reproducing, they produce microscopic gametophyte stages inside cones or flowers that are never seen by a casual observer. This lab will allow you to look at the details.

To set the stage for your lab work, a brief overview of the life cycle of seed plants will be given, starting with the sporophyte. Sporophytes produce reproductive structures that we commonly call cones or flowers. Inside these structures are sporangia that produce spores. Seed plants are **heterosporous** meaning that they produce unisexual spores that also differ in size. Male spores are typically smaller and are called **microspores** and will develop into male gametophytes. The larger female spores are called **megaspores** and will develop into female gametophytes. Both types of spores are produced by meiosis from diploid cells located in different sporangia respectively called **microsporangia** and **megasporangia**. The spores are never released from the sporangia, but are retained as they develop by mitosis into microscopic gametophytes.

The megasporangia producing megaspores are surrounded by one or two layers of sporophyte tissues, the **integument**. The integument, the enveloped megasporangium, and the megaspore are collectively called an **ovule**. Inside the ovule, the haploid megaspore divides by mitosis to produce a microscopic, but multicellular, female gametophyte. It will produce one or two eggs by mitosis and they will be retained in the ovule.

In microsporangia, haploid microspores will develop into microscopic male gametophytes, that at maturity, consist of only a few cells. These cells will be surrounded by a desiccation-resistant covering to form a pollen grain. Pollen, containing the male gametophyte, is released from the microsporangia and must travel by some means to the female gametopyte to complete the life cycle. Pollen is unique to seed plants, representing a mobile male gametophyte that can travel great distances through the air or on animals to fertilize distant eggs. Consequently, this adaptation promotes greater genetic variability.

If pollen successfully transfer sperm to an egg, the resulting zygote marks the start of a new sporophyte generation. The other female gametophyte tissues surrounding the zygote become a source of nutrition allowing the zygote to undergo limited development to form the embryo stage of the new sporophyte generation. After the embryo reaches a certain size, growth stops. The integuments surrounding the megasporangium then harden, encasing the embryo and nutritive tissues. This composite structure is the **seed**. When seeds are released into suitable environments, the embryos resume growth, producing mature sporophytes to complete the life cycle.

To summarize, the additional life cycle adaptations to a terrestrial environment that we see in seed plants are:

- Gametophyte stage is reduced to a microscopic stage dependent on the sporophyte.
- The male gametophyte is mobile and resistant to desiccation because it is encased in a pollen grain. Genetic outcrossing is promoted because the mobile male gametophyte can deliver sperm to distant females independent of water availability.

DERIVED CHARACTERISTICS OF SEED PLANTS 215

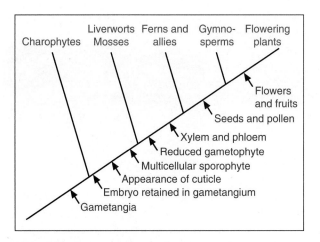

Figure 18.1   Cladogram of plant evolution showing the sequential development of adaptations to land.

- Retention of the female gametophyte embedded in the sporophyte and retention of the embryo in the female gametophyte allow seed development.
- Seeds provide an effective means for sporophyte dispersal.

This sequential progression of evolutionary events can be used to make a cladogram (**fig. 18.1**).

In addition to the life cycle strategy changes, seed plants exhibit other adaptations to a terrestrial environment. Having well-developed vascular tissues allowed seed plants to develop large aboveground bodies, with some species reaching 300 feet in height. This reduced competition for sunlight with low-growing bryophytes and ferns. Secondary growth allowed the stems to increase in girth so that they could carry large numbers of leaves and not break under the strain. Root systems also became highly developed to anchor the stems and to provide water needed for metabolic activities. Most seed plants also developed symbiotic relationships between their roots and soil fungi, enhancing the roots' ability to absorb water and minerals needed to supply the increased cellular mass of large bodies.

LAB INSTRUCTIONS

You will study the life cycles of conifers and flowering plants in this lab, learning to apply the concept of alternation of generations to these familiar plants. You will also learn how seeds and pollen represent adaptations to a terrestrial environment.

Gymnosperms

The term *gymnosperm* means "naked seed" and is used today to describe a clade of four phyla that all have this characteristic. It is not a taxonomic category. Present-day gymnosperms include: cycads, ginkgo, gnetophytes and conifers (**fig. 18.2**). All bear naked seeds at the base of the scales found in characteristic cones or similar structures of these plants. **Cycads** are tropical plants. They look somewhat like palms from a distance and have a stem that may be 18 m high bearing leaves in a cluster at the top, hence the common name for some species is "sago palms." Only one species of **ginkgo** survives today, although fossils dating back 270 million years are known. All plantings today are derived from a few trees found in the Imperial Gardens of China in 1690. It is widely planted as an ornamental in the United States. It has fan-shaped leaves with parallel venation and seeds are born on stalks with a fleshy coating. The **gnetophytes** are a diverse group found in moist tropical areas and arid areas, including North America. Some species are tree-like; others, shrubs; and yet others, vines. The **conifers** are by far the largest and most diverse group and include the familiar pines.

In this section you will look at pines as examples of the gymnosperms. The sporophyte stage is the conspicuous stage in the life cycle and the gametophyte stage is reduced to a few cells that give rise to eggs and sperm.

(a)

(b)

(c)

Figure 18.2   Examples of gymnosperms: (*a*) cycad; (*b*) ginkgo; (*c*) gnetae.

Phylum Coniferophyta (Conifers)

The conifers are an economically important group of over 500 species of plants. They include pine, fir, spruce, hemlock, redwood, sequoia, cypress, and cedar; many of the species are used for building lumber and as a source of fiber for paper. Most, but not all, have needlelike leaves, tree or shrublike body plans, well-developed vascular tissues, and extensive secondary growth. Cones are common reproductive structures, but fleshy structures perform this function in yews and junipers.

Sporophyte Stage of Pine You will start your study of conifers by looking at the adaptations of the sporophyte body to land environments.

1 ▸ *Leaves* Look at the demonstration specimens of conifer needles (pines, firs, spruces, yews, *etc.*). Needles are produced in bunches of 1 to 8, collectively called a **fascicle**, with the number depending on the species. Note the shiny appearance to the needles. This is caused by the presence of a thick, waxy **cuticle**, which prevents evaporation of water, adapting the plants to dry environments. The reduced surface area of the needles and the pyramidal shape of most conifers are adaptations that enhance shedding of snow in cold climates. Many conifer species hold their leaves year-round and is why they are called evergreens, although some angiosperms also do so. Needles are usually retained for 2 to 4 years, although the bristlecone pine may retain its for 45.

Stem Sequoias reach heights of over 300 feet. Large seed plants such as this must have stems that can withstand high wind forces, bear the weight of large branches and numerous needles, and conduct water to the cells in the leaves and food materials to cells in the roots. To see how they can do this, we will look at a cross section of a pine stem.

2 ▸ Examine a cross section of a pine stem with your compound microscope (**fig. 18.3**). The tightly packed cells making up most of the section are **secondary xylem**, the same type of cells you buy at the lumber yard when you specify pine lumber. Note how none of the cells contains cytoplasm. At one time a living cell was present but its function was to make the cell wall and die, leaving behind a fluid-conducting and structural element which is integral to the life of the plant. The cell walls contain a stiffening material called **lignin**. It strengthens the cell wall so that it can bear the stresses of added weight and wind shear. Furthermore, the hollow center of each dead cell can serve as a conduit for water conduction until it finally plugs with mineral crystals. Thus, this tissue is an adaptation that allows conifers (and other seed plants) to increase their size and outcompete those with no, or poorly developed, vascular tissues.

Surrounding the stem are the several layers of cells that comprise the **bark**. The outer bark protects the tree from the drying effects of the terrestrial environment,

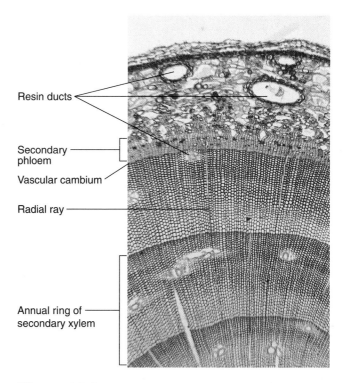

Figure 18.3 Cross section of a pine stem.

including some protection from fire. The inner bark has many resin ducts that seal wounds and protect against fungal invasion. It also contains another type of tissue, **secondary phloem.** Phloem cells are living cells. They transport photosynthetic products from the needles to the nonphotosynthetic cells in the roots, fueling their growth and role in water and mineral accumulation.

Between the inner edge of the secondary phloem and the outer edge of the secondary xylem is a single cell layer, the **vascular cambium**. Its cells divide to form new secondary xylem inwardly and secondary phloem outwardly, increasing the girth of the stem as it elongates from its tip. The coordinated growth in girth and height adds to the strength of the stem.

The roots of gymnosperm sporophytes are also adapted to their terrestrial life style and large body size. As the stem grows, so must the roots to provide an anchor point as well as the water and minerals needed in metabolism. Many gymnosperms have augmented their own roots' capabilities by forming symbiotic relationships with soil fungi.

Microsporangia and Male Gametophyte The familiar pine, fir, and spruce trees represent the sporophyte generation of conifers. To locate the sporangia and gametophytes, the reproductive structures must be examined. For pines, these are the cones. Pine trees develop two types of cones: small pollen (staminate) cones, which are on the tree for only a few weeks, and the larger, more familiar ovulate cones ("seed cones," "pine cones"), which can remain on a tree for two or more years.

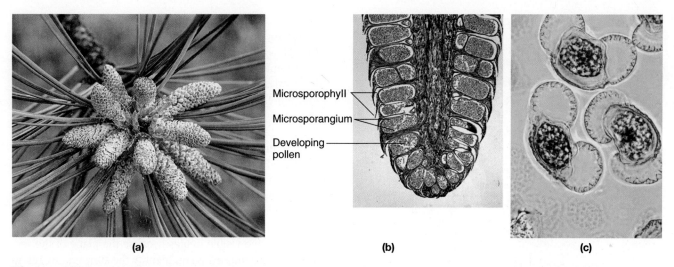

Figure 18.4 Pine male reproductive structures: (a) pollen (male) cones; (b) Longitudinal section of pollen cone; (c) characteristic pollen of pine.

The small male cones appear singly or in clusters at the branch tips (fig. 18.4a) in the spring. These **pollen cones** usually occur on lower branches and female cones are on upper branches. Since pollen falls when it is released, it is carried by the wind away from the higher female cones on the same tree. Something as simple as cone location is an adaptation that promotes genetic outcrossing.

3 Examine the pollen cones available in the lab. Each is composed of modified leaves, which each bear two **microsporangia** where pollen are produced. If the pollen cones are reasonably fresh, hold one over a drop of water on a microscope slide and tap the cone to release pollen into the drop. Add a coverslip and observe with your compound microscope. Alternatively, a prepared slide of pollen can be used.

Pollen is surrounded by a thick wall, the outer layer of which expands to produce two air-filled bladders (fig. 18.4c), making pine pollen resemble "Mickey Mouse" hats. Speculate on what might be the function of these two bladders so typical of pine pollen.

4 How is pollen produced? To answer this question, use your compound microscope to examine a prepared slide of a longitudinal section of an immature pollen cone (fig. 18.4b). It is composed of whorls of **microsporangia**, radiating from a central stalk (fig. 18.4b). Depending on the maturity of the microsporangium, it will contain pollen in sequential stages of development.

In young microsporangia, **microsporocytes**, also called microspore mother cells, divide by meiosis to produce four haploid microspores. Unlike in mosses and ferns, the spores are not released but are held in the sporangium and you can see them as the darkly staining structures that are undergoing further development to become pollen. Each microspore will become a multicellular **male gametophyte.** When released from the pollen cone, it will consist of only four cells; a **tube cell,** a **generative cell,** and two prothallial cells. When mature, the male gametophyte will consist of a tube cell, a sterile cell, and two sperm cells. Thus, spores develop to become gametophytes as in seedless plants, but the gametophyte is reduced to something that just qualifies to be called multicellular with its four cells. Note that no antheridia are produced by the male gametophytes of conifers. The sequence of events in the origin of the male gametophyte is summarized in figure 18.5. As the pollen matures, the microsporangia, now called pollen sacs, rupture, releasing enormous clouds of yellow pollen into the air, most of which never reaches a female gametophyte. After releasing pollen, the pollen cones fall off the tree.

The development of pollen is a major reproductive adaptation to terrestrial environments. With the appearance of pollen, male gametes became mobile and highly resistant to desiccation. No longer was a water film required for the sperm to swim from a male gametophyte to an egg produced by a female gametophyte. Pollen could deliver sperm through the air, increasing the likelihood of genetic outcrossing to female gametophytes beyond the immediate vicinity.

Megasporangia and Female Gametophytes The familiar pine cone is an ovulate cone, containing small reproductive structures we have not encountered before, **ovules.** At its early stages of development, the ovule contains a megasporangium that will produce a single haploid megaspore.

While still in the megasporangium, the spore develops into a microscopic multicellular female gametophyte that produces eggs inside an archegonium. Sperm fertilize the egg in place to form a zygote, the first cell of a new sporophyte generation. The zygote develops into an

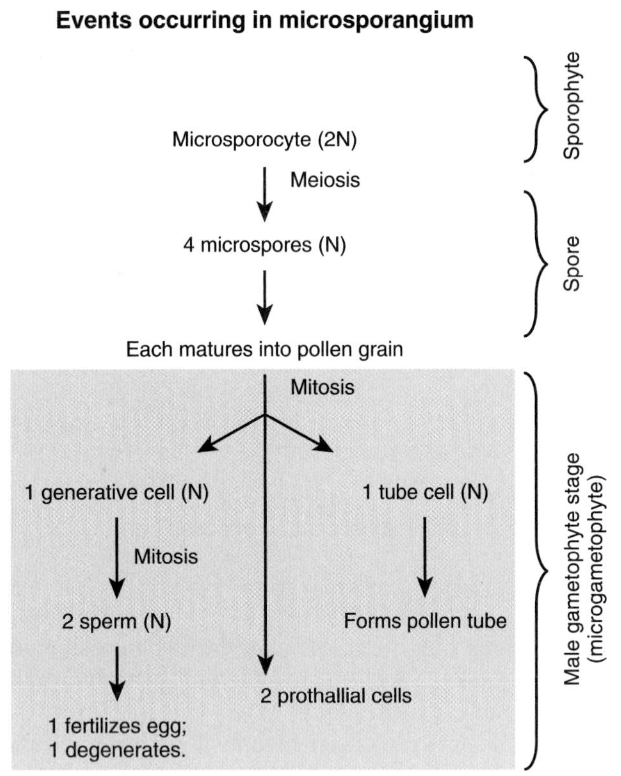

Events occurring in microsporangium

Microsporocyte (2N) — Sporophyte

↓ Meiosis

4 microspores (N) — Spore

↓

Each matures into pollen grain

Mitosis ↙ ↓ ↘

1 generative cell (N) | | 1 tube cell (N)
↓ Mitosis | ↓ | ↓
2 sperm (N) | 2 prothallial cells | Forms pollen tube
↓
1 fertilizes egg;
1 degenerates.

— Male gametophyte stage (microgametophyte)

Figure 18.5 Sequence of spore formation and development of male gametophyte generation in gymnosperms.

Ovules can be observed in young cones by cutting away the tip and about one-third of the scales. While observing through a dissecting microscope, remove a scale from the large remaining piece. Look at the base of the scale for two oval structures on the surface that faced the smaller end (upper surface when the young cone was on the tree). The name gymnosperm, meaning naked seed, describes the exposed nature of the ovules.

6▸ To study the microscopic structure of an ovule, obtain a prepared slide of a longitudinal section of an ovulate cone and look at it with the scanning objective of your compound microscope. After determining which end of the section is the tip of the cone and what structures are scales, find an ovule at the base of a scale (fig. 18.6). Not all scales on the slide will have visible ovules because the plane of the section may have missed them. Identify the surrounding **integument,** the protective outer layers of cells that will mature to become the seed coat. The contents of an ovule may be in one of many stages of development, depending on when the section was made. You may have to look at several ovules to see all of the sequential steps outlined below.

Inside a young ovule is the megasporangium. In it, a single diploid **megasporocyte** divides by meiosis to produce four haploid **megaspores,** three of which die. The remaining haploid megaspore develops into a haploid **multicellular female gametophyte** by repeated mitotic cell divisions. As the gametophyte develops, it draws nutrients from the surrounding tissue. At a later stage of development, it produces two **archegonia** at the end of the ovule near the central axis of the cone. Each produces a functional egg. Thus in pines, the female spores that are produced never leave the parental sporophyte's body and the gametophyte develops in place. The sequence of events in the development of the female gametophyte is summarized in figure 18.7.

7▸ **Fertilization and New Sporophyte Generation** For fertilization to occur, sperm produced by the male

embryo inside the ovule. As the ovule matures, its outer covering hardens to produce a seed, the defining characteristic of the seed plants.

5▸ Obtain an ovulate cone from the supply area. Ovulate cones are borne on the tips of young branches on the upper part of the tree. The cone forms with the pointed tip upward, but as it matures it inverts. Each cone is made up of whorls of heavy scales. Each scale bears two ovules.

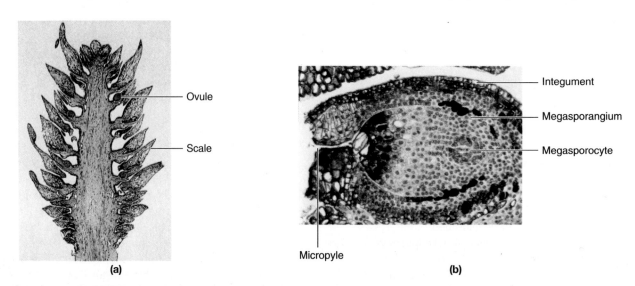

Figure 18.6 Structure of a pine cone. (a) Longitudinal section through an ovulate (female) pinecone. (b) Magnified section through ovule showing micropylar opening and megaspore mother cell.

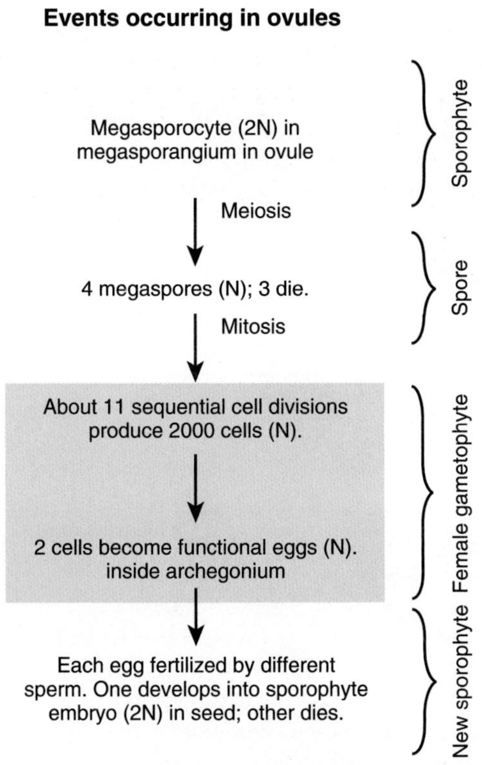

Events occurring in ovules

Figure 18.7 Sequence of spore formation and development of female gametophyte generation in female cone.

Sporophyte: Megasporocyte (2N) in megasporangium in ovule
↓ Meiosis
Spore: 4 megaspores (N); 3 die.
↓ Mitosis
Female gametophyte: About 11 sequential cell divisions produce 2000 cells (N).
↓
2 cells become functional eggs (N) inside archegonium.
↓
New sporophyte: Each egg fertilized by different sperm. One develops into sporophyte embryo (2N) in seed; other dies.

gametophytes in pollen must reach the eggs held in the ovules in the pine cones. How do the sperm reach the eggs in the ovules if each is surrounded by the protective integuments? Look again at the slide of the ovulate cone and examine several ovules carefully. When the plane of the section is correct, you should see an opening in the integument on the end facing the central axis of the cone (fig. 18.6b). Called the **micropyle,** this opening allows the pollen tube access to the female gametophyte.

Fertilization in pines involves a series of events that may take over a year to complete, whereas in spruces, firs, and hemlocks, it may be completed in a few weeks. **Pollination** is the transfer of pollen to the ovulate cone from the pollen cones. Pines produce incredibly large numbers of pollen grains that are randomly carried by wind. Most simply fall to the ground, but some land between the scales of upturned ovulate cones of the same species. Windborne pollen enters the cone openings and is trapped in a droplet of fluid secreted at the micropyle on the upper surface of the scales. As this droplet dries, pollen is drawn toward the micropyle. The scales of the cone now press tightly together (close) and over the next year fertilization and seed development gradually occur.

When pollen reaches the micropyle, it produces a pollen tube that grows down the canal toward the egg (fig. 18.8). Two sperm pass down the tube and enter the egg cytoplasm. The nucleus of one sperm fuses with the nucleus of the egg, signaling the start of a new diploid sporophyte generation. The second sperm nucleus degen-

erates. The second egg in the ovule also may be fertilized by sperm from another pollen tube, but only one of the zygotes will develop. The zygote divides to form an embryo that pushes into the surrounding tissue, drawing nutrients to sustain growth.

As the embryo develops, the integument covering the ovule hardens into the seed coat. It may take up to an additional year for the seed to develop. In essence, *the seed is a mature ovule* containing the embryo of the next sporophyte generation.

8▶ Look at a slide of a longitudinal section of a pine seed. These seeds are truly remarkable structures containing cells from three generations in the life cycle of the pine: (1) the parent sporophyte contributes the seed coat; (2) the stored food materials in the seed are haploid female gametophyte tissues; and (3) the embryo is the diploid sporophyte of the new generation.

9▶ Examine dry, mature pine cones for evidence of seeds. When air- or water-borne, or transported by an animal to a new location, the seed can germinate, dispersing the species. Thus in gymnosperms, when compared to mosses and ferns, seeds have replaced spores as the means of dispersal.

10▶ Analyze Observations You have now observed the stages of pines in the alternation of generation strategy of plant life cycles. In so doing, you saw how the gametophyte generation is greatly reduced compared to the familiar sporophyte generation. Nonetheless, it has the same function as gametophytes in seedless plants: gamete production by mitosis. You saw how the sporophyte stage produced spores as in seedless plants, but how, in addition, it also nutritionally supported the microscopic gametophyte stages arising from those spores. You also observed two evolutionary innovations seen for the first time in gymnosperms: ovules that perform multiple functions before becoming seeds and also pollen that house mobile male gametophytes.

Discuss with your lab partner what it is about the gametophyte and sporophyte stages of gymnosperms that represent adaptations to terrestrial environments.

Angiosperms

Angiosperms are a monophyletic clade of seed plants. As in gymnosperms, angiosperms have ovules that at first contain megasporangia, later female gametophytes, and then zygotes that mature into embryos as the ovules become seeds. Angiosperms also produce pollen containing male gametophytes. What sets them apart from gymnosperms is that they produce flowers and fruits. We will focus on these strutures as we explore how the concept of alternation of generations applies to angiosperms.

Angiosperms are the most diverse of the plant groups, living in mountain, desert, and freshwater and seawater habitats. Over 250,000 species have been described. They range in size from the tiny duckweeds about 1 mm tall to

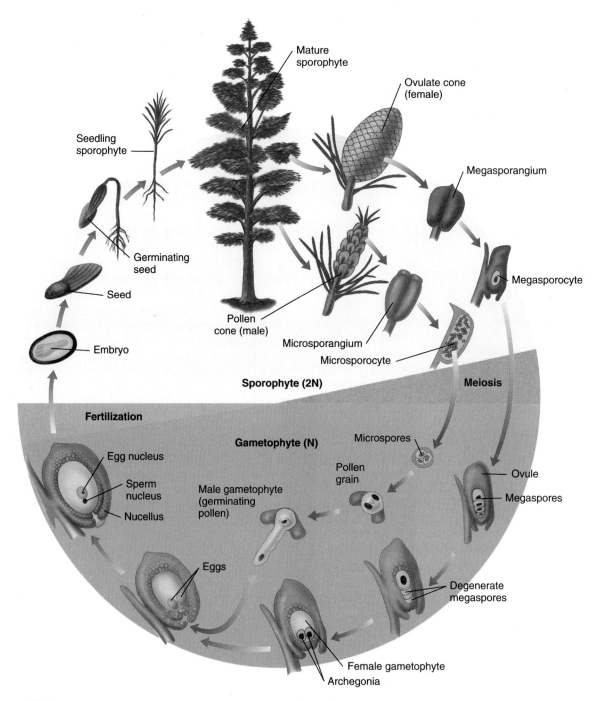

Figure 18.8 Life cycle of a pine, showing sporophyte and gametophyte stages. Gametophyte stages are microscopic.

the 100-m tall eucalyptus trees in Australia. All angiosperms are classified in phylum **Anthophyta,** but there are several clades in the phylum. The two largest are **monocots,** which include the grasses, lilies, bamboos, palms, bromeliads, and orchids; and **eudicots,** which include broad leaf garden flowers herbaceous plants, shrubs, and trees, vines, and cacti.

The sporophyte stage dominates the angiosperm life cycle as it did in gymnosperms, but unlike gymnosperms the ovules are contained in an organ you will see for the first time, an **ovary.** At first part of the flower, the ovary enlarges following fertilization to become a fruit containing the ovules that have matured into seeds. The name angiosperm comes from the Greek words *angeion* (=container) and *sperma* (=seed) and together emphasize the role of the fruit as a seed container. Like gymnosperm, the term *angiosperm* does not have taxonomic status and is used as a common name for the clade.

In many species, flowers attract animal pollinators, promoting genetic diversity through cross fertilization. Fertilization in angiosperms involves unique **double fertilization events,** where one sperm fuses with an egg to produce the zygote, and another sperm fuses with a second type of cell to produce a nutrient-laden tissue called the **endosperm** inside the ovule. The endosperm fuels the early development of the embryo and seedling during seed germination.

Sporophyte Floral Structure

The flower is a complex reproductive structure produced by the sporophyte. Male and female gametophytes develop from spores that are produced and retained within the flower. Without a microscope and carefully prepared slides, it is not possible to observe the gametophyte stages. Some species have flowers that contain both male and female gametophytes while other species have distinct male and female flowers. Those species producing separate male and female flowers may have individual plants with flowers of one sex or may have flowers of both sexes on the same plant. Flowers may be single or in clusters. A flower aggregation is called an inflorescence.

11▸ Go to the supply area and break a single *Gladiolus* flower from an inflorescence. *Gladiolus* produces both male and female gametophytes in the same flower.

Note how the floral parts arise from a green **receptacle** at the end of the pedicel. The brightly colored **sepals** and **petals** arise as whirls of modified leaves from the receptacle. Can you detect an odor from the flower? _____

The bright colors and fragrances of flowers attract pollinators such as insects, birds, bats, and small mammals. This adaptation promotes genetic outcrossing between distantly located individuals of the same species. Animals are attracted to specific flower species as a result of coevolution of visual and olfactory attractants in the plants and corresponding receptors and behaviors in the animals. Many flowering plants, such as the grasses, are wind pollinated. Consequently, their flowers lack attractants and are inconspicuous.

Remove the sepals and petals to reveal the sex organs inside the flower (fig. 18.9). The central stalklike structure is the **pistil** and represents the female portion of the flower. It may consist of one or more **carpels**. Carpels are thought to have evolved by the folding and fusion of specialized ancestral leaves that bore female sporangia along their margins. The swollen base of the pistil is a hollow chamber, the **ovary**, where the **ovule**(s) is (are) located. Some flowers, such as those of peaches, have one ovule in the ovary, while those of watermelons have thousands. As in gymnosperms, several events will occur in the ovule: spore production, female gametophyte

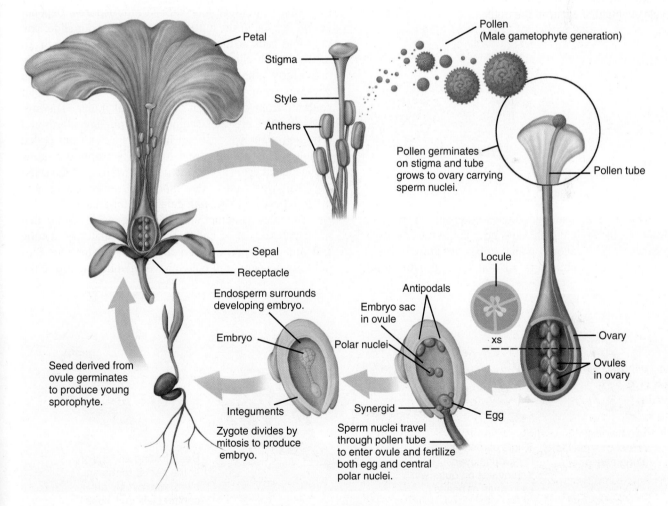

Figure 18.9 The sporophyte generation of an angiosperm bears flowers that produce a few cells that act as the male and female gametophytes. Male gametophytes are the pollen produced by the anthers of the stamens. Female gametophytes are the embryo sacs contained in the ovules located in the ovaries. Nuclei of the cells in the gametophyte generation are haploid. The diploid number of chromosomes is restored with formation of the new sporophyte generation at fertilization.

development, fertilization, and embryo development as the ovule matures to become a seed. The ovarian wall will enlarge to become the fruit.

Above the ovary, find the **style,** a stalk that ends in an expanded tip of glandular tissue, the **stigma.** It produces sweet, sticky material that attracts pollinators, traps pollen, and stimulates pollen tube growth. For pollen to transfer sperm to the eggs, the tube must grow through the tissue of the style to the ovarian region. Estimate the distance from the stigma to the bottom of the ovary in the *Gladiolus* that you are dissecting. How long must the pollen tube grow to deliver sperm to the egg? _____

12 Coordinate this part of the dissection with another person at your lab table so that one of you cuts cross sections of the ovary and of the style on one flower. The other person should cut a longitudinal section through all parts of the pistil from another flower. Compare the sections. You may want to use a dissecting microscope to see more detail.

As you look at the cross section of the ovary, note the three chambers called **locules** (fig. 18.9) that are derived from the ancestral fusion of three carpels (**fig. 18.10**). Projecting into the locules from the central shaft are little beadlike structures, the **ovules.** Name five significant life cycle events that occur in the ovule.

Now look at the longitudinal sections of the style and ovary. In the style region, note how the tissue is loosely organized. This allows pollen tubes to grow through a rather open architecture so that their growth is not impeded. Note that there are several ovules in vertical arrays in the locules. Try to count the number of ovules in one locule. How many did you find? _____ Estimate how many ovules are in the ovary of your plant. _____ How many seeds could this flower have produced? _____

Alongside the pistil were long filaments with expanded tips that you most likely removed to make your ovarian sections. These were the **stamens.** Find one and look at it closely. The structures at the tips are **anthers,** and contain four microsporangia that produce microspores. They develop into the male gametophyte stage (pollen). Use a razor blade to cut across the anthers to reveal the chambers and their contents. As in gymnosperms, no antheridium is produced. The anthers are thought to have evolved by the folding and fusion of specialized leaves that bore male sporangia along their borders.

13 *Microsporangia and Male Gametophyte* Use your compound microscope to examine a prepared slide of an anther cross section (**fig. 18.11**). Four **microsporangia** (pollen sacs) should be clearly visible. They contain thousands of, if not more, **microsporocytes** that each divide by meiois to produce four haploid **microspores.** Depending on the age of the anther used to make your slide, you may find cells at the end of meiosis I (in clusters of 2), meiosis II (in clusters of 4), or maturing pollen. During maturation, a heavy wall impregnated with sporopollenin develops around the cells. It protects the delicate cellular contents from desiccation and physical injury. The surfaces of pollen are often sculptured in unique patterns and a trained plant scientist can identify species from pollen samples. This is often used to identify plants that grew thousands of years ago as the pollen's walls (not the cells) are highly resistant to decomposition. Bogs, amber, and other deposits provide rich databases for analysis.

The male gametophyte developing inside each pollen grain will consist of only three cells at maturity: a **tube cell** and two **sperm.** Often the male gametophyte does not fully develop until pollen have been transferred to the

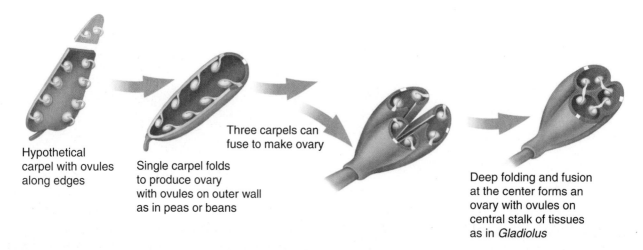

Figure 18.10 Hypothetical model for evolutionary development of ovary from carpels, ancestral leaves that bore ovules along edges.

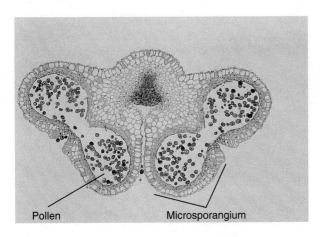

Figure 18.11 Photomicrograph of a cross section of microsporangia in an anther.

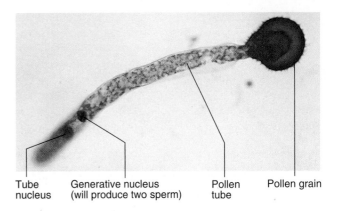

Figure 18.13 The germinating pollen grain is the male gametophyte of flowering plants. It begins to develop as shown when it is transferred to the stigma of a flower.

stigma. As in gymnosperms, the packaging of male gametophytes in pollen makes them mobile, capable of fertilizing eggs at distant locations. The steps in the formation of the male gametophyte are summarized in **figure 18.12**.

14▶ It is possible to induce pollen germination in the lab on a microscope slide, *i.e.*, to activate the male gametophyte and see a growing pollen tube. Your instructor may set this up as a demonstration or you may do the procedure. Take a depression slide and add a drop of germinating solution which contains 12% raw sugar made up in a dilute mineral solution. Choose a flower that is shedding pollen and lay an anther in the drop, teasing it apart to release pollen. Add a coverslip. Observe this preparation for evidence of pollen tube growth over the next two hours at approximately 15-minute intervals. Record your observations below as drawings with marginal time nota-

tions. Alternatively, you may look at a prepared slide of a germinating pollen tube. See **figure 18.13**.

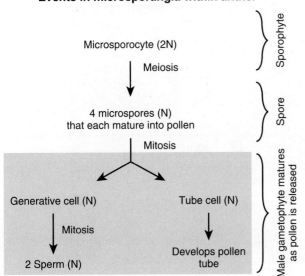

Figure 18.12 Sequence of spore formation and development of male gametophyte in anthophytes.

15▶ *Megasporangia and Female Gametophyte* Your instructor may have several slides made with cross sections of ovaries taken at different stages of maturation. If so, you can use them in sequence as you work through the stages below. Look at the slides first with the scanning objective. Orient yourself by finding the three ovarian locules that you saw in your dissection of the whole flower. Each locule should contain two ovules, although some ovules may be missing because the plane of the section did not cut through where they were located (**fig. 18.14**).

Look closely at an ovule. It is similar to those you observed in gymnosperms, except it is inside an ovary and not exposed. In young ovules, **integuments** begin to develop around a central **megasporangium** and will completely encase it except at the **micropyle** which allows later pollen tube access into the ovule. In the megasporangium, one diploid **megasporocyte** divides by meiosis to produce four haploid cells. Three die, leaving one functional **megaspore**. This is the first cell

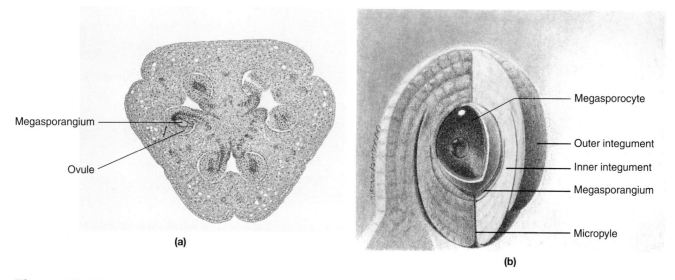

Figure 18.14 Structure of a lily's ovary: (a) photomicrograph of a cross section of an ovary; (b) structure of an ovule, containing the megasporangium in which the multicellular female gametophyte develops from a single megasporocyte; each gametophyte eventually produces one egg.

of the female gametophyte generation. What is its ploidy level? _____

The female gametophyte matures as the megaspore divides three times by mitosis to produce eight nuclei in one cytoplasm. This multinucleate structure is called the **embryo sac.** What is the ploidy level its nuclei? _____

Three of the nuclei, called the antipodals, then migrate to the end of the embryo sac away from the micropyle (**fig. 18.15**). Two, called the polar nuclei, migrate to its center. Three migrate to the end nearest the micropyle, where one enlarges to become the egg nucleus and the others are antipodals. Cell membranes form around each nucleus separately, except for the polar nuclei pair in the center. A single membrane forms around them to yield a binucleate cell. These seven cells with eight nuclei comprise the entire **female gametophyte** stage of the flowering plant. The events in the development of the female gametophyte are summarized in **figure 18.16**.

16 Fertilization

Transfer of pollen from its source to the stigma of a flower is called **pollination.** As in gymnosperms, it is a separate event from **fertilization,** the fusion of egg and sperm to produce a zygote. Pollination, although a prerequisite for fertilization, does not guarantee that it will occur because some complex events must happen. If a demonstration slide of double fertilization is available, look at it after reading the following paragraph.

Pollen grains transferred to or landing on the stigma are activated by carbohydrates, salts, and other biochemical components of the nectar secreted by the stigma's glandular tissues. Pollen tubes growing out from the pollen grain penetrate through the tissues of the stigma and style to reach the ovules in the ovary. Once in the locules of the ovary, the tubes must "find" the micropyle of the ovule and penetrate into the embryo sac. Fertilization occurs when two sperm cells travel through the pollen tube and enter the embryo sac. One sperm fuses with the egg to form a diploid zygote, the first cell of the next sporophyte generation. The second sperm combines with the two polar nuclei in the single cell at the center of the embryo sac to form a special triploid tissue, the

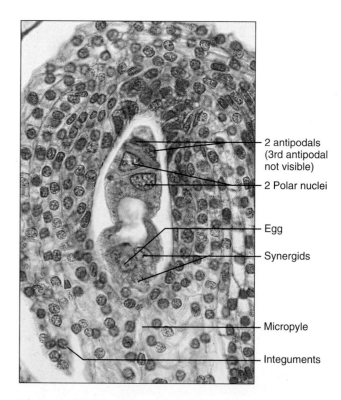

Figure 18.15 Mature female gametophyte of a flowering plant consists of eight cells in the embryo sac surrounded by outer integuments of ovule.

arrangements of the locules and the seeds in the sections. Estimate the number of seeds in some representative fruits, such as a pepper, bean, or pea (estimate the number in one locule and multiply by the number of locules). Record these numbers next to your drawings. Using your seed data, estimate the number of ovules that would be found in a flower of the same species. Are any of these numbers impressive?

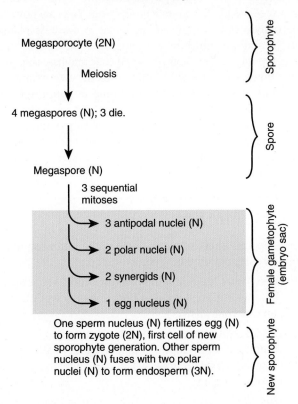

Figure 18.16 Sequence of spore formation and development of the female gametophyte in anthophytes.

endosperm. Rich in starch and fats, it will nourish the developing plant embryo. Anthophytes are said to have **double fertilization** because two sperm are involved, but only one diploid embryo is formed.

Seeds and Fruits

Following fertilization, the embryo develops inside the ovule as the ovule matures to become a seed. The integuments surrounding the seed harden into a seed coat that protects the embryo from physical injury and desiccation. As the seed develops, the ovarian wall begins to enlarge, often becoming a fleshy fruit, although some fruits can be hard nuts. Fruits are another example of how plants and animals have coevolved. Many fruits are colorful, sweet, and fragrant. They attract animals that eat them. The seeds encased in their protective coats pass through an animal's digestive system unharmed. They are often deposited along with a little fertilizer at some distance from the parent plant where they germinate and continue the life cycle. Thus, fruits and seeds are adaptations that promote geographic dispersal of a plant species, an interaction that will be studied in detail in lab topic 26.

17 In the lab are several fruits that have been sectioned either across or longitudinally. Study the fruits and make some quick sketches below in which you show the

18 Grass Flowers

Take a look at a demonstration specimen of the highly specialized flowers of grasses. Most have flowers arranged as inflorescences (fig. 18.17). Because grasses are wind-pollinated, they lack showy petals and fragrances that attract pollinators. The grasses known as cereals and grains are a

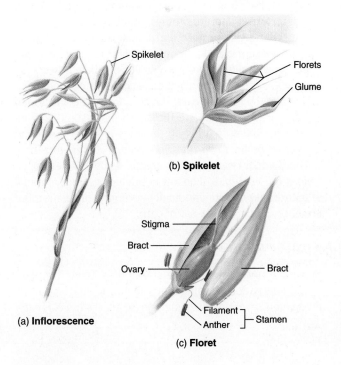

Figure 18.17 Structure of flowers in oats: (a) inflorescence; (b) side view of spikelet; (c) structure of a single floret.

major component of the human food supply and include corn, wheat, rye, and oats. Sorghum and sugarcane are also economically important grasses.

Look at the inflorescence of an oat in the specimen area. Oat flowers are encased in modified leaves called bracts. Branching from the main inflorescence are several spikelets. If a spikelet is removed and the bracts are dissected away, three florets are revealed. Each is a bisexual flower.

Look at a single floret with a dissecting microscope. This reproductive structure produced by the sporophyte is enveloped by two other bracts, one smaller and the other larger with a long extension. Spread the bracts to reveal the male and female sexual structures. Three stamens are present, terminating in anthers where pollen is produced as discussed earlier. When mature, the anthers split and release pollen into the air.

Find the central pistil that ends in two terminal feathery **stigmas.** Why do you suppose the stigmas are feathery?

At the base of the pistil, note the swollen ovary. A single ovule is located here. At maturity, it will contain a mature female gametophyte as in the other flowering plants. Following fertilization, the seed develops in the ovary and is harvested as grain, surrounded by remnants of the ovarian wall that are removed in processing as chaff.

Now take a look at the demonstration materials of corn flowers. Corn differs from other grains because its sporophytes bear separate male and female flowers. The male flowers are borne at the top of the plant as the **tassel** and release their pollen into the wind. The pistillate flowers are located lower on the plants at nodes on the stem and eventually develop into the **ears,** the corn fruit. The **silk** found in ears is what remains of the styles that extended from every egg. Pollen that fell on the ends of these styles had to develop pollen tubes that were several centimeters long. The individual kernels of corn are the seeds that developed from multiple ovules. Look at the demonstration material of a corn tassel, a longitudinally sectioned pistillate inflorescence, and a mature ear of corn.

19 Analyze Observations

You have now observed the stages of a flowering plant in the alternation of generations life cycle strategy. In so doing, you saw how the gametophyte generation is greatly reduced compared to the familiar sporophyte generation. Nonetheless, the gametophytes have the same functions as in seedless plants: gamete production. You saw how the sporophyte stage produced spores as in seedless plants, but in addition nutritionally supported the microscopic gametophyte stages. You observed that angiosperms produce flowers and fruits in addition to the two evolutionary innovations seen for the first time in gymnosperms: ovules that become seeds and pollen. Discuss with your lab partner what it is about the gametophyte and sporophyte stages of angiosperms that represent adaptations to terrestrial environments where animals have significant influence.

Lab Summary

20 On a separate sheet of paper, answer the following questions as assigned by your instructor.

1. Seed plants are said to be heterosporous. What does that mean? Give examples from your lab work to support your answer.

2. Describe one male and one female reproductive evolutionary development that is found in all seed plants and explain how they are an adaptation to terrestrial living.

3. Backed up by observations from your lab work, describe where you would look for the male and female gametophytes in a conifer and what the gametophytes would look like.

4. Backed up by observations from your lab work, describe where you would look for the male and female gametophytes in a flowering plant and what the gametophytes would look like.

5. Based on your lab work, what evolutionary developments are found in flowering plants but not in gymnosperms?

6. Describe how the sporophytes of seed plants are adapted to terrestrial living.

7. Comparing seed plants to seedless plants, describe differences in what life cycle stages are involved in dispersal of the species.

8. How does pollination differ from fertilization?

9. Summarize your observation from the previous lab (17) and this one by filling in **table 18.1** on page 228.

You may want to try the critical thinking questions that apply some of the knowledge you gained in doing this lab.

Internet Sources

The Wollemi pine (*Wollemia nobilis*) is a newly discovered (1994) pine. Forty individuals were found growing in canyon lands to the west of Sydney, Australia. Use a search engine such as Google to locate information on this newest discovery of an ancient group. Why is it considered to be a significant discovery?

Learning Biology by Writing

A major theme in this lab topic and the one before was the phylogenetic trend in the plant kingdom that allowed plants to inhabit terrestrial environments. In a short, two-page essay describe observations that you made in both labs that support or contradict the following hypothesis: there is strong evidence that plants can be arranged in a phylogenetic sequence that corresponds to increasing adaptations to a terrestrial environment. Indicate whether you accept or reject this hypothesis based on your observations.

Critical Thinking Questions

1. How does a seed differ from a spore?
2. If a species of pine has a diploid number of chromosomes equal to 40, how many chromosomes will be found in a cell from an embryo in the seed? In a pollen nucleus? In a megasporocyte? In a megaspore?
3. If a species of flowering plant has a diploid number of chromosomes equal to 30, how many chromosomes will be found in a cell from the embryo in a seed? In a pollen nucleus? In a megasporocyte? In endosperm tissue?
4. If all flowering plants depended on the wind to carry pollen from one individual to another, do you think there would be any colorful flowers and fruits? Explain your answer.
5. Describe coevolution and how the concept applies to flowering plants and many animals.
6. Why do grasses and conifers lack any showy, sweet, fragrant reproductive structures?

Table 18.1 Life Cycle Comparisons. Summarize Your Observations by Adding *Yes* or *No* to Indicate if a Feature Is Found in a Group.

Characteristics	Bryophyta (Mosses)	Pterophyta (Ferns)	Coniferophyta (Conifers)	Anthophyta (Flowers)
Airborne haploid spores				
Haploid gametophyte				
Egg and sperm produced by mitosis				
Flagellated sperm				
Water-dependent fertilization				
Diploid sporophyte				
Dependent sporophyte				
Independent photosynthetic gametophye sporophyte				
Dependent gametophyte				
Independent photosynthetic gametophyte				
Spores produced by meiosis				
Ovules				
Pollen				
Embryo protected by seed coat (seeds)				
Seed within fruit				
Vascular tissues				

Lab Topic 19

Investigating Fungal Diversity and Symbiotic Relationships

Supplies

Resource guide available on WWW at
http://www.mhhe.com/labcentral

Equipment

Compound microscopes
Dissecting microscopes

Materials

Living Specimens
- Mushrooms (from grocery store) or mushroom culture kit
- *Rhizopus* zygospore culture kit
- *Sordaria* culture kit
- Miscellaneous fungi samples: puffballs, bracket fungi, molds, others as locally available
- Lichen specimens (crustose, foliose, and fruticose) including *Cladina*

Field guide to mushrooms
Slides and coverslips
Loaf of bread with wrapper showing sodium propionate added
Fresh bread with no preservatives
Sealable plastic bags
Lactophenol stain
Microscope slides
- *Allomyces* haploid phase (gametophyte)
- *Allomyces* diploid phase (sporophyte)
- *Penicillium* conidiophores
- *Peziza* section
- *Rhizopus* combination slide with sporangia and zygospores
- *Coprinus* mushroom, section of gills

Mycorrhiza and root sections
Lichen thallus section
Lichen ascocarp section

Student Prelab Preparation

Before doing this lab, read the Background material and sections of the lab topic that have been assigned by the instructor.

Use your textbook to review the definitions of the following terms:

Ascomycota	lichen
ascus	monokaryotic
Basidiomycota	mycelium
basidium	mycorrhizae
Chytridiomycota	plasmogamy
coenocytic	ploidy
dikaryotic	saprophytic
fruiting body	septa (septum)
gametangium	sporangium
heterokaryotic	spore
hypha	Zygomycota
karyogamy	

Describe in your own words the following concepts:

Fungal life cycle from spore stage to spore stage

General body form of a fungus when growing and at the time of sexual reproduction

In addition, do some field work and bring fungi to the lab that you find on campus.

After finishing the prelab review, write any questions you have about terms, concepts, or techniques in the margins of this lab topic. The lab observations will help you answer these questions, or you can ask your instructor during the lab.

Objectives

1. To learn the body plans and life cycles of representative fungi from each phylum
2. To observe the symbiotic relationships that fungi enter into with various plants (mycorrhizae) and algae (lichens).

Background

Commonly called yeasts, mildews, molds, rusts, blights, mushrooms, and puffballs, fungi are found in all ecosystems. At one time, fungi were considered degenerate plants that had lost their photosynthetic capability. Now it is quite clear that the fungi are a separate kingdom, the

Fungi (Mycota). The oldest fossils of fungi are over 460 million years old.

Fungi are found everywhere from the tundra to the tropics in both aquatic and terrestrial habitats. It is estimated that an acre of forest soil ten inches deep can contain over two tons of fungi. About 5,000 species are pathogens of crops and 150 species cause animal and human disease. A 1984 study showed that 40% of the deaths from infections acquired while in hospitals were from fungal infections, not bacteria. Athlete's foot and ring worm (a misnomer) are common, but not lethal, fungal infections of the skin. Members of phylum Chytridiomycota have recently been implicated in the global decline of frog populations, causing lethal skin infections.

Some fungi that grow on foodstuffs produce powerful toxins. Aflatoxins produced by fungi that grow on stored grain are carcinogenic at concentrations of a few parts per billion. Ergot is an LSD-like hallucinogen produced by some fungi that grow on stored grains. Some historians think that the Salem witch hunts of the 1600s were induced by ergot, a fungal contamination of colonial grain supplies, that caused mass hallucinations.

Fungi are important components of the biogeochemical cycles for minerals. Their decomposing action releases inorganic compounds and carbon dioxide that would otherwise be tied up in dead plants and animals and unavailable to other organisms. Fungi also enter into symbiotic mycorrhizal relationships with higher plant roots, increasing the absorptive capacity of the host plants. Lichens (fungi in association with algae) are important colonizing organisms on rock faces, creating an environment for other organisms and building soil. A few species are important in manufacturing; yeasts are used in baking and brewing and others in manufacturing antibiotics, such as penicillin. Remarkably, the fungi also are able to colonize synthetic environments. For example, some fungi live in jet fuel, on photographic plates, and on other manufactured hydrocarbon materials. These types of observations suggest that they might be useful organisms for bioremediation, cleaning up the environment after organic chemical spills.

Except for the unicellular yeasts, fungi are multicellular. The typical body is composed of a mass of intertwined filaments and is referred to as a **mycelium.** The individual filaments called **hyphae** (fig. 19.1) are composed of cells joined end to end so that they are only one cell in diameter. No tissues are found in a mycelium: all cells are generalized to perform the functions of the entire organism. The familiar reproductive structures called mushrooms consist of hyphae that are tightly intertwined to form a solid aggregation. The hyphae of fungi are surrounded by cell walls composed of the polymer **chitin,** the same material found in the exoskeletons of several groups of animals. Chitin, along with DNA analysis, are considered good evidence for similar evolutionary origins of fungi and animals. Chitin is not found in the bacterial, protistan, or plant kingdoms.

In most species of fungi, the hyphae are divided by cross walls called **septa,** but the septa are perforated,

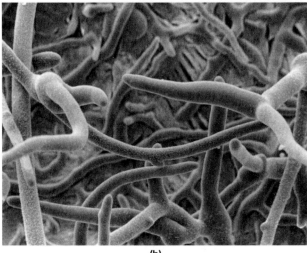

Figure 19.1 (a) Hyphae growing out from mycelial mass; (b) scanning electron micrograph of hyphae in a mycelium.

allowing cytoplasm to flow from one cell compartment to the next. This allows the rapid bulk transport of material from one part of the organism to another without crossing cell membranes. New hyphal growth is achieved as new cells form at the tips, increasing their length. Fungal growth is indeterminate and continues as long as food is available. Some individual fungi may be hundreds of years old, if not older. Some fungi lack septae and are said to be **coenocytic** (aseptate); meaning many nuclei are found in a single large cytoplasm.

Fungi are heterotrophic and they are completely dependent on preformed carbon compounds from other sources. Consequently, fungi live as saprobes, parasites, and symbionts of dead and living plants, animals, and protists.

Because of their chitinous cell walls, fungal cells cannot engulf food materials. Instead, the hyphae secrete digestive enzymes that break down food into small organic molecules that the hyphae absorb. Hyphae are never more than several micrometers thick; thus, they have a great surface-area-to-volume ratio for efficient absorption. Cytoplasmic streaming through pores in the septae, in addition to diffusion, provides efficient transport to nonabsorbing areas. Mycelial mats must continually grow outward to exploit new food supplies and may spread over several acres in undisturbed environments. Some parasitic fungi produce specialized hyphae called haustoria, which penetrate a host's cells and absorb food materials produced by the host.

Cell division in fungi is by mitosis or meiosis. Both are unusual in that the spindle forms inside the nucleus. After the chromosomes have segregated, the nucleus divides in two. Cytokinesis does not necessarily follow mitosis so cells may contain many nuclei.

Fungi reproduce both asexually and sexually by releasing **spores,** haploid cells encased by a tough cell wall. Spores are small enough to be airborne to virtually any location. If they land on suitable substrates, the spores germinate and produce haploid hyphae by mitotic nuclear divisions to continue the life cycle. Asexual spores are produced by mitosis. These spores are genetically identical to the parent and to each other. They are produced when conditions are good for growth. The life-cycle strategy can be stated something like this: If the parent genotype can grow under the present conditions, then asexual spores allow the fit genotype to produce genetic clones rapidly to take full advantage of the situation.

The process of producing sexual spores is more complicated and usually occurs in response to changing environmental conditions when growth slows. Sexual spores are always produced by meiosis, sometimes followed by mitosis. Consequently, they are genetically different as a result of crossing over and independent assortment. Review your work on *Sordaria* in lab topic 11 for a detailed explanation of crossing-over.

Sexual reproduction in most fungi (excluding the chytrids) involves two distinct processes: plasmogamy and karyogamy. **Plasmogamy** occurs when the cytoplasms from the terminal cells of hyphae from two parents fuse. Fungi do not have sexes. Instead there are **mating types** designated + and −. The differences are not obvious and involve the genetically determined surface chemistries of hyphae. When the cytoplasms of two mating types unite, a **heterokaryon** is formed, meaning that there is a single cytoplasm containing haploid nuclei from two different parents. If there are two nuclei per cell, then the heterokaryon is called a **dikaryon**. Dikaryotic hyphae can continue to grow for years with each type of nucleus dividing by mitosis during the growth period, producing new cells that have one of each type of nucleus. Because mitosis occurs inside the individual nuclei, there is no mixing of chromosomes and the genetic lineages of the nuclei from different parents remain intact.

Usually under conditions of poor nutrition, the haploid nuclei in a dikaryon will fuse to produce a single diploid nucleus, a process called **karyogamy.** This diploid cell is called a **zygote** and is the only diploid cell in the life cycle. It will divide by meiosis to produce four haploid nuclei. These nuclei may divide again by mitosis and eventually are packaged into spores that are released. Given suitable conditions, the spores will germinate, where they land, producing new hyphal growth. During the brief diploid phase and meiosis, genetic recombination occurs through crossing over and independent assortment so that sexually produced spores are genetically different from each other. This mixing of genes allows the organisms to test new genetic combinations against the natural selection factors in the environment.

Often the events of karyogamy occur in what is called a **fruiting body,** the familiar mushrooms, puffballs, and fungal cups that are apparent to the casual observer of nature. However, it is not a true fruit and has no relationship to fruits produced by flowering plants. The term is a hold-over from the days when fungi were thought to be plants.

Nearly 100,000 species of fungi have been described. It is estimated that over ten times that number await discovery. Mycologists traditionally group the fungi into at least five phyla and one nontaxonomic group. These are:

Phylum Chytridiomycota—Chytrids: aquatic organisms that share characteristics with both the protists and fungi; saprobes and parasites; only fungi with flagellated cells in their life cycle and only fungi that produce both haploid and diploid nuclei by mitosis in hyphae; about 1,000 species.

Phylum Zygomycota—Zygomycetes: fungi living in soils, or on decaying plant and animal matter, or as parasites of insects, coenocytic (aseptate) cells; about 1,100 species.

Phylum Ascomycota—Sac fungi living in a variety of habitats and in symbiotic relationships; septate, dikaryotic cells in most of life cycle; sexual spores produced in an ascus, a structure that gives the group its name; about 60,000 species.

Phylum Basidiomycota—Club fungi; important in decomposing wood and in forming mutualistic and parasitic relationships with plants; septate, dikaryotic cells in most of life cycle stages; sexual spores produced in a basidium, a reproductive structure that gives this group its name; about 25,000 species.

Phylum Glomeromycota—Glomales; a group of over 150 species formerly classified within the Zygomycota, but, as of 2002, placed in a provisional, new phylum based on DNA analysis; coenocytic (aseptate) cells; only asexual reproduction has been observed; produce large asexual spores that contain hundreds of haploid nuclei; all species form mycorrhizal relationships with plant roots. This phylum will not be studied separately in this lab topic, but you will see an example when you look at mycorrhizae.

Deuteromycetes—a nontaxonomic group; imperfect fungi with asexual reproduction only. This group will not be studied in this exercise; about 17,000 species.

Recent DNA sequencing and comparison studies suggest that some of the organisms traditionally placed in these phyla, and also in protist groupings, should be grouped in their own clade. These clades are Microsporidia, Neocallimastigomycota, and Blastocladiomycota. We will continue to follow the five phylum traditional approach.

Traditionally, the lichens are studied with the fungi. Lichens, however, are not single organisms. Instead, they are a unique association of two species: one an alga (or a cyanobacterium) and the other a fungus. While you will study their structure, their taxonomy will not be discussed.

LAB INSTRUCTIONS

In this lab topic, you will examine the body plans and life cycle stages of several representative fungi as well as observe several symbiotic relationships.

Observation of Field Samples

1 You should start this lab by looking at locally collected samples of fungi. On your way to class, look under bushes on campus or in the back of your refrigerator for fungi. Bring them to the lab.

Look at your samples under a dissecting microscope and tease them apart. Make slides of small pieces and look at them with your compound microscope. As you work through this material, review the definitions for mycelium, hypha, and fruiting body in the Background section.

Make some sketches of your sample. Write any questions you have about how it grows, what it feeds on, or how it reproduces. Use the space below.

After this investigative phase, you will study prepared slides of fungi from four of the common phyla. As you do so, some of your questions should be answered and you may be able to assign your specimen(s) to one of the phyla.

Phylum Chytridiomycota

Once considered protists, members of the phylum Chytridiomycota are now classified as fungi based on the occurrence of chitin in their cell walls and similarities in their DNA nucleotide sequences. Among the fungi, chytrids are unusual because they are the only fungi to produce flagellated reproductive cells, called **zoospores**. This suggests that the origins of fungi were from an ancestral aquatic flagellated protist. It is interesting to note that the origins of animals are also thought to have been similar protists. This flagellated stage of the life cycle limits chytrids to aquatic and wet soil environments. Among the fungi, chytrids are also unusual because they have both diploid and haploid hyphae as you will soon see.

2 Get a prepared slide of an *Allomyces* diploid phase (also called a sporophyte) from the supply area. Alternatively, a live culture may be available in a petri dish. If so, make a wet-mount slide by adding some hyphae from the dish into a drop of water on a microscope slide. Add a coverslip. Look at the slide first with the scanning objective, adjusting lighting to improve contrast. Then use the 10× objective, going to high power only to see detail.

Body Plan Note that the hyphal filaments though only a single cell wide, are several cells long with numerous branches of the filament (**fig. 19.2**). An individual filament is a **hypha,** and collectively they form a **mycelium.** The filaments are structurally supported by surrounding cell walls, consisting of the polymer chitin. Look along the length of a single hypha using high power and note the cross walls called **septae** (sing., **septum**). The cross walls have several pores in them so that cytoplasm of one cell is connected to that of the next. Draw some of the hyphae in the circle below.

Allomyces is **monokaryotic,** meaning it has one nucleus per cell, but you will not be able to see this on your slide because special staining techniques are

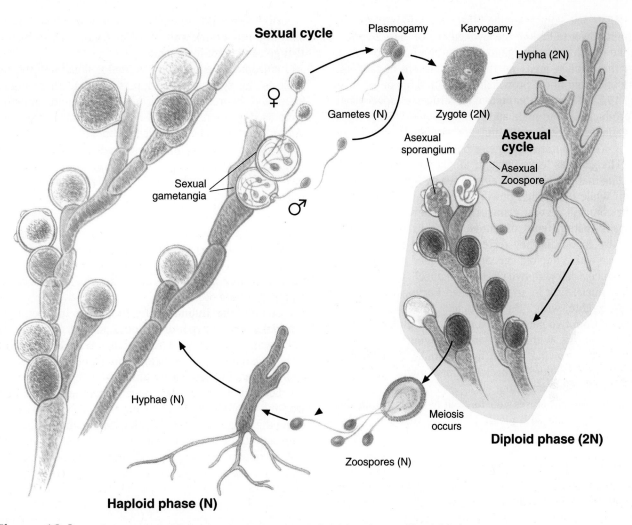

Figure 19.2 Life history of *Allomyces,* showing asexual diploid and sexual haploid phases.

required. Fungi grow by adding cells to the ends of the hyphae, pushing forward into new areas of their environment. They do not grow in girth to become several cells wide. As a hypha pushes forward, enzymes are secreted at the tip to digest new food that is encountered. The products are readily absorbed across the extensive surface area of the thin filaments.

Asexual Reproduction Chytrids are unusual for fungi in that some hyphae have diploid nuclei, while others have haploid nuclei. You cannot distinguish between the two except when reproductive structures are present.

3   Look at a wet-mount slide from a culture of a diploid phase or a prepared slide of the diploid phase (sometimes called a "sporophyte," another botanical term holdover). Focus on the swollen tips found on some hyphae. These swollen cells are asexual **sporangia** (sing. **sporangium**). There are two types: one type is relatively thin-walled and tan in color (stains may obscure color) while the other is darkly colored with a thick wall. The thin-walled type functions in asexual reproduction. Intranuclear mitosis in the sporangium with cytokinesis produces motile, **zoospores,** cells with one diploid nucleus in a small amount of cytoplasm and a flagellum. When released from the sporangium, the asexual diploid zoospores swim away from the parent and populate new areas by settling and undergoing intranuclear mitosis to produce new diploid hyphae (fig. 19.2) that are genetically equivalent. If you have not already done so, modify your drawing in the circle to include some sporangia. Chytrids must live in wet habitats; can you think of a reason why?

Sexual Reproduction The darkly colored, thick-walled sporangia, which should also be visible on the slide, function in sexual reproduction. In these sexual sporangia, zoospores are produced by intranuclear *meiosis, not mitosis*. These zoospores will have haploid nuclei that are genetically variable as a result of crossing-over and independent assortment. After they are released and settle, the sexual zoospores grow into multicellular haploid hyphae by mitosis (fig 19.2). These haploid hyphae play a critical role in sexual reproduction.

4   To see the structures involved in the next steps of sexual reproduction, you will need another slide of the *Allomyces* haploid (or gametophytic) phase. It will have been made using haploid hyphae.

Again use your scanning objective, working through to high power to look at the swollen tips of these hyphae. Unlike the single sporangium seen on the first slide, note that the swollen areas on these hyphae are in tandem, with one often tinted brownish orange and the other clear (fig. 19.2). These are **gametangia,** structures specialized to produce haploid male and female gametes. Both types of gametes have flagella and are motile, but eggs are larger. You will not see gametes on a prepared slide.

The gametes leave the gametangia through a circular opening that you should be able to see by focusing up and down while using high power. Note that each gametangium contains several gametes. After they are released, opposite mating types of gametes are attracted to each other by chemicals they secrete. A haploid male gamete will fuse with a haploid egg, an event called **plasmogamy** to produce a cell with two nuclei. In that cell, the male and female nuclei fuse in an event called **karyogamy** to produce a **zygote,** with a diploid nucleus containing genetic material from two parents. The zygote will undergo intranuclear *mitosis* without an accompanying cytokinesis, producing a growing diploid hypha similar to what you saw on the first slide. Go back to your drawing and add a hypha with gametangia to it.

Label the structures you have drawn, indicating the following: hyphae with ploidy levels, asexual and sexual sporangia, gametangia.

5 ▸ *Summarize Observations* Compared to other fungi, name two characteristics that make organisms in the phylum Chytridiomycota different.

Fill in the information required in column 1 of **table 19.1** on page 246.

Pair off with a lab partner. One of you should explain asexual reproduction in chytrids while the other critically comments on the description, and then trade roles as the other explains sexual reproduction.

Return your slides to the supply area.

Phylum Zygomycota

The 1,100 members of this phylum are saprophytic, terrestrial fungi living in the soil on dead plants or animals. The zygomycetes do not produce flagellated spores nor do they produce a distinct egg or sperm as you saw in the Chytridiomycota. In sexual reproduction, two mating types of hyphae undergo plasmogamy to produce a **zygosporangium,** a thick-walled sexual reproductive structure that gives the group its name.

Rhizopus, the black bread mold, illustrates the life cycle of the zygomycete group (**fig. 19.3**). These organisms were a problem for bakers before mold inhibitors were used in commercial breads. Now sodium propionate and other chemicals are added to bread to prevent mold spoilage. This fungus also grows well on fruits and vegetables, causing spoilage during storage. The "fuzz" that develops on old strawberries is composed of hyphae from this fungus.

6 ▸ **Body Plan** Several days before lab, two mating types of *Rhizopus* (+ and −) were inoculated onto a sterile potato dextrose agar in a petri plate. Observe the culture through your dissecting microscope. Alternatively, you may view a prepared slide of *Rhizopus.*

Identify the **hyphae** spreading from the points of inoculation to form **white mycelial** mats. Scoop up some of the white hyphae with a probe and make a wet-mount slide. Wicking a drop of lactophenol stain under the coverslip will make them easier to see.

Look at the slide with your compound microscope, working through the magnifications to high power. Scan along the length of a single hypha. As seen in the Chytridiomycota, the hyphae are only one cell thick, giving each cell maximum exposure to its environment. Why is this an advantage to the cells? (Consider how fungi acquire food.)

Note that unlike what you saw earlier in *Allomyces,* the hyphae lack cross walls (septa). All members of this phylum are **coenocytic;** their hyphae lack cross walls (a condition known as **aseptate**). The single, large cytoplasm contains many nuclei. You cannot tell by looking at them but all are haploid. Streaming of the cytoplasm along the length of the hypha distributes food and metabolic products. In the circle below, draw some of the hyphae.

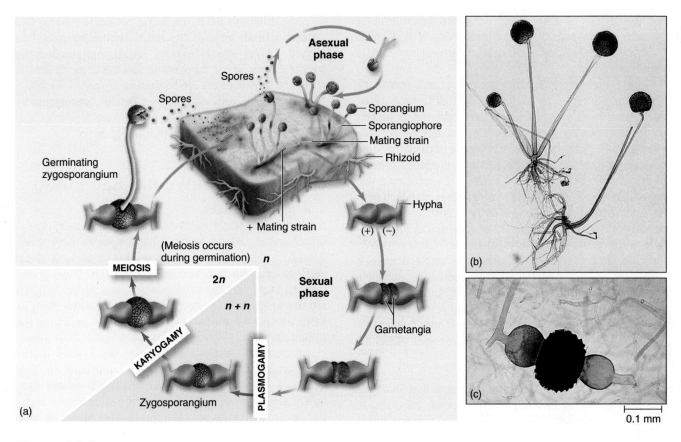

Figure 19.3 *Rhizopus,* the black bread mold: (*a*) sexual and asexual life cycles; (*b*) asexual sporangiophores; (*c*) sexual zygosporangium.

7 *Asexual Reproduction* Extending above the surface of the petri dish and possibly flattened on your slide, find some erect hyphae called asexual **sporangiophores** (fig. 19.3). These erect structures speak to the strength of the chitinous cell walls surrounding the living cells. At the tips of the hyphae, find the small black, spherical bodies called **asexual sporangia**. These are asexual reproductive structures that produce thousands of asexual spores by mitosis. These **spores** are single cells surrounded by heavy chitinous cell walls that protect the delicate cell from drying and mechanical abrasion. When released, they are wind-carried to new locations where they germinate to produce new mycelia.

If the spores in these structures are produced by mitosis from haploid nuclei, what is their **ploidy** level (= the number of complete sets of chromosomes in any one cell)? Are they genetically the same or different?

Add some sporangiophores with sporangia to your drawing.

8 *Sexual Reproduction* *Rhizopus* also reproduces sexually. Two mating types are found, differing in surface chemistry designated as + or −. When + and − hyphae meet in the environment, small swellings grow toward each other and differentiate into **gametangia,** literally meaning gamete containers (fig 19.3). Each gametangium contains many haploid nuclei from the parent hypha. A septum forms at the base of each gametangium, separating its cytoplasm and nuclei from the parent. **Plasmogamy** occurs as the two gametangia fuse to produce a single large cell, a **zygosporangium.** It will contain many haploid nuclei from both mating types and is said to be heterokaryotic, literally meaning with multiple different nuclei.

In the culture plate zygosporangium formation should be visible as a black zone at the region where the two mycelial mats overlap. Use a dissecting needle to carefully remove some hyphae with these structures from the black line and make a wet-mount slide. If a black zone is not evident, prepared slides of *Rhizopus* zygosporangia may be substituted. Look at the slide with the low-power objective on your compound microscope.

In the developing zygosporangia, the many haploid nuclei from the two parents fuse in pairs during multiple **kayogamy** events to form half as many diploid nuclei. Thus, in these fungi, a zygote in the usual sense is not formed because the resulting zygosporangium has many diploid zygotic nuclei rather than one. In addition, the spores it produces contain many diploid nuclei rather than one haploid nucleus. Inside the zygosporangium, a thick cell wall develops around the cytoplasm and nuclei to produce a zygospore. It can withstand drying and harsh conditions. When favorable conditions are encountered, it germinates.

Just prior to germination, meiosis occurs within all of the diploid nuclei, forming genetically different haploid nuclei as a result of crossing-over and independent assortment. The resulting haploid nuclei may divide by intranuclear mitosis, building up large numbers of nuclei. Spores develop when cell membranes form around a small amount of cytoplasm and each haploid nucleus as a resistant outer cell wall is laid down. As these events occur, a stalked **sexual sporangiophore** grows out of the zygosporangium. At the tip of the sporangiophore is a **sexual sporangium** containing the newly formed haploid spores. When released, they are wind-carried to new locations where they may germinate or die depending on local conditions and their genetic content. Haploid hyphae growing from the spores will eventually reproduce by asexual and sexual means.

Add a drawing of two gametangia and a zygosporangium to your earlier drawing of this species. Label all structures in the drawing.

9 *Summarize Observations* You have seen that *Rhizopus* produces spores by two different processes, one asexual and the other sexual. Why have both? Sexual spores differ from those that either parent alone could produce because of biparental inheritance and the genetic recombination events that occur during meiosis. Thus, new combinations of genes are produced and are tested when the spore germinates to produce hyphae. In a changing environment, this increases the fitness of the fungus. If only asexual spores were produced, all offspring would be the same. This would be fine in an unchanging environment, but could be problematic in a changing one. Why?

Make a list of what you think are the characteristics that distinguish members of the phylum Zygomycota from those in phylum Chytridiomycota.

Pair off with a lab partner. One person should explain asexual reproduction to the other. Switching roles, the second partner should explain sexual reproduction. Be sure to discuss when mitosis and meiosis occur in the life cycle and which stages are haploid or diploid.

Summarize your observations on the Zygomycota by completing column 2 of table 19.1 on page 246.

Return your materials to the supply area.

Phylum Ascomycota: Sac Fungi

This phylum includes such diverse forms of fungi as the molds, powdery mildews, and the cup fungi found on decaying wood. About 60,000 species have been described. About half the species associate with green algae or cyanobacteria to form lichens. The blue, green, and red molds on spoiled foods are ascomycete fungi as are the tree diseases, chestnut blight and Dutch elm disease, which have made these trees virtually extinct in the United States.

Ascomycete hyphae are septate with pores in the cross walls allowing cytoplasm and nutrients to flow from one compartment to another. Fungi in this phylum share a common diagnostic characteristic: the **ascus** (pl. **asci**), a saclike sexual reproductive structure that produces haploid spores. The common name, sac fungi, emphasizes the uniqueness of this structure. In lab topic 11, you looked at *Sordaria*, a species in this phylum, and observed the results of meiosis in the asci. Review pages 126 to 127. Many, but not all, species in this group also produce specialized asexual reproductive spores called **conidia**.

10 In the supply area, there is a culture of *Sordaria* growing in a petri dish. Make a wet-mount slide of a small amount of hyphae. View the hyphae under low, medium, and then high power of your compound microscope. Reduce the light intensity so that you have good contrast and scan along the length of a hypha looking for **septa**. Can you see them? If not, try wicking some lactophenol stain under the coverslip. Draw a few hyphae in the circle below.

Molds and Mildews The general terms mold and mildew describe discolorations and odors caused by fungi growing in moist areas. Housekeepers use the term to describe things that grow in the back of refrigerators or on damp basement walls. Gardeners use the term to describe over a 1,000 species of parasitic fungi that grow on the leaves of ornamental plants, often with a white powdery appearance. All molds are not members of this phylum because the term describes a growth form in much the same way that the term tree does. Trees can be flowering plants or conifers each in a different phylum. However, to illustrate molds we will use *Penicillium*, a common mold assigned to this phylum.

Widely distributed in nature, members of the genus *Penicillium* are common organisms causing food spoilage in homes that can be casually identified by their greenish color (**fig. 19.4**). Some members of the genus produce the antibiotic penicillin that is widely used to treat bacterial infections of animals and humans. In nature, species produce this compound to reduce the competition with bacteria for food.

Whether *Penicillium* is a member of the phylum Ascomycota is a bit controversial. *Penicillium* has never been observed to reproduce sexually. No one has seen *Penicillium* produce asci. Why then are they included in the group? *Penicillium* reproduces asexually by producing conidia, a secondary characteristic of the phylum that you will now see.

11 ***Asexual Reproduction*** Obtain a prepared slide of *Penicillium* from the supply area. Scan to the ends of the hyphae and then use high power to study the highly branched broomlike structures found there. These are **conidiophores,** specialized sporangia that produce asexual spores called **conidia** by mitosis. Ascomycete molds spread asexually when the conidia are released and carried by air currents to new locations. The conidia are highly resistant to desiccation and can remain dormant in soil samples for years, essentially providing a safe haven for the genetic complements that are being carried. If your microscope is good, you will also be able to see that the hyphae are **septate.**

Cup Fungi You have probably seen the cup-shaped **ascocarps** (fruiting bodies) produced by members of phylum Ascomycota. Found on decaying logs and leaves (**fig. 19.5c and d**), ascocarps represent the sexual phase of the life cycle. The inside surface of the cup is lined with asci that produce sexual haploid spores.

12 ***Body Plan*** Look at the samples of ascocarps available in the lab. The relatively large ascopcarp would have been connected to an extensive mycelium beneath it. The fleshy body of the ascocarp is composed of intertwined hyphae packed closely together. The hyphae, if you could see them, would be septate and contain two genetically distinct haploid nuclei, depending on the stage of the life cycle. Like the hyphae of other fungi, a chitinous cell wall would surround the cytoplasm. The hyphae would be a single-cell thick and grow by increasing in length, not girth, allowing the immobile organism to reach out and take advantage of fresh food materials. The growing hyphae would secrete enzymes that digest wood. The resulting small molecules would be absorbed across the extensive surface area of the thin filaments and be transported throughout the mycelium by cytoplasmic flow through the pores of the septa.

13 ***Sexual Reproduction*** To see the cellular organization of an ascocarp, obtain a prepared microscope slide of a section of *Peziza* and examine it using the scanning objective. Compare what you see to the artist's conception in figure 19.5a. One edge of the section will be darkly stained. This is where the asci are located.

Figure 19.4 *Penicillium* mold often develops on stored foods such as on these tangerines. White areas are hyphae. Blue-green areas are where asexual reproductive structures called conidiophores have developed.

Look first at the loosely organized area beneath this layer to see the hyphae that are cut at all sorts of angles and packed tightly together to form the fleshy ascocarp. Though you cannot distinguish among them, there are three types of hyphae based on the types of nuclei they contain. **Monokaryotic** hyphae have one haploid nucleus per cell. Because they grew by mitosis from germinating spores of either the + or − mating types, they represent two types of hyphae. Such monokaryotic hyphae make up the bulk of the ascocarp. In the center of the mass, a third type occurs. It formed when + and − hyphae underwent **plasmogamy** (cytoplasmic fusion) to produce **dikaryotic** hyphae with two genetically distinct nuclei. The nuclei remain separate. Because one of the strains donates its nucleus into the cytoplasm of the other, the participating hyphae are given names: an **antheridium** is a nuclear donor and an **ascogonium** a recipient.

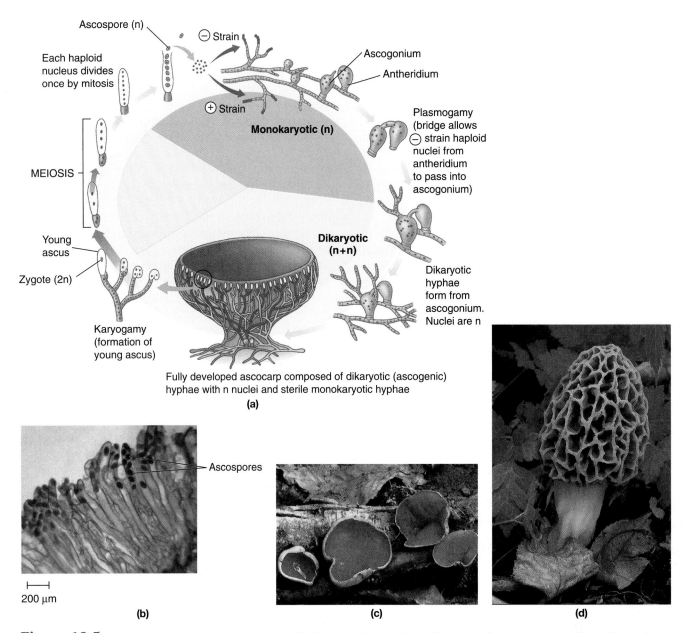

Figure 19.5 Ascomycete cup fungus structure. (*a*) Hyphae of opposite mating types fuse to produce dikaryotic hyphae. These grow to produce an ascocarp fruiting body in the shape of a cup. Nuclear fusion occurs in certain cells lining the cup. The diploid zygote divides by meiosis and then by mitosis to produce eight spores held in a saclike structure, the ascus. Spores are released and germinate to produce haploid hyphae. (*b*) Photomicrograph of ascospores in ascus sacs. (*c*) Living cup fungus *Sarcoscypha coccinea*. (*d*) Morel mushrooms are also cup fungi.

The two nuclei in the dikaryotic cells undergo intranuclear mitosis and cytoplasmic division to produce two cells each having a pair of + and − nuclei. The dikaryotic hyphae continue to grow upward to the surface of the cup. Here the + and − nuclei in the terminal cells finally undergo **karyogamy** (nuclear fusion) to produce a cell with a diploid nucleus, a **zygote**. It has a short lifetime that ends when it divides by meiosis to produce four haploid nuclei. In *Peziza* and many other species, each of the four nuclei will then divide by mitosis to form eight haploid **ascospores** inside a saclike **ascus**. The inside of the cup is lined with thousands of these asci. What two processes occurring during meiosis make the spores found in different asci genetically different?

Look at your slide and find the layer of darkly stained asci (fig. 19.5*b*). Inside each ascus are sexual spores called **ascospores**. Confirm the number of spores given above. Is it correct? _____ Draw an ascus in the following circle.

In some species, ascospores are explosively released into the air as water is taken up into the ascus and can travel over 2 meters. If they land in suitable locations, the spores germinate and grow by mitosis into haploid mycelia of the + or − type.

14 *Summarize Observations* Make a list of what characteristics distinguish members of the phylum Ascomycota from those in phylum Zygomycota.

Summarize your observations of this phylum by completing column 3 of table 19.1 on page 246.

Pair off with someone in the lab and explain the body plan and asexual life cycle of an ascomycete fungus. Trade roles and the other person should explain the sexual life cycle.

Return your materials to the supply area.

Phylum Basidiomycota: Club Fungi

This is a varied phylum of saprophytic, mutualistic, and parasitic fungi. They play a central role in the decomposition of plant litter, accounting for up to 2/3 of the biomass found in the soil, excluding soil animals. Smuts and rusts that infect various grain crops and vegetables are found in this group. About 25,000 species have been described. Puffballs, mushrooms, toadstools, and bracket fungi are the fruiting bodies, called **basidiocarps**, **(fig. 19.6)**. They bear unique microscopic club-shaped sexual reproductive cells, called **basidia** that each produce four haploid spores. A single mushroom may produce a billion spores. Asexual reproduction occurs far less commonly in this phylum than in the others you have studied.

Many mushrooms are edible, while others are among the most poisonous organisms known. The *Amanita* mushrooms, known as "destroying angels" (fig. 19.6a), contain a chemical that blocks the function of RNA polymerase in human cell nuclei, meaning that no RNA is made. One bite of an *Amanita* mushroom contains enough poison to kill an adult human and this is only one example of several that could be mentioned. *Beware: There are no known antidotes for fungal poisons and there is no easy method for distinguishing the edible from the poisonous species.*

(a)

(b)

(c)

Figure 19.6 Three examples of basidiomycota fungi: (a) poisonous mushroom *Amanita muscaria;* (b) shelf fungus growing on side of dead tree; (c) a puffball releasing spores.

Body Plan As in other phyla, spores germinate and grow by intranuclear mitosis to produce haploid hyphae. Composed of monokaryotic cells joined end to end, they form the thin filaments in mycelial mats that are essential to food gathering. The cells have chitinous cell walls and are always septate with pores in the cross walls that allow transport between cells. Mushrooms grow quickly, often in a day or so, because materials can be rapidly transported through the mycelium to the growing basidiocarp.

When + and − hyphae meet, the terminal cells undergo **plasmogamy** to form a dikaryotic cell with two nuclei derived from different genetic lineages (**fig 19.7**). Dikaryotic hyphae grow from this cell with new haploid + and − nuclei being produced by intranuclear mitosis. Thus, as cells are added to the growing hypha, each new cell is dikaryotic with one + and one − nucleus. Such growth may continue for years as the mycelial mat spreads through its food source.

Under appropriate conditions, the dikaryotic hyphae will aggregate, producing fleshy basidiocarps. In these structures, the terminal cells of the hyphae differentiate into basidia, where the true sexual phase of the life cycle takes place. In the basidia, **karyogamy** occurs, producing the only diploid nucleus in the life cycle. It unites the genetic material, which was long ago derived from two parents, into one nucleus. This nucleus almost immediately undergoes intranuclear meiosis to produce four haploid nuclei that are genetically different because of biparental inheritance, independent assortment, and crossing-over. Each of the four nuclei migrates to the periphery of the basidium where it pinches off with a bit of cytoplasm and is surrounded by a rather impervious spore wall. When released from the **basidium**, the spores, called **basidiospores** (fig 19.7*b*) to convey their origins, are wind-carried until they fall in random locations. Only those landing in suitable locations for the species will germinate to continue the life cycle.

In Oregon an individual basidiomycete, dubbed the "humongous fungus," was found to extend throughout nearly 10 square kilometers of high mountain forest soil. It is estimated that it had grown from a single spore deposited over 2,000 years ago and had slowly spread as innumerable interconnected hyphae representing one mycelium.

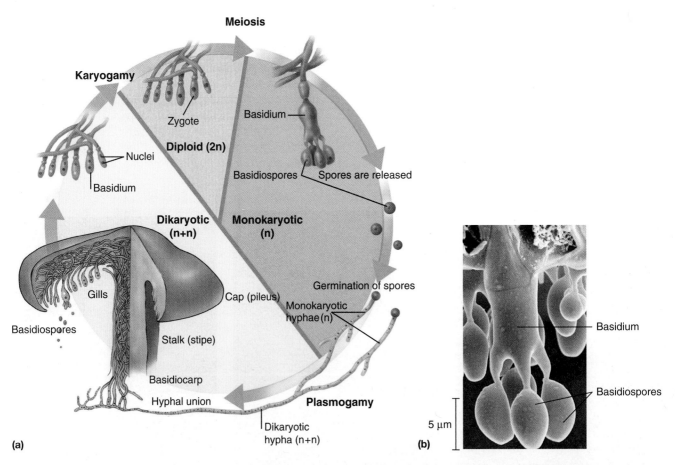

Figure 19.7 Life cycle of a mushroom. (*a*) Haploid hyphae from two mating types fuse to produce a dikaryotic mycelium; each cell has two nuclei, one derived from each parent. This dikaryotic stage will form a basidiocarp, where nuclear fusion occurs in the basidia located on the gills. The zygote thus formed divides by meiosis, producing four basidiospores which germinate and form haploid hyphae. (*b*) Scanning electron micrograph of a basidium, showing basidiospores forming.

15. Get a common edible mushroom (*Agaricus bisporus*) from the supply area. Note its fleshy consistency produced by tens of thousands of tightly interwoven dikaryotic hyphae that have associated in such a way as to form the **cap** (pileus) with the **gills** on its undersurface and the supporting **stalk** (stipe). It is interesting to think about how these hyphae have communicated with each other during the mushroom's development to achieve this structure. Chemical signaling and use of different genes at different times are most likely involved.

Split your mushroom longitudinally through the cap and stalk. Look at the cut surface with a dissecting microscope. Can you see the interwoven hyphae? Use a razor blade to cut an extremely thin section of the stalk, put the slice in a drop of water on a slide and tease it apart with dissecting needles and add a coverslip. Look at with your compound microscope. Sketch some of the hyphae below.

16. *Sexual Reproduction* The basidia where karyogamy, meiosis, and spore formation take place are located on the gills found under the cap. To see them, take a piece of a gill and try to cut a thin section with a razor blade on a slide next to a drop of water. Put the section in the water, add a coverslip and look at it with the scanning objective of your compound microscope. Along the edges of some of the gills you should see spores, called basidiospores, adhering to the cells or perhaps floating free.

17. To see greater detail, obtain a prepared slide of a mushroom's gills and look at it with the scanning objective of your compound microscope. Some slides may be a cross section of the cap with the stalk at the center and the gills radiating away from it. Other slides may be longitudinal sections of a single gill. Both will show the detail we want.

Look at the section closely using scanning, and then medium power. First, note how the interior of the gill is made from interwoven hyphae cut at various angles. Now look at the gill surfaces. Several darkly stained cells will be seen with "bumps" protruding from their external surfaces. Look at these with high power and count the maximum number of "bumps" you can see on a single cell. How many? _____ Does this suggest what type of cell division may be involved in their formation? _____

The cells with the "bumps" are **basidia** (sing. **basidium**) and are the terminal cells of dikaryotic hyphae. Karyogamy of the + and − haploid nuclei occurs in these basidia, forming a short-lived diploid zygote nucleus. It quickly divides by meiosis, producing four haploid nuclei that migrate to the outer surface of the basidium. The "bumps" seen earlier are the **basidiospores** that bud off each basidium containing a small amount of cytoplasm, one of the haploid nuclei, and a spore wall. Each basidiospore is attached to its parent basidium by a thin stalk of cytoplasm that will break as it dries, releasing the spore. When released, these spores will be wind-borne to new locations where, if conditions are favorable, they germinate to produce either + or − hyphae. Draw a basidium with attached basiodiospores in the blue circle where you drew hyphae. These structures are a diagnostic characteristic for the phylum.

18. About a day before lab, your instructor placed a cap of a mushroom with the gill opening facing downward onto a piece of white paper and left it there. If you lift the cap, you can see the huge numbers of dark spores that were released in a day. The resulting image is called a spore print. Mycologists, people who study fungi, use the pattern and color to identify fungal species.

19. *Summarize Observations* Make a list of what you observed that distinguish members of the phylum Basidiomycota from members of the phylum Ascomycota.

Be sure to summarize your observations by providing the information requested in column 4 of table 19.1.

Pair off with someone in the lab and have them explain to you the sexual life cycle of a mushroom why you critically comment on the description.

Return all materials to the supply area and dispose of trash as directed.

Fungal Associations

Lichens

Lichens are composite organisms consisting of a fungus growing in close association with a photosynthetic green alga or, in some cases, with cyanobacteria (**fig. 19.8**). The fungus and alga together produce a unique "superorganism"

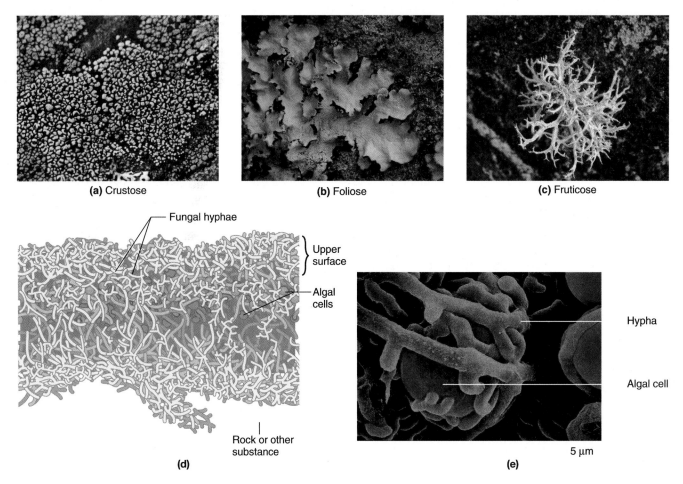

Figure 19.8 Lichens: (a) *Rhizocarpon geograpicum;* (b) *Cetrelia chicitae;* (c) *Evernia mesomorpha;* (d) artist's conception of a foliose lichen in cross section; (e) artificially colored scanning electron micrograph shows close association between the green algal cells and hyphae of fungus.

that can colonize environments where neither could live alone. The fungus contributes to this mutualistic, symbiotic association by absorbing minerals and moisture from the environment. The algae are nestled among the fungal hyphae, where they benefit from the absorptive processes, and, in turn, contribute carbohydrates and other organic molecules to the fungus. About 5 to 10% of the lichen's dry weight is due to algal cells. The fungal component in 99% of the lichens is a member of the phylum Ascomycota, although a few have a basidiomycete component.

Lichens are found from the arctic to the tropics, growing on rocks, trees, and soils. About 15,000 kinds of lichens have been described. They are often given genus and species names, but strictly speaking, this name applies only to the fungal component. A single species of green algae can be involved in many distinct lichens. They range in color from black and white to delicate shades of green, yellow, brown, and red. The different pigments protect the algae from the intense sunlight they experience on exposed surfaces. Lichens produce acids that gradually break down the rocks they grow on, contributing minerals to the buildup of soil. Lichens that contain cyanobacteria are able to fix nitrogen, contributing nitrogen compounds to developing soils.

Lichens are able to live in harsh environments because they can survive long periods of desiccation. When it rains or fog rolls in, a lichen can absorb upto 35 times its own weight in water. As water is absorbed, the algae become photosynthetically active until a dry period begins. Because lichens are dry most of the time, they have very slow growth rates, increasing in diameter from less than 1 to 10 millimeters per year. On this basis, some individual lichens may be thousands of years old.

20▶ Body Plan Examine the examples of whole lichens available in the lab. Note the three general forms (fig. 19.8). (1) **Crustose** lichens grow as a crust on surfaces. (2) **Foliose** lichens are more or less leafy in appearance. (3) **Fruticose** lichens are shrublike with branching and intertwined fibrous parts.

21▶ Ask your instructor if you should dissect a lichen. If so, obtain a small sample of the lichen, *Cladina.* Place it in a drop of water on a microscope slide. Using the blunt end of a dissecting needle, smash the specimen. Use forceps and a dissecting needle to tease it apart into very small bits, removing any large clumps. Add a coverslip and view with your compound microscope. Can you see both fungal and green algal cells? Make notes and sketches.

INVESTIGATING FUNGAL DIVERSITY AND SYMBIOTIC RELATIONSHIPS   243

These are the sexual reproductive structures of the fungal member of the lichen association. What phylum is the fungal member from? _____

Mycorrhizae

Many fungi live in a mutualistic symbiotic relationship with the roots of living plants and contribute to their growth. If pine trees are planted in sterile soil, their growth is slow. If a small amount of soil from a natural pine forest is added, growth promptly increases. Why? Fungal mycelia in the added soil form associations with the pine tree roots. The hyphae spread out from the roots and aid in mineral and water absorption, greatly increasing root efficiency (**fig. 19.9b**). In turn, the fungi obtain carbon compounds from the roots of the plants. Over 90% of the vascular plants, including many crops, form such associations. Fossil roots over 400 million years old contain mycorrhizal fungi, attesting to a long and stable coevolution.

Body Plan Two types of mycorrhizal relationships occur, defined by how the fungal hyphae interact with the roots. The **arbuscular mycorrhizae,** also known as endomycorrhizae, are more common and are found in 70% of all plant species. Fungi from the new phylum Glomeromycota, once considered to be Zygomycota, associate with approximately 200,000 species of plants. The roots of any individual plant that you see in nature may have associations with several species of glomeromycete fungi. The hyphae of these fungi penetrate not only into the roots, but also through the cell walls where the hyphal tips flare out into structures called **arbuscules.** The fungus does not actually penetrate into the living cell. Instead,

22 Obtain a slide of a cross section of a foliose lichen. Look at it with your compound microscope. Compare the general organization to figure 19.8d. Are the algae scattered or concentrated in layers? _____ Add any details seen through your microscope to your drawing.

Reproduction Lichens do not reproduce sexually, although the fungus may do so individually. Fragmentation of the lichen asexually propagates the association. Some lichens form special structures called **soredia,** which are minute fragments of fungus and algae that may be wind-carried to new locales.

23 If a slice of a lichen ascocarp is available, look at it with your scanning objective working through medium and high power to see details. Projecting from the upper surface, you should see flat cone structures that are darkly stained with a brownish upper edge. Look at this closely.

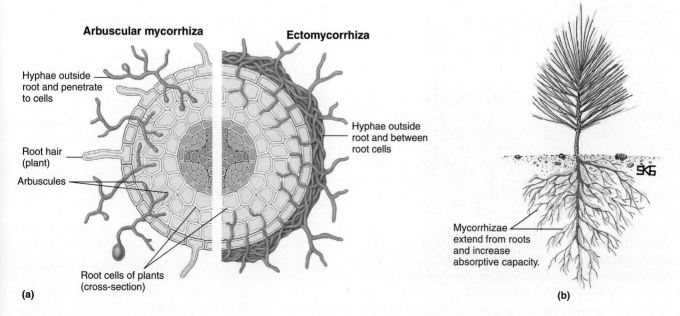

Figure 19.9 (a) Comparison of two types of mycorrhizal relationship; (b) Young pine tree with ectomycorrhizae enlarging the roots' absorptive zone.

the arbuscule cell membrane presses closely against the plant cell membrane, forming a highly folded interaction area that allows the fungal hypha to draw nourishment from the plant cell as the hyphae provide water and ions to the root cells (fig. 19.9a).

Beeches, birches, oaks, and pines, as well as several other species, do not enter into arbuscular relationships. Instead, they form **ectomycorrhizal** ones where the fungal hyphae surround the roots and pass between cells in the roots (fig. 19.9a). A few fungal species from phyla Ascomycota and Basidiomycota tend to enter into this relationship.

24▸ Your instructor will have set up some demonstration slides of cross sections of plant roots that have arbuscular or ectomychorrizae. The ectomychorrizal roots have a halo of hyphae surrounding them with some penetration into the root as the hyphae pass between cells. The arbuscular mychorrizal roots lack the halo but if you look closely at the cells in the outer half of the root, some cells have darkly stained structures as well as cytoplasm in them. The darkly staining structures are arbuscles.

In each half of the following blue circle make a sketch of half of each kind of root.

Fun with Fungi (Optional)

25▸ *Identifying Unknowns* Now that you have studied all of the phyla and several associations of fungi, see if you can apply your new knowledge. At the beginning of the lab, you looked at specimens that you and your classmates had collected. Look at them again and see if you can name the phyla that the specimens represent.

Have your instructor check your identifications.

Fungus Amon-gus Fungal spores are everywhere, floating in the air. When they land on suitable substrates, they germinate and produce hyphae, eventually a mycelial mass, and finally reproductive structures. Chances are that you never observed a complete life cycle. This will be your opportunity.

26▸ Your instructor will have some small sealable sandwich baggies and loaves of bread without added preservatives. You will use the bread to trap and grow some of these fungi. Take a slice of bread and go outside and wave it through the air several times or drop it on a dusty spot on the ground or a window ledge. Place it in the baggie along with 20 drops of distilled water. Seal the bag and take it back to your room. Poke a few small holes in the bag to allow for air exchange. Place the bag in a dark warm place.

Each night look at the bag and keep a journal of the changes that occur. Note such things as when you first see mycelial growth, what percent of the slice is covered by fungi (at each observation period), and when fruiting bodies appear. Bring your bread mold and journal to class next week.

Lab Summary

27▸ On a separate sheet of paper, answer the following questions as assigned by your instructor.

1. What makes a fungus different from a plant or an animal? From the various protists?
2. List the distinguishing characteristics of the four phyla of fungi studied in this lab.
3. If you were given a sample to identify, what characteristics would you use to determine if it was a lichen or a pure fungus?
4. In general terms, describe a sexual life cycle in fungi. What stages are haploid and which ones are diploid? When is nuclear division by mitosis and when by meiosis?
5. Why do fungi produce both sexual and asexual spores? Does one have advantages over the other? Explain.
6. Using table 19.1 as a guide, list those ancestral characteristics that are shared by all fungi.
7. Using table 19.1 as a guide, list those derived characteristics that are unique for each of the phyla studied.
8. Compare the life cycles of fungi from the phylum Ascomycota with those from the phylum Basidiomycota. How are they similar? How do they differ?
9. What is it about lichens that allows them to live under harsh as well as lush conditions?
10. Pine trees grown in sterilized soil grow slowly. If some soil with fungal hyphae from a natural forest is added, they begin to grow much faster. How do you explain this?

You may want to try the critical thinking questions that apply some of the knowledge you gained in doing this lab.

Internet Sources

Considerable information is available on the World Wide Web on the topic of lichens. A starting place for locating information is the Internet Resources for Bryologists and Lichenologists. Use Google to locate the URL as it changes from time to time. Check out the Images and Information section. An interesting site about fungi in the news (with pictures of the fungus of the month!) is Dr. Tom Volk's site at http://botit.botany.wisc.edu/toms_fungi

Learning Biology by Writing

Write a short, 200-word essay describing the shared and unique features among the four phyla of fungi.

Critical Thinking Questions

1. The *Suillus lakei* mushroom is commonly found growing around the base of conifers, especially the Douglas fir. Offer a possible explanation of this association of a fungus and plant.
2. DNA testing revealed that an *Armillaria bulbosa* fungus growing in the Upper Peninsula of Michigan covered 15.9 hectares (38 acres). Explain how this basidiomycete could get to be so large. Would you expect to find larger examples?
3. Crustose lichens growing on rock surfaces may increase in diameter only a millimeter or so each year. If a lichen started growing on the day you were born, how many millimeters in diameter would it be now? How big is that in inches?
4. If an asexual spore from an Ascomycota fungus mildew has 21 chromosomes, how many would you expect to find in a sexual spore?
5. Imagine that you find a fungus that has septate hyphae and forms a fruiting body visible to the naked eye. What information do you need to decide what phylum the fungus is in?

Table 19.1 Summary Table for Fungal Diversity

Characteristic	Chytridiomycota	Zygomycota	Ascomycota	Basidiomycota
Common names?				
Names of unique structures?				
Fruiting body name?				
Ploidy of hyphal nuclei?				
Septate or aseptate hyphae?				
Monokaryotic or dikaryotic?				
Instant or delayed karyogamy?				
Where meiosis occurs?				
Asexual of sexual spores or both?				
Ploidy of spores?				
Associates with what other organisms?				

Lab Topic 20

FROM BASAL TO BILATERAL ANIMALS

SUPPLIES

Resource guide available on WWW at
http://www.mhhe.com/labcentral

Equipment

Compound microscopes

Dissecting microscopes

Saltwater aquarium with living demonstration specimens of sponges, jellyfish, and anemones

Materials

Preserved specimens

    Leucosolenia

    Gonionemus

    Taenia (tapeworm—demonstration)

    Assorted sponges

Prepared slides

    Leucosolenia, whole mount (Flinn Sci.)

    Grantia or *Scypha,* c.s or l.s.

    Hydra, longitudinal section

    Obelia, whole mount

    Cnidocyte, demonstration slide

    Dugesia, cross section and whole mount

    Clonorchis sinensis or other fluke, whole mount

    Tapeworm, scolex and mature proglottids composite

Living *Dugesia*

Living *Hydra*

Raw liver meat or boiled egg yolk

Watch glasses or screw-top vials

Spring water

5% acetic acid

STUDENT PRELAB PREPARATION

Before doing this lab, read the Background material and sections of the lab topic that have been assigned by the instructor.
    Use your textbook to review the definitions of the following terms:

archenteron	body plan
bilateral symmetry	Cnidaria
blastula	diploblastic
ectoderm	Platyhelminthes
endoderm	Porifera
gastrula	radial symmetry
mesoderm	triploblastic

Describe in your own words the following concepts:

The advantages of cellular specialization to form tissues and organs

The advantages of bilateral symmetry and a body cavity

How body plans become more complex through the sequence sponges to flatworms

    After finishing the prelab review, write any questions you have about terms, concepts, or techniques in the margins of this lab topic. The lab experiments should help you answer these questions, or you can ask your instructor during the lab.

OBJECTIVES

1. To study the functional anatomy and life cycles of representatives from three basal animal phyla
2. To illustrate the organizational differences between animals with and without tissues
3. To illustrate asymmetry, radial symmetry, and bilateral symmetry
4. To collect evidence that tests the following hypothesis: Animals in the three phyla (studied during this lab) can be arranged in a sequence from simple to complex.

BACKGROUND

What makes animals different from other organisms? Animals are heterotrophic multicellular, eukayotic organisms. Most feed by ingesting food and digesting it extracellularly. Their cells lack cell walls and are held together by external structural proteins such as collagen and by specialized cellular junctions. Most animals have excitable tissues, muscles and nerves that allow coordinated movements. Most are sexually reproducing and the diploid phase of the life cycle is predominant. The zygote usually undergoes a series of developmental stages starting with cleavage and progressing through blastula and gastrula stages. Many animals have independent larval stages that are often self-feeding, and then metamorphose into sexually mature adults.

The evolutionary relationships in the animal kingdom are topics of debate as new data become available. Most agree that animals originated from an ancestral stock of colonial flagellated protists (supergroup Ophistokonta), sequentially diversifying over time into the many forms seen today. About 40 phyla of animals are recognized, including two new phyla described since 1984. Vertebrates, animals with backbones, are found in only one of these phyla and account for less than 5% of all named animal species. All others are invertebrates and may be less familiar. In this book 10 of the 40 phyla are discussed. These phyla are shown in **figure 20.1**.

Traditionally, the 40 phyla have been grouped into what we now call clades that share common characteristics. Until recently, these clades were based on answering four rather simple questions about an animal's anatomy and embryology. These are:

1. Does the animal have tissues?
2. What is the symmetry of the body: radial or bilateral?
3. During embryonic development, how does the egg initially divide into many cells and how do the mouth and anus form?
4. Does the animal have a body cavity?

In recent years the basis for forming clades has shifted emphasis from only morphology to include molecular characteristics. The techniques of molecular biology allow the nucleotide sequences in genes or the amino acid sequences in proteins to be determined. Once sequences are known, new techniques in bioinformatics can be used to compare them, using computers and statistical tests. The analyses allow scientists to show that two kinds of animals share a particular percentage of their molecular characteristics, and assess the probability that those two kinds are more closely related to each other than to other kinds of animals. The underlying reasoning in systematics is that if two animals are related through evolution, there will be greater similarity in their genes and proteins. Therefore they should be grouped in the same clade. Analysis of two types of genes have proven very useful: those coding for ribosomal RNA and those for a group of genes controlling embryological development of body plan, called *Hox* genes.

When the composition of clades based on molecular data are compared to clades based on the traditional anatomical data, many of the groupings remain the same, thus validating earlier hypotheses that certain animal groups are related. For example, the clades based on tissue organization and body symmetry are supported as are the

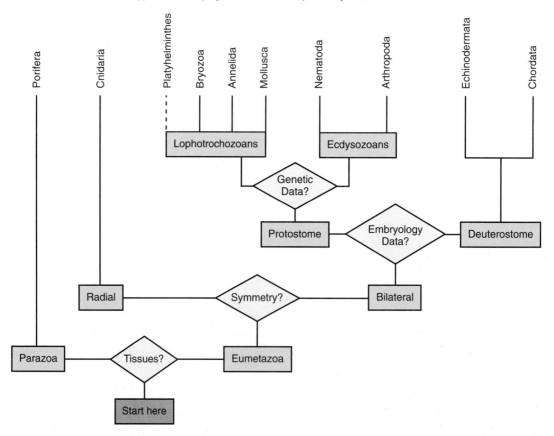

Figure 20.1 Ancestral and derived characteristics of major clades of animals. Clade names are shown in green boxes. Animals can be assigned to clades on the basis of answering four questions in yellow diamonds: (1) Does animal have tissues? (2) What is animal's symmetry? (3) What are the embryological developmental patterns? (4) Are there similarities in the gene or protein sequences?

clades based on embryological data. However, significant disparities have also been found where the traditional approach suggests one relationship but the molecular data suggest another. The earlier hypotheses based on traditional data are then rejected because they have been falsified by new evidence. What have not been supported are the clades based on body cavity (coelom) development. The coelom may have appeared in different lineages or disappeared, and it is clear that one type of cavity did not evolve into another over evolutionary time.

Figure 20.1 can be used to show the usefulness of these cladistic ideas. Let's assume that you were given data about an unknown organism that met the basal criteria for being an animal. You are asked to assign it to one of the ten phyla you will study in this course. Figure 20.1 tells you that you should first ask, "Does it have tissues"? If it does not, then it is a member of the clade Parazoa which contains the phylum, Porifera. On the other hand, if it has tissues, then it is a member of the clade Eumetazoa which contains most animals. If you establish that it is a Eumetazoan, you must now determine its symmetry. If it is radially symmetrical, then it must be a member of the phylum Cnidaria. If it is bilaterally symmetrically, as most animals are, more information will be needed to narrow down the choices.

Assuming your unknown is a bilaterally symmetrical animal, you will now need information about its embryological development. Figure 32.1 in lab topic 32 shows the early developmental stages of a starfish. Most animals go through similar stages as they develop from zygotes. Look at the last photo in the sequence. This is called a gastrula stage. The digestive system develops at the gastrula stage when a tube, called the archenteron, grows inward from one side of a spherical group of cells called a blastula. In one clade, called the **protostomes,** the opening of this tube to the external environment will become the mouth. In the other clade, called the **deuterostomes,** the opening will become the anus. There are additional differences between the two clades but these will be discussed in later lab topics. Let's assume that the data you have allows you to determine it is a protostome, as most bilaterally symmetrical animals are.

Note that the protostomes are divided into two clades based on molecular data: lophotrochozoans and ecdysozoans. Access to such data would allow you to decide what clade your unknown animal belonged to, but that would take some insights to interpret. Alternatively, you could make use of certain traits that correlate with the molecular data. Animals in the lophotrochozoan clade use subcellular structures called cilia to move water over their outer surfaces for feeding or for locomotion. Those in the ecdysozoan clade lacks cilia. All ecdysozoans are covered by a cuticle, an external skeleton which protects the body and is periodically shed as the animal grows. Assuming that your animal had ecdysozoan characteristics, you now know that it is a member of either the phylum Nematoda or the phylum Arthropoda. Now use the phylum characteristics for the final assignment to a phylum. For example, if it was wormlike with no appendages and the exoskeleton was made of collagen, it would be a nematode.

The example you have just worked through shows the utilitarian value of grouping organisms into clades. Using only five pieces of information, you identified an animal to the phylum level. However, there is more to it. Applying the underlying philosophy of systematics, grouping animals in a clade indicates genetic similarity and relatedness through evolution.

Molecular data clearly shows animals are monophyletic, evolving from colonial flagellated protists nearly 700 million years ago (M.Y.A.). We are not sure what this animal looked like as it is probably extinct today. However, we can hypothesize that it most likely exhibited the previously described basal characteristics of animals but lacked tissues. From these animals evolved a group that had tissues, cells specialized for particular functions, allowing the associations of tissues that would lead to organ formation. This primitive eumetazoan could have been radially or bilaterally symmetrical, although it is hypothesized that radial symmetry is the ancestral condition and bilateral symmetry developed later.

The bilaterally symmetrical group gave rise to most types of animals we see today. Consequently, the development of bilateral symmetry as a shared characteristic is considered a major event in animal evolution. Bilateral symmetry most likely affected nervous system development. Bilaterally symmetrical animals move through their environment anterior end first. This caused selection for clustering of sensory receptors with associated nerves at the anterior and nerve tracts leading to other body regions. The general adaptability of this body plan is supported by the number of animals having it. During the Cambrian Explosion from about 565 to 525 million years ago, the three major clades with bilateral symmetry appeared in the fossil record. These are Lophotrochozoans, Ecdysozoans, and Deuterostomes. The origins of most of the animals in the 40 phyla we see today are traceable to this time.

In this and subsequent lab topics on animal diversity, you will look at only 10 phyla. The intent is to illustrate the derived characteristics that allow animals to be assigned to clades and to give you some common experience with animals that you might never have seen before. As you study the animals, there will be many new anatomical and taxonomic terms to learn. Flashcards and memorization sessions will help you. However, do not neglect to comprehend the big picture. The big picture is gained by comparing and contrasting one animal group with another. For example, as each new animal group is studied ask which of its characteristics:

- Is ancestral and shared with other clades
- Is derived and not shared with other clades

In addition, you should be able to describe how each phylum studied has solved the problem of performing:

- Feeding and digestion
- Exchanging respiratory gases
- Distributing nutrients to remote cells

- Maintaining salt and water balance as well as eliminating metabolic wastes
- Locomotion and support
- Sensing and reacting to the environment
- Reproduction

If you are able to do this, you will achieve a sense of diversity that is fascinating rather than bewildering.

LAB INSTRUCTIONS

In this lab topic, you will study representatives from three phyla: Porifera, Cnidaria, and Platyhelminthes. These animal phyla were chosen for the first lab topic on the animal kingdom because they clearly illustrate differences in tissue organization, symmetry, and development of organ systems.

Before starting this lab, we will propose a hypothesis for you to test, to falsify if you can. The hypothesis is: *"There is no evidence that animals in the phyla studied here can be arranged in a sequence from simple to complex."* If you find evidence that they can, then you have falsified the hypothesis and must accept an alternative. Formulate an alternative hypothesis and state it below.

Metazoa

Available molecular and morphological evidence indicates animals, also called Metazoa, are monophyletic derived from the supergroup Ophistokonta which includes a traditionally protist clade, the choanoflagellates. Choanoflagellates are characterized by having a collar of microvilli, fine cytoplasmic surface extensions that surround a single flagellum (see figure 16.14). The beating of the flagellum drives water through the sieve of microvilli filtering out fine particles that are then ingested. Some choanoflagellates are colonial, and it is from similar species that we think metazoans evolved by becoming multicellular. Multicellular organisms differ from colonial ones because all cells in multicellular organism do not have the same structure and function. In the hypothetical evolutionary scenario to explain the appearance of sponges, some collar cells lost their flagella and collar. They then took on functions related to support, body shape, and reproduction, but remained dependent on other feeding collar cells for sustenance. When such events occurred, animals appeared for the first time on earth. Subsequent evolution of these first animals yielded the diversity of animals seen today.

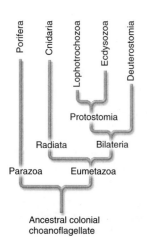

Parazoa

Basal animals such as sponges are clearly multicellular with specialized cells, but those cells remain totipotent, meaning that each type of cell can change into the other types. Consequently, no one cell type is traceable to a group of cells in a developing embryo as is the case for the true tissues found in other animals. Multicellularity also requires that the cells of an animal's body: (1) remain joined together as a functional whole, usually as a result of specialized proteins on the cell's surface, (2) have specialized cell junctions, and (3) secrete a collagen-type protein to form an extracellular matrix. Sponges have some of the specializations characteristic of animals, but not all. For these reasons, the animal kingdom (Metazoa) is traditionally divided into two branches: Parazoa and Eumetazoa. Parazoa, meaning almost animals, recognizes that sponges have most characteristics of other animals, but not all. Eumetazoa, meaning true animals, is a category containing those animals that have true tissues.

Phylum Porifera

About 8,300 or so species of animals are sponges. Although classified into a single phylum, molecular analyses of genes and proteins strongly suggest that the sponges are polyphyletic and should be split into at least four clades. Consequently, in the not too distant future, the phylum Porifera may no longer be used and animals in several new phyla will commonly be called sponges, but for now we will group them together. There are also molecular indications that all other animals are derived from one of the sponge clades, suggesting that sponges are not a side branch in the evolutionary tree of animal life (Parazoa) as has long been thought, but part of the main trunk.

Body Plan All sponges are aquatic with most species being marine, although some live in fresh water. They have asymmetric body plans, varying from encrusting growths to globular aggregations that can be greater than 2 meters across. Except for their motile larval stages, sponges are sessile, remaining in one place throughout their lives.

1▶ Look at the demonstration specimens in the lab to see the variety of body shapes. For dried specimens, you are looking at skeletal remains, but in preserved specimens you see the skeleton with adhering cells. In most specimens, you can see that the body surface is perforated by numerous openings, hence the phylum name **Porifera,** meaning pore bearing. If you dove to a shallow ocean reef, you would see many other shapes and the sponges would be vividly colored. The colors are not produced by the sponges themselves. Instead, pigmented endosymbiotic dinoflagellates, algae, and cyanobacteria live inside the sponge's cells, yielding a rainbow of colors. In this mutualistic association, the protists and cyanobacteria gain ammonia and carbon dioxide from their hosts while contributing oxygen and photosynthetic products.

2 To see the body structure in more detail, obtain a whole-mount slide of *Leucosolenia* (**fig. 20.2a**). Use only your scanning objective because the slide is thick and the other objectives may smash the coverslip.

The body wall is closed at the base, but open at the tip, creating a vase-shaped body outline. This is a colonial species and small budding colonies may be present. Water, containing oxygen and food particles, enters a central cavity, the **spongocoel,** through very small pores in the body wall and exits through the **osculum,** the opening at the end of the body. The spongocoels of individuals in a colony may or may not be interconnected.

The body wall consists of two layers of cells: one covering the surface and another lining the spongocoel. In-between is a loosely organized region called the **mesohyl,** consisting of gelatinous proteins, fibers, and secreted mineral elements. A sparse population of amoeboid cells wanders through the mesohyl. If you focus up and down while looking at one edge of the specimen, you will see the mineral elements of the skeletal system. The wandering cells in the mesohyl make these three-pronged **spicules** by secreting calcium carbonate around collagen fibers. Some species make spicules from silica and others, especially natural bath sponges, do not produce mineralized spicules. Instead they secrete a protein similar to collagen called **spongin** that forms a framework supporting the sponge's body. The familiar natural bath sponge comes from sponges with this type of skeletal system. The chemical composition of the skeletal system is used as a diagnostic character to place sponges in traditional taxonomic classes which we will not discuss.

Three types of body wall arrangements are found in sponges. Asconoid sponges, such as *Leucosolenia,* have the simplest bodies (fig. 20.2a). The other arrangements can be derived from this shape by folding of the body wall, best seen in the syconoid body type (fig 20.2b).

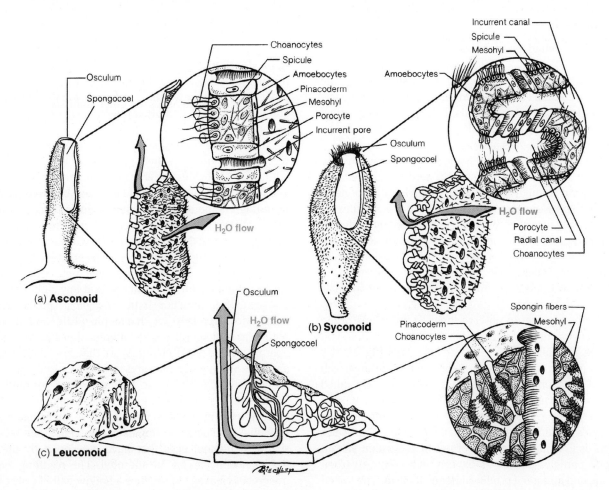

Figure 20.2 Cells in sponges' bodies are organized into one of three body types, depending on species. Longitudinal sections through three body types showing the relationship of spongocoel to choanocytes and canals; (*a*) asconoid body type; (*b*) syconoid body type; (*c*) leuconoid body type.

In the leuconoid plan (fig. 20.2c), not only is the body wall folded, but the spongocoel is branched to produce a labyrinth of interconnected chambers, as seen in bath sponges.

3 *Functional Anatomy* To view cellular organization in the body wall, obtain a slide of a longitudinal section of *Grantia* (=*Scypha*), a small sponge with a syconoid body plan. Look at it first with the dissecting microscope to orient yourself. Note (1) how the body wall is folded, (2) the open central spongocoel and (3) possibly the osculum, depending on where the specimen was sectioned. Note how the body wall is uniform without areas specialized into organs.

Switch to your compound microscope to study the cell organization in one of the body wall folds. How many cell layers do you see between the outside and the central cavity of the spongocoel? _____

Examine some cells of the outer surface to see the flattened **pinacoderm** cells. They lack the specialized cellular junctions found in cells on the surfaces of other types of animals and are only loosely held together to form the animal's outer covering. If a sponge is vigorously shaken in calcium-free seawater, many outer cells will come loose, indicating the absence of firm cell junctions as found in other animals. If you look closely at the pinacoderm in one of the folds, you should find **porocytes,** cells that are shaped like cylinders spanning the body wall. Water is drawn into the fold and then passes through the porocyte opening into the spongocoel. Cells surrounding the porocytes contain contractile proteins. Though not muscle cells, as found in other animals, these cells can contract closing the pores. All contractions are local as there are no nerves coordinating the activity.

Beneath the pinacoderm is the gelatinous mesohyl. In life, amoeboid cells in the mesohyl secrete spicules that often extend through the pinacoderm giving the sponge's surface a spiky appearance. You will not see spicules on this slide because they were chemically removed during its preparation so that the tissue could be sliced without tearing. Spicules may offer some protection against predators, but are not the primary defense. Sponges produce a number of unpleasant toxins that deter other animals from feeding on them.

Use high power to observe the cells lining one of the spongocoel's outward folds, Although difficult to find, you may see the collar cells, called **choanocytes** (fig. 20.2b), that are important in feeding. Representing the inner cell layer of the sponge body wall, they are collectively called the **choanoderm.** The hundreds of thousands of flagella beating in the choanoderm draw water in through the porocytes and drive it out through the osculum. This water stream is essential because it brings in a constant stream of food particles as well as oxygen and removes carbon dioxide and ammonia produced in metabolism. The folded body wall seen in syconoid sponges like this one greatly increase the surface area of the choanoderm compared to asconoid sponges (fig. 20.2). Leuconoid sponges, with their many chambers containing choanocytes, have the greatest ability to capture food. Respiratory gas exchange and excretion occur by diffusion between the water stream and individual cells that are rarely far from it. Only the cells of freshwater sponges have any specialization for excretion. They contain an organelle, a contractile vacuole, that periodically accumulates water and expels it from the cytoplasm, preventing osmotic swelling.

Sponges are **filter feeders.** They feed by removing bacteria, protists, and small bits of detritus from the water passing through their bodies. Studies have shown that a 1 cm^3 piece of sponge can pump 20 liters (~5 gallons) of water in a day, capturing about 80% of the suspended organic matter. Food particles are caught in mucus secreted by the collars and settle to their base to be engulfed by the choanocyte, forming food vacuoles. Digestion starts as the food vacuoles move from the tip of the choanocyte to the base, where they are transferred to wandering amoebocytes in the mesohyl that, in turn, transfer them to pinacocytes and other cells. Based on this information, would you say that the spongocoel is the same as a digestive system?_____ A study done by administering fluorescent dye labeled bacteria indicated that dye is seen in food vacuoles within 30 minutes, within amoebocytes at 60 minutes, and in the water leaving the osculum after 24 hours, providing a time estimate for complete digestion.

Reproduction Sponges can reproduce asexually by fragmentation and budding. Small pieces of sponge that are broken off will grow into new individuals. Hurricanes and heavy wave action often fragment sponges in shallow waters over reefs.

Freshwater sponges produce specialized asexual reproductive bodies called **gemmules.** At the onset of winter, amoebocytes congregate in the mesohyl, accumulate food from wandering amoebocytes, and secrete a surrounding mantle of spongin and spicules that seals them in a capsule. Gemmules are very resistant to freezing and drying, allowing the sponges to overwinter in harsh conditions. In spring, the amoebocytes emerge, grow, and divide to produce a completely new sponge with the same genetic composition as its parents and any siblings.

Sexual reproduction is the way most sponges reproduce. Most sponges are **hermaphroditic** and produce both eggs and sperm but at different times in their lives. Sperm develop from choanocytes that lose their flagella and migrate into the mesohyl where they undergo meiosis before being released in a cloud. Sperm are randomly drawn into a second sponge by its feeding current. They are captured by choanocytes that lose their collar after ingesting the sperm. Acting as carrier cells, they transfer sperm to haploid eggs that have developed by meiosis from amoebocytes in the mesohyl. The resulting zygote undergoes cleavage to produce a hollow ball of cells called a blastula. Its cells may continue dividing, filling the internal cavity of the hollow ball while flagella develop on the surface cells. The free-swimming larva bursts out of the parent sponge and may swim for a few days before settling and developing into an adult sponge. As a result of biparental inheritance, all sponges produced in this way are genetically unique.

4 *Summarize Observations* Before beginning your study of the next group, organize your observations of sponges by filling in **table 20.1** on page 264. With your lab partner, discuss the body plan of sponges. Explain how a sponge feeds, exchanges respiratory gases, and gets rid of excretory products.

Eumetazoa

A revolutionary step in the evolution of animals was the development of true tissues, groups of cells with specialized structures and functions traceable back to a specific location in a developing embryo. A critical specialization was the development of epithelial tissues, cells that are tightly bound to each other by special proteins to form leak-proof cellular junctions on both internal and external surfaces. Absent from sponges, epithelial tissues are found in all animals grouped in the clade Eumetazoa. What makes this development noteworthy is that it made extracellular digestion possible, allowing animals to digest prey larger than the single cells on which sponges feed. Having a leak-proof seal meant that enzymes could be secreted into a digestive cavity and not be lost, or worse, leak into other spaces and digest the very animal doing the secreting.

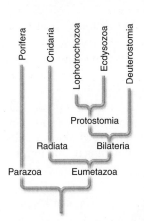

Leak-proof cellular junctions alone did not result in extracellular digestion. Body plans had to change providing a digestive chamber containing enzymes. One theory suggests that in a blastula stage (see fig. 32.1) of a small group of ancestral sponges, surface cells grew inward to form a saclike chamber. An analogy may help you to visualize this. If you take a soft balloon and press your finger against its wall, you form a tube that now projects inside the balloon. When such a chamber developed in a sponge ancestor and its cells developed the ability to secrete digestive enzymes, a simple digestive system was derived.

It is intriguing to note that all eumetazoans go through this hypothetical evolutionary sequence during embryological development. Zygotes divide to produce a hollow ball of cells, a blastula. At the end of the blastula stage, a tube, called the archenteron or primitive gut, forms by invagination to produce a gastrula stage. It consists of two layers of cells: an outer epidermis, called the ectoderm, and the wall of the archenteron called the endoderm. All other cells in eumetazoans arise from mitotic cell divisions in these germ layers. For example, nervous tissues arise from ectodermal cells and tissue lining the gut from the endoderm. Animals having these two layers are said to be diploblastic, another derived characteristic that separates eumetazoans from the sponges.

Radiata

Symmetry is a characteristic of all eumetazoans whereas sponges lack any symmetry to their typical body plans. Radially symmetrical animals have the general body form of a cylinder with body parts arranged around a central axis passing lengthwise through the cylinder, in much the same way that an umbrella is symmetrical around its shaft. Body parts arranged around this axis confront the environment equally in most directions, unlike in bilaterally symmetrical animals where the head and tail, left and right, and dorsal or ventral surfaces set up unequal axes of confrontation. The origins of radial symmetry are not well understood and it may be that both radial and bilateral symmetrical animals are derived from the same group, not one from the other. Radial symmetry is found in aquatic sessile and drifting animals. Having body parts arranged in radial symmmetry means that, more or less, all sides are equally likely to encounter and react to stimuli from the environment.

Phylum Cnidaria

The phylum Cnidaria includes about 10,000 species commonly known as jellyfish, sea anemones, sea fans, and corals. Molecular data indicate that the group is monophyletic and all species share the following common derived characteristics:

- True tissues derived from the embryonic endoderm and ectoderm
- Radial symmetry
- A digestive system, allowing extracellular digestion
- A hairnet-like nervous system, allowing sensing and coordinated movements, but no major nerve tracts or brain
- *Hox* genes, absent from sponges but found in all other animals
- Unique stinging cells called cnidocytes that allow large prey to be captured

The phylum Cnidaria contains four taxonomic classes: **Hydrozoa,** which includes *Hydra, Gonionemus, Obelia,* and the *Portuguese man-of-war;* **Scyphozoa,** which consists mainly of marine jellyfish; **Cubozoa,** which consists of a group of very toxic jellyfish called box jellies; and **Anthozoa,** which includes sea anemones and corals. Of these, the anthozoans are considered basal to the others.

Most cnidarians are marine, but some freshwater species are known. Individuals are often found as members of large colonies as in coral reefs. Many cnidarians contain endosymbiotic dinoflagellates or green algae. The spectacular colors seen on coral reefs are due to these protists living inside the cells of corals, anemones, and other animals. During periods of heat stress, the dinoflaglates sometimes evacuate the coral's tissues, a phenomenon known as coral bleaching.

Body Plans Three basic body forms are found among cnidarians: a planula larval stage and two adult forms—a sedentary **polyp** stage and a free-swimming **medusa** (jellyfish) stage. Life cycles may involve one or both of the adult stages.

The bodies of cnidarians consist of two well-defined epithelial tissues, an outer **epidermis** covering the body, and an inner **gastrodermis** lining the digestive system. Between these cell layers is a layer of gelatinous material called **mesoglea** in which widely scattered amoebocytes are found. In some species the mesoglea is very thin but in others quite thick. In this phylum, you will look at three model species, illustrating both adult body plans.

A Polyp: Hydra *Hydra* is a genus of hydrozoans that live attached to stones and vegetation in freshwater streams and ponds. Most are brown or green colored because of symbiotic algae. *Hydra* is unusual in that it has only a polyp stage in its life cycle.

5) Use an eye dropper to get a living *Hydra* from the supply area and put it with water into a small screw-top vial. Put on the top and look at the animal with your dissecting microscope at low light intensity. When disturbed, *Hydra* tends to contract into a ball. Let the animal sit for several minutes while you watch it intermittently. It should extend its tentacles and body (**fig. 20.3**).

Once extended, note the radial symmetry when viewed from the top. This is a classic example of a polyp stage with a central body column terminating in a mouth surrounded by tentacles. Do you see any evidence of specialized organs? _____

Gently roll the vial back and forth or poke it with a probe. From your observations, would you conclude that *Hydra* has sensory and nervous systems? Muscles?

Polyps such as *Hydra* can slowly crawl along a substrate by extending the pedal disc, attaching it, and pulling

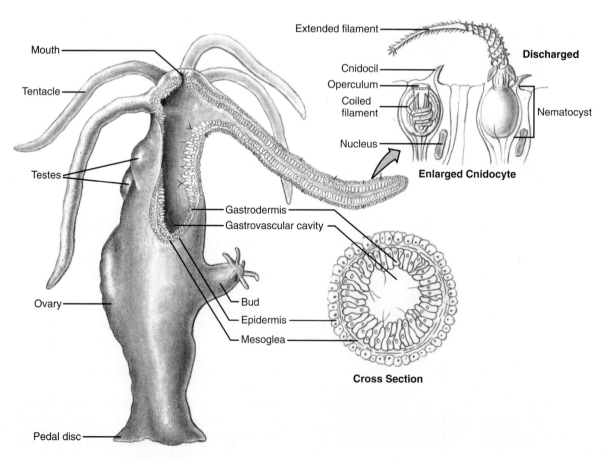

Figure 20.3 Anatomy of *Hydra*.

themselves forward. Some polyps will actually do somersaults, bending over, attaching the tentacles, loosening the pedal disc, and flipping over.

Your instructor may ask you to feed your *Hydra*. With an eye dropper add a *Daphnia* next to the *Hydra*. If you are lucky, it will immobilize and ingest one as you watch.

6 *Hydra* immobilizes its prey with unique stinging cells, called **cnidocytes,** on its tentacles. To see these, put a drop of water on a clean microscope slide. Transfer your *Hydra* to the drop. Gently add a coverslip and observe with the scanning objective or low power objectives.

Scan along the tentacles to see if you can find evidence for cnidocytes. Adjust light intensity and condenser height to get the best views. Gentle tapping on the slide may cause them to discharge. If not, add a small drop of 5% acetic acid beside the coverslip and wick it under while watching. If you are lucky, you will see a cnidocyte discharge a long thread. These threads are often tipped with a hardened point and contain potent toxins that immobilize prey. All cnidarians have them. Box jellies found in tropical oceans have toxins so powerful they can kill a human. Fortunately, most jellies are not as lethal and their toxins cause only temporary discomfort.

7 *Functional Anatomy* To see the tissues of the body wall, use your compound microscope to look at a prepared slide of a longitudinal section of *Hydra* (fig. 20.3). How many well-defined cell layers do you see in the body wall? _____

The outer epidermal cells are tightly joined to one another to form an epidermis. These columnar cells have a basal–lateral extension, giving an isolated cell somewhat the appearance of an inverted letter T. The basal extension, called a myoneme, contains contractile proteins, that can shorten the body or reduce its width. Other cell types are also found in the epidermis but will not be discussed. Widely scattered nerve cells (that you will not be able to identify) underlie the myonemes and coordinate their contractions. The mesoglea in *Hydra* is very thin. The gastrodermis includes a number of types of cells that form leak-proof seals with adjacent cells. Some secrete mucus that lubricates the gastrodermis. Others secrete digestive enzymes into the gastrovascular cavity when food is present. Some have myonemes and contribute to body movements.

Study the epidermis of the tentacles closely to see the structure of the cnidocytes. The threads that you saw earlier are coiled up in unique organelles, called **nematocysts,** that are not found in other kinds of animals. A small projection from the nematocyst is a trigger. When stimulated it causes the nematocyst vesicle to take on water, raising the internal pressure to the point that the thread is rapidly forced out, uncoiling in the process.

8 If available, look at the demonstration slide of discharged cnidocytes. Compared to the cell, how many times longer is the extended thread? _____

Prey caught on the tentacles are moved to the mouth and stuffed into the gastrovascular cavity. *Hydra,* as all cnidarians, has a sacklike digestive system. It lacks an anus. Anything that is not digested is forcefully regurgitated by contractions of the body wall. Because there is no anal opening, it is called an **incomplete digestive system.** Note how the gastrovascular cavity extends into the tentacles from the body column. As digestion proceeds, food molecules can pass into these spaces and nourish the surrounding cells.

Realize that you observed no specialized organs other than the gastrovascular cavity. How do you think that *Hydra* performs the important physiological functions of gas exchange, circulation, and excretion?

Hydra has a hydraulic skeleton which results from the interaction of the nervous system with the myonemes in the body wall and the fluid in the gastrovascular cavity. When the mouth is closed and the myonemes contract, the body stiffens. By controlling which myonemes contract and which relax, different regions of the body (column or tentacles) can be extended or retracted as you observed with your live *Hydra*.

Reproduction Hydra, like most solitary polyps, can reproduce asexually by budding. Buds form as an outgrowth of the parent's body wall that lengthens and forms tentacles to become a small replica of the parent. It eventually breaks off and lives independently.

Other polyps, such as sea anemones, may actually lose fragments of their pedal disc as they crawl along substrates. These fragments are able to reorganize into small versions of the parent polyp. In colonial species such as corals, new individuals resulting from budding remain attached to the adjacent parent, so that a spreading mat of individuals develops. All polyps asexually produced are genetically identical and represent a clone.

Sexual reproduction is often triggered by harsh environmental conditions. In species having only a polyp stage in their life cycles, gonads develop as clusters of gamete-producing cells in the body wall of polyps. These temporary gonads are seen as swellings. Those near the base are ovaries and will produce eggs. Those near the mouth will produce flagellated sperm. Your instructor may have set up demonstration slides of *Hydra* at sexual maturity so that you can see the histology of the gonads.

Sperm swim to the eggs and fertilize them in position in the body wall. The resulting zygote divides to produce a blastula that, in turn, invaginates to form a gastrula. This diploblastic stage develops into a free-swimming planula larval stage that leaves the parent. Cilia on its surface cells allow it to feebly swim at the mercy of water currents. After a few hours to a few days, the larva settles to develop into a new polyp. Because of biparental inheritance, polyps produced by sexual reproduction differ genetically.

9 ***A Medusa: Gonionemus*** Commonly called jellyfish, the medusa is the second alternative body form found in adult cnidarians. Some cnidarians have only a medusa and no polyp stage in the life cycle. Others have both polyp and medusa stages. Members of the genus *Gonionemus*, found in shallow bays on both coasts of North America, have both polyp and medusa stages, but the polyp is quite small and will not be studied here.

Obtain a preserved specimen of *Gonionemus* in some fluid in a small dish from the supply area.

Body Plan The animal is obviously radially symmetrical. Find the upper and lower surfaces, respectively called the **umbrellar** and **subumbrellar** surfaces. The body consists of an outer epidermis and an inner gastrodermis with the types of cells examined in *Hydra*. In-between the two cell layers is a very thick **mesoglea** layer composed of collagen-like proteins that are hydrated, giving the body a jellylike consistency. About 95 to 98% of the jellyfish's weight is due to the water hydrating the mesoglea. This gives them buoyancy nearly equal to that of seawater; therefore, the animal expends little energy in swimming to remain suspended as it hunts for food. Jellyfish feed on zooplankton, small fish, and larvae of other animals.

Functional Anatomy As you study your specimen, compare it to **figure 20.4**. Identify the **manubrium** hanging from the center of the subumbrellar surface. At its end is the **mouth** which opens into the **gastrovascular cavity**. It connects to four **radial canals** that pass to the periphery of the bell-shaped body where they connect with a **circular canal** passing around the circumference. What do you think is the function of this canal system?

Medusae can actively swim and respond to stimuli. When epitheliomuscular cells in the bell contract making the bell smaller, water is expelled. A thin shelf of tissue, the **velum,** passing around the rim narrows the cross-sectional area for the water to escape, thus increasing the velocity obtained from a single contraction allowing the animal to "jet" along. Jellyfish typically swim upward and then slowly sink capturing unwary prey as they descend. Although you will not be able to see them, small sensory organs of equilibrium called statocysts are located at the base of some of the tentacles. They contain small calcareous "stones" surrounded by nerve cells. If a medusa tilts in the water, the statocyst shifts and contacts nerves on one side. Outputs from these nerves cause compensatory swimming movements so the animal remains more or less upright when swimming in calm water.

Tentacles hanging from the edge of the bell contain batteries of cnidocytes that can stun and capture prey. The Lion's Mane jellyfish common in cooler polar seas have bells upto 2m in diameter and tentacles several meters long. They capture prey by swimming to the surface and then slowly sink with the tentacles outstretched. Small fish are stunned when brushing against the tentacles. The tentacles move prey into the mouth, and to the gastrovascular cavity. Digestion products are carried through the radial and circular canals to remotely located cells. Despite its complicated canal structure, the gastrovascular cavity is a sac with only one opening, the mouth. Anything that cannot be digested must be regurgitated.

The gonads are best seen from the subumbrellar view, attached to the radial canals. Sexes are separate. Eggs and sperm are released into the sea and fertilization is external. When life cycles have both polyp and medusa stages, the polyp is the asexual reproduction stage and the medusa is the sexual stage. You will look at sexual reproduction in more detail in the next specimen.

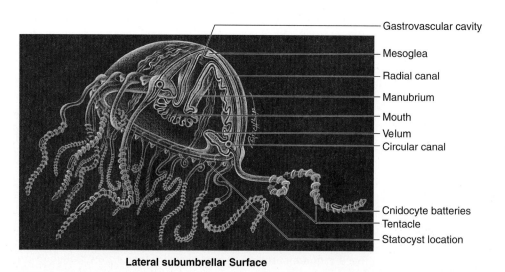

Lateral subumbrellar Surface

Figure 20.4   The anatomy of the jellyfish *Gonionemus* sp.

How do you think an organism like *Gonionemus* performs the important functions of gas exchange, excretion, and circulation?

A Colonial Form: Obelia Cnidarians in the genus *Obelia* are small marine colonial animals that attach to seaweeds and pilings in shallow waters along the Atlantic and Pacific coasts of North America. *Obelia* is studied because it has a life cycle with both polyp and medusa stages that are easily observed. The polyp stage is a colony of several individuals and reproduces asexually. The medusa stage is solitary and reproduces sexually with separate male and female individuals. The medusa stages are often short lived compared to the polyps.

10 Get a prepared microscope slide of a colonial polyp stage of *Obelia* from the supply area. Look at it with the scanning objective before switching to medium power.

Body Plan At first this colony might look more like a plant to you than an animal. There is a central stalk with many branches. The colony is sessile and attaches by a basal holdfast to suitable substrates. At the tip of each branch there is a polyp (**fig. 20.5**). You should see two types of polyps: feeding polyps called **hydranths**, with tentacles, and reproductive polyps called **gonangia**, which lack tentacles. The gonangia asexually bud off small medusa which sexually reproduce. The reproductive polyps are dependent on the feeding ones for nourishment. The similarity of a hydranth's anatomy to *Hydra* should be easy to recognize. Note how the body is surrounded by a translucent noncellular covering, the **perisarc** made of the polysaccharide chitin, a common structural molecule in both animals and fungi. It serves as an external skeleton supporting and protecting the living part that is collectively called the **coenosarc**.

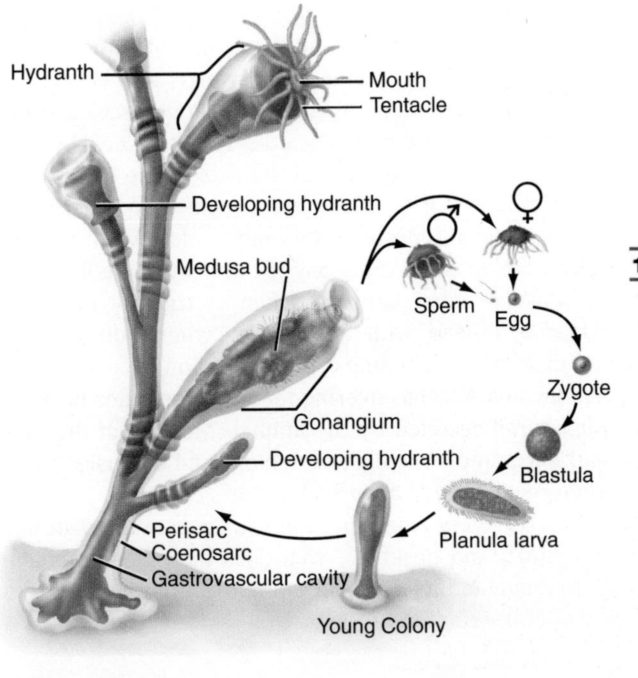

Functional Anatomy Hydranths gather food in much the same way that *Hydra* does. The tentacles have **cnidocytes** that immobilize prey that are brought into a **gastrovascular cavity** through the **mouth**. Digestion is extracellular.

What is interesting is that the gastrovascular cavities of all individuals in the colony are interconnected. Scan the colony and note the continuous chamber in the coenosarc from one individual to the next. This means that if one hydranth captures prey, the digestion products are shared.

Reproduction *Obelia* reproduces sexually. Small medusae develop on the buds of the reproductive polyps. They break off and escape into the surrounding water. If a second slide of *Obelia* medusae is available, look at it and note the similarities to *Gonionemus*.

Both male and female medusae are produced. This form of asexual reproduction allows a single sessile colony to produce many mobile reproductive individuals, increasing the chances of genetic outcrossing. Fertilization is external and the zygote develops into a ciliated planula stage that will swim and drift before settling on substrate where it produces a new colony.

If you look carefully at this organism you will not see any organs for gathering oxygen or releasing carbon dioxide or ammonia. There is no circulatory system. How does a small animal like this perform those functions?

11 *Summarize Observations* Stop your lab work for a moment. Pair off with another student and explain how Cnidaria feed and digest, obtain oxygen and get rid of carbon dioxide and ammonia, move and respond, and reproduce. Compare these processes in Cnidarians to the same ones in sponges. Similarities? Differences? Fill in the information required in table 20.1 at the end of this lab topic.

Bilateria

A majority of the animal phyla have bilateral symmetry, attesting to the success of this type of body organization. These animals all have an anterior–posterior axis, a left–right axis, and a dorsal–ventral axis. The *Hox* genes seen first in cnidarians increase in number in bilateral animals and control regional specialization along the anterior–posterior

Figure 20.5 Anatomy and life cycle of *Obelia*, showing vegetative and reproductive polyps and sexual medusa stages.

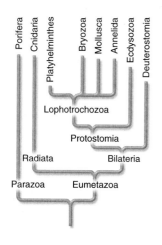

axis. Accompanying bilateral symmetry is the organization of the nervous system from the simple nets seen in cnidarians to complex tracts, often with anterior clusters of neurons forming integrating centers.

A major derived characteristic of bilaterians was the evolution of a triploblastic embryo, in contrast to the diploblastic one of cnidrians. In a triploblastic animal, a third germ layer develops during the gastrula stage. It forms when some endodermal cells migrate from the archenteron into the space between the endoderm and ectoderm (fig 21.1). These cells divide and specialize to produce striated muscle and connective tissues which make up the bulk of the bodies of bilaterally symmetrical animals. Consequently, these animals are able to move faster, prompting selection for greater nervous system evolution. Connective and muscle tissues also provided frameworks for organ development which prompted development of respiratory, circulatory, and excretory systems to handle the increased metabolic loads and body sizes. The traditional phylogeny of the animal kingdom divides the bilateral animals into groupings based on what type of body cavity was present. Molecular studies have not supported this phylogeny.

On the basis of DNA and protein analyses, embryological characteristics, and other anatomical characteristics, the clade Bilateria is divided into three groups: Lophotochozoa, Ecdysozoa, and Deuterostomia. You will study these in detail in the next three labs. Here we will look at one phylum from the lophotochozoan clade, Platyhelminthes, to illustrate the changes in animal form associated with bilateral symmetry. We will explore the other derived characteristics of the lophotrochozoans in the next lab topic.

Phylum Platyhelminthes

Approximately 20,000 species are found in the phylum Platyhelminthes, commonly called flatworms. The phylum contains four traditional taxonomic classes: Turbellaria—free-living flatworms; Trematoda and Monogea—parasitic flukes; and Cestoidea—tapeworms.

Compared to cnidarians, their major derived characteristics are:

- Bilateral symmetry
- Increased numbers of *Hox* genes
- Three germ layers, triploblastic, with true muscle and connective tissues
- Nervous system development with tracts and integrating centers
- Presence of an excretory system
- Dorsal–ventral flattening of body
- Parasitic lifestyle in many, but not all, species

Body Plan No doubt when you look at these animals, their anatomy will seem more familiar because of their bilateral symmetry. Flatworms differ from other bilateral animals in that they are compressed dorsoventrally and lack appendages. They have well-developed organs for digestion, excretion, coordination, and reproduction.

Unlike most bilaterally symmetrical animals, flatworms have a gastrovascular cavity and are acoelomate, meaning that they do not have a body cavity. Instead, the internal organs are embedded in surrounding tissues called parenchyma.

Most animals in this group are parasitic, living in the digestive systems of vertebrates. Consequently, you will see that many organ systems are reduced, while the reproductive system is greatly enhanced. For example, tapeworms lack a digestive system because they can absorb digestion products from the host's intestine. Furthermore, the nervous and muscle systems are not highly developed because they live a rather protected life where they need not find prey or escape predators.

While studying sponges and cnidarians, we did not make a distinction among taxonomic classes. Here we will, because the body organization and biology of flatworms varies greatly across the classes and it makes sense to study them one at a time.

Class Turbellaria Members of the genus *Dugesia*, commonly called planarians, are free-living freshwater flatworms found worldwide. They live under rocks and in streams and lakes. Other members of the class live in marine environments.

12 **Functional Anatomy** Obtain a living planarian, place it in a petri dish in some spring water, and observe its locomotion. What type of symmetry do you see? Can you identify the anterior end? The dorsal surface? Left side? On **figure 20.6**, add labels to indicate these three body axes.

Although not visible, there are clusters of nerve cell bodies at the anterior end, forming a primitive integrating center. Nerve cords run along the animal's length.

Observe what happens when you gently touch the planarian's head with a probe or when you turn the organism over. A gliding-type locomotion occurs because cilia located in cells covering the ventral surface beat in a mucus trail secreted by the animal. Muscles in the body wall allow the twisting, shortening, and extensions seen when the animal is disturbed.

13 Place a very small piece of raw liver in the dish near the animal and observe. What does the planarian's ability to move and its behavior tell you about its muscle and nervous systems?

cavity is highly branched. What is the advantage of this branching in an animal that lacks a circulatory system?

Planarians are carnivorous scavengers. The mouth is on the ventral surface about halfway along the body. A tubular **protrusable pharynx** connects to it. The tube can be everted through the mouth, turning inside out as it extends and presses against a food source. Enzymes released through the pharynx partially digest any food, and the resultant slurry is sucked into the gastrovascular cavity where digestion is completed.

Flatworms lack an anus. Do they have a complete or an incomplete digestive system? _____
If an animal lacks an anus, it must regurgitate undigested residues. In so doing, it loses digestive enzymes and any recently consumed new food.

Though reasonably well developed, the excretory and nervous systems will not be seen in your specimen. Special staining techniques must be used to make them visible. The nervous system consists of an anterior ganglion and two nerve tracts extending the length of the body with several cross connectives. The excretory system consists of several structures called protonephridia located laterally along the length of the body. They seem to function more in osmoregulation than in excretion. The reproductive system will not be studied in detail in this organism. No specialized respiratory gas exchange organs are present nor is there a circulatory system. How does the small dorsal ventrally flattened body plan with a branched gut facilitate the performance of these necessary functions?

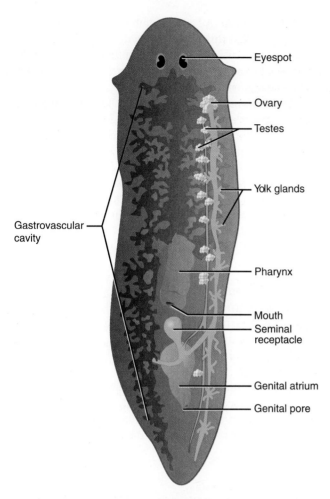

Figure 20.6 Internal anatomy of planaria, a flatworm. Nervous and excretory systems are not shown and will not be visible on your slide. Diagram shows digestive system on left and reproductive system on right. Both systems are found on both sides in real animals.

15> Now obtain a microscope slide of a cross section of a *Dugesia*. Compare the slide to figure 20.7. Most of the interior is filled with cells in contrast to the the thin layers found in sponges and cnidarians. This is characteristic of triploblastic animals. The openings that you see are the cavities of the digestive system that have been cut in cross section. The tissue organization is an obvious feature.

Can you see well-defined layers of cells? _____
The ventral surface of the worm can be identified by carefully looking for the **ciliated cells** on the surface. **Gland cells** on the anterior ventral surface secrete mucus, which aids in locomotion. Some cells on the dorsal surface contain darkly staining **rhabdites.** When provoked, planarians release the sticky contents of the rhabdites, producing a repellent slime.

Beneath the outer surface find the ends of the **longitudinal muscles.** What happens to the animal's shape when they contract? _____
Also find the **circular muscles,** which pass around the animal's body. What happens when they contract? _____
_____ **Dorsoventral muscles** should

Note the two light-sensitive, but not image-forming, **eyespots** on the anterior end. If you cover part of the dish with cardboard so that a dark shadow falls on the planarian, does it preferentially move into the light or stay in the shadow? What is the advantage of this behavior to the animal in its normal habitat?

14> Return your live animal to the supply area and get a stained whole-mount slide of *Dugesia,* a common planarian, and look at it with your compound microscope.

The branched gastrovascular cavity should be clearly visible (fig. 20.6). Each lobe of the saclike gastrovascular

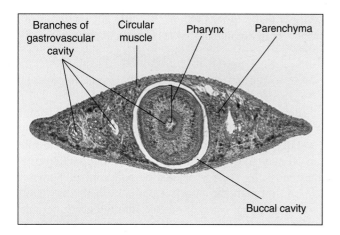

Figure 20.7 Cross section through pharyngeal region of planarian. Body is solid tissues except for openings corresponding to branched gastrovascular cavity and buccal cavity surrounding the protrusible pharynx.

Class Trematoda All adult members of class Trematoda, commonly called **flukes,** are parasitic in vertebrate hosts. After the eggs are shed, they go through very complex life cycles involving larval stages that infect intermediate hosts, usually fish or snails.

Clonorchis sinensis is a fluke that lives in human bile ducts. Eggs released into bile flowing to the intestine are voided with the feces. **Figure 20.8** shows the life cycle of this organism. Eggs develop into five sequential larval stages that first inhabit snails and then fish as intermediate hosts. Eating raw or improperly prepared fish containing *Clonorchis* larvae encased in cyst walls causes infection in humans. The adults mature in the human body, producing eggs that start the cycle anew.

16 Obtain a slide of *Clonorchis sinensis*. Look at it with low power and find the structures shown in **figure 20.9**.

also be visible passing from the dorsal to the ventral surfaces. When they contract, what happens to the shape of the worm? _____

Return your slides and materials for planaria to the supply area. If you compare a planarian to a sponge or a *Hydra,* which would you say is more complex? Why?

Functional Anatomy Like its free-living relative you just studied, *Clonorchis* has a dorsoventrally flattened body, but it lacks eyespots and a protrusable pharynx. At the anterior end is an **oral sucker** which the fluke uses to grasp the bile duct wall. The **mouth** at the center of the sucker ingests cells lining the bile duct. A short **esophagus** leads to two branches of the **gastrovascular cavity.** A second sucker is located on the ventral surface. Would you expect to find an anus in this animal? _____
How are nutrients distributed throughout the organism?

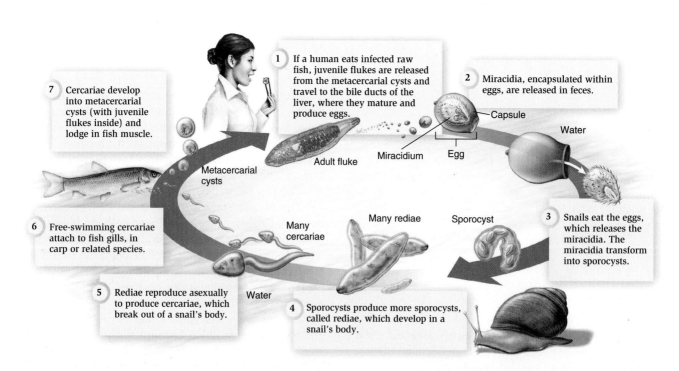

Figure 20.8 The complete life cycle of a trematode. As an example, this figure shows the life cycle of the Chinese liver fluke *Clonorchis sinensis.*

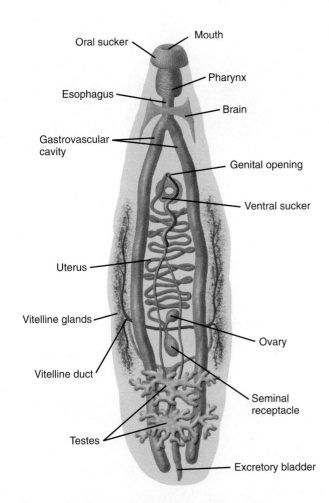

Figure 20.9 Anatomy of an adult human liver fluke. Although a brain and excretory bladder are shown, they will not be visible on your slide.

Would you expect to find a body cavity in a flatworm like this?

The brain and excretory bladder shown in figure 20.9 will not be visible.

Reproduction The reproductive system occupies a substantial part of the body. These animals are hermaphroditic with both male and female reproductive organs, but self-fertilization does not occur.

Locate the paired **testes** in the posterior third of the worm. Small ducts lead from the testes to the **genital opening** located near the ventral sucker.

The **ovary** is anterior to the testes. It produces eggs that pass into the **uterus** where they are fertilized by sperm from previous copulations held in the **seminal receptacle**. Yolk from the **vitelline glands** combines with the fertilized egg and a shell is formed before the eggs are released through the genital opening.

As is the case with most parasitic organisms, the nervous and muscular systems of *Clonorchis* are reduced, and the reproductive system is enlarged. This ensures that large numbers of fertilized eggs are produced to compensate for the great odds against successfully completing a life cycle. Each egg ingested by a snail will asexually produce several larvae, amplifying the reproductive potential even more. Further development depends on the larvae infecting a fish and ultimately being eaten by a human. Many larvae simply die, but the more larvae there are, the greater the chances of survival for the species.

In what ways is a fluke similar to a planaria? How does it differ?

Class Cestoidea There are about 1,000 species of tapeworms; all are parasites living in the intestines of vertebrate hosts.

Body Plan The general body form is similar to a long ribbon of theater tickets. It is made up of small repeating units called **proglottids**. In 1991, a tapeworm 37 feet long was pulled from the mouth of a woman in Mississippi. It consisted of thousands of proglottids and had migrated from her intestine!

Tapeworms are highly adapted to a parasitic way of life. The anterior end has special holdfast structures (**fig. 20.10***a*) which anchor the animal to the intestinal wall of its host. They lack a digestive system and absorb nutrients from the host's intestinal fluids, sometimes causing malnutrition and weight loss in the host. There are no specialized organs for circulation or respiration.

17 From the supply area, obtain a composite or separate slides showing a scolex, mature, and gravid proglottids. From these representative samples taken from different body regions, you should be able to piece together an understanding of a tapeworm's anatomy. There may also be whole tapeworms on display so that you can see the body regions that the slides were made from.

The **scolex** is at the anterior end. Note the hooks and suckers that anchor the tapeworm in the intestine. Behind the scolex is the **neck** region. Here repeated cell divisions produce new very small proglottids by mitotic cell divisions and differentiation. As new ones are formed and grow, older ones are displaced toward the posterior.

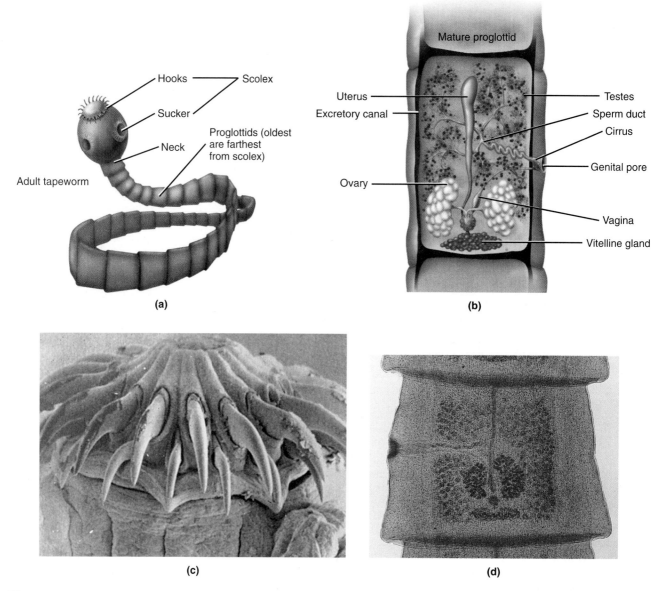

Figure 20.10 Anatomy of a tapeworm: (a) external anatomy of whole worm; (b) internal anatomy of a mature proglottid; (c) scanning electron micrograph of hooks on scolex; (d) photo of mature proglottid.

Reproduction Tapeworms do not asexually reproduce. Look at a representative mature proglottid. As you study the mature proglottid, refer to figure 20.10b to identify the organs. Most of the proglottid is devoted to reproduction.

Tapeworms are hermaphroditic and each proglottid contains both male and female reproductive organs which both empty their gametes into a common genital atrium. Internal, self-fertilization is common. Fertilized eggs develop in the uterus as a desiccation resistant shell forms around the developing zygote. Proglottids that are filled with mature eggs are called **gravid** proglottids and are found near the posterior end of the animal. A single gravid proglottid may contain up to 100,000 eggs. Gravid proglottids are shed with the host's feces and break open, releasing their eggs to contaminate an area. Thus, the bulk of the tapeworm's body is virtually a serial reproductive machine, producing a huge number of eggs. This is most likely related to the risk inherent in a parasite's life cycle which depends a lot on chance happenings.

In the United States, 1% of the slaughtered beef cattle are infected with the larval stages of beef tapeworm, *Taenia saginata*. It can live for years encysted in the muscles of cattle. When a human eats undercooked beef (rare) containing the encysted larvae, they are activated and attach to the intestinal wall. Gravid proglottids are shed with the feces and can contain 50,000 to 100,000 fertilized eggs that remain viable for months. If fecal material contaminates the vegetation eaten by cattle, the eggs develop into larvae that bore out of animal's intestine and migrate to the muscles where they encyst. Federal inspection of beef at slaughterhouses removes infected carcasses, but about 20% of beef consumed is not inspected.

An estimated 40 million people worldwide carry the parasite, especially in countries where meat is not inspected, raw meat is consumed and human fecal material is used as fertilizer. Various "worming" drugs eliminate an infection. Other species of tapeworms have similar life cycles with fish as the intermediate hosts and are becoming a problem as fish farming increases.

18 *Summarize Observations* Organize your observations of flatworms by filling in table 20.1. With your lab partner, discuss how the body plan of flatworms is different from those seen in sponges and cnidarians. What is similar among the three phyla studied? What is different?

Lab Summary

19 On a separate sheet of paper, answer the following questions as assigned by your instructor.

1. What are the basal characteristics of animals?
2. Think generally about the animals you saw in the lab work. If you compare animals in the clade Parazoa with those in the clade Eumetazoa, what similarities and differences come to mind?
3. If you compare animals with radial symmetery to those with bilateral symmetry, what similarities and differences come to mind?
4. What do you think are the advantages of having a body that is bilaterally symmetrical compare to one radially symmetrical?
5. What are the basic requirements for an animal to have a separate system of digestive organs? What is the advantage of a digestive system over intracellular digestion?
6. Each one of the phyla studied today had a unique feature. What were they?
7. When animals have tissues, they also have organs. Why are these two evolutionary developments correlated?.
8. None of the animals studied in this lab had organs specialized for respiratory gas exchange or circulation of body fluids. How do these animals perform these important physiological functions?
9. At the beginning of this lab, the following hypothesis was made: Animals in the phyla studied here can be arranged in a sequence from simple to complex. What observations have you made today and summarized in table 20.1 that can be used to test this hypothesis? What is your conclusion?

You may want to try the critical thinking questions that apply some of the knowledge you gained in doing this lab.

INTERNET SOURCES

Many of the species in phylum Platyhelminthes are parasites. Scientists who study these animals belong to professional societies that promote research and dissemination of information. Use your browser to connect to American Society of Parasitologists at http://asp.unl.edu/. What did you find most interesting at these sites?

LEARNING BIOLOGY BY WRITING

Write an essay summarizing your observations that support or fail to support the null hypothesis that these three phyla can be arranged in a phylogenetic sequence from simple to complex. Be sure to recognize those characteristics related to parasitism as secondary developments and not as major phylogenetic trends.

CRITICAL THINKING QUESTIONS

1. Radially symmetrical animals lack a brain and nerve cord. Why?
2. Sponges are sessile and often green in color. Why aren't they considered plants?
3. To control parasitic infections by flukes, snails are often removed from ecosystems. Why?
4. Brushing against some jellyfish can kill you. Why?
5. Why is nervous system development coordinated with bilateral symmetry?

Table 20.1 Summary of Characteristics in Basal Animal Phyla

	Porifera	Cnidaria	Platyhelminthes
Common names			
Tissues?			
Symmetry?			
Embryological germ layers?			
Name organ systems present			
Intra- or extracellular digestion?			
Herbivore, filter feeder, predator, or parasite?			
Unique features			
Describe body plan			

Lab Topic 21

LOPHOTROCHOZOA: INCREASED COMPLEXITY

SUPPLIES

Resource guide available on WWW at www.mhhe.com/labcentral

Equipment
Compound microscopes

Materials
Preserved specimens for demonstrations
   Polychaetes
   Chiton
   Snail
   Various mollusc shells
Live leeches and polychaetes for demonstration
Live *Lumbriculus variegatus* culture kit
Live earthworms (night crawlers) for class
Preserved clams for students
Frozen (thaw) whole squid for students
Prepared slides
   Bryozoan zooid, *Cristatella*
   Earthworm, cross section
   Clam, cross section
Dissecting trays and instruments
Capillary tubes, 0.7 mm × 75 mm

Solutions
70% ethanol

STUDENT PRELAB PREPARATION

Read the Background material and sections of the lab topic that have been assigned by the instructor.
   You should use your textbook to review the definitions of the following terms:

Annelida	lophophore
blastopore	Lophotrochozoa
Bryozoa	mesoderm
clade	Mollusca
coelom	protostome
ectoderm	spiral cleavage
endoderm	trochophore larva
gastrula	

Describe in your own words the following concepts:
What a clade is and why it is important
What a gastrula is and why it is important
The general body plan of an annelid
The general body plan of a mollusc

After finishing the prelab review, write any questions you have about terms, concepts, or techniques in the margins of this lab topic. The lab activities should help you answer these questions or you can ask your instructor during the lab.

OBJECTIVES

1. To observe a lophophore and trochopore
2. To illustrate the development of complexity in bilaterally symmetrical animals
3. To learn the functional anatomy of a bryozoan, earthworm, clam, and squid

BACKGROUND

In the previous lab, you explored how body plans are more complex in animals that have tissues, bilateral symmetry, organs, and triploblastic embryological development. These four ancestral traits are shared by animals in 31 of the 40 animal phyla. In this lab topic we will look at bilaterally symmetrical animals in detail and will continue to do so for two more labs.

Bilaterally symmetrical animals are divided into two clades based on molecular similarities that correlate with embryological characteristics (see fig. 20.1). Understanding the gastrula stage of embryonic development and the development of the digestive tube in bilaterally symmetrical animals is the key to understanding the basis for these clades.

In one clade, the **protostomes,** the initial opening (blastopore) of the digestive tube in the gastrula will become the mouth in the adult animal. In the other group, the **deuterostomes,** the blastopore becomes the anus (**fig. 21.1**). In either case, the development of the digestive tube establishes the anterior–posterior axis, but the polarity is opposite in the two clades.

Correlated with the fate of the blastopore in these two groups is a second embryological characteristic: development of the mesoderm. Bilaterally symmetrical animals are **triploblastic** with three distinct layers of cells in their

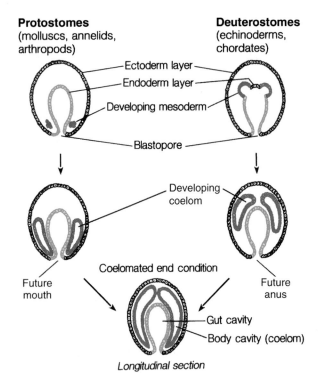

Figure 21.1 Two embryological development patterns are found in bilaterally symmetrical animals. In the protostome pattern, the blastopore of the gastrula becomes the mouth and mesoderm develops from a splitting of cell masses. In deuterostomes the gastrula's blastopore becomes the anus and mesoderm forms from cells splitting off the developing gut. The end results look very similar, but the processes producing the results are very different.

gastrula stages. All other cells found in the adult animal arise from these layers, but not in a random pattern. They follow very strict lines of development. The outer layer of the gastrula, the **ectoderm,** will form the outer covering of the animal and the nervous system. The cells lining the digestive tube of the gastrula, the **endoderm,** develop into the lining of the digestive system and digestive glands. The remainder of the adult animal's body (muscles, skeleton, connective tissues, reproductive tissues, *etc.*) comes from the third layer of cells, the **mesoderm.**

In addition, most bilaterally symmetrical animals have a tube-within-a-tube body plan with a body cavity (**coelom**). You have seen a coelom if you have ever cleaned a fish or a turkey before cooking. It is the cavity in which the organs are found. A coelom has many functions: (1) organs contained in it can enlarge or move independent of the body wall, such as the stomach extending while feeding and later emptying; and (2) fluids in the cavity can cushion internal organs from injury, serve as a primitive circulatory system, or act as a hydrostatic skeleton.

The embryological origin of the coelom differs between the protostome and the deuterostome clades. In **protostomes,** the mesoderm originates as a solid mass of cells that grows inward from the endoderm near the blastopore (fig. 21.1). This mass then splits to form a sac that enlarges to line the body cavity. The term schizocoelomate describes how the body cavity develops in protostome animals by this splitting of mesoderm buds.

In **deuterostomes,** which include the echinoderms and chordates (lab topic 23), the mesoderm arises as an outpocketing of the endoderm to form sacs at the end away from the blastopore (fig. 21.1). These sacs grow out to line the coelom. The term enterocoelomate describes the origin of the coelom from the digestive lining in these animals.

While the protostome-deuterostome dichotomy seems very logical and easy to apply, it is often difficult to make the distinction in practice. When observing the embryology of many animals, it is difficult to observe the fate of the blastopore, let alone to determine how the mesoderm and the coelom develop. In fact, the embryological developmental sequence of many animals is yet to be described. See lab topic 32 for a discussion of embryology. Over the years as new embryological evidence has been discovered, there have been debates over whether a phylum should be considered a protostome or deuterostome. Only recently, with the advent of molecular techniques and genetic analysis, has a new and easier method appeared.

Genetic analysis supports the protostome/deuterostome dichotomy among the bilaterally symmetrical animals. About four phyla are considered deuterostomes, the rest, some 30 phyla, are protostomes. The molecular data indicate that within the protostome group, there are two additional clades based on genetic differences, a group called the **lophotrochozoans** and another called the **ecdysozoans.** In the previous lab, you studied phylum Platyhelminthes, which are basal lophotrochozoans, to illustrate bilateral symmetry. In this lab you will look at some additional representative lophotrochozoans and in the next lab you will study the ecdysozoan phyla.

Although the lophotrochozoan clade is based on genetic similarities, it is a diverse group. The name, lophotrochozoan is a made-up word, new to the language of biology. It attempts to capture two morphological characteristics seen among members of the clade. Not all of the included phyla have both characteristics. About four have a feeding structure called a lophophore which consists of ciliated tentacles. Several, but not all, of the other phyla have a larval stage in their life cycle called a trochophore. Some phyla in the clade have neither a lophophore nor a trochophore, but are included because of genetic and molecular similarities.

LAB INSTRUCTIONS

After looking at a lophophore and trochophore that give the unusual name to this clade of protostomes, you will study the functional anatomy of an earthworm, clam, and squid.

Clade Lophotrochozoa

The derived characteristics of animals in the clade Lophotrochozoa include:

- Genetic and protein similarities;
- Protostome embryo developmental patterns:
  - Triploblastic embryos
  - Spiral, determinate cleavage
  - Blastopore that becomes mouth
- Ciliated cells on surfaces of larvae or adults
- Most, but not all, have:
  - Lophophore feeding structure or trochophore larval stage
  - Complete complex muscle, sensory and nervous systems
  - Complete digestive systems
  - A coelom
  - Organ systems for circulation and excretion

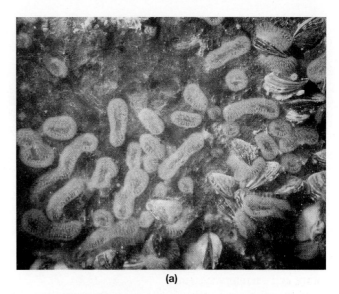

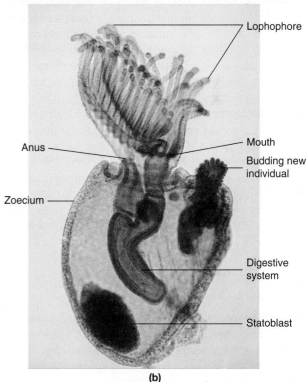

Figure 21.2 An example of a lophotrochozoan from the phylum Bryozoa. (*a*) A colony of *Cristatella* sp; (*b*) an individual zooid with a prominent lophophore.

Within what we are calling the lophotrochozoans, some biologists think that phylum Platyhelminthes, examined in the last lab, and a few other phyla should be moved into a new clade called the Platyzoa. The new clade would be a sister clade to Lophotrochozoa within a broader grouping called the Spiralia. We have not adopted this analysis because there is no general consensus on the new scheme.

We start by looking at some animals that demonstrate the distinguishing characteristics: the lophophore feeding organ and the trochophore larva.

Phylum Bryozoa (Ectoprocta)

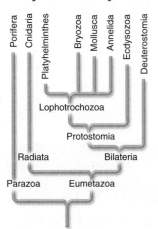

To illustrate lophophores we must look among three or four phyla of lesser-known animals. Bryozoans, also known as moss animals, ectoprocts, and lace corals, have been chosen to illustrate this group because study materials are readily available. Of the 5,000 species, most are marine but about 50 species live in fresh water as encrusting colonies. At one time they were considered deuterostomes but modern molecular evidence puts them clearly in the protostome lophotrochozoan group.

1. Obtain a prepared whole mount slide of *Cristatella* sp., a freshwater species. Look at it with your scanning objective. We are not interested in studying cellular detail so high magnification is not required.

Body Plan Most bryozoan species are colonial. An individual is called a **zooid** (fig. 21.2). How many zooids are on your slide? _____ Colonies develop when zooids asexually bud to form new individuals that remain attached to the parent. Zooids are bilaterally symmetrical, have a well-defined body cavity, and tissues and organs despite a microscopic size.

Most colonies are flat encrusting mats on surfaces or floating gelatinous masses. The animals live inside an exoskeleton, called a **zoecium,** made of the polymer chitin. It can be hardened by calcium carbonate deposition.

The geological record shows that ancestral species formed extensive marine reefs as do modern-day corals. *Cristatella* produces a jellylike colonial matrix with thousands of zooids embedded in the surface "facing" outward (fig 21.2a).

Functional Anatomy Note the crown of ciliated tentacles. This is the **lophophore** and is used in filter feeding. The beating of cilia on the tentacles moves bacteria, protists, and small bits of suspended organic matter towards the mouth at the base of the crown. Particles are caught in mucus that passes into the mouth. The tentacles also function in gas exchange and excretion. Bryozoans have a complete digestive system which is U-shaped. The anus is located external to the lophophore so that fecal material does not foul the feeding apparatus. Bryozoans lack circulatory and excretory systems.

Asexual reproduction is either by fragmentation of the parent colony or internal budding to form **statoblasts.** Composed of cells surrounded by two chitinous valves, statoblasts are able to overwinter and survive freezing while the parent colony dies. Growth can be rapid; a colony in Yugoslavia was observed to grow to 38,000 individuals over a five-month period.

Sexual reproduction also occurs. Individuals are hermaphroditic. Motile sperm are released at the tips of the tentacles and caught in the lophophores of other individuals. Eggs are brooded inside the exoskelton. The zygote develops into a motile larval stage that drifts in currents before settling to produce a new colony.

Return your slide to the supply area. Fill in the information for phylum Bryozoa in **table 21.1** at the end of this lab topic.

Phylum Annelida

2 The 15,000 species in the phylum are placed into three taxonomic classes. Look at the demonstration specimens in the lab as you read the class descriptions.

As you look at the demonstration specimens, you will see that the body plans across the three taxonomic classes are remarkably similar. Externally, they look like a long segmented tube with little apparent anterior to posterior specialization, although internally there are definite head and tail ends. If you cut through the body wall of any specimen, except leeches, you would find a cavity, a **coelom.** It allows movement of the internal organs somewhat independent of movements in the body wall, *e.g.,* allowing the gut to expand after a meal. The coelom is divided into chambers corresponding to each segment by "bulkheads" of membranous tissue, called **septa,** that function as part of the hydrostatic skeleton. A digestive tube, large blood vessels, and nerve cords run the length of the animal, passing through the septa, setting up a tube-within-a-tube body plan that will be evident in many other animals you will study. In each segment, the body wall muscles are separate and there are separate excretory organs and nerve centers. Segmentation offers two advantages to a bilaterally symmetrical animal. First, it allows complex locomotory movements because muscles in each segment can be controlled individually. Second, it provides a redundancy in some organ systems so that damage to a segment is not fatal.

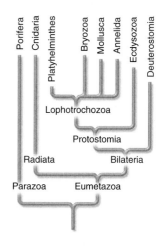

Class Polychaeta is the most diverse, containing about 11,000 species that are all marine (**fig. 21.3a and** *b*). Some are carnivorous, others burrow, ingesting organic matter, and some are filter feeders. All have a tubular segmented body that bears lateral fleshy extensions of the body wall called parapodia that are used in locomotion and gas exchange. Some make tubes of chitin, mucus, and bits of sand. The life cycles of polychaete worms include a trochophore larval stage. These worms have a distinct coelom.

The 1,000 or so species of leeches in the **Class Hirudinea** (fig. 21.3) live in fresh water, although some are terrestrial, living in moist rain forests. Leeches are carnivores and scavengers. Some feed on invertebrates, others are blood-sucking parasites and others consume organic detritus. The coelom of a leech is reduced and is not divided into segments.

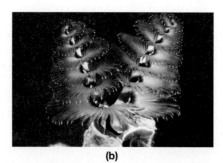

(a)   (b)   (c)

Figure 21.3 Representatives of phylum Annelida: (*a*) polychaete marine annelid; (*b*) the Christmas tree worm is a polychaete fan worm; (*c*) a freshwater leech.

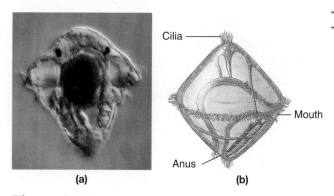

Figure 21.4 (a) A trochophore larval stage of a marine polychaete; (b) artist's interpretation of larval anatomy.

Annelids in the **class Oligochaeta** are commonly known as earthworms, although several species live in freshwater and marine habitats. They feed on organic debris and are important in turning over soil and as members of food chains. You will dissect an earthworm to develop an appreciation of annelid functional anatomy.

Recent molecular studies have led some scientists to suggest that leeches and earthworms should be grouped together in a new clade called Clitellata.

Larval Anatomy A trochophore larva is found in the life cycles of polychaetes and marine molluscs. It has become one of the icons, for the clade lophotrochozoa, even though it is not found in the life cycles of any freshwater or terrestrial species.

Slides of these larvae are not readily available from supply houses, so **figure 21.4** will have to suffice as your introduction. Trochophore larvae hatch from fertilized eggs, usually within 24 hours following fertilization. The larva is fully functional except that it lacks a reproductive system. A band of cilia allows it to swim and to collect food. There is a fully functional gut. It has light sensory structures, touch receptors, and muscle bands. Following a period of drifting and feeding in the sea as plankton, the trochophore will metamorphose into an adult in annelids and, in molluscs, into a second larval stage called a veliger.

Most terrestrial and freshwater species have direct development. The zygote develops into a miniature version of the adult and there is no distinct larval stage.

Live Aquatic Oligochaetes

We acknowledge the ideas of the late Charlie Drewes used to create this activity. You can read more about these worms at: www.eeob.iastate.edu/faculty/DrewesC/htdocs/Lvgen4.htm.

Your instructor may have live blackworms or mudworms (*Lumbriculus variegatus*) for you to start your study of annelids. These small oligochaetes are found in organic sediments in shallow water near the shores of lakes, ponds, and marshes throughout North America and Europe. They are easily raised for classroom use and their somewhat transparent bodies allow organ systems to be observed in action.

To observe the worms, they must be placed in a capillary tubes. This can be accomplished in two ways. If capillary tubes are filled with water, added to a petri dish containing these worms, covered with water and left overnight, many will crawl into the tubes and be ready for lab the next day. Alternatively, you can use a disposable pipet to transfer a blackworm to a petri dish in a drop of water. Use a capillary tube with a suction bulb and draw the animal into the capillary tube. This involves some trial and error.

Attach the tube to a microscope slide with tape and observe it with the scanning objective of your compound microscope. Reduce the light intensity so the heat does not kill the worm. As the light passes through its body, you should be able to see the digestive system, major blood vessels, and other organs.

We will use *Lumbriculus* to document blood-flow annelids. Note that the blood is red due to hemoglobin which aids in oxygen transport. Unlike vertebrates, the hemoglobin is dissolved in the plasma of the blood and is not contained in cells. The circulatory system is closed so that blood always flows in vessels and capillaries in the various tissues and organs. Note the waves of peristaltic contraction that sweep along the dorsal blood vessel driving blood toward the anterior end. The dorsal vessel branches off smaller segmental vessels that are also contractile. Blood flows from anterior to posterior in a ventral blood vessel that is not contractile. Make a sketch below of a whole worm and indicate the flow pattern of blood in the major vessels.

How often does the dorsal vessel contract per minute? _____

Return your blackworm to the supply area.

Now think back to your observations of *Hydra* and *Planaria* in the previous lab. If you compared *Lumbriculus* to these animals, how would you characterize:

Movement and behavior?

Body fluid circulation?

Digestive systems?

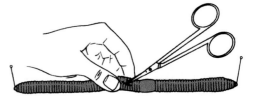

(1) Pinch skin with fingers. Cut through body wall (off center) from the clitellum to the anus. Do not damage internal organs by jabbing points downward.

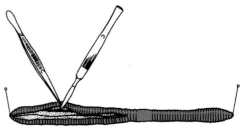

(2) With scalpel, cut through septa on both sides. Pin body wall to tray.

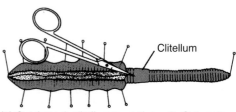

(3) Cut through clitellum toward mouth. Cut septa and pin to end.

Figure 21.5 Procedure for opening an earthworm.

Earthworm Dissection

Live earthworms can be obtained inexpensively at bait stores and are preferable for dissection. They can be anesthetized by submersion in tobacco water made with a crushed cigarette and then placed in 70% ethanol for a few minutes prior to dissection.

5 *External Anatomy* External structures in the earthworm are usually located by reference to the segments numbered from the anterior end. You can identify the anterior end by locating the **clitellum,** a swollen band covering several segments in the anterior third of the worm (see **fig. 21.5**). The clitellum secretes a cocoon around the eggs when they are released. Earthworms are hermaphroditic, having both male and female organs, so every worm has a clitellum.

The mouth is located in the first segment, and overhanging the mouth is a fleshy protuberance. The anus is in the last segment. There are no obvious external sense organs at the anterior end. The dorsal surface is identifiable by the **dorsal blood vessel,** which appears as a dark red line. Having established the anterior–posterior axis and dorsal–ventral axis, it is quite easy to see that this is a bilaterally symmetrical animal.

The surface of the worm has an iridescent sheen because light is refracted by the **cuticle,** a thin noncellular layer of collagen fibers. They are secreted by the cells of the underlying epidermis. Mucus secreting cells in the epidermis lubricate the body surface and keep it moist.

Run your fingers back and forth along the sides of the worm and feel the projecting chitinous bristles called **setae.** Separate sets of muscles allow these to be extended or retracted. Each segment has two pairs of setae on the ventral surface and two pairs on the side. How would the setae help the earthworm in locomotion?

Find the **excretory pores** located in the lower half of each segment (ventrolaterally), except for the first few and the last one.

In the region of segments 9 to 15, the paired reproductive system openings are found on the ventral surface. Depending on the species, the **male pores** are usually in segment 15 surrounded by fleshy lips, and the **female pores** are in segment 14 just anterior to the male pores. In the grooves between segments 9, 10, and 11 are the small paired openings of the seminal receptacles, part of the female reproductive system. They store sperm following copulation. Note the **sperm grooves** extending from segment 15 to the clitellum.

6 ▶ *Internal Functional Anatomy* To examine the internal anatomy of the worm, lay it in a dissecting pan dorsal side up, pin it through the fleshy lip, stretch it slightly, and pin it through the last segment. Open the worm as in figure 21.5 by first cutting from the clitellum toward anus. Cut to one side of the dorsal blood vessel and keep points of scissors up against the inside of the body wall to avoid cutting internal structures.

After finishing the posterior incision, cut from the clitellum forward. In segments 1 through 5 be very careful, because if you cut too deep you will destroy the "brain" and pharyngeal area. After the incision is made, pin the body open by putting pins at a 45° angle through every fifth segment of the body wall. This will provide a quick reference as you try to find various structures. When the specimen is opened and secure, add enough water to the pan to cover the open worm. This prevents the tissues from drying, reduces light reflection, and floats organs and membranes for easier viewing.

Note the internal organization of the worm with its obvious tube-within-a-tube body plan. Observe the anterior concentration of reproductive organs (**fig. 21.6**). Also observe how the body cavity (**coelom**) is divided by **septae,** forming compartments corresponding to the body wall segmentation.

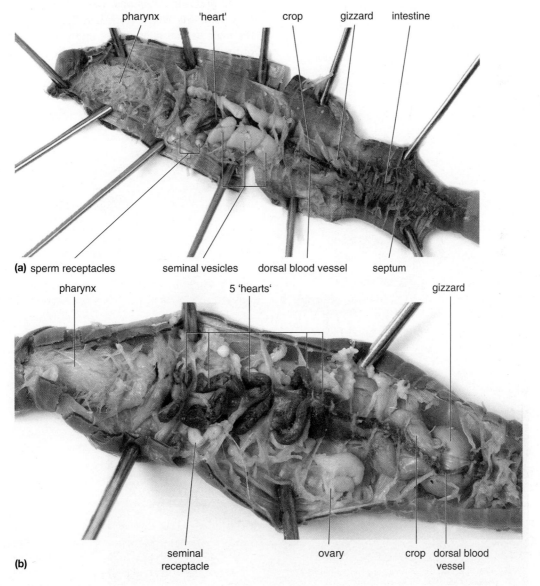

Figure 21.6 Dorsal view of the internal anatomy of an earthworm: (*a*) appearance when first opened; (*b*) removal of reproductive organs reveals structures beneath.

Each compartment is normally filled with fluid, which acts as a hydrostatic skeleton. Well-developed **circular** and **longitudinal muscles** in the body wall exert pressure on this fluid and allow a variety of movements.

If only the circular muscles contract, how will the shape of the worm change? How does the shape change when the longitudinal muscles contract?

The septae prevent coelomic fluid from sloshing from one end of the worm to the other when body wall muscles contract. By confining fluid and muscle contraction to one segment, each can independently expand and contract, greatly increasing the range of motion of the whole body.

Reproductive System Because the reproductive organs obscure other systems, we will study reproduction first. Earthworms are hermaphroditic, having both male and female reproductive organs, but self-fertilization does not occur (**fig. 21.7**).

The male system consists of three pairs of large **seminal vesicles,** which span body segments 9 to 13 lying close to the esophagus. They surround the two pairs of tiny **testes** that you will not be able to see. Sperm pass into the seminal vesicles where they mature before being released via very small sperm ducts that open to the outside in grooves on the ventral surface of segment 15.

The female system consists of two pairs of **seminal receptacles** in segments 9 and 10, which receive sperm from the partner during copulation. The sperm are temporarily stored here and later released when eggs are released. A small pair of **ovaries,** located on the septa between segments 12 and 13 (fig. 21.7), release eggs into the **oviducts,** which carry them to openings located in grooves on the ventral surface of segment 14.

To copulate, earthworms approach one another head on, pressing their ventral surfaces together. They stop when about 1/5 of their bodies overlap with their anterior ends facing opposite directions. Each clitellum secretes mucus that holds the worms together. Sperm are released from the male genital openings of one worm into the openings of the seminal receptacles of the other, the process occurring simultaneously for each partner. They then separate,

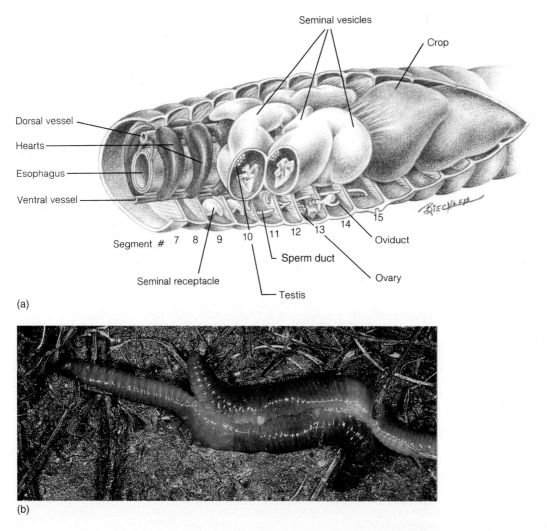

Figure 21.7 Reproductive system. (*a*) Lateral view of reproductive organs in an earthworm; (*b*) Copulating worms

each taking away sperm from the other, but fertilization has not occurred and may be delayed for days.

Fertilization occurs when the clitellum secretes a ring of mucus surrounded by a protective layer of chitin. The ring moves forward from the clitellum, passing over the openings of the oviducts in segment 14 where unfertilized eggs are released into the mucus. As the mucus and eggs continue forward, they pass over the openings of the seminal receptacles between segments 9 and 10. Stored sperm are released and now fertilization occurs in the mucus. The ring slips off the anterior end of the worm and its ends close to form a cocoon. While in the protective cocoon, the fertilized eggs develop directly into worms similar to adults.

Digestive System Find the tubular digestive tract running from the mouth to the anus (fig. 21.6). Having studied the reproductive organs they now can be removed to see the digestive system. Behind the mouth is the muscular **pharynx.** It appears to have a "fuzzy" external surface because several dilator muscles extend to the body wall. Contraction of these muscles expand the pharynx which sucks in particles of soil and detritus. Peristaltic waves of muscle contraction sweep ingested material down the **esophagus** into a thin-walled storage area, the **crop.** From it, ingested material passes into the muscular **gizzard,** which grinds it into a fine pulp. Digestion and absorption of organic material take place in the **intestine.** Any undigestable soil mineral particles pass through the system unchanged and are voided through the **anus** as castings. What is the longest part of the digestive system? How does its length relate to its function?

Compared to a *Hydra* or planarian, how is the digestive system of an earthworm different? Is it a complete or incomplete digestive system?

Circulatory System Earthworms have **closed circulatory** systems in which the blood circulates from arteries to capillaries to veins and back again to arteries. The blood is red because it contains the oxygen-carrying protein hemoglobin. Respiratory gases are exchanged by diffusion with the capillaries near the surface of the body wall. Earthworms have no special respiratory structures. This is one reason why earthworms come to the surface during heavy rain storms. Water percolating into their burrows carries much less oxygen then air and the worms will suffocate if they do not escape.

Find the **dorsal blood vessel** overlying the digestive tract. (fig. 21.6). It collects blood from the capillaries in each segment. From your observations of living blackworms, does blood flow anteriorward or posteriorward in this vessel?

Surrounding the esophagus are five pairs of pulsatile arteries, called the **"hearts."** They pump blood from the dorsal

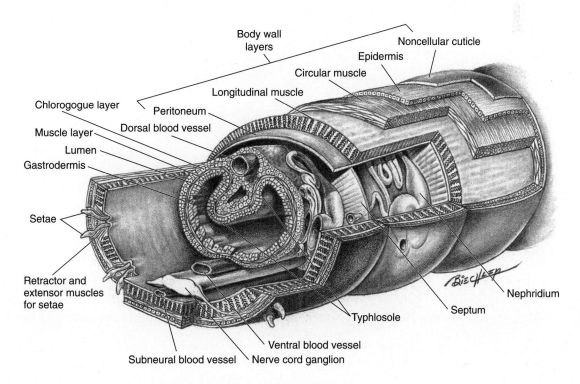

Figure 21.8 Segmental anatomy in posterior half of earthworm.

vessel into the **ventral blood vessel.** Blood in the ventral vessel flows posteriorward.

Excretory System Rather than a pair of large kidneys, earthworms have microscopic excretory organs in each segment except for the first few and the last. **Nephridia,** the excretory organs, are found laterally in the septae of each segment (**fig. 21.8**). These organs filter coelomic fluid and remove waste products. To see a nephridium, try this. Put a drop of water on a slide. Use forceps to remove a septum from the side of the intestine and mount it in the drop. Add a coverslip and observe with your compound microscope. Adding a stain such as methylene blue or neutral red may improve viewing.

The nephridium has a funnel-like opening into the coelom. Coelomic fluid, enters the nephridium and passes through a long tube which opens to the outside through the segmental nephridiopores. As the fluids move along the tubes, salts, nutrients, and water are reclaimed as needed. Anything not reabsorbed is excreted from the body.

Nervous System Refer to figure 21.8 and note the position of the ventral nerve cord and ganglia. Carefully remove the digestive system tube from the posterior third of the worm and look for the cord lying on the floor of coelom with **ganglia** in each segment. Ganglia are clusters of nerve cell bodies. They have long cytoplasmic processes that extend outward to muscles and sensory structures as well as to other nerve cells. The ganglia are connected to each other by a nerve cord.

How many lateral nerves do you see coming from each ganglion? _____

Trace the nerve cord up into segments 1 through 5. This will be difficult. It may be possible to reveal the **subpharyngeal ganglion** lying beneath the pharynx. Trace the nerve cord forward to the **circumpharyngeal connectives** that connect to a pair of **cerebral ganglia** lying above the pharynx. Considered the worm's "brain," these clusters of nerve cells process sensory inputs from touch, and chemoreceptors on the anterior segments. The ganglia control the complex muscles of the pharynx. It is not a brain in the mammalian sense because many of the body movements are handled by the segmental ganglia.

7▶***Histology*** Obtain a prepared slide of a cross section of an earthworm. Using the scanning objective of your microscope, compare the section to **figure 21.9**.

First note the obvious. The muscular body wall surrounds an open space, the **coelom** or body cavity. This arrangement allows the organs to move independent of body wall movements. Note the composition of the body wall. Use figure 21.8 to identify:

Cuticle—noncellular consisting of secreted collagen fibers.

Epidermis—cells that produce cuticle and secrete mucus to keep it moist.

Circular muscles—when they contract fluid in segmental components causes elongation.

Longitudinal muscles—when they contract body shortens.

Peritoneum—smooth-surfaced cells that line body cavity and grow outward to form septa.

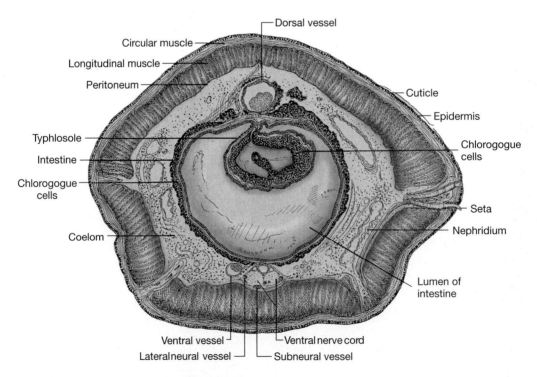

Figure 21.9 Microscopic cross section of earthworm.

You may see some **setae** and their associated musculature in the body wall. When extended, the setae anchor the worm in its burrow, allowing it to draw its body through the soil.

Now direct your attention to the digestive system in the center of the section. Observe the **typhlosole,** a dorsal infolding of the intestine, which greatly increases the area of absorption. Surrounding the intestine, find the **chlorogogue layer,** which functions like the vertebrate liver, storing glycogen and fat and metabolizing amino acids. Finally, note the muscles in the wall of the digestive tube that allow it to move independently of the body wall.

Three major longitudinal blood vessels should be visible: the **dorsal, ventral,** and **subneural vessels.** Finally, identify the **ventral nerve cord.**

You are finished with your dissection of the earthworm. Dispose of the animal according to the directions of your lab instructor.

8 ▶ *Summarize Observations* Turn to table 21.1 at the end of this lab topic and fill in the information for phylum Annelida. Turn to your lab partner and discuss what organ systems you saw in annelids that were not present in bryozoans. Think back to your work with flatworms in the last lab and discuss how annelids differ from flatworms in terms of organ systems present. What common characteristics do animals in all three of these phyla share?

Phylum Mollusca

The 150,000 species of molluscs are soft-bodied animals, although many produce a hard shell to protect the body. They live in marine, fresh water, and terrestrial environments. Marine species have trochophore larvae in their life cycles which is evidence of their relation to the annelids. However, molluscs lack segmentation and the coelom is not prominent, usually being confined to a cavity around the heart. Common representatives are chitons, snails, slugs, clams, oysters, octopi, and other less well-known animals. Their great diversity is reflected in the number of species and the eight taxonomic classes. Four of the classes are represented by few species and will not be discussed here.

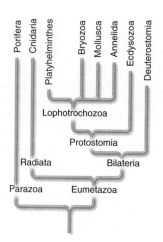

Body Plan Unlike annelids that have rather constant body plans, molluscs exhibit a variety of shapes, sizes, and features. Small molluscs may be only a millimeter or so in length and large ones, like the giant squid, can weigh in at 250 kg, making it the largest invertebrate.

Figure 21.10 provides outline drawings of the basic body types found in adult molluscs. Each type usually consists of four parts: visceral mass, foot, mantle, and shell. The **visceral mass** is composed of soft tissue and contains the organs of the digestive, excretory, reproductive, and nervous systems. The dorsal aspect of the visceral mass is surrounded by the tissues of the **mantle.** It performs a number of functions. If a calcareous shell is present, the mantle secretes it in such a way that the shell grows with the animal never being shed. Outgrowths from the mantle called **ctenidia** function in respiratory gas exchange and sometimes in food gathering. The **muscular foot** is involved in locomotion, sometimes food capture, and in attachment. The **shell** is a $CaCO_3$ exoskeleton that supports and protects the soft parts of the body. Slugs and some marine molluscs have no shell, and squids have internal skeletons made from chitin, not $CaCO_3$.

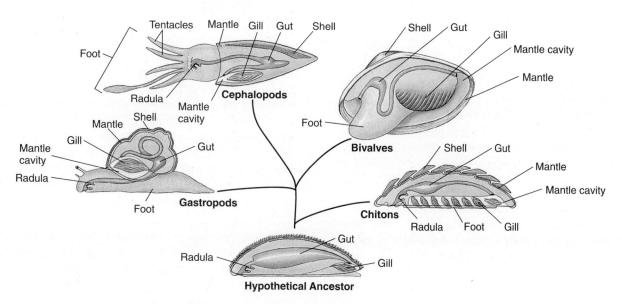

Figure 21.10 Body plans among the molluscs.

Molluscs have well-developed organ systems. The digestive system is complete with mouth and anus. Except the bivalve molluscs (clams and related species), molluscs have a radula, a tonguelike feeding structure whose surface is covered by chitinous teeth that works like sanding paper to erode the soft tissues of prey. The circulatory system is open, meaning that there are no capillaries, and blood flows directly from vessels into tissue spaces, except in squids and octopi which have closed systems. Respiratory systems involve separate organs for gas exchange. Muscles and nerves are well developed, allowing many complex behaviors. There are distinct excretory organs. Sexes are usually separate, except in some freshwater and terrestrial snails. Marine species have a trochophore larval stage.

9▶ Several specimens of different molluscs are on display in the laboratory. They will be arranged by taxonomic class, and you should write brief descriptions of each under the appropriate following headings. As you look at these specimens, be sure to identify these four regions of the body: **visceral mass, foot, mantle,** and **shell.** Also note the type of symmetry exhibited.

1. Class *Polyplacophora.* These marine animals, commonly known as chitons, live in intertidal areas, although a few species live at depths of more than 7,000 meters. Their oval bodies have eight dorsal calcareous plates which protect the soft body. They have a broad, flat ventral foot which allows them to creep and which can act like a suction cup, holding them to firm surfaces. The mouth is ventral with a radula. Most species graze on encrusting algae. Write your description of the body regions below.

2. Class *Gastropoda.* Commonly known as snails, slugs, limpets, conchs, and whelks, the gastropods are the largest class in the phylum. All gastropods were bilaterally symetrical as embryos but their bodies undergo torsion and spiraling during embryological development. Torsion involves the 180° rotation of the visceral mass so that the digestive and nervous systems have roughly a U-shape. Spiraling involves the coiling of the visceral mass inside a shell. Most have a shell, except slugs and nudibranchs. The well-defined head usually has tentacles that bear chemo-, mechano-, and light receptors. Write your descriptions of body regions below. What is the adaptive advantage of the shell?

3. Class *Bivalvia.* This group includes clams, oysters, mussels, and scallops. They are sedentary, filter-feeders characterized by a laterally compressed body surrounded by a two-piece shell. The shell is lined by the mantle. The head is indistinct. A large muscular foot can be extended and is used in burrowing. You will dissect a clam in the lab and do not need to write a description.

4. Class *Cephalopoda.* Commonly known as nautiluses, squids, and octopi, these animals are active marine predators. Nautiluses have an external shell, but in others the shell is internal or absent. You will dissect a squid later and do not need to write a body plan description.

Class Bivalvia

10▶ *External Anatomy* Obtain a specimen of a freshwater clam. Bivalves have no distinct head, so it is a little difficult to become oriented to the body plan. Note the two **valves** that make up the halves of the shell. The shells are hinged on the **dorsal** surface of the animal (**fig. 21.11**). The ventral surface is opposite the hinge.

In some preserved specimens, the muscular **foot** will be extending between the valves. It extends anteriorward. At the posterior, toward the dorsal surface, darkly pigmented **siphons** may also be visible in the gap between the shells. Holding the animal dorsal up and anterior away from you, identify the right and left valves. A cut through the hinge and following the gap between the valves would divide the animal into equal halves, *i.e.,* it is bilaterally symmetrical.

Now study the external surface of one valve. Use a scalpel to scrape the surface. The outer layer that you are removing is the **periostracum.** Composed of the insoluble protein conchiolin, it protects the calcium carbonate portion of the shell from chemical and physical erosion. Think back to your chemistry class, is $CaCO_3$ eroded by acids or bases? _____ The lines on the shell represent periods of growth, but are not annual growth lines. The shell and its periostracum are secreted by cells at the edges of the mantle. The hinge region is oldest and the regions at the edge were just added.

11▶ *Internal Functional Anatomy* To open the clam for study, you must reach in through the gap between the valves and cut the anterior and posterior **adductor muscles,** which connect the left and right shells. See figure 21.11 for their locations.

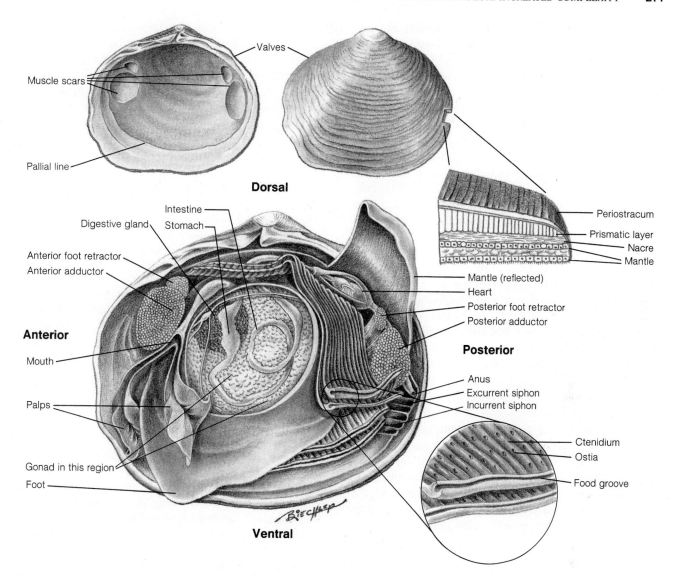

Figure 21.11 Anatomy of a partially dissected freshwater mussel.

> **CAUTION**
> Be careful: Never wrap your fingers around the shells and cut toward them. A slip of the scalpel will give you a nasty cut.

Holding the clam in your left hand in such a way that you will not jab yourself, carefully insert a scalpel through the gap between the shells. Press the point along the inside of the left valve until you feel the muscle and then slice. *As you slice, be aware of where your fingers are so that you do not slice them as well.*

After opening, free the soft mantle adhering to one valve and cut the elastic ligament that forms the hinge. You should now be able to remove one valve leaving the soft body parts in the other.

Note how the soft tissue of the **mantle** forms an envelope around the visceral mass. The mantle performs three important functions: (1) it secretes the shell; (2) it has cilia on its internal surface, which help draw in water; (3) it is a major site of respiratory gas exchange along with the gills. Look for the **pallial line** about 1 cm from the edge of the valve where the mantle attaches.

Examine the dorsal–posterior edge of the mantle and find the thickened, pigmented edge of the mantle. If you match up the left and right mantles, they form two tubes, a ventral **incurrent siphon** and a dorsal **excurrent siphon.** Cilia on the surfaces of the siphons, mantle, and ctenidia (gills) draw water into the mantle cavity over the ctenidia and force it out the excurrent siphon. This water brings a stream of oxygen and suspended, microscopic food to the animal and carries away wastes. In some marine clams, the siphons may be a third of a meter long, allowing the clam to live buried in sand or mud.

Lay back the mantle to expose the **visceral mass** and muscular foot. When circular muscles in the foot contract, it extends in much the same way that a water-filled balloon elongates when squeezed at one end. Once extended into mud or sand, the tip of the foot expands to anchor itself and the anterior and posterior retractor muscles (fig. 21.11) pull the body toward the anchored foot, providing a reasonable form of locomotion. Some marine clams can burrow a foot or more in water-saturated sand in a minute.

Feeding and Digestion Remove the upper mantle. Most of the visceral mass is covered by two flaps of tissue, the **ctenidia,** often called gills. Although they also function in respiratory gas exchange, the primary function of these structures is gathering food. Clams are filter-feeders, obtaining food by filtering suspended algae, detritus, and bacteria from water entering the mantle cavity through the incurrent siphon.

12 ▶ The structure of the ctenidia is best seen in microscope slides of cross sections of bivalves (**fig. 21.12**). Get your microscope and a prepared slide. Find the visceral mass hanging like a pendant with curtains of tissue surrounding it on both sides. The outermost curtain is the mantle, the next two are the ctenidia, and the larger central mass is the visceral mass. Because the shell was removed before the slide was made, you will not see it.

Look carefully at one ctenidium. Note that it is hollow with a **water tube** passing up the center. Close examination will reveal that the surface of the ctenidium has small porelike openings, the **ostia.** Water entering the mantle cavity passes through the ostia into the water tubes and then upward to the **suprabranchial chambers** (fig. 21.12). From there the water flows to the excurrent siphon.

Any particles contained in the water stream are filtered out at the ostia where they are trapped in mucus on the surface of the ctenidia. Cilia on the surface of the ctenidia move the mucus down to the food groove that passes along the edge of a ctenidium. Find the **food groove** on your slide.

13 ▶ Now return to your dissection specimen. Look at the tip of the ctenidia where the food groove is located. Follow the food groove anteriorward to the region of the cut anterior adductor muscle. In this region between two flaps of tissue called **palps,** you should find the **mouth** of the clam. Food particles trapped in mucus enter the digestive system here and pass to the **stomach.**

To see the stomach, remove the ctenidia covering the visceral mass. Scrape and pick away the solid tissue of the visceral mass to find the digestive tract. Tissue of a large, greenish-brown digestive gland surrounds the stomach and a lighter gonad surrounds the intestine. The intestine winds its way dorsally through the visceral mass and then posteriorly to the anus. It is located near the excurrent siphon in the region of the posterior adductor muscle. With a partner discuss how a clam feeds. Start with food gathering and then, using your dissection, trace how food enters the mouth.

Circulatory System Clams have open circulatory systems. Animals with open circulatory system have no capillaries. **Hemolymph,** the circulating fluid, enters the heart directly from the surrounding tissue spaces, the **hemocoel,** and is pumped through arteries to the tissues. At the tissues, the hemolymph does not enter capillaries. Instead, it is released directly into the tissue spaces from where it eventually percolates back to the heart.

To see the clam's heart, look at the dorsal surface of the visceral mass just anterior to the posterior adductor muscle (fig. 21.11). Remove the thin membrane in this area to reveal the **heart** sitting in the **pericardial cavity.** The heart is three-chambered: one **ventricle** and two **atria.** As the intestine travels from the visceral mass to the anus, it passes through the ventricle. Two **aortas** leave the ventricle, one passing anteriorward and the other posterior.

Excretory System Beneath the heart, are a pair of dark brown excretory organs called **nephridia** (kidneys). They remove nitrogenous wastes from the hemolymph. Urine passes from these to a pore in the suprabranchial chamber above the ctenidia. Water flowing through this chamber carries wastes out the excurrent siphon.

You will not study the nervous or reproductive systems of the clam. Dispose of your clam as instructed.

Class Cephalopoda

While clams are rather sedentary, squid have streamlined bodies and a jet propulsion mechanism that make them fast-moving predators. Large eyes, grasping arms and tentacles, and a beaklike structure for biting and tearing make them even more formidable. Depending on the species, squids can range in size from 2 cm up to 15 meters for giant squids.

14 ▶ Squid obtained from seafood suppliers make the best materials to dissect as the tissues look fresh and retain a natural color and consistency. Get a squid in a dissecting pan from the supply area along with dissecting instruments.

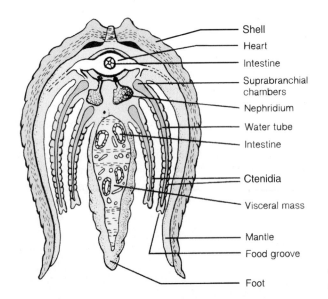

Figure 21.12 Cross section of a freshwater mussel. Shell will not be included on microscope slide.

External Anatomy Squids offer an excellent lesson in the adaptability of the basic molluscan body plan: mantle, visceral mass, and foot. The class name Cephalopoda means head foot and describes how these two body regions have joined to form one region. The arms and tentacles represent the foot but they are joined to the head with its pair of eyes and central mouth.

The **collar** marks the beginning of the muscular **mantle** (fig. 21.13) which covers the visceral mass. The mantle contains specialized pigmented cells called **chromatophores**. Each has tiny muscle fibers attached to it, and the animal can expand or contract its chromatophores to change colors in seconds. The mantle tapers at one end and has two muscular **fins**. These are like diving planes on a submarine and, when the animal is swimming, can be used to steer up or down as well as to the side.

Because the tentacles and arms, 10 in all, are modifications of the foot region, they represent the theoretical ventral surface. The end of the animal bearing the fins is the hypothetical dorsal surface. A midline cut passing through this axis divides the animal into equal halves, *i.e.*, it is bilaterally symmetrical.

Look at the suckers on the arms and note the small teeth around the periphery of each. They help hold prey. Spread the arms and look into the circle made by their bases to see the **mouth**. A horny **beak** should be seen protruding from the mouth.

Study the collar area and find the **siphon** (sometimes called the funnel) which is very important in locomotion. If you hold the animal up with the arms ventral, then the surface on which the siphon is found is considered the anatomical posterior surface. Squids swim at speeds up to 20 mph by water jet propulsion. Note the interlocking cartilages on the mantle and siphon. They support the funnel and mantle in this area. When muscles in the mantle are relaxed, water flows into a chamber beneath the mantle (mantle chamber) through the collar. Circular muscles then contract, first in the collar area to prevent water escape, and then in the rest of the mantle. As pressure builds, the water jets out the siphon. Slit open the funnel to see the fleshy valve which regulates water flow through the funnel. By a rapid cycling of relaxation and contraction phases, a continuous jet of water is produced. Muscles attached to the siphon can turn it in different directions which, along with the fins and arms, helps the squid steer very complex paths as it pursues fish.

The well-developed **eyes** are image forming. There is an iris that regulates light entering the eye and behind it a spherical lens which focuses the light. The lens is focused by muscles that move it back and forth. It does not bend as do the lenses of vertebrates. Remove an eye and cut the lens free to see it.

Functional Internal Anatomy To see the visceral mass, the mantle has to be cut to reveal the **mantle cavity.** Do so by laying the animal in the pan with its posterior side up. How do you tell which side is posterior?

Cut lengthwise from the collar on one side of the siphon to the apex and pin the tissue folds back. The mantle cavity is not a coelom. It is a cavity that is continuous with the external environment. It contains the gills and visceral mass. **Figure 21.14** shows the internal anatomy of a male and female squid.

In the region of the siphon, note the large retractor muscles running through the mantle cavity. Some of these adjust the position of the siphon and others can retract the squid's head. The **anus** will be situated near the inner opening of the funnel. Its location assures that fecal material will pass out of the mantle cavity and not foul the gills. The digestive system, therefore, forms a large loop, passing from the mouth down into the visceral mass before curving back to the anus.

Near the midline, below the base of the siphon find the dark **ink sac.** It actually empties into the rectum and out through the anus. When disturbed, the pigment melanin is released from the sac into the mantle cavity and is expelled as water is released by the siphon. The dark cloud in the water confuses predators and can mask a squid's escape.

If the sac has ink in it, remove it by snipping it at both ends and put it in a dish. Take a dissecting needle and dip it in the ink and write something on paper. This is what artists call sepia ink.

Respiratory System In the mantle cavity you should see two long feathery **gills.** They function in gas exchange. They are constantly bathed by water entering the mantle cavity during locomotion. Their delicate structure is protected by the surrounding mantle.

The organs of the visceral mass are covered by a thin transparent tissue (membrane). If you carefully remove

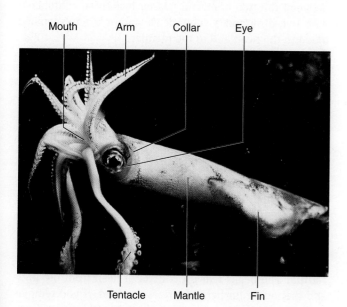

Figure 21.13 External anatomy of a squid.

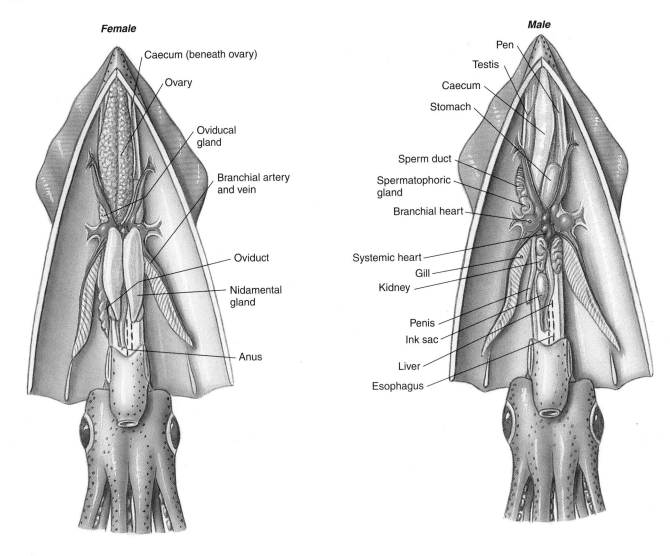

Figure 21.14 Internal anatomy of squid: female on left; male on right.

it, you will be able to find the organs. If your specimen is a female, two large **nidamental glands** may be present covering the other organs. They are part of the reproductive system and produce a jelly which covers the eggs as they are shed. Remove the glands to reveal the structures beneath.

Circulatory System Unlike other molluscs, cephalopods have a closed circulatory system with blood traveling from the heart to capillary beds and draining back to the heart through veins. The blood contains the respiratory pigment hemocyanin. Like hemoglobin, it increases the blood's oxygen transport ability, but unlike hemoglobin it is blue in color. Look at the base of a gill on each side and identify small yellowish **branchial hearts.** They receive blood returning from the mantle and head and pump it into the gills through an branchial artery. Essentially, these are auxiliary pumps that assure blood flow through the gill where respiratory gas exchange takes place. Blood returning from the gills drains into branchial veins which carries it to a larger **systemic heart** located on the midline between the two branchial hearts. It is three chambered with two atria and a ventricle. It pumps the oxygenated blood to the head and to the mantle.

Excretory System A pair of triangular shaped **kidneys** in a single nephridial sac can be found toward the head from the branchial hearts near the midline. Nitrogenous waste exchange takes place in the kidneys. Urine is voided into the mantle cavity and is expelled through the siphon.

Digestive System Most of the digestive system is not readily visible as it passes in a long U-shape from the mouth to the anus. However, some careful tracing will allow you to see most parts. Remove the siphon and make a median incision in the mass below it to expose the muscular **buccal bulb** which bears the two horny **beaks.** Open the beaks and look into the buccal cavity to see the **radula.** The beaks tear off chunks of flesh from the prey and the radula further pulverizes the food. If you remove a portion of the radula and look at it with your microscope, you will see its minute teeth.

The thin walled esophagus passes through the triangular, yellowish **liver,** which sits on the midline, before opening into the **stomach.** Extending from the stomach to the apex is a **caecum.** Food passes into the caecum for storage and flows back into the stomach as digestion proceeds. It can be small or large depending on whether your animal fed before being caught. The **intestine** leaves the stomach near the point where the esophagus entered and travels toward the collar ending in the anus identified earlier.

Nervous System Cephalopods have a highly developed nervous system. Push the buccal bulb and its retractor muscles aside to reveal two white **stellate ganglia** on the back wall of the mantle. Giant nerve fibers should be seen radiating from them. These control the contractions of the mantle in swimming. There are other ganglia in this area, but you need not look for them.

Reproductive System The sexes are separate in cephalopods and the gonads are not paired. The gonads extend from the apex of mantle cavity to about midway to the head.

If you had a female, you removed the whitish **nidamental glands** as you looked at the internal organs. The single **ovary** extends from mid-body away from the head to the apex. It sheds eggs into the coelomic cavity and they enter the **oviduct.** Move the ovary and caecum aside to find the oviduct underneath them. Close to the left branchial heart, the oviduct enlarges to form the **oviducal gland.** It secretes a shell around the egg. The oviduct continues toward the head and enlarges to form the **ostium** that opens into the mantle cavity. Eggs pass out of the siphon in strings. Find a student who dissected a male. Explain to each other the anatomy of the reproductive systems in your specimens.

In males, the single, elongate and light-colored **testis** is in a similar location to the ovary in females so that the same organs will have to be moved to reveal it. Sperm are shed into the coelomic cavity from an opening in the testis. They enter a coiled sperm duct that conveys them to the region of the left branchial heart. As they pass they are formed into aggregations of cells called spermatophores or sperm packets. They are stored in a spermatophore sac and are released from the **penis.** During mating, the male inserts one of his arms into his mantle cavity to remove a spermatophore that was released and transfers it to the mantle cavity of a female. Find a student who dissected a female and explain to each other the reproductive anatomy of your specimens.

Skeletal System Remove the visceral mass and feel the back mantle wall. The hard structure is the squid's **pen,** a longitudinal stiffening element that is analogous to a shell although it is made of chitin. Cut into the wall and dissect it out.

You are now finished with the dissection of your squid. Dispose of the remains as directed by your instructor.

16 *Summarize Observations* Turn to table 21.1 at the end of this lab topic. Fill in the information for phylum Mollusca. Turn to your lab partner and discuss the common characteristics shared by clams and squid. Then describe how they differ.

Lab Summary

17 On a separate sheet of paper, answer the following questions assigned by your instructor.

1. What are the shared characteristics of protostome animals that you have studied so far (Platyhelminthes, Annelids, Molluscs)?

2. Describe in your own words what a lophophore is. Do the same for a trochophore larva. Why are animals with these structures grouped in the same clade?

3. Describe the body plan of an annelid. Briefly describe how an earthworm feeds/digests, exchanges respiratory gasses, circulates fluids, excretes wastes/salts, and moves.

4. What are the diagnostic features of the phylum Mollusca?

5. Explain how a clam is a filter feeder. Draw a diagram that shows the pathway of food particles from water into the mouth of a clam.

6. Describe what anatomical and physiological adaptations to a predatory way of life you saw in squid.

7. What evidence can you present from your observations to indicate that annelids and molluscs are more complex than cnidarians or flatworms?

You may want to try the critical thinking questions that apply some of the knowledge you gained in doing this lab.

Internet Sources

Use the WWW to discover what kind of discussions are occurring among scientists about Lophotrochozoa. Type the term into Google. You will get lots of "hits." Many will be class notes from other colleges. Scan the list looking for articles written by scientists who are arguing a point or reporting research. Read a few of these written by people working at reputable institutions or published in journals. Write a couple of paragraphs summarizing what they are saying.

Learning Biology by Writing

You have had the opportunity to study the general anatomy of an earthworm, clam, and squid as well as animals in lab topic 20. As you consider the representatives of the six animal phyla that you have studied, what evidence do you have to accept or refute the hypothesis that annelids and molluscs are more complex than the sponges, cnidarians, and flatworms? Summarize your arguments in a one-page essay.

Critical Thinking Questions

1. Based on your knowledge of an earthworm's anatomy, reproductive behavior, and physiological systems, explain why large earthworms come to the surface after a heavy rain.
2. Clams and oysters are often eaten raw. The diner cuts the adductor muscles as you did, removes the top valve, and slurps the live animal. Clams and oysters are also filter feeders. Why is this a potentially dangerous meal?

Table 21.1 Summary of Lophotrochozoan Characteristics

	Phylum Bryozoa	Phylum Annelida	Phylum Mollusca
Common name(s)?			
Tissues?			
Symmetry?			
Lophophore?			
Trochophore?			
Name organ systems present			
Type digestive system?			
Type circulatory system?			
Type skeletal system?			
Describe body plan			

Lab Topic 22

Ecdysozoa: Simple and Complex

SUPPLIES

Resource guide available on WWW at
http://www.mhhe.com/labcentral

Equipment

Compound microscopes
Dissecting microscopes

Materials

Dissecting pans and instruments
Preserved Specimens
    Acorn and gooseneck barnacles
    Ascaris
    Horseshoe crabs
    Spiders
    Miscellaneous insects (demonstration)
Live Specimens
    Crickets or roaches
    Crayfish
    Daphnia
    Vinegar eels (nematodes)
Prepared slides
    Pin worm
    Trichinella encysted
    Mosquito mouthparts
    Fly mouthparts
Spirulina flakes

STUDENT PRELAB PREPARATION

Before doing this lab, read the Background material and sections of the lab topic that have been assigned by the instructor.
    Use your textbook to review the definitions of the following terms:

Chelicerata	Ecdysozoa
chitin	exoskeleton
coelom	Hexapoda
collagen	Nematoda
Crustacea	open circulatory system
cuticle	

Describe in your own words the following concepts:
The general body plans of nematodes and arthropods
The diversity of the Phylum Arthropoda

After finishing the prelab review, write any questions you have about terms, concepts, or techniques in the margins of this lab topic. The lab activities should help you answer these questions, or you can ask your instructor during the lab.

OBJECTIVES

1. To illustrate the body plans and functional anatomy of wormlike Ecdysozoans
2. To illustrate the diversity of body plans and functional anatomy of segmented Ecdysozoans

BACKGROUND

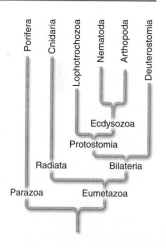

One of the biggest surprises coming from the molecular analysis of animal evolutionary relationships is the discovery of two clades within the protostomes: the Lophotrochozoa which you studied in the last lab and the Ecdysozoa. The ecdysozoans are a significant group of animals. About 75% of all known animal species are ecdysozoans, primarily because the arthropods are in the group. The eight phyla include such animals as roundworms, water bears, velvet worms, barnacles, crabs, spiders, and insects as well as many other such creatures. Repeated attempts based on the analysis of over a thousand genes have failed to falsify the clade Ecdysozoa, suggesting instead that it is a natural, monophyletic grouping of animals.

While molecular similarities established the clade, biologists sought a directly observable characteristic exhibited by all members of the clade to use as a more intuitive diagnostic character. The first to be noticed was that all of the included animals have an exoskeleton that is shed as the animal grows. Another word for shedding is **ecdysis,** hence the clade name **Ecdysozoa** (=animals that shed). A second characteristic is animals in the ecdysozoan clade lack surface cilia both in their larval and adult stages. Ecdysozoans do not use cilia for locomotion nor do they use cilia for generating feeding water currents as do many lophotrochozoans and deuterostomes.

The ancestor of the ecdysozoans is visualized as a small, wormlike, bilaterally symmetrical protostome animal that was a burrower. The presence of an exoskeleton would have protected the animal's tissues from abrasion but would have precluded the animal from having cilia because they are extensions of living cells on surfaces. If the surfaces were covered by a nonliving layer, the cilia would be useless. The animal had a terminal mouth at the end of an protrudable proboscis and a complete digestive system. It is not certain whether the ancestor was segmented or this was a later derived characteristic.

The eight ecdysozoan phyla are divided into two additional clades based on shared derived traits. One clade, called the Cycloneuria, is characterized by having:

- Nerve ring surrounding pharynx from which clade name is derived.
- No segmentation or appendages.
- Wormlike body.
- Collagen-based exoskeleton.
- Mouth at anterior terminus.
- No circulatory system.

also

- Nematodes are best-known examples, but clade includes four other phyla.
- Most likely paraphyletic.

The other clade, the Panarthropoda is characterized by:

- Segmented bodies.
- Segmentally repeated appendage bearing terminal clawlike structures.
- Chitinous exoskeleton.
- Specialized organs for gas exchange.
- Ventral ganglionated nerve cord.
- Open circulatory system with dorsal heart.

also

- Arthropods are best known examples, but clade includes two other phyla.
- Most likely monophyletic.

LAB INSTRUCTIONS

The intent of this lab is to introduce you to the diversity of members of the clade Ecdysozoa, by looking at representatives of both branches of the clade: wormlike animals and segmented animals. You will dissect a nematode worm and then several very different animals that reflect additional evolutionary lineages within the phylum Arthropoda.

Phylum Nematoda

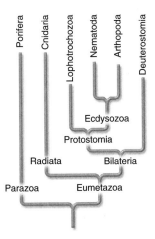

The 80,000 + species of nematodes, commonly called roundworms, live in marine, fresh water, and terrestrial environments from the poles to the equator, in deep ocean trenches to the highest mountains. About 20% are parasites of plants or animals. The numbers of individual nematodes living on earth is huge: it is estimated that a spade full of organic earth can contain over a million microscopic worms. Size ranges from less than 1 mm to 5 cm, although a few species are larger, such as a parasite of sperm whales that can be over 1m long.

In 2002 the Nobel Prize was shared by researchers who worked on *Caenorhabditis elegans,* a tiny nematode easily grown in the lab that has proven to be a fantastic model for studying animal development. During its development, 1,090 cells are produced, but 131 die, leaving 959 cells in the adult animal. The fate of every cell from the two-cell stage onward is known and the complete nervous system has been mapped. The prize-winning researchers were able to identify the genes that regulated the developmental processes giving rise to tissues and organs.

Body Plan Nematodes have an unsegmented, bilaterally symmetrical cylindrical body tapering to points, blunter at the anterior. The external body wall is covered by a noncellular cuticle composed of collagen that functions as an exoskeleton. It is rather nonelastic and must be shed for the animal to grow. Ecdysis occurs four times during development in free-living species. Beneath the body wall is a body cavity, called a pseudocoel, containing the organs. A complete digestive system runs from the mouth to the anus, giving the animal a tube-within-a-tube body plan. There are no organ systems for gas exchange or circulation. Embryos are triploblastic and follow the protostome pattern of development.

Ascaris will be used to demonstrate the functional anatomy of nematodes. It is similar in appearance to most free-living nematodes, except it is much larger. It is a parasite living in the intestines of humans and swine. The U.S. Center for Disease Control indicates that 1.5 billion people were infected in 2002, especially in tropical and subtropical areas where sanitation is poor. Infection in the

United States is rare. If a person ingests eggs, the worms develop in the intestines and new eggs are shed with the feces. It is treatable with drugs.

1   Obtain a dissecting pan, instruments, and a preserved *Ascaris* from the supply area.

Identify the external features, using **figure 22.1**. Note the proteinaceous, noncellular **cuticle** secreted by underlying cells. It prevents rapid drying and acts as an exoskeleton, supporting and protecting soft tissues, as well as allowing locomotion. Identify the anterior terminal **mouth** and the posterior **anus**. Three lips surround the mouth with the largest indicating the dorsal aspect of the worm. Internal tubules of the excretory system passing down the left and right sides can be seen as light-colored lines. Although you really cannot see them, chemical and mechanical receptors are found on the anterior and posterior ends.

Sexes are separate. Females have a rather straight body while males have a definite hook on the posterior end, often with two protruding spicules that guide sperm during copulation. In males, both the reproductive and digestive systems open to the outside through a common opening, the cloaca, while females have separate openings for both systems. The female system opens on the ventral surface about one-third of the body length from the anterior tip. What is the sex of your specimen? _____ Be sure to look at a worm of the opposite sex.

Functional Anatomy

Pin the animal dorsal side up in your dissecting pan with two pins. Cautiously make a single longitudinal cut along the middorsal line, being careful to not cut deeply and damage internal organs. Pin the body wall down. Flood your dissecting tray with enough water to make the organs float. This will make structures easier to see. Before you start to identify organs, answer this question: what type of symmetry does a nematode have?_____

Note how all internal organs are exposed because they lie in a body cavity. As fluids ebb and flow during body movements, the fluid-filled cavity acts as a circulatory system, distributing nutrients and removing wastes from remote cells. The fluids also serve as a hydroskeleton by transmitting pressure changes from muscle contraction in one body region to another. The body cavity also allows internal organs to move independently of other organs or the body wall. Last, the body cavity allows organs such as the stomach to expand and store food or gonads to enlarge during breeding season. The body cavity in nematodes, as well as in several other animals, is called a pseudocoelom because it is not lined by a peritoneum, a type of connective tissue. In pseudocoelomate animals, no muscles are found in the wall of the digestive tube and there are no membranous tissues, called mesenteries, holding the organs in place. Gently lift some of the tubular organs to confirm this.

Gently tease the coils of the reproductive organs to the side to reveal the straight digestive tube running from

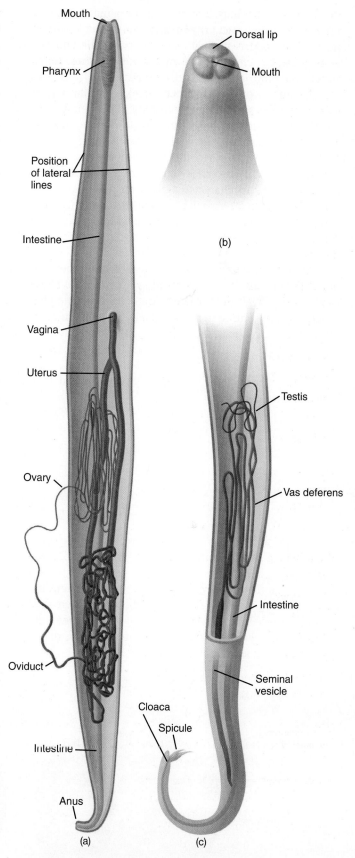

Figure 22.1   Anatomy of *Ascaris*: (*a*) internal anatomy of female; (*b*) oral view of female; (*c*) posterior end of male.

the mouth to the anus. Just behind the mouth, note the fuzzy-appearing muscular **pharynx.** Unlike our digestive tube where food is pushed along by peristaltic contractions of muscles in the wall of the tube, *Ascaris* must pump food into its digestive tube because it lacks wall muscles and is only a single cell thick. The pharynx is able to expand, sucking food in through the open mouth, and then to contract when the mouth closes to pump food along the tube. There is no stomach and the pharynx connects to the intestine where digestion and absorption take place. Undigested material is voided through the anus.

The nerve ring, characteristic of the ecdysozoan worm subclade is difficult to see. It surrounds the pharynx about half way between the mouth and the end of the pharynx. Two very small nerve cords run the length of the body on the internal dorsal and ventral surfaces.

The excretory system consists of excretory tubes that pass along the inside of the left and right body wall. These connect to difficult-to-see large excretory cells in the pharyngeal region. This system most likely functions in regulating osmotic concentrations more than in excretion. The system empties to the outside through two lateral excretory pores near the pharynx. In smaller free-living nematodes, the canal system is replaced by specialized cells called renette cells.

Nematodes lack any structures for respiratory gas exchange. How do they perform this necessary function?

Reproductive System The male reproductive system is a single long, highly coiled tubule. The distal end is the **testis** where sperm are produced by meiosis. It joins a vas deferens that conveys sperm to an enlarged **seminal vesicle** where they are stored. During mating two spicules are inserted into the female genital pore as sperm are released through the **cloaca,** a common opening of the reproductive and digestive systems. Nematode sperm are amoeboid, lacking flagella as in other animals.

The female reproductive system is Y-shaped. The tips of the arms of the Y are the highly coiled **ovaries** where meiosis begins. The eggs pass from the ovaries into the coiled **oviducts** which continue on to form two rather straight **uteri.** The uteri join together to form the **vagina** with a separate opening to the outside. Fertilization is internal, usually in the ovary region, and then shells are laid down around the zygotes as they pass toward the vagina. A mature worm can produce hundreds of thousands of eggs.

Eggs pass out of the host's digestive system with the feces. When ingested by a suitable host, the juveniles hatch and burrow to the liver and lungs where they mature before migrating back to the intestine.

Dispose of the carcass as directed. Be sure to wash your hands and dissection instruments thoroughly. There is a slight possibility that all of the *Ascaris* eggs were not killed by the preservative; you would not want an infestation.

2 *Cross-Sectional Anatomy* Obtain a cross section of an *Ascaris* and look at it first with the scanning objective of your compound microscope to see details not evident in gross dissection. Compare what you see to **figure 22.2**.

Note the open body cavity with the free-floating digestive and reproductive systems unattached to the body wall and surrounding tissues because of a lack of mesenteries. Several tubules fill the body cavity, representing sections through the round reproductive tubules. Each tube is filled with gametes in some stage of development. If you have a female, the largest tube will be filled with fertilized eggs that have a shell around them. One large tubule will be out of round, have a thin wall, and no contents. This is the digestive tube. How many cell layers can you count in the digestive tube wall? (*Hint:* look for nuclei because each cell has one nucleus.) _____ Identify the two lateral canals of the excretory system, cut in cross section and located 180° from each other against the inside of the body wall.

Study the **cuticle** covering the animal and note the absence of cells. Beneath it, find cellular **epidermis** that

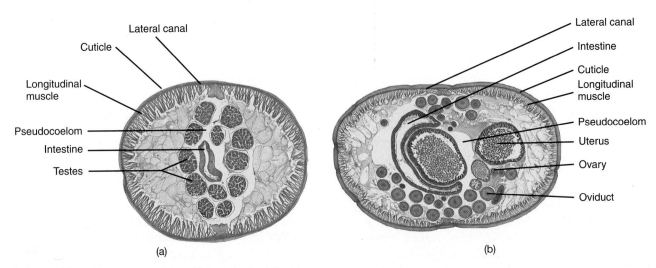

Figure 22.2 Cross section of *Ascaris:* (a) male and (b) female.

secretes the cuticle. Beneath the epidermis is a **longitudinal muscle** layer, composed of cells that run anterior to posterior, but are now cut in cross section. No circular muscles are found in the body wall. When the longitudinal muscles contract, the semirigid cuticle prevents the body from shortening. If muscles on one side contract, the animal bends. When they relax, the elasticity of the cuticle and the pressure of fluids in the body cavity straighten the animal. Unlike earthworms that can extend, contract, and twist their bodies, roundworms can only move forward by thrashing movements. These do not allow the animal to swim effectively, and it is confined to movement along surfaces.

3. Live Nematodes

To finish your study of nematodes, you will observe living nematodes. Vinegar eels are small nematodes that live in rotting vegetation and naturally fermented, unpasteurized vinegar. They are not parasites and cause no harm if ingested. Often soils from greenhouse pots that were not sterilized or from undisturbed wood-lands will contain other small nematodes. Make a slide by putting a drop of the culture on a slide. Remember the nematodes are at the bottom of the culture because they cannot swim.

Add a coverslip and look at the nematodes through your compound microscope. Note the thrashing movements due to the interaction of the longitudinal muscles and semirigid cuticle and hydraulic skeleton. Look inside the translucent animal and try to identify the organs you saw in *Ascaris*.

4. Rogue's Gallery

While most nematodes are free-living, many species are parasites of animals and plants. In the lab are two demonstration slides that you should take a look at: pinworms and *Trichinella*.

Pinworm *Enterobius vermicularis* is a common parasite of children both in the United States and abroad. Living in the large intestine, mature females migrate to the anal region to lay up to 16,000 eggs. Migration causes irritation and itching in the perianal area. A child scratching the anal area contaminates his/her hands. As children often put their hands in their mouths, the cycle is completed as new eggs are ingested. The U.S. Center for Disease Control estimates that 40 million U.S. citizens (13%) are infected, making it the most common worm parasite in this country. Take a look at the demonstration slide of this nematode.

Trichinella *Trichinella spiralis* is a serious parasite of rats, pigs, and humans as well as many wild mammals (boar, bear, seal, and canines). Infection is called trichinellosis. Fortunately, public health practices have made it rather uncommon in the food supply.

Adults live embedded in the intestinal lining of carnivores. Females will bear up to 1,500 juveniles. The immature forms enter the circulatory system and are distributed throughout the host's body. They burrow into the muscles where they are encapsulated by reactions of the host's tissues, although they sometimes end in the brain or eyes during this migration phase. The life cycle ends there unless the host is eaten by another carnivore. If contaminated muscle is ingested, the juveniles leave the cyst and take up residence in the new host's intestine, continuing the cycle.

If humans eat raw pork, they risk infection. Public health officials have broken the cycle in two ways. First public education has resulted in pork being served thoroughly cooked. Second, it used to be the practice to feed garbage scraps to hogs. If the garbage contained raw pork muscle, the parasite continued in pig herds. Laws now prohibit feeding uncooked garbage to hogs. During the last decade, in the U.S.A., an average 12 cases were reported per year, usually related to eating improperly cooked wild game.

The demonstration slide of this nematode is a section of infected muscle. What you will see in the muscle tissue is a cyst with a small worm coiled inside waiting to be eaten!

Filarial Worms Filarial worms are parasites that live in the blood and lymphatic systems of mammals and birds. Your dog may take drugs for heartworm, a filarial worm that lives in the circulatory system, often accumulating as large masses in the atria of the heart, blocking their function. New animals become infected because young worms, called microfilaria, hatch from eggs and circulate in the blood. Mosquitoes feeding on an infected dog pick up the young worms and transfer them to a new host.

An estimated 120 million people living in 80 tropical or subtropical countries are infected with another filarial worm. Also transmitted by mosquito bites, the adults take up residence in a human's lymphatic ducts in the legs and arms. As they grow, they can plug the ducts. Our lymphatic system is important in draining fluids from the intercellular spaces. If blocked, the accumulating fluid causes severe swelling, leading to a limb becoming grotesquely large. The condition is known as elephantiasis. There is no treatment other than surgery or waiting 6 to 7 years for the worms to die.

5. Summarize Observations

Summarize your observations of nematodes by filling in **table 22.1** at the end of this lab topic. Turn to your lab partner and discuss how nematodes perform the important functions of gas exchange, digestion, circulation, excretion, and movement involving both muscles and the exoskeleton.

Phylum Arthropoda

Approximately a million species of arthropods have been described and the world population level of arthropods is estimated at a billion billion (10^{18}) individuals. Crustaceans, insects, millipedes, spiders, and ticks are but a few

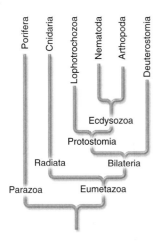

examples of this diverse phylum that has representatives in virtually every environmental habitat from the ocean depths to mountain tops, including aerial environments.

Body Plan As in the annelids, the bodies of arthropods are segmented. This similarity led many to conclude that the annelids and arthropods were closely related. However, molecular evidence places them in different clades: annelids in the Lophotrochozoa and arthropods in the Ecdysozoa. In the arthropods, the segments are often fused to make distinct body regions, such as the **head, thorax,** and **abdomen.** Some have a **cephalothorax** in which the head and thorax regions are fused. Appendages are always jointed.

Arthropods are encased by a noncellular **cuticle** composed of chitin, a complex polysaccharide stiffened by secretions of calcium salts and cross-linked proteins. Acting as an exoskeleton, the cuticle protects and supports, but has joints where thin regions of the cuticle allow bending. For terrestrial species, the cuticle is effective at preventing desiccation and may explain why the arthropods have been so successful on land. In order for individual arthropods to grow, they molt their exoskeletons, and most species have several developmental stages in their life cycles. Muscles occur in distinct bundles rather than being part of the body wall and allow a variety of movements, especially in combination with the jointed appendages.

Because the exoskeleton limits diffusion, gas exchange in large arthropods is through special respiratory structures, such as gills, book lungs, or small tubules called tracheae. The digestive system of arthropods is more or less a straight tube leading from the mouth to the anus. Arthropods have an open circulatory system with a dorsal artery and heart that pumps hemolymph anteriorward. The nervous system consists of two ventral cords with ganglia serving as integrating centers, allowing complex behaviors and locomotion. They have well-developed sensory capabilities with organs of sight, hearing, smell, and balance. Arthropods have distinct excretory organs that remove nitrogenous wastes from body fluids and which function in electrolyte balance. The sexes are usually separate.

Five evolutionary lineages can be seen in the arthropods. These are:

Clade Trilobitamorpha: about 4,000 species of extinct trilobites known only from fossils, disappearing about 250 million years ago during the great Permian extinction. Bodies were segmented with similar appendages on most segments. Modern arthropods tend to have specialized appendages on different segments.

Clade Chelicerata (= Cheliceriformes): about 74,000 species of horseshoe crabs, mites, spiders, scorpions, and ticks; name reflects the feeding appendages called chelicerae. You will look at a horseshoe crab and spider as representative of this clade.

Clade Crustacea: about 45,000 species of crabs, crayfish, barnacles, and many others; primarily aquatic although there are terrestrial species (pill bugs); have two pairs of antennae, compound eyes, and appendages that branch. You will dissect a crayfish, observe the anatomy of a microcrustacean (*Daphnia*), and observe the external anatomy of barnacles as representatives of this clade.

Clade Myriapoda: about 13,000 species of centipedes and millipedes; mouthparts are mandibles; definite head; numerous body segments bearing one or two pairs of walking legs; one pair of antennae. You will not study this group.

Clade Hexapoda: about a million species of insects; mouthparts are mandibles; body consists of three regions of fused segments, head, thorax, and abdomen; three pairs of legs and usually two pairs of wings. You will look at the anatomy of a cricket as a representative of this clade.

There is some debate among biologists about how these lineages are related. Some consider each lineage a subphylum of the Arthropoda. Others suggest Myriapoda and Hexapoda be lumped together in a clade called Uniramia because they have one pair of antennae and non-branching appendages. Others argue that the Myriapoda, Crustacea, and Hexapoda be lumped together in a clade called Mandibulata based on the fact all have jawlike mandibles. All of this is confusing for the student who simply wants to know: What should I learn? Unfortunately, you need to know a little about all of this. Currently, there is not an agreed-upon answer to the taxonomic relationships within the arthropods. On the other hand, this controversy makes it an exciting time to become a biologist. There is much work to be done to solve these problems. Will you be among those who do?

Clade Chelicerata (= Cheliceriformes)

Most living chelicerates are now terrestrial animals, although at one time in geological history they rivaled the crustaceans for predominance in marine ecosystems. Chelicerates include horseshoe crabs, scorpions, spiders, ticks, and mites as well as many extinct forms.

The derived characteristics of the group include:

- Body consists of a cephalothorax and an abdomen.
- Cephalothorax bears 6 pairs of appendages: 2 pairs of mouthparts, and 4 pairs of walking legs.

- Mouthparts include a pair of chelicerae, diagnostic for the group.
- No antennae.

Horseshoe Crab

Horseshoe crabs are marine scavengers that consume molluscs, worms, and plant material as they slowly move over the sandy bottoms of shallow bays and oceans. Despite its name, it is not a true crab because it lacks mandibles and has chelicerae.

6▶ Look at the horseshoe crab on demonstration in the laboratory. From a dorsal view, the exoskeleton is fused into a thick, protective **carapace** that covers the 7 segments of the cephalothorax and the 12 segments of the abdomen. A hinge joint of thin cuticle marks the dividing line between the body regions and allows the animal to bend. A tail-like **telson** extends from the tip of the abdomen. Despite its threatening appearance, the telson is used more in recovering, when the animal turns upside down, rather than in defense. At the telson's base is the **anus**. A pair of **compound eyes** are prominent on the anterior of the cephalothorax and two simple eyes **(ocelli)** are found on either side of a small spine located on the midline anterior to the compound eyes.

Turn the animal over and look at its ventral surface (fig. 22.3). Six pairs of jointed appendages surround the medially located mouth. Note how they are jointed, allowing them to extend and bend, so the animal can move food to the mouth and walk. The small first pair are **chelicerae** used in manipulating food. All but the last pair end in **chelae,** pincerlike structures used to grasp food. The base of the appendages bear spiny **gnathobases,** which macerate food and force it into the mouth as the appendages are moved in a kneading fashion.

The ventral surface of the abdomen is covered by six pairs of flat appendages. The first of these is the **genital operculum.** The openings of the reproductive system lie beneath the operculum. Under the remaining five plates are **book gills,** so-called because in life the gills fan back and forth as you would fan the pages of a book. They also can be used to propel the animal when it swims.

What type of symmetry does this animal have? _____

Spider

7▶ Look at the demonstration specimen of a preserved garden spider with a dissecting microscope. Most specimens will be female because males are smaller and not routinely sold by biological supply houses.

Note the two body regions, the **cephalothorax** and the **abdomen,** which are connected by a slender **pedicel,** a modified first abdominal segment (fig. 22.4). Locate the eight **ocelli,** simple eyes, on the anterior dorsal surface of the cephalothorax. It does not have compound eyes.

Turn the spider over and examine the ventral surface. Six pairs of appendages should be visible. The first are the **chelicerae** with hollow, hardened **fangs** that inject a paralytic poison into prey. The second pair of appendages are the **pedipalps,** which serve a sensory function in feeding. Behind the pedipalps are four pairs of **walking legs,** each composed of several segments.

On the ventral abdominal surface, find the paired slits, **spiracles,** that open into the **book lungs.** Hemolymph circulates in the plates and exchanges respiratory gases with the air.

On the median line of the abdomen near the spiracles, find the platelike **epigynum,** which covers the female genital opening. The openings of three pairs of **spinnerets**

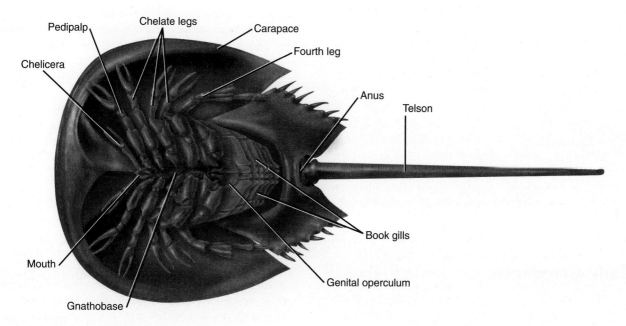

Figure 22.3 Ventral view of horseshoe crab.

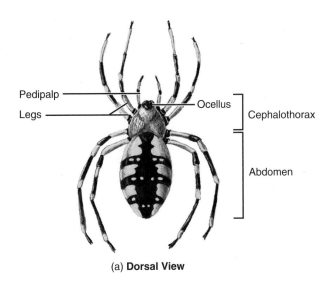

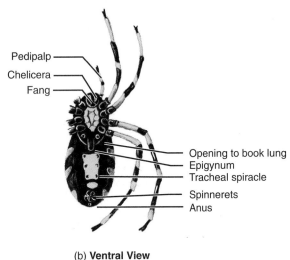

Figure 22.4 Spider: (a) dorsal view; (b) ventral view.

should be visible on the ventral surface near the tip of the abdomen. Silk glands beneath the openings exude a viscous solution of silk protein that hardens when exposed to air.

Find the small paired **spiracles** on either side of the midline of the ventral surface of the abdomen. These open into a system of tubules, the **tracheae,** that supplement the gas exchange capabilities of the book lungs. How does this system of gas exchange differ from that of the earthworm?_____

Fill in the information required in the second column of table 22.1 at the end of this lab topic to summarize your observations on Chelicerates.

Clade Crustacea

No other phylum contains as many marine species as there are crustaceans. Many other crustacean species live in fresh water and there are even a few terrestrial species such as the pill bugs that live in damp terrestrial habitats. Crustaceans range in size from a millimeter or so long to lobsters that can exceed 40 pounds. Crustacea are a very diverse group as your study of barnacles, freshwater *Daphnia,* and crayfish will demonstrate.

What distinguishes the Crustacea from the rest of the arthropods?

- Head develops from the fusion of five body segments each bearing a pair of appendages.
- Head can be fused with the exoskeleton of the thorax to form a carapace. Some species are encased in lateral valves.
- Abdominal segments are usually distinct and the terminal segment bears a telson.
- Appendages are biramous, meaning they branch.
- Gas exchange usually across gill surfaces.
- Most have both simple and compound eyes.

Barnacles

Barnacles are sessile marine crustaceans that live attached to rocks, ships, whales, turtles, and seaweeds. Some barnacles have a long stalk, the gooseneck barnacles (*Lepas*); and others have a compact body, the acorn barnacles (*Balanus*) (**fig. 22.5**). Examples of both are on demonstration in the lab.

8 Obtain a preserved barnacle in a small dish of water so you can observe it with a dissecting microscope. Although barnacles superficially resemble molluscs because of their calcified outer exoskeletons, dissection reveals that the animal inside has a segmented body, jointed appendages, and a chitinous exoskeleton, characteristics diagnostic of the phylum Arthropoda.

To understand barnacle anatomy, visualize a "shrimp-like" arthropod lying on its back within the surrounding plates. The animal is attached to the basal plates by the back of its head. Its large thorax, bearing several appendages, and small abdomen are free. Flaps of tissues, called a mantle, grow out from the back of the head and line the internal side of the plates creating a mantle cavity. The animal's body and the inside of the mantle are covered by a thin chitinous cuticle. Muscles in the mantle can open and close the surrounding plates, protecting the body from predators or from drying when in intertidal zones. When the plates are open, the animal can extend itself by doing the equivalent of our lying on the floor and lifting our legs up and curling them over our heads. If you look at your barnacles from the side, the taller end is the posterior of the animal, and shorter anterior; dorsal is down and ventral up.

The procedure for opening the plates to reveal the animal's body, depends on whether you have an acorn or gooseneck barnacle. For acorn barnacles, pry open the ventral opercular plates to reveal the mantle cavity. Do not cut with scissors or they will be ruined by the calcium plates. If you have a gooseneck barnacle, the long stalk

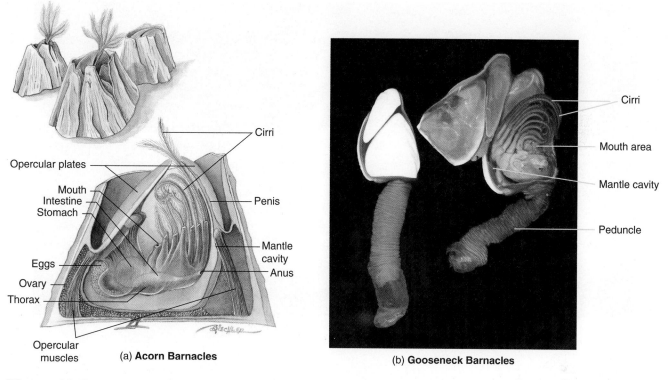

Figure 22.5 Body plans of barnacles. (a) acorn barnacle; (b) gooseneck barnacle.

is the peduncle that attaches the animal to its substrate and the calcified region is the capitulum. A longitudinal slit runs anterior to posterior along the free ventral edge of the capitulum's largest plates. Pull the plates apart to enter the mantle cavity and cut the upper one free with a scalpel.

Note the brownish mantle lining the cavity and the animal's body. The thorax bears six pairs of appendages called **cirri,** which have fine comblike projections. The animal feeds by curling its body up, extending its cirri, and sweeping them through the water, capturing microplankton. The mouth is just anterior to the first pair of cirri and is surrounded by small mouthpart appendages. The digestive system runs the length of the body to the anus located between the last pair of cirri.

Barnacles are hermaphrodites. The single long penis (6× longer than the body) arises just anterior to the anus. The two openings of the female reproductive system are inconspicuously located at the base of the first pair of cirri. In a population of barnacles, individuals acting as females secrete chemical signals that activate the male systems of neighbors. Being sessile, barnacles cannot move to find mates. Instead, a stimulated individual extends its penis and probes the surroundings. Sperm are deposited in the mantle cavity of a receptive neighbor who then releases eggs. Fertilization is external and the developing eggs are brooded until young motile larvae hatch and disperse. Few barnacle species are capable of self-fertilization. Sperm exchange is not reciprocal at the time of mating, although sex roles may be reversed at another time.

If you have a gooseneck barnacle, use a scalpel to cut the peduncle lengthwise to reveal a small cavity containing the ovaries and associated ducts. The wall of the peduncle consists of an outer layer of circular muscles that elongate the peduncle when they contract, and an inner layer of longitudinal muscles that shorten it.

Clean up your barnacle dissection as directed by your instructor.

Daphnia

9 Commonly known as the water flea, this microcrustacean inhabits ponds and lakes in North America and Europe.

Make a slide of living *Daphnia* by catching one in a pipette from the lab culture and transferring it with some water to a slide. Use your dissecting microscope to watch it swim. Using **figure 22.6,** describe what body parts are used for propulsion.

Add a coverslip and observe it through your compound microscope. *Daphnia* are covered by a transparent **carapace,** an exoskeleton that wraps around the body. The head bears a conspicuous, single compound **eye,** two large **antennae,** and two smaller antennae that are difficult to see.

These animals are filter-feeders. Legs inside the carapace terminate in comblike **setae,** which sieve bacteria, algae, and

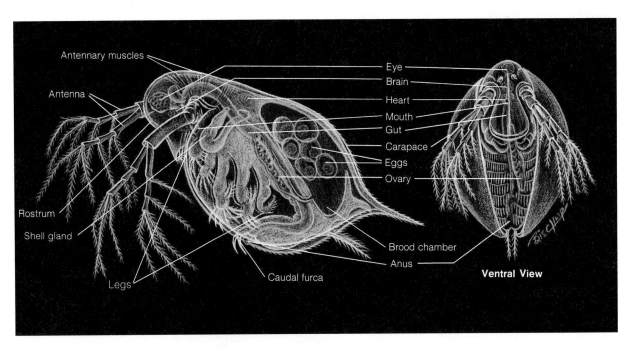

Figure 22.6 Internal anatomy of *Daphnia* in lateral and ventral views.

small particles from the water. Food balls are transferred to the **mouth** and take about one-half to three hours to pass through the digestive system. Often the gut appears green because it contains algae sieved from the culture.

10▶ Feeding behavior can easily be observed. Remove the coverslip from your depression slide. Dip a dissecting needle into a thick suspension of *Spirulina,* a blue-green alga, and remove a small drop. Transfer this to the *Daphnia* slide while watching through your dissecting microscope. You should be able to see the sieving action of the leg combs and blue-green algae should be taken into the digestive tract.

Most likely the sex of the *Daphnia* you are observing is female. *Daphnia* can reproduce both asexually and sexually. Most natural populations of *Daphnia* are almost entirely female, produced by a type of asexual reproduction called **parthenogenesis.** When reproducing this way, females produce diploid eggs by mitosis, and the eggs will develop into adults. Therefore, the eggs, and individuals that develop from them, have genomes equal to the mother's and are clones. Under optimum conditions, a female may produce up to a dozen or more eggs per brood, repeating every three days or so with up to 25 broods produced in a lifetime. The eggs are released into a brood chamber beneath the carapace and develop into small replicas of the parent. When the female sheds her carapace during the molt cycle, young adult asexual females are released and within a few days will themselves produce parthenogenic eggs. This allows for rapid population growth when conditions are favorable.

When conditions change to less than optimum due to low oxygen, inadequate food supply, high population densities, drying of a pond or puddle, or cooling temperature, sexual reproduction occurs. The asexual mother will now produce parthenogenic eggs that develop into sexual males and sexual females. The ability to do this raises fundamental questions about how sex is determined and is currently the subject of much research. The genome of *Daphnia* is being closely studied by researchers collaborating in the *Daphnia* Genomics Consortium. (See http://daphnia.cgb.indiana.edu/). Sex determination seems to be controlled by the mother and may involve some type of hormone that influences development. Much is yet to be learned about how this occurs.

Sexual males and females produce haploid sperm and eggs by meiosis. The males copulate with the sexual females who produce only two eggs. The resulting zygotes are retained in the brood chamber but are encased in a protective jacket. When the sexual female molts her carapace, the eggs stay in the carapace and sink to the bottom. They are resistant to drying, freezing, and even passage through the guts of other animals. The embryo remains in a quiescent state until conditions improve. At that time, asexual females escape the protective case and the cycle repeats. Researchers have found that some encased embryos remain viable for up to 20 years.

Return the *Daphnia* to the supply area and clean your slide.

Crayfish

You will use a crayfish for your detailed dissection of a crustacean. Fresh crayfish can be obtained from restaurant suppliers frozen and then thawed. Fresh material is better to dissect because the tissues retain natural colors. Obtain a crayfish, dissecting instruments, and a dissecting pan from the supply area.

11▶ *Body Plan* Examine the external anatomy of the crayfish (fig. 22.7). The body is divided into two major regions: the

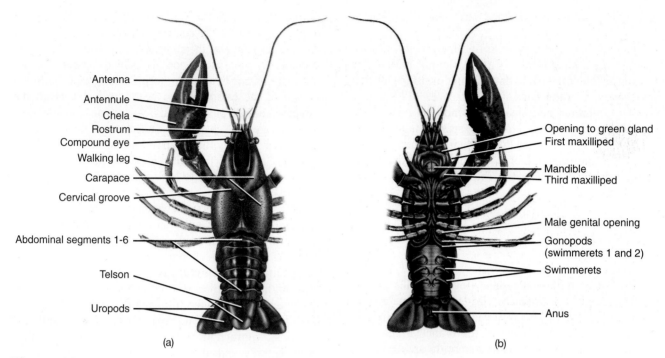

Figure 22.7 External anatomy of male crayfish: (a) dorsal view; (b) ventral view.

anterior **cephalothorax,** covered dorsally by the **carapace** and the posterior **abdomen,** composed of 6 segments. The abdomen ends in a flaplike structure, the **telson.** Thirteen fused segments make up the cephalothorax. The **cervical groove,** a cross body groove in the carapace, marks the separation of the head from the thorax. The head ends anteriorly in the pointed **rostrum.**

Lay the animal on its back and examine the appendages. On the last abdominal segment are two flattened lateral appendages, the **uropods.** Contraction of the ventral abdominal muscles draws the telson and the extended uropods under the body, allowing the crayfish to swim rapidly backwards. On the other abdominal segments are small **swimmerets.** If your specimen is a male, the swimmerets on the first abdominal appendages will be pressed closely together and pointed forward. They serve as sperm transfer organs, receiving sperm from openings at the base of the last pair of walking legs. If your specimen is a female, these swimmerets are small. Be sure you observe both males and females in the lab.

The large appendages of the cephalothorax are the **walking legs.** The first pair of legs are called **chelipeds** because they bear **chelae** (pincers), which are used in grasping and tearing food as well as in defense.

Several appendages surround the mouth. It is instructive to remove them from one side of the mouth. This can be done by grabbing them with forceps near the base and pulling. If you do this, arrange them on a paper towel from anteriormost to last as you remove them. Note that all appendages except for the long antennae are biramous, meaning that they branch from a single base segment (**fig. 22.8**). The heavy **mandibles** are the jaws that grind food. There are also four other pairs of

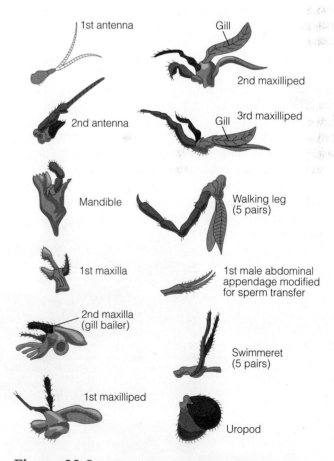

Figure 22.8 Biramous appendages of crayfish removed to show basal and two distal segments for each appendage. Arranged in order from most anterior to most posterior as you scan down the first and second columns.

mouthparts that handle food and carry sensory organs for taste and touch.

Examine the mouthparts to find the second maxilla, which bears a large elongated plate, the **gill bailer.** The sculling action of the bailer draws water over the gills that are located under the sides of the carapace.

The **antennae** and smaller **antennules** are the last appendages on the head. They have sensory functions.

The gills are contained in branchial chambers, located laterally in the cephalothorax. What are the advantages and disadvantages of having the gills enclosed by the exoskeleton?

Cut away the lateral carapace over the gills on one side of the crayfish. Gradually clear away the material to reveal the feathery gills. Water is drawn in through an opening at the posterior end of the branchial chamber, passes over the gills, and exits anteriorly as a result of the action of the gill bailer. Cut off a gill and mount it in a drop of water. Add a coverslip and look at it with your compound microscope. Sketch what you see below.

Hemolymph, crustacean blood, circulates through the gills and comes in close proximity to water passing over the gills, where carbon dioxide and oxygen exchanges occur.

12 *Functional Anatomy* You will now study the internal organs. With scissors, cut very carefully to one side of the dorsal midline of the carapace from where it joins the abdomen forward to a region just behind the eyes. Do not point the scissor tips down. Keep them against the inside surface of the carapace. The heart lies just beneath the carapace and will be destroyed if you are not careful. Use a probe to carefully separate the carapace from the underlying tissue that secretes the carapace. Completely remove the carapace. Now cut through the exoskeleton covering the abdomen from the first to last segments. Carefully remove the exoskeleton to expose the underlying structures. Remove the gills and the membrane separating them from the internal organs. Flood the tray with water so that the organs float.

Compare the opened crayfish to **figures 22.9** and **22.10**. Starting in the abdomen, note the two thin longitudinal bands of **abdominal extensor muscles** that extend forward into the thorax. Contraction of these muscles straightens the abdomen from a curled position. Beneath the extensors, find the large segmented **abdominal flexor muscles.** Contraction of these muscles bends the abdomen ventrally and allows the crayfish to swim rapidly backward in escape reflexes. Given the relative size of these two sets of muscles, which ones would you say generates more force? How is this related to locomotion?

Passing along the dorsal midline of the muscles is the **intestine** with the overlying **dorsal abdominal artery.** If you are dissecting an injected specimen, this artery should contain colored latex. Carefully remove the long bands of the extensor muscles to reveal the organs beneath them.

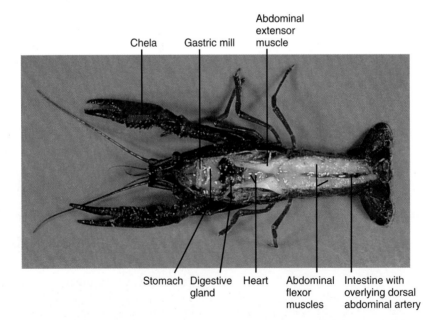

Figure 22.9 Dorsal view showing internal organs in a partially dissected crayfish.

openings, the **ostia,** which should be visible. When the heart contracts, small flaps of tissue seal the ostia from inside the heart. Hemolymph in the heart is forced out into difficult-to-see arteries. As figure 22.10 indicates, these include the single **ophthalmic artery,** the **paired antennary arteries,** paired **hepatic arteries,** and a single **sternal artery.** Try to locate one or more of these and show your instructor.

Arthropods have open circulatory systems. When hemolymph is pumped away from the heart in arteries, it leaves the arteries and enters the spaces in tissues at the periphery of the animal. There are no capillaries as are found in animals with closed circulatory systems (annelids some molluscs, and vertebrates). Hemolymph in the tissues is displaced by the new hemolymph and percolates back to where it enters the heart through the ostia. Hemolymph contains the respiratory pigment hemocyanin. When it is pumped through the sternal artery it flows into the gills before returning to the heart. This circuit allows respiratory gas exchange.

Reproductive System From your specimen's external anatomy, you should know whether you have a male or female. In females, the paired ovaries lie ventral and lateral to the heart. If collected in breeding season, the ovaries may be enlarged and filled with eggs. In males, the paired testes should be visible on either side of the midline beneath the heart. Be sure to look at the internal anatomy of both males and females by locating someone who has an animal of the opposite sex from yours.

Digestive System Trace the digestive system from the mouth through the **esophagus** to the **stomach** and into the **intestine.** Find the anus ventrally at the base of the telson. Locate the **hepatopancreas,** a large gland that secretes digestive enzymes into the stomach. Cut the esophagus and intestine to remove the stomach. Open the stomach and observe the hardened areas of the wall, the **gastric mill** that grinds food. Remnants of its last meal may be present. At the posterior end of the stomach, fine bristles prevent large pieces of ingested food from entering the intestine.

Nervous System Remove all organs and look on the floor of the body cavity to find the ventral nerve cord. You should see **segmental ganglia,** collections of neurons that function as integration centers. Lateral nerves leave the ganglia and innervate the appendages. If you follow the nerve cord anteriorly, you will find ganglia beneath and above the stub of the esophagus.

Many arthropods have **compound eyes.** Such eyes are made up of hundreds of individual light-sensitive units called **ommatidia.**

Cut the surface off one of the crayfish's eyes and mount it in a drop of water on a slide. Look at the slide with your compound microscope. Sketch what you see. Compound eyes give an animal mosaic vision. What does this mean?

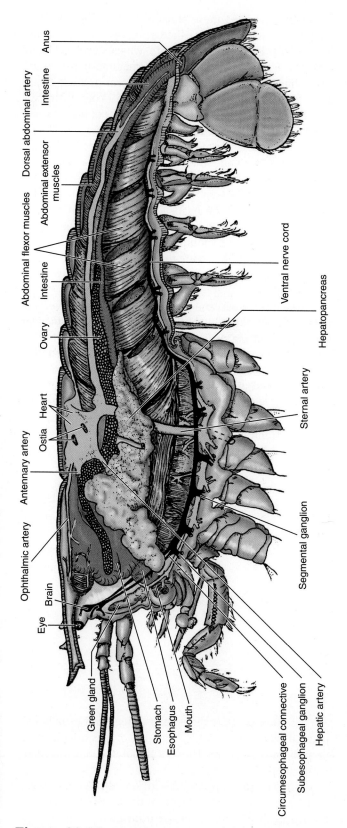

Figure 22.10 Sagittal section of crayfish showing positions of internal organs.

Circulatory System Trace the dorsal abdominal artery anteriorward to where it joins the delicate membranous **heart** located in the posterior third of the cephalothorax. Hemolymph surrounding the heart enters it through small

Excretory System Remove the organs from the cephalothorax. Find the **antennary glands** at the base of the antennae. These are the crayfish's excretory organs and also are known as green glands, though they are not green in color.

Dispose of your crayfish as directed by your instructor. Pair off with another student in the lab and explain to each other how a crayfish gets oxygen to its tissues and the pathways of circulation of hemolymph.

To summarize your observations on crustaceans, fill in the second column of table 22.1 at the end of this lab topic.

Clade Hexapoda (=Insects)

The characteristics of the hexapods include:

- Body with three regions; a head, thorax, and abdomen.
- Appendages do not branch at ends as in crustaceans.
- Head appendages include one pair of antennae, two pairs of maxillae, one pair of mandibles.
- Have a unique tracheal gas respiratory system.
- Have a unique malpighian tubule excretory system.

There are about 28 orders of insects and nearly a million species, making them the most diverse group of animals. The adaptability of the insect body form has developed over 400 million years and allowed insects to be successful in virtually every ecosystem, except in marine environments where the crustaceans predominate. Insects are the most common animals in terrestrial environments, attesting to the exoskeleton's ability to prevent water loss as well as to support the animal. Insects are among the relatively few animals that fly and this may contribute to their success by allowing them to escape predators and to disperse widely. Ecologically, they are important members of food chains, are extremely important in the pollination of eudicot flowers with which they have coevolved, and can be transmitters of diseases caused by viruses, bacteria, and protists. The organ systems of insects are highly developed and some are unique as your dissection work will demonstrate.

13 ***Body Plan*** Obtain a cricket, cockroach, or grasshopper from the supply area.

Note your specimen's noncellular exoskeleton composed of chitin and proteins which are secreted by underlying cells. As these animals mature, they undergo a series of molts (ecdysis) increasing in size and changing slightly in morphology; *e.g.*, wings develop. Many other insects undergo a complete metamorphosis, radically changing the body form from larval to adult stages as in the caterpillar-to-moth transition. Your instructor may have on display some living examples of the stages in insect development. Be sure to look at them.

For your insect, note the three distinct body regions: head, thorax, and abdomen. The **head** bears a single pair of **antennae** which function in chemoreception (odors) and as touch receptors. Compound and simple (ocelli) eyes also occur on the head. Grasshoppers and crickets have chewing mouthparts. Use your dissecting microscope to observe the mouthparts of your specimen. Find the heavy **mandibles** with the hardened edges. Which way do the mandibles move to chew food, vertically or horizontally?_____

Note the upper and lower "lips" which cover the mouth. Small **palps** attached to **maxilla** help maneuver food to the cutting surfaces of the mandibles.

14 Other insects have mouthparts specialized for biting or lapping. Microscopes in the demonstration area have slides of mosquito and fly heads to demonstrate their mouthpart specializations. Mosquitos have biting-sucking mouthparts and the house fly has lapping mouthparts. Make diagrammatic drawings of each below.

The **thorax** bears three pairs of **legs** and usually two pairs of **wings** (fig. 22.11). One pair of legs originates on each thoracic segment. The legs are jointed, allowing agile movements.

In male grasshoppers and crickets the last leg is also used to produce sound. Teeth on the largest segment are rubbed against the edge of the front heavy wing to produce the familiar stridulation or chirping sounds of these species.

The wings originate from the thoracic segments, but are not appendages. The forewings are heavy and the hindwings membranous. Some insects, such as fruit and house flies, have only one pair of wings. The second pair is reduced to small knob-like projections.

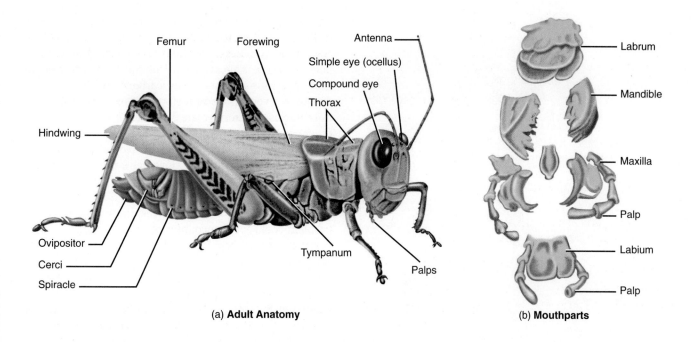

Figure 22.11 Grasshopper: (a) external anatomy; (b) anatomy of head and chewing mouthparts.

The **abdomen** consists of 11 body segments and sometimes fewer in other species. Count the number in your specimen. Look along the side of the abdomen and note the 10 respiratory spiracles. These open into a system of tubes that pass throughout the body (see fig. 22.12) and are part of the tracheal respiratory gas exchange system. The last three segments are modified in the different sexes, either for copulation or egg laying. A pair of sensory **cerci** can be found on the eleventh segment. On the first abdominal segment, find the **tympanum** that functions in sound reception.

As you have studied the external anatomy of your insect, you should have noticed how the exoskeleton is arranged in plates called **sclerites,** some with rigid junctions, called **sutures,** others with flexible junctions. Pigments in and under the exoskeleton give many insects their distinctive colors.

Functional Anatomy Preserved insects are usually not good materials to dissect because the exoskeleton prevents rapid penetration of the preservative so that the organs are often digested by enzymes escaping from the digestive system. For this section, you will use a freshly killed cricket or cockroach. They can be killed by exposing them to ethyl acetate in a closed chamber or placing them in a freezer.

Insects are compact animals with many organs and tissues crammed into a small space. Given the relatively small size, this makes dissection a challenge. In this section, you will have an opportunity to develop your microsurgery techniques as the dissection should be done while viewing through a dissecting microscope.

15 Obtain a freshly killed cricket or roach, a petri dish filled with wax or a dissecting mat, and four small pins. Remove the insect's wings by cutting them off at the base. Cut off the legs as well. Fasten the animal to the wax, dorsal side up, by putting a pin through the head and the last abdominal segment. Now comes the crucial step—opening the animal.

Using fine, sharp scissors and forceps, loosen a dorsal sclerite near the tip of the abdomen and insert the tips of the scissors. Cut along the lateral edge forward to the thorax. Raise the dorsal flap and use a dissecting needle to work it free of underlying tissues. Be careful not to damage the dorsal aorta which lies underneath. Now continue your cut forward through the thorax. When at the front of the thorax, make a new cut across the body to the other side. Make a similar cut in the last abdominal segment. Now work the dorsal flap of the exoskeleton free and lay it over and pin it down. Flood the tray with water to float the organs and keep them from drying. You will now systematically work your way down through the animal, exposing different organ systems. Use **figure 22.12** as a guide to identification.

Circulatory System Insects have open circulatory systems, as do all arthropods. Vessels move a circulating fluid called hemolymph, from one body region to another where the fluid enters the spaces between tissues and gradually percolates back.

If you removed the exoskeleton carefully, you should see the **tubular heart** on top of the body mass in a small depression. Hemolymph enters the heart from the surrounding spaces through minute lateral openings called **ostia** and is pumped forward by peristaltic waves through

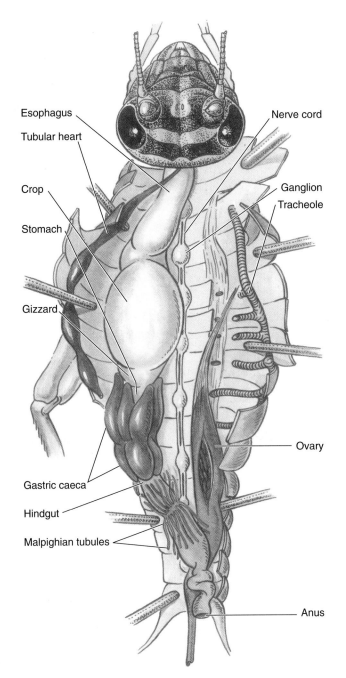

Figure 22.12 Internal anatomy of a cricket. Fat body is not shown for clarity.

a **dorsal aorta** to the head. There it leaves the aorta and enters the tissue spaces, collectively called the **hemocoel**. Hemolymph gradually flows back to enter the ostia and completes the circuit.

As you look at the organs, a chalky white **fat body** may fill the hemocoel. In females, the fat body can be replaced by swollen oviducts holding up to 600 eggs. Remove this material carefully to reveal the respiratory and digestive systems.

Respiratory System Insects (and some spiders) have a unique respiratory system, the **tracheal system** which conveys oxygen directly to the tissues.

As you work through the fat body, you should occasionally see glistening whitish tubes. These are **tracheoles** of the tracheal system. They branch from main **tracheae** coming from the spiracles that you saw on the external lateral surface. This system of tubules branches into fine tubes that directly take away carbon dioxide and bring oxygen to every tissue in the body. The hemolymph does not serve as an intermediate carrier as blood does in many other animals. In larger insects, air sacs associated with the tubes help to ventilate them (see fig. 29.1). When an insect uses muscles for movement, the sacs are compressed and air is forced out and drawn in as the muscles relax.

An interesting sidebar on respiratory systems is a problem beekeepers encounter. A parasitic mite (a chelicerate) parasitizes bees. It invades and lives just inside the spiracles in the tracheal system. This interferes with the respiratory system and can kill a whole bee colony.

Digestive System Insects have a complete digestive system. As you reveal the digestive system, note the different regions along its length. The foregut has three regions: the **esophagus, crop** (a storage area), and the **gizzard** where chitinized plates on the inner wall grind the food into a fine pulp. Food enters the **stomach** with six fingerlike **gastric caeca** that secrete digestive enzymes. The hindgut or **intestine** leaves the stomach and passes to a **rectum** and **anus**. The rectum is a water-reclamation organ and produces a relatively dry fecal material.

Excretory System The excretory system of insects is unique and consists of structures called **malpighian tubules**. These can be seen radiating like threads from the middle of the hindgut. The cells in the walls of these tubules absorb nitrogenous waste materials from the hemolymph that bathes them, forming uric acid. Crystals of uric acid enter the hindgut and are compacted with fecal material before defecation.

Reproductive System The sexes are separate in insects and show sexual dimorphism (sexes look different). Based on your observation of external anatomy, you should know if your insect is male or female. If the gonads were not enlarged and filling the body cavity, remove the digestive tube to find them.

Nervous System After removing the digestive and reproductive systems, you should see the internal floor of the body cavity. Passing from anterior to posterior on the midline, are two whitish **nerve cords** which form three paired ganglia in the thorax and five smaller ganglia in the abdomen. **Ganglia** are collections of nerve cell bodies and are where integration of nerve signals takes place. If this is not apparent, pour the saline out of the dish and add a few drops of methylene blue stain to the body cavity. Let it sit for a few minutes and then wash it out. The nerve cords should stain blue. Additional ganglia are found in the head but will not be observed in this dissection.

You are now finished with your insect dissection. Dispose of the carcass and clean your dissecting instruments as directed.

Summarize your observations of this group by filling in the required information in the third column of table 22.1 on the last page of this lab topic.

Keying Insects (Optional)

16 If you are to do all of the dissections included in this lab, there probably will not be enough time to also do this portion of the lab. However, your instructor may choose to skip some dissections and use this activity to introduce you to the diversity of insects.

There are over 28 taxonomic orders of insects. For example, butterflies are in the order Lepidoptera while beetles are in the order Coleoptera. With close to a million species of insects in so many orders, experts turn to taxonomic keys to help them identify newly collected specimens.

If you are to do this activity, your instructor will have a collection of unknown insects in the lab. They will be numbered so that you can check your identification later against a master list. Keys to taxonomic order will also be available and you will be given directions in their use.

17 *Summarize Observations* You have looked at several arthropods exhibiting a wide range of body plans. What are the common characteristics of all arthropods that you saw in your lab work? Discuss these characteristics with your lab partner to see if you both agree. Then take turns and describe the alternative solutions arthropods have to solving the problems of gas exchange, circulation, feeding, and locomotion.

Lab Summary

18 On a separate sheet of paper, answer the following questions assigned by your instructor.

1. What are the derived characteristics of the clade Ecdysozoa? Phylum Nematoda? Phylum Arthropoda?
2. What are the general differences between the three living subphyla of arthropods studied in lab today? What are the similarities?
3. Describe the circulatory and respiratory systems of a crayfish. Compare them to the systems found in an insect and to those in a nematode.
4. What are the similarities between annelids, molluscs, nematodes, and arthropods? What are the differences?
5. Arthropods live in marine, fresh water, and terrestrial environments where they encounter the full range of osmotic challenges from overhydration to desiccation. Discuss these challenges and what it is about their body plans that makes them so adaptable.

You may want to try the critical thinking questions that apply some of the knowledge you gained in doing this lab.

INTERNET SOURCES

Check the WWW for information about arthropods as vectors of diseases. Use your Internet browser to connect to Google at www.google.com. Type in Center for Disease Control. When connected to CDC find the search function. Enter the name of an arthropod-borne disease such as: Chagas; filariasis; Lyme disease; malaria; sleeping sickness; or West Nile virus. Read about the disease and answer these questions: (1) Name of disease? (2) Causative agent? (3) Arthropod vector? (4) World region(s) where common? (5) Estimated number of people affected?

CRITICAL THINKING QUESTIONS

1. Why are there no really large arthropods? What could be engineering-type factors limiting body size?
2. How does the segmentation of an arthropod compare with the segmentation of an annelid?
3. Arthropods are the most diverse (greatest number of species) and the most common (greatest number of individuals) of all the animals. List several reasons why you think they have been so successful.
4. If you were offered a generous serving of wild boar that was only singed on the barbecue so that the center was raw, why should you refuse it?

LEARNING BIOLOGY BY WRITING

Write an essay that discusses the ancestral and derived traits of the Ecdysozoa. Based on the evidence gathered in this lab, explain the derived traits of Nematoda and Arthropoda, speculating why these phyla have so many and such diverse species.

Table 22.1 Summary of Ecdysozoan Characteristics

	Nematoda	Arthropoda		
		Chelicerata	Crustacea	Hexapoda
Symmetry?				
Cuticle composition?				
Skeletal system type?				
Any ciliated cells?				
Terminal mouth?				
Segmented?				
Appendages?				
Muscle in gut?				
Locomotion mode?				
Type digestive system?				
Type circulatory system?				
Heart present?				
Type respiratory system?				
Describe nervous system				
Describe excretory system				
Describe body plan				
Unique features?				

Lab Topic 23

Deuterostomes and the Origins of Vertebrates

Supplies

Resource guide available on WWW at
http://www.mhhe.com/labcentral

Equipment

Compound microscopes
Dissecting microscopes

Materials

Preserved or fresh specimens
    Demonstration specimens of the five classes of echinoderms
    Sea star
    Branchiostoma (amphioxus or lancelet)
    Adult sea squirt (*Ciona* or *Molgula*)
    Perca flavescens (perch); fresh trout from fish farms are often less expensive
    Sea urchins, sand dollars, and other echinoderms
Prepared slides
    Sea star arm, cross section
    Ascidian tadpole larva
    Branchiostoma, whole mount
    Branchiostoma, cross section
Dissecting pans and instruments
Demonstration fish skeleton

Student Prelab Preparation

Before doing this lab, read the Background and sections of the lab topic that have been assigned by your instructor.
    Use your textbook to review the definitions of the following terms:

bilateral symmetry	Echinodermata
cephalization	radial symmetry
Cephalochordata	Urochordata
Chordata	Vertebrata
deuterostome	water vascular system

    Describe in your own words the following concepts:
The general body plan of an echinoderm
The general body plan of a chordate
The organ systems you expect to find in the animals to be studied.
    After finishing the prelab review, write any questions you have about terms, concepts, or techniques in the margins of this lab topic. The lab activities should help you answer these questions, or you can ask your instructor during the lab.

Objectives

1. To illustrate the functional anatomy of echinoderms
2. To trace the development of the chordate body plan in primitive chordates
3. To study the functional anatomy of a bony fish
4. To collect evidence to test the hypothesis that all chordates share derived characteristics that suggest an evolutionary relationship

Background

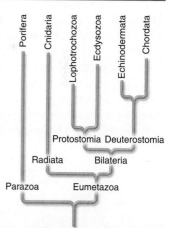

Bilaterally symmetrical eumetazoans can be placed into two major evolutionary lineages: the protostomes and the deuterostomes. Most animal phyla are in the protostome lineage and you sampled their diversity in the last two labs. In this lab, we turn our attention to the deuterostomes that include such diverse organisms as starfish in the phylum Echinodermata and mammals in the phylum Chordata. The clade is small with only about 60,000 living species in four phyla, representing 5% or less of the known animals, but they are the most familiar because vertebrates, animals with a backbone, are included.

    Several shared derived characteristics distinguish deuterostomes from other bilateral animals. These include:

    First cell divisions of zygote are by radial cleavage (see fig. 32.1).

    The blastopore of the gastrula stage becomes the anus and the mouth develops later, the opposite of the sequence observed in protostomes.

Mesoderm forms as buds from the distal end of the archenteron.

Similarities in nucleotide sequences in genes and amino acid sequences in proteins.

Mesodermally derived endoskeleton.

It is interesting to note that the new molecular data do not refute the relationships based on morphological data. In the strict logic of science, the hypothesis that Echinoderms and Chordates as well as two other minor phyla are related is not falsified. The hypothesis is understood to be true until someone can devise other independent observations to falsify it. This is what makes science so interesting: the constant suggesting of new ideas and then the logical testing of those ideas to determine their validity.

LAB INSTRUCTIONS

You will observe the anatomy of four deuterostome animals: an echinoderm (a sea star) and three chordates (a tunicate, an amphioxus, and a fish).

Phylum Echinodermata

The approximately 7,400 species of this phylum are spiny-skinned, slow-moving, or stationary marine animals. There are no freshwater or terresterial representatives. Adults share two unique traits not found in any other animal clade. The bodies of adult echinoderms exhibit **pentamerous radial symmetry**. Their bodies can be divided into five, or multiples of five, equal parts. Despite their unusual symmetry, echinoderms are included in the ancestral Bilateria clade because their larvae are bilaterally symmetrical (**fig 23.1**), suggesting their radial symmetry evolved later rather than first. Pentamerous symmetry develops when the larva metamorphosis into an adult. The second shared characteristic is the **water vascular system**, absolutely unique in the animal kingdom. It performs a variety of physiological functions, the most obvious being locomotion.

Many of the physiological systems we have seen in other bilaterally symmetrical animals are absent in Echinoderms. Separate excretory and osmoregulatory organs are not found, a trait that may have limited them to marine habitats. Likewise, distinct systems for body fluid circulation and respiration do not occur in these animals. Furthermore, there is no brain or main nerve cord: the nerves are arranged more like a net that extends through the body, receiving information from numerous sensors and controlling numerous muscles.

Echinoderms have a complete digestive system with a mouth and anus. Adults have an internal, calcareous skeletal system composed of separate ossicles that allow the body to bend, except in sea urchins where the ossicles fuse into solid plates. The sexes are separate but difficult to distinguish. Fertilization is external and eggs develop into larvae that metamorphose into adults. Some species can reproduce asexually by detaching body parts that grow into smaller replicas of the parent.

Echinoderm Diversity

1. Living echinoderms are grouped into five taxonomic classes, and 15 or more are recognized among the nearly 13,000 fossil species dating back to the Cambrian. Go to the area where the display specimens are laid out and find the representatives of each class (**fig. 23.2**).

    Crinoidea (sea lilies and feather stars) have:
    - Featherlike arms that radiate from the body.
    - A body that looks like a cup with mouth upward.
    - Sea lilies have body at end of stalk; feather stars can be free swimming.
    - Open ambulacral grooves on arms bear tube feet that filter food from water.

    Ophiuroidea (brittle stars and basket stars) have:
    - Body is stellate with five thin arms joining a central disc.
    - Mouth opening on undersurface.
    - Ambulacral grooves closed with tube feet lacking suckers.
    - Browse on seafloor or filter feed.

    Echinoidea (sea urchins and sand dollars) have:
    - Globe- or disc-shaped bodies with ossicles fused into a solid test.
    - No arms.
    - Closed ambulacral grooves.
    - Mouth opening on undersurface.

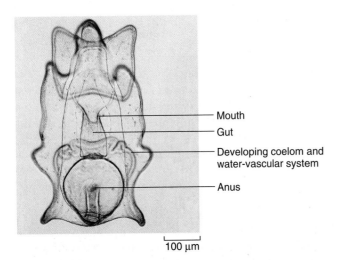

Figure 23.1 Bilaterally symmetrical larva of a sea star before it metamorphoses into a pentaradial adult.

Figure 23.2 Diversity of echinoderms: (*a*) sea lily; (*b*) feather star; (*c*) brittle star; (*d*) sea urchin; (*e*) sea cucumber.

- Movable spines.
- Feed by scraping encrusting algae from surfaces.

Holothuroidea (sea cucumbers) have:
- Soft and cucumber-shaped body.
- No arms or spines.
- Skeleton reduced to widespread ossicles.
- Mouth opening at one end and anus at other.
- Circle of feeding tentacles around mouth.
- Tube feet reduced and sometimes absent.
- Suspension and deposit feeders.

Asteroidea (sea stars)
- You will dissect a sea star and observe its anatomy in detail.

Asteroidea

Sea stars, starfish to some, are among the easiest to recognize invertebrates. The 1,800 or so species are distributed worldwide in benthic coastal and reef habitats in tropical areas, although there are a number of species in deeper, cooler subpolar areas. They are carnivorous, feeding on other invertebrates, especially molluscs and corals, but also on dead animal matter. A few are suspension feeders catching plankton in mucus secreted on their body surfaces.

2 *External Anatomy* Obtain a sea star and a dissecting pan. Note the pentamerous radial symmetry with five arms radiating from the central disc. The external and internal anatomy of each arm is identical, a feature that will allow for a unique approach to dissecting the animal.

Identify the **oral** surface of the sea star with the mouth at the center of the **central disc** and **ambulacral grooves** passing along the centerline of each arm (**fig. 23.3**). If you bend back the covering spines, the fleshy **tube feet** can be seen in the groove.

Place the animal in a dissecting pan, mouth side down. The upper surface is the **aboral** surface. Find the lighter-colored **madreporite** near the edge of the disc, a small sievelike opening into the water vascular system.

Add enough water to the pan to just cover the specimen. Use a dissecting microscope to examine the aboral surface of the central disc. Near the base of the spines find several small pincers called **pedicellaria** (fig. 23.3, surface detail). These may extend and contract to grab and remove small encrusting organisms from the sea star's skin. This is why sea star in aquaria always look pristine in comparison to many other invertebrates that are encrusted with algae. Also on the skin, note the many soft, blisterlike **dermal gills** (sometimes called papulae). They connect directly with the body cavity and function,

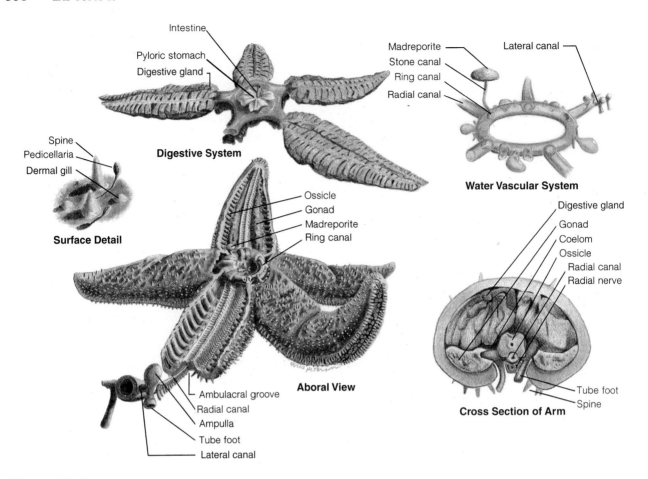

Figure 23.3 Anatomy of a sea star. Animal is dissected by layers in each arm to show different internal organ systems.

along with the tube feet, in respiratory gas exchange and excretion of nitrogenous wastes.

3 If a prepared microscope slide of pedicellaria is available, look at it with a compound microscope. Muscles under nervous system control can bend and extend the stalk and operate the pincers. Draw what you see below.

Internal Anatomy You will use the animal's pentamerous symmetry to your advantage as you study the identical internal anatomy of the arms. Each arm can be dissected in a different way to reveal the internal organs. The central disc, however, does not have this symmetry, so you will have only one opportunity to see what is there.

The body wall contains muscles and the skeletal system. It is made of thousands of calcareous plates called **ossicles** held together by flexible connective tissues. Muscles attached to the ossicles allow the body to flex and extend in a surprising number of ways.

4 *Cross-Sectional Anatomy*

Cut off one arm about halfway between the base and tip. Note the numerous ossicles that had to be cut. Examine the cross section and compare it to a cross section of the arm in figure 23.3. The major organs in each arm are in a distinct body cavity (=coelom). More than half the cavity toward the aboral surface is filled by a pair of greenish-colored **digestive glands** (=pyloric caecae). They connect to one of two stomachs located in the central disc. The remainder of the cavity may be filled with paired gonads or may be quite small if the animal was not captured during the breeding season. In the ambulacral groove area, note the supporting ossicles, spines, and tube feet.

5 Greater detail can be seen by looking at a prepared slide of a cross section of an arm with scanning power of your compound microscope. After orienting yourself, note the many ossicles that have been cut in the body wall. On the surface you may see occasional pedicellaria or dermal gills. Internally, glandular and gonadal tissue occupies the coelom. Look most carefully at the ambulacral groove area and note the hollow tube feet. Above the tube feet you should see the specialized ossicles of the water vascular system. Organized into a tubelike structure, they form the radial canal that passes down each arm from the central disc to the tips (fig. 23.3, main illustration). The tube feet are connected to this canal by a short lateral canal, allowing fluids to enter and leave as the tube feet extend and contract.

Put your microscope aside and return to your dissection.

6▶ On another arm, cut off the aboral surface from the tip to the base, being careful not to cut into the central disc. Carefully lift up and free the body wall to reveal the large digestive gland.

On a third arm, cut off the aboral surface as before, protecting the central disc area from stray cuts. In this arm, remove the digestive gland to reveal the paired gonads. These may be quite small if the animal was not captured during the breeding season.

On a fourth arm, cut off the aboral surface as before, but this time remove both the digestive gland and the gonad to reveal the floor of the coelom with its ossicles and components of the water vascular system.

Note how the pentamerous radial symmetry allowed you to reveal the internal anatomy of the arms in five simple steps. Now comes the more difficult dissection: removal of the aboral surface of the central disc. The difficulty involves three fragile connections: (1) from the stomach in the central disc to the digestive glands in the arms; (2) the small stone canal that passes from the madreporite on the aboral surface of the central disc to the rest of the water vascular system; and (3) the very short intestine that passes from the stomach in the central disc to the anus on the disc's aboral surface (fig. 23.3).

Very carefully cut from the edge of the central disc nearest the madreporite to the madreporite, keeping the handles of the scissors down and the points up against the body wall to avoid cutting internal organs. When about 3 mm from the madreporite, start cutting around it, so that it sits in the center of a 6-mm circle of isolated body wall. Leave everything in place for the moment.

Functional Anatomy

Feeding and Digestion The digestive system is basically a very short tube running from the mouth to a small anus on the aboral surface of the central disc. You will observe the digestive system in reverse order from anus to mouth. Carefully lift the aboral surface away from the central disc as you peer under from the side. If you did not cut the **intestine** in your earlier dissection, you should see it as small tube attached to the interior of the body wall where the anus voids material to the outside. Some species will have small sacs extending from the intestine that aid in absorption. Cut the intestine and remove the body wall, leaving behind the small circle surrounding the madreporite. The **pyloric** stomach is now revealed. Look along its edges to find the paired ducts leading to the still intact **digestive gland** in one arm. Although the term digestive glands conveys the idea that they contribute enzymes into the stomach, this is a bit of a misnomer. Food particles and digested materials actually flow from the stomach into the glands, where digestion is completed, absorption occurs, and excess nutrients may be stored. Beneath the pyloric stomach is a second stomach, the **cardiac stomach.** It connects to the **mouth** via a very short **esophagus.**

Sea stars are voracious carnivores. They feed on bivalve molluscs, corals, and other sedentary animals. How do they do it? If we use a clam as an example, the sea star approaches and straddles the clam with its hinge (dorsal) side downward. The sea star attaches to the clam's shells with its tube feet and pulls. When the shells open a little, say one millimeter, the starfish everts its cardiac stomach through its mouth and into the opening. The stomach enters the clam inside out so that its glandular cells press against the clam's flesh and secrete digestive enzymes. As the enzymes do their work, the clam dies and the starfish no longer needs to struggle to keep the shells apart. Bits and pieces of digested clam are drawn into the digestive glands of the pyloric stomach where digestion is completed. When a starfish feeds on coral polyps, it simply everts its stomach and presses it against the living surface of the coral, digesting away the living layer, before moving on to a fresh spot.

Reproductive System On the floor of the coelom in each arm lie two **gonads**. Depending on the breeding cycle at the time the sea star was caught, the gonads may either fill the arm or be quite small. **Gonoducts** lead from each gonad to very small **genital pores** located at the periphery of the central disc on the aboral surface between the arms. You will not see them. Though the sexes are separate, they are difficult to distinguish except by microscopic observation of the gonad contents. Fertilization is external. Fertilized eggs show the deuterostome pattern of development. You can see the developmental stages of a sea star in fig 32.1

Water Vascular System The unique water vascular system is an internal hydraulic system associated with the functioning of the tube feet in locomotion and feeding. Earlier you observed some of the components. Here you will look at it as an integrated system.

Start by looking at the arm from which you removed the digestive gland and gonads. On the floor of the coelom, find the ossicles encasing the **radial canal** that passes from the central disc to the arm's tip. Short **lateral canals** branch from it, passing to the pairs of **tube feet.** Look at the arm that you cut off in cross section. Depending on where you cut, you may see lateral canals connecting to the tube feet. If not, make a fresh cut. Find the swollen top of a tube foot, the **ampulla.** Contraction of the ampulla forces water into the tube foot extending it. Water does not flow back into the lateral canal because of a valve arrangement. When the tip of the tube foot presses against a surface, it flattens. Muscles draw the center of the contact area upward, creating a suction similar to a suction cup. With the tube foot's tip thus attached, contraction of longitudinal muscles in the tube foot along with others in the body wall pulls the animal forward. As described earlier, the animal can also use the system to capture prey.

Now trace the radial canal back into the central disc. Carefully remove the stomach to reveal where the radial canal joins the circular **ring canal** passing around the periphery of the central disc. The radial canals from all five arms join in similar way (fig 23.3). Passing from the ring canal to the madreporite is the **stone canal,** so-named because of its high calcium salt content. Water entering the madreporite fills the various canals of the water vascular system.

Nervous System The nervous system of the sea star lacks cephalization, as is characteristic of radially symmetrical animals. Though you will not be able to see them, a **nerve ring** surrounds the mouth with **radial nerve cords** passing into each arm. At the junction of the ring and radial nerves are ganglia. The only differentiated sense organ is a very small eye-spot at the tip of each arm, though there are other sensory cells located throughout the epidermis.

Clean up your area and dispose of your starfish as directed.

7▸ *Summarize Observations* Note that in your complete dissection of this animal that you did not see a heart or blood vessels, any organs for respiration, or any excretory organs. How do you think the animal performs the functions of circulation, gas exchange, and excretion?

Explain to your lab partner how the water vascular system is used in movement. Have your partner explain to you how a sea star feeds on a clam.

Phylum Chordata

The approximately 56,000 species of chordates are quite diverse, ranging from simple invertebrate animals such as sea squirts to complex animals such as birds and mammals. Chordates share ancestral traits such as bilateral symmetry and a deuterostome pattern of development with echinoderms, but it is four shared derived characteristics that make them unique. At some stage in their life, all have:

- A longitudinal stiffening rod, called a **notochord,** or its derivative, running the length of their bodies.
- A hollow nerve cord located dorsal to the notochord.
- Embryonic pharyngeal gill pouches that often develop into slits in the short muscular tube that connects the mouth to the esophagus.
- A postanal tail so that the anus is not located at the very posterior of the body as in most invertebrates.

The phylum is split into three subphyla: Urochordata (about 3,000 species of tunicates, salps, and sea squirts), Cephalochordata (23 species of lancelets), and Vertebrata (about 53,000 species of fish, amphibians, reptiles, birds, and mammals). The first two subphyla contain animals that lack a backbone although they have notochords. They are invertebrates along with all of the animals studied so far. In the vertebrates, the segmented vertebral column provides a more flexible longitudinal stiffening structure and the notochord is only present as remnants represented by the discs between the vertebrae. In this part of the lab, you will look at representatives of all three subphyla.

Subphylum Urochordata

Depending on the species, these marine invertebrates are sessile filter feeders (sea squirts) or drifting pelagic filter feeders (salps and larvaceans) that may form colonies several meters long. The distinguishing chordate characteristics are found only in the larval stages and disappear in the adults. If only adults were studied, the relationship to the chordates would not be apparent.

8▸ *Observation of Derived Chordate Traits* We will start our study of the urochordates by looking at a slide of the larval stage to see the notochord and the other derived traits characteristic of chordates.

Obtain a slide of a tunicate tadpole larva and look at it only with low to medium power. Note the general resemblance to a frog tadpole, hence the common name (**fig. 23.4c**). Larvae develop from fertilized eggs and drift and swim in the ocean for a day or so before settling down to metamorphose into the adult form. At first, they are attracted to light and use their tails to swim to the surface where they may be caught in surface currents, an obvious dispersal mechanism. This positive phototropism is reversed and the animals sink to the bottom, attaching to available substrates by its anterior suckers. During the dispersal phase, they do not feed.

Look at the tail, for it is here that you find the **notochord**. Made from cartilage-like material, it stiffens the tail so that when surrounding longitudinal muscles contract the tail bends rather than shortens. This allows swimming movements. Dorsal to the notochord you may be able to make out a shadow of the **nerve cord,** although this is difficult to see and there is no indication that it is hollow. The anus is located at the posterior end of the intestine. Clearly, the **tail extends postanally.**

The main part of the body consists of an outer covering, the tunic surrounding a large pharyngeal basket and various internal organs. The **pharyngeal slits** in the basket wall should be visible as a repeating pattern. We will talk about how the basket is used in feeding when we discuss adult anatomy.

When a larva settles on a suitable substrate, it undergoes a change in body form. The tail is reabsorbed with its notochord and nerve cord and the chordate characteristics are lost as the adult body develops. As these changes occur, the animal develops into a sedentary, filter-feeding adult.

Of the four shared derived chordate characteristics, which one did you not see in the tadpole larva?

9 ▶ Adult Functional Anatomy Adult tunicates are available in the lab for you to observe. Look at them with your naked eye and through a dissecting microscope (fig. 23.4a and b). The outer covering of the animal is the

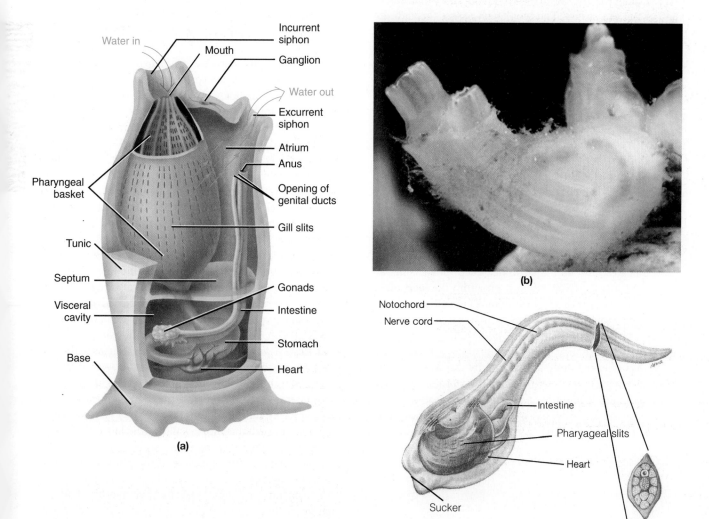

Figure 23.4 Stages in the life cycle of a tunicate: (*a*) internal anatomy of an adult; (*b*) adult sea squirt; (*c*) anatomy of a tadpole larva.

tunic and is the basis for the common name tunicate. Made from a cellulose-like material, it gives the body a semirigidity. The body shape can change when underlying muscles contract. When they contract, water often shoots out of the siphons, hence their second common name, sea squirts. An internal septum divides the body into an atrium and a visceral cavity containing the organs.

Feeding and Digestion Opposite the basal end, find the two **siphons.** One opens into the **pharynx** and the other into a surrounding chamber called the **atrium.** In some species the tunic is translucent and allows you to observe the enlarged pharynx. It has slits in its walls, forming what is called a **pharyngeal basket.** The beating of cilia on cells lining the pharynx draws water through the incurrent siphon and mouth into the pharynx. The water passes through the slits into the atrium and then out through the excurrent siphon. Suspended materials (such as algae, protists, and small larvae) cannot pass through the slits and are retained in the basket. Curtains of mucus produced by the pharynx flow downward and trap the food particles. Entrapped particles pass on to an **esophagus** at the base of the basket. The esophagus passes through the septum connecting to the stomach located in the visceral cavity. As food is digested, it passes into the intestine that loops around 180° to enter the atrium. Nondigested material is released through the **anus** located near the excurrent siphon. The water stream exiting here carries fecal material away.

Other Organ Systems Tunicates have no specialized respiratory system. Gas exchange occurs by diffusion across body surfaces that are exposed to the water stream passing through the animal. No special excretory organs are present and nitrogenous wastes are thought to be lost by diffusion. Tunicates have an open circulatory system. A tubular heart is located in the visceral cavity below the stomach.

Tunicates are hermaphroditic, each possessing an ovary and a testis lying in close proximity to each other near the loop of the digestive system. These closely placed gonads are shown as one structure in figure 23.4a. Genital ducts pass from the gonads to the region of the excurrent siphon where they open near the anus. When eggs or sperm are released, they are carried by the water stream into the environment. Fertilization is external. Development is rapid, exhibiting a deuterostome pattern, and tadpole larvae settle within a day to metamorphose into the adult body form. What is the advantage of having the anus and genital ducts open near the excurrent siphon?

Tunicates are bilaterally symmetrial. Where must the plane of the section pass to create equal halves?

Given this plane, what end of the animal is anterior and which sides are left and right?

Subphylum Cephalochordata

The subphylum Cephalochordata contains a few dozen species of small (5 cm long) fishlike species commonly known as lancelets. They live in shallow marine environments throughout the world. In some places, lancelets occur in sufficient numbers to be used as a human food source. The common American species is in the genus *Branchiostoma*, though it is often called amphioxus, the former genus name for the group.

10 ***Observation of Derived Chordate Traits*** Lancelets are best studied using a dissecting microscope with transmitted light to view specimens that have been stained and mounted on microscope slides or embedded in plastic. There is no need for high magnification.

This is a bilaterally symmetrical animal. Determine which end is anterior and which posterior (**fig. 23.5**): likewise, which side is dorsal and which ventral, as well as left and right sides. Note the general similarity to the shape of a fish, although these animals are not fish.

At the anterior end, find the **rostrum.** Beneath this overhanging hood is the oral opening surrounded by a fringe of oral cirri that have sensory and filtering functions. Running along the dorsal surface is the **dorsal fin.** The cross striations seen in the dorsal fin are storage chambers for food reserves such as glycogen, fat, and protein. The body ends in a **caudal fin.** The **anus** is located just before the caudal fin on the ventral surface, indicating these animals have a **postanal tail.** About one fourth of the body length forward from the anus is the **atriopore,** equivalent in function to the excurrent siphon of tunicates. Running forward from the atriopore on the ventral surface are two folds of tissue that function in respiratory gas exchange.

No doubt what is most visible in the lancelet is a series of about 50 diagonal structures, starting just behind the rostrum. These are the gill bars of the **pharyngeal basket** with slits between them. This observation provides the data needed to say these animals have another diagnostic characteristic of chordates, pharyngeal slits. We will discuss its role in feeding later.

If you focus up and down on the lancelet, you will see a chevron (<<<<) pattern in the body wall. These are blocks of segmented muscles. Each block is a **myomere.**

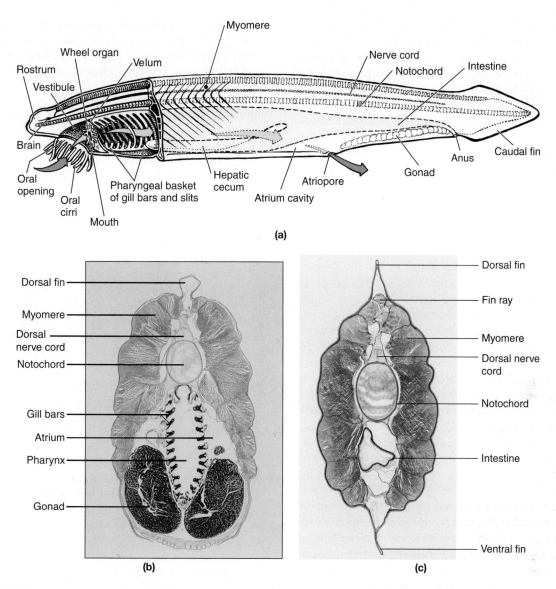

Figure 23.5 Lancelet anatomy: (*a*) internal organs; (*b*) cross section in region of pharyngeal basket; (*c*) cross section in posterior half.

Segmented muscles such as this also characterize vertebrates, chordates with segmented backbones.

Dorsal to the pharyngeal basket, find a line of darkly pigmented cells. These pigment cells shield the **nerve cord** that lies ventral to them. The anterior of the nerve cord is slightly enlarged to form a primitive brain. If you focus up and down, you will pick up the faint outline of the nerve cord and be able to trace it back to the tail.

Now focus up and down and adjust the light intensity while looking ventral to the nerve cord. You should see a rodlike structure with cross striations. It extends from the rostrum to the tail. This is the **notochord,** a third trait found in all chordates. This animal has a **nerve cord that is dorsal to the notochord,** suggesting that it exhibits all of the chordate's characteristics, but chordates have a hollow dorsal nerve cord. To determine if the cord is hollow, it must be seen in cross section.

11▶ Get a prepared slide of *Branchiostoma* and look at it with your compound microscope. If your cross section was taken near the pharyngeal basket, it will look like figure 23.5*b*. If it was taken near the tail, it will look like 23.5*c*. In either case, locate the notochord on the slide and look at the tissue mass dorsal to it. Do you see a clear area lacking cells at its center, the central canal? If so, this is your final piece of evidence. Lancelets have a **dorsal hollow nerve cord.** You now have all the evidence needed to say that lancelets are chordates.

12▶ *Functional Anatomy* We will now investigate the biology of lancelets. As you read about the various functional systems, locate the structures indicated in bold type.

Feeding and Digestion Mature lancelets burrow tail first into soft sands and gravels to a depth where their oral cirri lie just above the sand. Water is drawn through the cirri into the vestibule by the beating of ciliated cells, collectively called the **wheel organ.** The stream of water brings in bacteria, protists, and small bits of detritus, the lancelet's food. At the back of the vestibule, water enters the **mouth** and flows to the **pharyngeal basket.** Water exits the basket through the slits in its walls flowing into the **atrium,** a space surrounding the basket. From the atrium it leaves the animal through the **atriopore.**

Food particles are caught in mucus secreted by cells in the vestibule. The mucus stream does not pass through the basket slits. Instead it is directed to the **esophagus** at the posterior of the basket. From there, it passes on to the stomach and intestine where its contents are digested and nutrients absorbed. The **hepatic cecum,** located ventral to the stomach, contributes enzymes and stores nutrients much as our liver does. Undigestible material is released through the anus.

Support and Movement The notochord is the primary longitudinal stiffening structure. You probably notice its faint cross striations. The notochord is not a solid rod of cartilage. Instead, it is made of several discs stacked end to end much like a stack of coins. The discs consist of transverse groups of muscle that are innervated from the dorsal nerve cord. A tough membrane surrounds the discs organizing them into a flexible rod. When the discoidal muscles contract, the interior space of the notochord decreases, raising the pressure of fluids within the notochord and stiffening it.

When the myomeres in the body wall contract, the notochord prevents the animal from shortening. Instead the animal bends side to side as muscles on one or the other side alternatively contract and relax, allowing a primitive swimming locomotion. Because the stiffness in different regions of the notochord can be controlled, the animal does not simply bend side to side but can bend in serpentine patterns to facilitate swimming.

Circulation You will not be able to see the components of the circulatory system in your specimen. It is a closed system with arteries connected to veins through capillary beds. There is no heart. Circulation is from an aorta located ventral to the pharyngeal basket. It contracts in peristaltic waves, driving blood through vessels in the gill bars to branchial arteries dorsal to the gill bars. Blood then flows to the tissues, generally in a posterior direction with major vessels supplying the muscles and intestine. Veins collect blood from the intestine and body wall, returning it to the aorta.

Respiration The gill bars do not function in respiratory gas exchange to any extent. Carbon dioxide and oxygen diffuse across the body surface. The blood does not contain cells or respiratory pigments to facilitate gas transport.

Excretion The excretory system will not be seen. It consists of numerous clusters of protonephridia that empty via ducts into the atrium.

Nervous and Sensory Systems The dorsal nerve cord has a small swelling at its anterior end that functions as a rudimentary brain. Pigmented cells, also found at the anterior end, shield underlying photosensitive receptors, providing the animal with an ability to detect the direction of light. They do not form images and only sense light, dark, and shadow.

Reproduction Sexes are separate. The multiple **ovaries** or **testes** are located posterior to the hepatic cecum and ventral to the intestine along the walls of the atrium. Eggs or sperm are released into the atrium, and pass out through the atriopore with feeding water currents. Fertilization is external. Development follows the ancestral deuterostome plan, resulting in a free-swimming larval stage that gradually changes to the adult form.

If you reflect on the general anatomical plan of lancelets, their general similarities to a fish are obvious. Except for the lack of segmented backbone, jaws, and appendages, its general body plan is that of a fish. The similarity to a tunicate tadpole larva also should be apparent. This similarity probably means that both lancelets and cartilaginous fish are related ancestrally through some organism much like a tadpole larva. One theory suggests that lancelets and fish may have originated from mutant tadpole larvae that failed to metamorphose into adults, and became sexually mature as juveniles. They produced offspring with a new set of derived characteristics, eventually resulting in a new group of organisms.

Return your specimens to the supply area.

13 *Summarize Observations* Discuss with your lab partner, what is similar and what is different in how a tunicate and lancelet feed.

What organ systems are present in the lancelet that you did not see in the tunicate?

Provide evidence from your lab work that both lancelets and tunicates are chordates.

Subphylum Vertebrata

Among the ancestral chordates, the head was the next major character to evolve, giving rise to the craniate clade. In craniate animals the anterior end of the dorsal nerve cord is enlarged, forming a brain encased in a skull. Additional derived traits include:

- Clustering of sensory organs near the brain.
- The development of gills from the gill bars of the pharyngeal basket.
- More complex musculature.
- Higher metabolic rates.
- Development of a heart with at least two chambers.
- The development of kidneys.

About 30 species of marine hagfishes (=slime eels) are the only surviving examples of animals with the fewest derived craniate characters.

From the ancestral craniate clade, a new group of organisms evolved during the Cambrian period. Called vertebrates, they had the following derived traits:

- A segmented vertebral column which encased and protected the dorsal hollow nerve cord and has only remnants of the notochord.
- Development of an endoskeleton consisting of cartilage or bone accompanied by the subsequent appearance of skeletal elements supporting two pairs of appendages.

About 40 species of lampreys are the only surviving examples of animals with the fewest vertebrate characters.

During the Silurian Period (417 million years ago) a new clade arose from the ancestral vertebrate stock, the jawed fish. They shared all of the ancestral vertebrate traits, but had a new derived trait: jaws. Jawed vertebrates include all living vertebrates, except lampreys. The evolution of jaw bones is traceable to modifications of the gill arches, derived from the pharyngeal basket.

You will dissect a perch or trout, as a representative of the 30,000 species of bony fish in the class Actinopterygii. In later labs, you will study vertebrate anatomy in a mammal, the fetal pig.

Fish are familiar animals and you can immediately recognize the bilateral symmetry and general body regions: head, trunk, and tail. As later dissection will demonstrate, fish are craniate animals because the head contains the brain surrounded by a protective skull and has several clustered sense organs, the eyes being the most prominent. It is obvious that they are jawed animals.

14▶ *Fish External Anatomy* Obtain a fish, dissecting instruments and dissecting pan from the supply area. Preserved perch are often used for dissection but fresh fish (trout) from a fish farm can be used. They have the advantage of fresh tissues and natural color and may actually cost less than preserved perch. The anatomical drawings can be used with most species. If using a preserved fish, rinse it with tap water to remove the preservative.

> **CAUTION**
>
> **Be careful:** the dorsal fin of a perch (but not trout) contains spines that can give you a painful puncture wound.

On the head above the mouth, find the four **external nares.** These are openings into small blind sacs containing chemoreceptor cells, giving fish a sense of smell. Water enters through the anterior apertures and passes out through the posterior ones, providing a steady stream over the sensors. These are not like our nostrils which open into the pharynx. Salmon can detect certain "odors" at a concentration of one part in a billion. West Coast migratory salmon use this sense when retuning to home streams to spawn after years in the open Pacific Ocean. They navigate to the general region of the stream using electromagnetic senses, and then find the right stream by its chemical signature. Fish also have an ability to taste, with small sensory structures on their lips and in their mouths.

Fish have ears, but there are no external openings. They are located behind the eyes and consist of two parts, as do ours. One is a semicircular canal system that provides a sense of balance and acceleration–deceleration. The second part senses sounds transmitted through the water and their body tissues.

Find the **dorsal fins** and the ventral **anal fin** (fig. 23.6). These fins help the fish maintain an upright position. The thin membranes of the fins are supported by skeletal elements, the **fin rays.** Cut the tips of the dorsal fin rays off with scissors to protect your hands during the subsequent dissection. Trout have a fatty adipose fin located posterior to the dorsal fin. The **caudal fin** or tail is the main propulsive fin in fast swimming. Note the anus located on the ventral surface at the base of the tail. What trait of all chordates have you just observed?

Note the paired anterior **pectoral** and posterior **pelvic fins.** Used in a sculling motion, they allow the animal to move slowly. These fins are appendages and are supported by the bones of the appendicular skeleton deeper in the body wall. If you have eaten a whole fish, you are quite aware of the main axial skeleton consisting of vertebrae, ribs, and the skull. What derived characteristic of vertebrates would you have just observed?

15▶ If a demonstration specimen of a fish's skeleton is available in the lab, take a look at it to observe the axial skeleton and the appendicular skeleton supporting the appendages. Note the segmented vertebral column. The bones of the vertebrae surround and protect the nerve cord which is located dorsal to the centerline of the column. If separate vertebrae are available, look at them and identify the dorsal space through which the nerve cord passes. What derived chordate characteristic have you just observed?

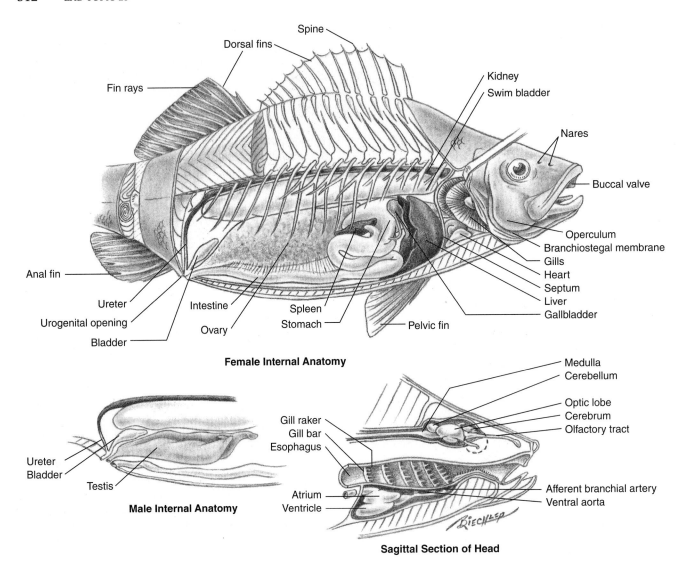

Figure 23.6 Internal anatomy of a perch.

16 The body is covered with scales, larger in the perch than in the trout. This is considered to be a dermal exoskeleton. Remove a few scales and mount them in a drop of water on a slide. Add a coverslip and observe. Scales provide fish with a thin, flexible armor. The rings that you see in the scales correspond to growth spurts and can be correlated with age, but they are not annual growth rings. Lipsticks and nail polishes containing "pearl essence" are made from silvery materials found in fish scales, primarily from herring. Underlying the scales are many mucus glands whose secretions protect the fish from microbial infections as the mucus is constantly added and washed off by normal motions. Sketch a scale below.

On each side of the body, find the **lateral lines,** visible as a row of small openings connecting to an underlying tubular canal. The canal contains sensory cells, called neuromast cells, that are sensitive to pressure changes. They respond to vibrations and water currents and aid the fish in swimming, locating prey, or avoiding predators. Some fish have cells that generate or are sensitive to changes in electrical fields, giving fish a sixth sense of their environment. Other sensory cells are found in the skin and respond to temperature differences as well as other stimuli.

Pry open the jaws and look beyond the mouth into the pharynx. Note how the gill arches form a pharyngeal basket that allows water entering the mouth to exit the pharynx. What derived characteristic of chordates have you just observed?

Fish Functional Anatomy

Respiratory System Examine the internal edge of the upper jaw to see the **buccal valve,** a small flap of tissue that seals the mouth opening like a gasket when the jaws are closed. On each side of the head are the **opercula,** large plates that protect the gills and function as a pump. Along the ventral, posterior edges of the opercula, find the **branchiostegal membranes** that act as seals.

Lift up the operculum and reach in with scissors and cut one of the gill bars loose with its gill filaments. Remove it and place it in a small dish with enough water to float the gills. Look at this with your dissecting microscope. The feathery gills have a tremendous surface area. Why is this important?

Fish pump water over their gills by the following mechanism. As the opercula are raised with the mouth open, the branchiostegal membranes close off the posterior gill openings so that water flows through the mouth into the pharynx. The mouth is then closed and sealed by the buccal valve. The opercula are then lowered, forcing water out the gill openings. Some fish with high metabolic rates, such as tuna, must constantly swim with their mouths open to ventilate the gills adequately.

17 ***Reveal Organs*** To expose the internal organs, lay the fish on its left side and insert the tips of your scissors just anterior to the origin of the anal fin. Cut forward along the midventral line to the pectoral fins. Make a second cut parallel to the first, but above the lateral line on the same side. Now connect the two cuts with a third at the anterior edge. Lay the right body wall back, free it from underlying tissues, and remove it.

Large fat bodies may fill the coelom in farm-raised fish. Fish do not store fat in their muscles as do mammals. Carefully remove the fat bodies to reveal the organs.

Note the various organs suspended in the coelom by **mesenteries,** thin, transparent outgrowths from the shiny **peritoneum,** a tissue that lines the body cavity. Dorsal to the other organs is the long balloon-like **swim bladder** (fig. 23.6). The air content of the bladder is actively regulated to maintain the buoyancy of the fish, allowing it to remain motionless in the water without settling to the bottom or floating to the top. In some fish, the air bladder is connected by a short duct to the pharynx. Air can be added to the bladder by gulping at the surface or released by "burping." More derived fish, such as perch and trout, lack the duct. Clustered in the wall of the swim bladder are cells forming a gas gland that can add or absorb gases from the blood or the bladder.

Feeding and Digestion Remove the right operculum to expose the gills attached to the **gill arches** of the pharyngeal wall. Insert the tips of your scissors into the mouth and cut through the angle of the jaw back through the gills to expose the inside of the pharynx. Remove the structures you have cut through.

Examine the inside of the pharynx and visualize how water entering the mouth flows through the slits in the pharyngeal wall, over the gills on the gill bars, and out through the external gill openings. On the inside surface of the gill bars, note the **gill rakers,** which act as a crude sieve in this species and direct ingested food toward the opening of the esophagus. In species that feed on suspended algae and detritus, the gill rakers may be highly developed, resembling a fine-tooth comb. Such fine openings are very effective in removing small particles from water passing through the pharynx. In predatory species, the gill rakers are widely spaced.

Pass a blunt probe through the mouth and down the **esophagus** into the **stomach** located dorsal to the liver. At the juncture of the stomach and the small intestine, note the many, and often large, **pyloric caecae.** These finger-like structures are outpocketings of the intestine and are thought to serve as temporary storage areas and possibly in absorption. Observe how the intestine leaves the stomach toward the anterior, producing a pouch. This is typical of predatory fish. This arrangement allows the stomach to expand to accommodate large prey and to hold such prey for long periods of digestion. Trace the **intestine** from the stomach to the **anus.** The **spleen** (not part of the digestive system) is the triangular-shaped organ located in the region between the stomach and the small intestine. Lift the lobes of the **liver** to find the globular, dark green **gallbladder.** What is its function?_____

Cut open the stomach to determine what your fish had eaten. List the contents in the following space.

Circulation Fish have a closed circulatory system. Find the **two-chambered heart** anterior to the **septum** in front of the liver, and ventral to the pharynx (fig. 23.7). The anterior chamber is the muscular **ventricle,** which pumps blood to the gills via the **ventral aorta** and **afferent branchial arteries.** Blood rich in oxygen and low in carbon dioxide leaves the gills via the **efferent branchial**

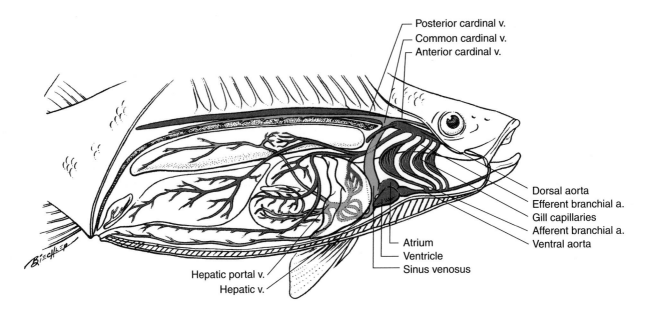

Figure 23.7 Circulatory system of perch. Major arteries are in red and veins are in blue.

arteries. Find the **dorsal aorta** dorsal to the gills. Blood collected from the gills flows via this major vessel to the rest of the body.

Blood is collected from the tissues via the **cardinal vein,** which runs just ventral to the vertebrae and from the digestive system via the **hepatic portal vein.** The latter flows into capillaries in the liver and is collected by the **hepatic vein.** The cardinal vein and hepatic vein flow into the **sinus venosus,** which returns blood to the **atrium** of the heart.

Excretion The excretory system consists of the paired **kidneys,** which are located dorsal to the swim bladder under the peritoneum covering the dorsal wall of the coelom (fig. 23.6). Carefully move the swim bladder and find the dark brown organs firmly attached to the dorsal wall of the body cavity and extending almost its entire length. Along the median border of the kidneys are the **urinary ducts.** As they pass posteriorly, they unite into a single duct, which passes ventrally to the urinary opening. A urinary **bladder** branches from the duct's distal end.

Reproduction System Perch have distinct males and females. Fish-farm-raised trout are usually all females. Females have a single large **ovary** lying between the intestine and swim bladder (fig. 23.6). If your fish is a female and was reproductively "ripe" when collected, the ovary will be filled with yellowish eggs. Eggs leave the posterior of the ovary via an oviduct that leads to a urogenital opening posterior to the anus.

If your fish is a male, paired **testes** will be visible caudal to the stomach. A **vas deferens** passes caudally in a longitudinal fold in each testis. These fuse to form a single duct that passes to a genital pore posterior to the anus.

18 *Nervous System* The ancestral craniate traits of vertebrates will now be investigated. Use a scalpel or razor blade to shave the skull bones off the dorsal surface of the head to reveal the brain. Work carefully and avoid cutting brain tissue as you approach the cranial cavity. Try to expose the brain from the first vertebra to the anterior tip. Note that you are being slowed down by a key feature of craniates: the cranium!

Note how the brain has specialized regions along its length (fig 23.6). Neurons entering from the spinal cord synapse with brain neurons for the first time in the single **medulla oblongata** located on the animal's midline at the posterior end of the brain. It appears roughly triangular as it expands anteriorward. Overlying it is the **cerebellum** that may have been damaged in the dissection. It has an important role in balance. Just anterior to the medulla are the large, paired roughly spherical **optic lobes.** Their size speaks to the importance of visual information in the lives of fish. The paired, elongated cerebral (cerebrum) lobes are anterior to the optic lobes. They seem to expand anteriorward into the large paired **olfactory lobes** and tracts, indicating the importance of chemical sensing to the animal. If we were to examine the brains of other vertebrates, we would see that the size of the olfactory processing area decreases while the size of the other areas increases.

To see the spinal cord in its protective vertebral casing, cut your fish in half, crossways. Look closely at the vertebral column and find the nerve cord. If you were to make a prepared microscope slide of the column cross section, you would see that the nerve cord is hollow. Does the location of the nerve cord and the fact that it is hollow have any phylogenetic significance? Explain.

You are now finished with your fish dissection. Clean your instruments and dispose of the carcass as directed.

19 *Summarize Observations* Discuss with your lab partner the organ systems seen in your dissection. Explain how a fish feeds and ventilates its gills. What organs did you see that were not found in tunicates or lancelets?

Lab Summary

20 On a separate sheet of paper, answer the following questions as assigned by your instructor.

1. What ancestral characteristics do echinoderms and chordates share?
2. What derived characteristics set echinoderms apart from chordates?
3. Describe the water vascular system of a sea star.
4. Briefly describe how a sea star performs the physiological functions of: (1) digestion; (2) gas exchange; (3) circulation; (4) excretion; and (5) body support.
5. Why are tunicates considered to be chordates?
6. Why is *Branchiostoma* considered an important evolutionary link among the chordates?
7. Have you observed any evidence that would allow you to refute the hypothesis that all chordates have a notochord, postanal tail, gill slits, and a dorsal nerve cord? Explain.
8. Using your dissected fish as an example vertebrate, list all of the derived characteristics you observed.
9. Explain how a fish ventilates its gills.
10. Describe how a large predatory fish swallows its prey and trace the pathway the food follows from mouth to anus.
11. Trace a drop of blood as it flows from a fish's ventricle to the intestinal wall and back to the ventricle.

You may want to try the critical thinking questions that apply some of the knowledge you gained in doing this lab.

INTERNET SOURCES

Use the WWW to locate information about current research being done on cephalochordates. Use your browser to connect to www.google.com. Type in the search terms, *Branchiostoma* and research. Scan several of the sites that are listed and write a few paragraphs summarizing what you think are the most interesting sites. List the URLs at the end of your summary.

LEARNING BIOLOGY BY WRITING

All general biology textbooks indicate that chordates have a set of shared derived characteristics: (1) a notochord; (2) a dorsal hollow nerve cord; (3) pharyngeal slits at some stage of the life cycle; and (4) a postanal tail. In a short essay, describe the evidence you have from your lab work that supports or refutes this generalization.

CRITICAL THINKING QUESTIONS

1. What ancestral traits suggest an evolutionary linkage between tunicates and lancelets?
2. Why are animals as different as a sea squirt and an eagle classified into the same phylum?
3. The crown of thorns starfish is causing problems on the Australian Great Barrier Reef by feeding on corals. How can a starfish feed on an encrusting animal?
4. Fish that are caught at depths of over 100 feet and then are rapidly brought to surface often look bloated and rapidly die even if kept in water. Why?
5. You have studied ten animal phyla in the last four labs. Describe the trends you observed as you moved from sponges to vertebrates.

Lab Topic 24

Investigating Plant Cells and Tissues

SUPPLIES

Resource guide available at
http://www.mhhe.com/labcentral

Equipment
Compound microscopes
Dissecting microscopes

Materials
African violet, geranium, or sycamore leaves
Apples, unpolished
Carrot
Celery
Natural fibers (sisal rope, linen, or burlap)
Pear
Peppers, green and red
Pine twigs
Potato
Red onion
Specimen house plants with waxy, unpolished leaves
Young sunflower plants
Zebrina leaves
Razor blades and glass plates as cutting surfaces
Prepared slides
    Cross and longitudinal sections of *Cucurbita* stem
    Endosperm of persimmon
    Pine wood macerate or use fresh material
    Pumpkin stem macerate (Triarch)

Solutions
Blue food coloring
IKI Stain
Phloroglucinol stain
Methylene blue stain
Water in dropper bottles
45% acetic acid

STUDENT PRELAB PREPARATION

Before doing this lab, read the Background material and sections of the lab topic that have been assigned by your instructor.

Use your textbook to review the definitions of the following terms:

cell wall	plastid
collenchyma	phloem
companion cell	sclerenchyma
dermal tissue	sieve tube member
fiber	tracheid
ground tissue	vacuole
meristem	vascular tissue
parenchyma	vessel element
plasmodesmata	xylem

You should be able to describe in your own words the following concepts:

Primary and secondary cell walls

Plant tissue systems

Primary and secondary growth

After finishing the prelab review, write any questions you have about terms, concepts, or techniques in the margins of this lab topic. The activities should help you answer these questions, or you can ask your instructor during the lab.

OBJECTIVES

1. To study the unique structures found in plant cells
2. To learn the characteristics of the basic types of cells and tissues found in plants.

BACKGROUND

Approximately 500 million years ago, some ancestral filamentous green algae made the transition to land. The plants we see today have descended from that ancestral stock. Plants dominate most terrestrial ecosystems and have influenced the evolution of bacteria, fungi, and animals that depend on them either directly or indirectly for food and shelter. Plants also influenced the physical environment, contributing to the oxygen and carbon dioxide geochemical cycles of the global ecosystem.

Plant cells exhibit the eukaryotic cell plan and, in addition, contain unique structures. These include cellulosic cell walls, cytoplasmic connections between cells, large central vacuoles, and plastids (fig. 24.1). In this lab, you will observe these unique structures in a variety of plant cells.

Collectively, the cytoplasm and nucleus of a plant cell are called the **protoplast.** It produces a cell wall and together these constitute what we call a plant cell. During the life of a plant, protoplasts in certain tissues often make cell walls and then die, leaving the cell walls to function either as supporting members or as conduits for the transport of water and minerals. Thus, when looking at mature plant material, you will see that some cells may have living protoplasts with cell walls and others will be only a cell wall left by a protoplast that has died.

Plant cell walls are composed of cellulose microfibrils and other polysaccharides. During cytokinesis, protoplasts produce a **middle lamella** and **primary cell wall.** Subsequently some cells, especially those that will be structural support cells, produce a **secondary cell wall** on the inside of the primary cell wall. It is much thicker and is impregnated with **lignin,** a phenolic polymer that stiffens the wall (fig. 24.2). After cellulose, lignin is the second most common molecule found in vascular plants.

Many protoplasts in plant bodies are not completely separate, although it may appear so through a light micro-

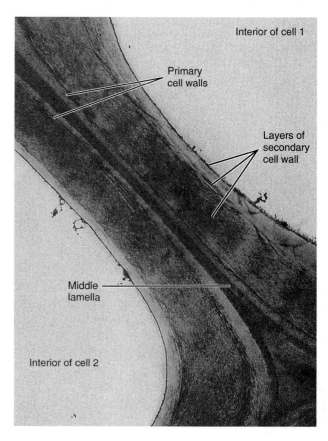

Figure 24.2 Transmission electron micrograph of cell wall showing middle lamella, primary wall, and secondary cell wall.

scope. Many, maybe even most, of the living cells in a plant are interconnected to adjacent cells by small bridges of cytoplasm. Called **plasmodesmata,** these bridges pass through openings in the cell walls and allow direct exchange of materials between cells (fig. 24.3).

Plant cells contain unique organelles called the **plastids.** Involved in photosynthesis and storage, these organelles have a variety of shapes and sizes. They are described in terms of the types of pigments they contain. Chloroplasts contain chlorophyll and are the organelles in which photosynthesis occurs. Chromoplasts contain yellow and orange-red pigments (carotenoids and xanthophylls) and are responsible for colors in many flowers and fruits. Leucoplasts lack pigments and are organelles that store starches, proteins, and sometimes lipids.

Mature plant cells often contain a large **vacuole** (fig. 24.1). Besides storing water and minerals, the vacuole plays an important role in the growth of plant cells.

The vegetative body of vascular plants consists of three regions: roots, stems, and leaves. Although these regions differ substantially in their form and function, all contain similar basic cell types and tissues. The differences are in the relative proportions of different kinds of cells and how they are organized into functional groupings. Four tissue systems are recognized in vascular plants: the meristematic systems, which are responsible for

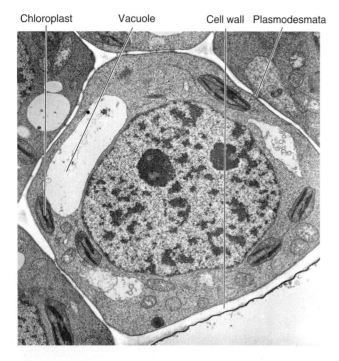

Figure 24.1 Transmission electron micrograph of a sunflower (*Helianthus*) leaf primordial cell showing the cellular structures that make plant cells unique: cell walls, plasmodesmata, chloroplasts, and a large vacuole.

INVESTIGATING PLANT CELLS AND TISSUES   319

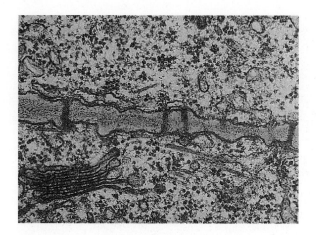

Figure 24.3 Transmission electron micrograph showing plasmodesmata passing through middle lamella and primary cell wall of adjacent cells. These passageways allow materials to pass readily from cell to cell.

growth; the dermal system, which comprises the covering of the plant body; the ground system, which represents the bulk of the body of many plants; and the vascular system, which conducts materials through the plant.

The **meristematic tissues** are the active regions of cell division in plants (fig. 24.4). **Apical meristems** are located at the tips of roots and stems where cell divisions lengthen the plant body. This is called **primary growth.** All other cell types arise by differentiation from the daughter cells of these divisions. In woody plants, additional meristematic tissues are responsible for what is termed **secondary growth,** growth in thickness or girth, of stems and roots.

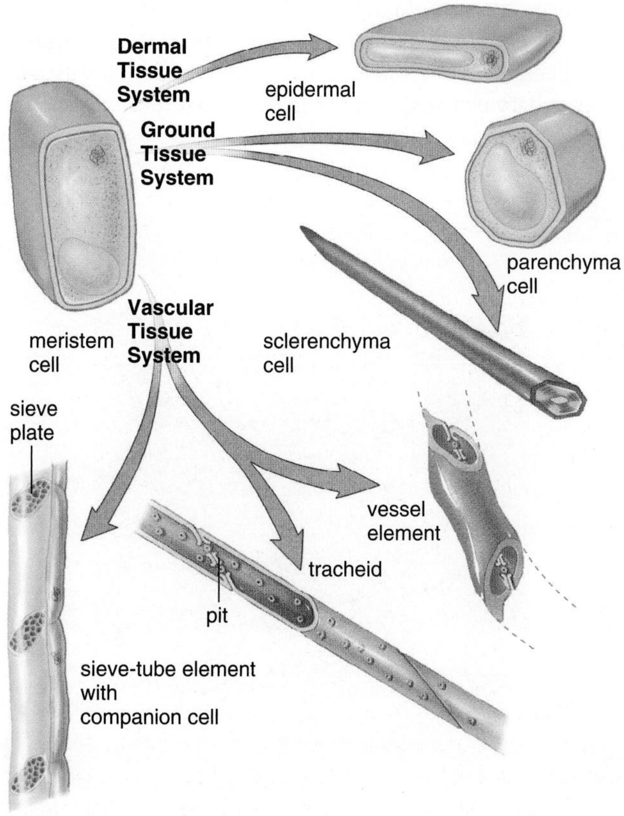

Some cells derived from the apical meristem develop into the cell types of **dermal tissue** that cover the surfaces of young root and shoot systems. These cells protect against invasion by microorganisms, prevent desiccation, allow for moisture and mineral absorption, allow gas exchange, and may produce structures or secretions that discourage herbivores.

Most of the living cells in plants are in **ground tissues.** Far from being filler in the plant body, ground tissues are among the most metabolically active cells. They are where photosynthesis occurs and where starches and other materials are synthesized and stored, and they are also active in producing reproductive structures.

The **vascular tissue system** allows materials to be transported from one part of the plant body to another: water from the roots to the above-ground parts, and products from photosynthetic cells to non-photosynthetic cells elsewhere in the body. In addition, vascular tissues often provide support to the shoot, allowing it to reach considerable height.

The specific types of cells found in each type of tissue are described in **table 24.1.**

LAB INSTRUCTIONS

In this lab, you will study the basic types of cells found in plants. Where possible, you will find the cell type in live materials. After learning to recognize many of the basic cell types, you will study how they are organized into the tissues of the living plant.

Investigating Unique Structures in Plant Cells

In this section, you will view slides from a variety of plant materials to observe those structures that are unique to plant cells: cell walls, plasmodesmata, vacuoles, and plastids.

Cell Walls

All plant cells produce a primary cell wall following mitosis and cytokinesis. The newly formed protoplasts first produce a **middle lamella** composed of gel-like pectins, a polysaccharide material that holds plant cells together. Cellulose is then deposited between the middle lamella and the plasma membranes of the protoplasts to form the **primary cell wall.** Enzymes (cellulose synthase) for producing this cellulose are found in the plasma membrane

Figure 24.4 Three basic types of primary tissues differentiate from cells produced by apical meristems.

Table 24.1 Tissue Systems, Tissues and Cells Found in Plants

Tissue Systems	Description of Tissues and Cell Types
Meristems	Cells that continue to divide and grow, forming tissues described below; apical meristems at root and stem tips allow growth in length (called primary growth); lateral meristems (also called cambiums—vascular and cork) found in roots and stems allow growth in girth (called secondary growth).
Dermal tissue	Includes living epidermal cells, trichomes, root hairs, and guard cells; protects internal cells and functions in water absorption and transpiration; arises from apical meristem; during secondary growth, epidermis is replaced by periderm arising from cork cambium.
Ground tissue	**Parenchyma.** Component of ground tissue; most common living cells in plant body and found in leaf mesophyll, herbaceous stems, fruits, and rays of woody plants; functions in storage, secretion, and wound healing; arises from apical meristem and retains capability to divide (typically no secondary cell wall) and differentiate into other tissue types. **Collenchyma.** Component of ground tissue; living cells forming distinct strands beneath epidermis in stems and petioles that support the primary plant body by producing unevenly thickened primary cell walls; arises from apical meristem. **Sclerenchyma.** Component of ground tissue; includes fibers and sclerids both of which are dead when they carry out their function in the plant; thick secondary cell walls with lignin provide support and mechanical protection; found throughout the plant, often in association with xylem and phloem; arises from apical meristem or vascular cambium.
Vascular tissue	**Xylem.** Component of vascular tissue; includes tracheids and vessels; functions in transport of water and minerals from roots to shoots; primary xylem arises from apical meristem and secondary xylem from vascular cambium; protoplast produces thick secondary cell wall and then dies; lumen of dead cells provides pathway for water transport. Parenchyma and sclerenchyma cells are also found in xylem. **Phloem.** Component of vascular tissue; includes living sieve tube members and companion cells; functions in distributing photosynthetic products and storage compounds throughout plant; primary phloem arises from apical meristem and secondary phloem from vascular cambium. Parenchyma and sclerenchyma cells are also found in phloem.

of the protoplasts. They use thousands of activated glucose molecules from the cytoplasm to form the cellulose polymer that the enzyme complex extrudes on the cell surface as a cellulose microfibril. The microfibrils are laid down in patterns (**fig. 24.5**) that control the final shape of the cell.

Primary cell walls are dynamic structures. They are extensible allowing cellular expansion, they undergo reversible changes in thickness, and they usually are penetrated by plasmodesmata. Many plant cells produce only a primary cell wall and these cells have a living protoplast at maturity. Other cells, which usually lack a protoplast at maturity, add a secondary cell wall before the protoplast dies.

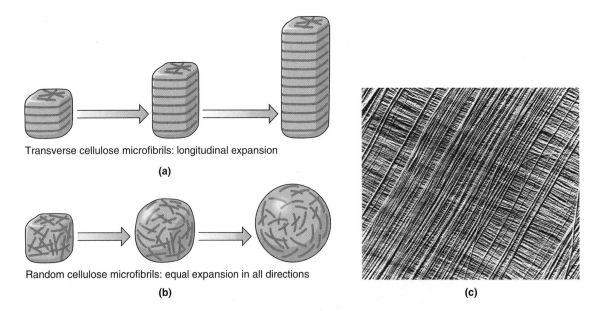

Figure 24.5 Patterns of cellulose deposition in primary cell wall determine how the cell can expand. (a) When cellulose is laid down transversely to long axis of cell, the cell can elongate to form a tubular cell. (b) When laid down randomly, the cell will expand equally in all directions until it contacts other cells that may alter its shape. (c) Cellulose in secondary cell walls is laid down in layers oriented at right angles to each other and is impregnated with lignin. This forms a rigid casing that prevents the cell from expanding.

1 ▶ ***Plasmodesmata and Primary Cell Walls*** Obtain a prepared slide of endosperm from the seed of a persimmon (*Diospyros*). This tissue supplies nutrients to the developing embryo in the seed.

As you look at this slide first with the medium power and then the high power objective, note the exceptionally thick primary cell walls. If you look at the wall midway between two adjacent cells, you should be able to see a faint line parallel to the surfaces of the cells. This is the **middle lamella,** a gel-like polysaccharide that holds the cells together. As you study the cell walls, you should also be able to see fine lines that seem to radiate from each cell. These are canals through which numerous plasmodesmata connect adjacent cells. Compare what you see to **figure 24.6**.

Name what you think are the advantages to having connections between the cytoplasms of adjacent cells.

Secondary Cell Walls In **secondary cell walls,** cellulose is laid down in ordered patterns with one layer oriented almost at right angles to the layer beneath it. The layers are impregnated with **lignin.** Secondary cell walls are stiff compared to unlignified primary cell walls. Secondary cell walls do not allow cellular expansion, they do not undergo reversible changes in thickness, and they lack plasmodesmata. Protoplasts make the cell walls and die leaving the walls behind to function as supporting members or conduits for water movement. Secondary cell walls are characteristic of many cells in xylem and sclerenchyma tissues and in some cells in phloem tissues. Other tissues commonly have cells with only primary cell walls.

Plant anatomists use the stain phloroglucinol to differentiate between primary and secondary cell walls. Phloroglucinol stains lignin bright red. Since lignin is found only in secondary walls, the stain is a chemical test to distinguish secondary from primary cell walls.

2 ▶ The flesh of a pear contains some cells with primary cell walls only and other cells with secondary walls. To show this, put a drop of phloroglucinol stain (*caution:* this stain is made with hydrochloric acid) on a slide. Use a razor blade to make a very thin slice of the pear's flesh and add it to the stain. After a few minutes add a coverslip and observe with your compound microscope.

Most cells on the slide will not be stained. These are parenchyma cells with only thin primary cell walls. They store the carbohydrates that make the fruit sweet. Scan around the slide until you find groups of cells that are red. These cells, sclereids, have thick secondary cell walls containing lignin which reacts with the stain. You may have noticed hard, gritty lumps whenever you have eaten a pear. The grit is sclereids with their heavy secondary cell walls. Similar cells are found in seed coats and in the walls of nuts, providing some protection from herbivores.

Use high magnification to study some of the sclereids. Note the small cavity, called the **lumen,** where the protoplast lived. Observe the radiating lighter areas in the thick cell walls. These pit canals were once filled with plasmodesmata.

Vacuoles

Much of a living plant cell's volume is taken up by one or more large membrane-bound vacuoles (fig. 24.1). The membrane surrounding the vacuole, the **tonoplast,** regulates what passes into or out of the vacuole. Vacuoles store water containing dissolved salts. They function as organelles of growth and as hydrostatic support structures. Following mitosis, many plant cells take water into their vacuoles as the primary cell wall is laid down. This stretches the primary cell wall, increasing the size of the cell. Many structures in herbaceous plants are supported by hydrostatic pressure. When the vacuole takes on water, it presses against the surrounding cell wall, often just a flexible primary cell wall. The composite structure is stiffened, much the same as a tire stiffens when air pressure is increased in the inner tube.

3 ▶ To see large vacuoles, cut a red onion in fourths and remove one of the fleshy layers (leaves). Snap the layer backward with a twisting motion. Use your fingernail or forceps to peel a cellophane-thin layer from the ragged edge and mount it in a drop of water on a slide. Add a coverslip and observe. Can you see a large central vacuole? _____ What color is it? _____ The water-soluble red pigments are called anthocyanins.

Figure 24.6 Endosperm cells from persimmon have very thick primary cell walls that are penetrated by plasmodesmata. The middle lamella is also visible as a faint line about halfway between adjacent cells.

Where is the cytoplasm of the onion cell? Draw a few onion cells below.

Plastids

4) Chloroplasts are the sites of photosynthesis in plants. Non-photosynthetic cells lack chloroplasts, but may have other types of plastids. In the lab there are green and red peppers. Your task is to make very thin sections of these specimens, using a razor blade, and to determine whether the entire cell is colored or whether the color is localized in plastids inside the cells. This will take some practice and you should attempt to get sections that are as thin as cellophane. Only in thin sections will you be able to see the structures. Make wet-mount slides of the sections as you get them. After you have finished viewing a slide, trade with someone else who has a slide of another tissue so that prep time is minimized.

Describe what evidence you have that the color is localized in structures in the cells. Include sketches of a few of the cells you observed. Be sure to label the drawings.

5) Leucoplasts are colorless plastids that often store starch. Potatoes contain substantial amounts of starch. Cut a very thin slice of a potato and make a wet-mount slide. Look at it to see if you can see leucoplasts in the cells. The stain I_2KI stains starch purple. Add a drop of the stain next to your coverslip and wick it under. Look at your specimen again. What color are the leucoplasts? What do they contain?

6) Summarize your observations by making a list of the unique structures found only in plant cells.

Investigating Plant Tissues

Meristematic Tissue

Located at the tips of roots, stems, and branches, as well as below the surface of the stems and roots of woody plants, meristematic tissues produce all other cells of the plant body. There are two types of meristems. **Apical meristems** are found at the tips of roots and stems where they add cells to a plant's length during primary growth. **Lateral meristems,** found only in the main shafts of stems and roots of woody eudicot plants, add the cells that increase the girth during secondary growth.

We will not look at meristematic tissues in this lab. The cells are small and rather uniform in appearance. In the next lab, you will observe apical and lateral meristems in roots and stems.

As cells produced by the apical meristems mature, they become specialized in tissues that perform particular functions in the plant. This process is called **differentiation.** The cells of the meristem produce three types of primary meristem cells: protoderm, procambium, and ground meristem. From these transitional meristems, the tissues differentiate to form the plant body.

Although the morphology of root and shoot systems vary considerably among angiosperms, all contain the same tissues and cell types. If you compared a root to a leaf or a monocot to a eudicot, you would not find different tissues. Instead, you would discover that there are different arrangements and proportions of the basic tissues. These tissues are the dermal tissue system, the ground tissue system, and the vascular tissue system. In the next sections of the lab, you will study the cells characteristic of these systems.

Dermal Tissue System

The **epidermis,** usually one cell thick, originates from the protoderm produced by the apical meristem. It covers the plant body and mechanically protects the tissues beneath it, reduces desiccation by secreting waxes and oils, allows for gas exchange, and may be involved in water absorption. As you will see, some epidermal cells are specialized for particular functions.

Woody eudicots, undergo secondary growth that destroys the primary epidermis covering the stems and roots. Secondary growth adds new cells beneath the epidermis. Stop a moment and consider the geometry of this situation. What would you hypothesize would happen if cells were added beneath an epidermis that is already a tight covering? It seems logical to expect that the epidermis would come under tension and eventually split, exposing other tissues in the stem or root to attack by microorganisms and to the drying effects of air. Plants have evolved a mechanism that prevents this from happening. As the plant expands, a secondary dermal system, called the **periderm,** develops. The outer layers of bark are examples of this system. When you are out and about campus, look carefully at the young twigs on a tree. You will see how the youngest twigs have a smooth epidermis, but the older ones have rough textured bark.

When you study roots and stems in the next lab topic, you will learn how the primary epidermis is replaced by the periderm. In this lab, we will concentrate on the primary epidermis and the types of cells found in it.

Epidermal Cells Basic epidermal cells tend to be flattened cells and often have cell walls that form interlocking joints with adjacent cells. Lacking chloroplasts, their function is protection.

7▶ In the lab are several plant specimens (fruits, stems, leaves) for you to use as study materials. Your challenge is to remove the surface layers from these plant specimens and make slides to see the structure of the epidermal cells. To illustrate basic epidermal cells, you should work only with materials that have a smooth (not hairy) outer surface. Before you make any slides of the materials, be sure to take a look at the specimens through a dissecting microscope so that you see the big picture.

To strip the epidermis from stems or leaves, fold the material over and then tear in such a way that the epidermis on the inside of the curvature is lifted off the stem or leaf. The piece of stem or leaf with the adhering epidermis should immediately be placed in a drop of water and the thin epidermis cut off. Add a coverslip and observe.

If you coordinate with others in the lab, each can prepare a slide from a different kind of plant or a different region and then slides can be exchanged. The object is to develop a comparative knowledge of basic epidermal cell structure.

As you study the specimens, answer these questions: Do these cells have chloroplasts? What is their shape? Are there spaces between the cells or do they butt against all adjacent cells? Do you see any unusual structures in the cells?

Study at least three different examples of basic epidermal cell types and fill in the comparative data required in **table 24.2**.

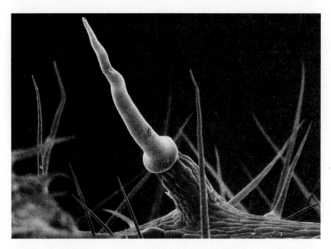

Figure 24.7 Trichomes on the surface of stinging nettle leaves (*Urtica dioica*) have pointed tips hardened with silica. When the tips penetrate the skin of an animal, they break off and a toxin flows from the cell into the skin causing a painful reaction that discourages herbivory.

How does the shape of epidermal cells reflect their function?

All epidermal cells do not have the flattened shapes you saw in the previous specimens. You will now look at some specialized epidermal cells.

Guard Cells Specialized epidermal cells called **guard cells** are found on leaves and stems where they surround a slit-like opening called a **stoma** (pl. **stomata**). Guard cells change shape to regulate water loss through stoma from the moist tissues below the epidermis. Carbon dioxide used in photosynthesis, and oxygen that is produced, move into or out of the leaf through the stomata.

8▶ Prepare a leaf peel of a *Zebrina* leaf from the lower epidermis where guard cells are more common. After mounting the peel in a drop of water, use your microscope to look at it. What organelle is found in guard cells that was not

Table 24.2	Epidermal Cell Characteristics			
Specimen	#1 =	#2 =	#3 =	#4 =
Chloroplast present?				
Shape?				
Edge spaces?				
Other				

found in other epidermal cells? _____
Draw a pair of guard cells below and label the stoma.

Trichomes These hairlike specialized epidermal cells give a fuzzy appearance to stems, leaves, and flowers. Their functions are varied. They can protect the plant from herbivores. Some trichomes, such as those of the North American stinging nettle (*Urtica dioica*), have sharp hypodermic points (**fig. 24.7**) which can penetrate skin. The cells produce a toxin that both stings and irritates so that a "smart" animal soon learns to browse elsewhere. Trichomes also reduce airflow directly over the surface and can reduce drying. Their reflective surfaces also can keep the leaf from overheating.

9▸ Any leaf or stem with a fuzzy surface has trichomes. African violet, geranium, or sycamore leaves are common examples. Your job is to figure out a way to make a slide so that you can see trichomes through your compound microscope. Before you make any slides of the materials, be sure to take a look at the specimens through a dissecting microscope so that you get the big picture about what you are studying.

Make a slide and draw some trichomes below. If there is more than one species with trichomes, be sure to look at several examples. Be sure to include labels identifying the source of the trichomes.

10▸ In the lab are some apples or some greenhouse plants. The surfaces are not bright and shiny and seem to be covered by a dull whitish layer. Take a cloth and rub the surface of an apple or leaf. Why are they now shiny? How does this observation relate to the function of the epidermis?

Root Hairs Epidermal cells near the tip of roots can produce root hairs. These tubular cytoplasmic extensions greatly increase the surface area for water and mineral absorption. Older root hairs are abraded off by the soil but new ones keep forming as the root's apical meristem adds cells that differentiate into protoderm and eventually root hairs. Thus the absorptive zone of the root keeps advancing as the root elongates. You will study root hairs in lab topic 25.

11▸ Summarize your observations by making a list of the type of cells found in epidermal tissue systems.

Ground Tissue System

Arising from the cell divisions in the ground meristem of the primary plant body, ground tissues are the most common types of cells found in plants. They function in basic metabolism, storage, and support in plants. Three basic cell types are found: parenchyma, collenchyma, and sclerenchyma (**fig. 24.8**).

Parenchyma These are the most common cells in the primary plant body. Parenchyma cells may function in photosynthesis, storage, secretion, and a variety of other tasks in leaves, stems, roots, flowers, and fruits. They retain their protoplasts at maturity and are morphologically characterized by large vacuoles and thin primary cell walls, although they sometimes will develop a secondary cell wall. Although you cannot see them with a compound microscope, plasmodesmata commonly interconnect the cells. Parenchyma cells can be almost any size or shape but are most often 14-sided polygons. Parenchyma tissue is usually loosely organized with spaces between adjacent cells (fig. 24.8a). Parenchyma cells retain the ability to divide even at maturity and can differentiate into other cell types.

When you looked at cells containing plastids at the beginning of this lab, you were looking at parenchyma cells found in leaves, roots, and fruits. Did the parenchyma cells stain with phloroglucinol? _____ What does this tell you about the composition of their cell walls?

12▸ Here you will look at parenchyma cells found in the stem of a non-woody herbaceous plant. Young sunflower (*Helianthus*) plants were cut from their roots before lab and the cut ends were immersed in a blue food color solution. This will preferentially stain the xylem, a vascular tissue, which you will study shortly. For now, you will look at the cells that make up the bulk of the stem.

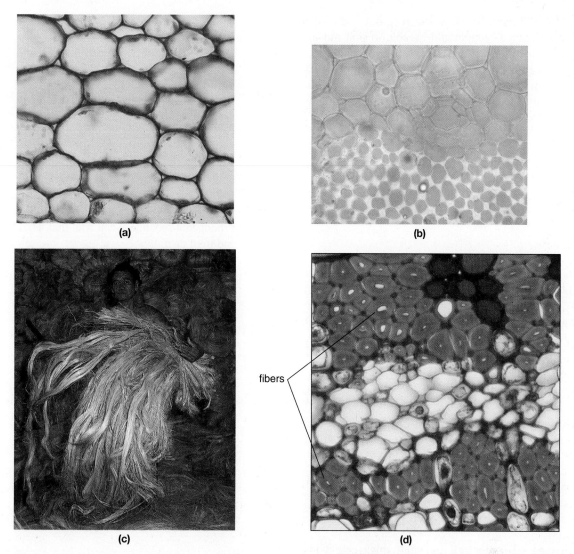

Figure 24.8 Examples of three types of ground tissue cells. (a) Thin, primary-walled, round parenchyma cells are common cells in stems, roots, and leaves where they are specialized for various metabolic functions. (b) Collenchyma cells (lower half of image) are characterized by unevenly thickened primary cell walls and provide support in growing parts of a plant. (c) Long sclerenchyma fibers are harvested to make rope and cloth. (d) Sclerenchyma fibers seen in cross section of phloem show how the lumen of the cell is almost completely filled with secondary cell wall material.

Take the stem and cut a very thin cross section and mount it in a drop of water on a microscope slide. Add a coverslip and observe.

How would you describe the shape of the cells? Are the cells tightly packed or loosely organized? Draw a few of the cells below with notes to the side.

13▸ Now prepare a longitudinal section of the stem so that you can observe the parenchyma in the other dimension. Add to your notes about cell shape. Are the cells tubular or more compact? _____

Collenchyma Like parenchyma, collenchyma cells are alive at maturity. The cells tend to be longer than wide and are characterized by primary cell walls of uneven thickness so that the cell walls appear thick at the corners (fig. 24.8b). In stems, petioles, and flowers, collenchyma occur as bands of cells. Because they have only primary cell walls that are not lignified, they are flexible, yet they provide support by a hydrostatic mechanism. In stems, the tendency of parenchyma cells to take on water forces them against the collenchyma bands making the stem semi-rigid.

14▸ The celery stalk (*Apium*) that we eat is a leaf petiole containing collenchyma strands just beneath the epidermis. Also in the petiole to the inside of the collenchyma will be parenchyma cells and vascular bundles of xylem and phloem. Obtain a piece of celery and strip out one of the strands. Once started, you should be able to pull out an entire strand from the length of the petiole.

Try to break an isolated strand to get an idea of its tensile strength. Then use a razor blade to cut a section of parenchyma from inside the petiole. How does its tensile strength compare to that of collenchyma? Write your impressions below.

17 Now obtain a prepared slide of a cross section of a *Tilia* stem. Find the heavy walled fibers in the phloem tissue of the stem. Note the very small lumen of the fibers. The protoplast produced a secondary cell wall that nearly filled the lumen and then died.

Given the small size of the lumens in fibers, do you think they would make good conduits for water transport? _____ If not, what could their function be?

15 Prepare two thin cross section slides of celery petiole: one with water and the other in phloroglucinol stain. Let them sit five minutes. Locate the collenchyma by looking for shiny cell walls of uneven thickness (fig. 24.8b). Note their location. Do not confuse the vascular bundles with collenchyma strands. Do the cell walls stain with phloroglucinol? Do collenchyma cells have secondary cell walls? Draw a few collenchyma cells below.

18 Summarize your observations by making a list of the types of cells found in ground tissue systems and describe where they would be found.

Sclerenchyma Characterized by rigid, thick cell walls composed of both primary and lignified secondary cell wall material, sclerenchyma cells usually lack protoplasts at maturity. They are usually found in mature regions of a plant's body that have stopped elongating. Two types of sclerenchyma cells are found in plants: fibers and sclereids. Fibers are long slender cells forming clumped strands often in association with vascular tissues. Their hard cell walls support the plant and can protect the phloem from the probing mouthparts of insects. Fiber strands of some species are economically important. Linen, burlap, jute, and sisal are examples of commercially important sclerenchyma fibers used to make cloth and ropes. Sclereids have very thick cell walls and tend to be cuboidal (fig. 24.8d). They are found in layers in seed coats, such as walnut shell, where they protect the embryo.

16 At the beginning of this lab, you observed sclereids when you stained stone cells in pears. Here you will look at fibers. In the lab are some pieces of sisal rope, burlap, or linen. Take a few fibers from these samples and mount them on a microscope slide in a drop of phloroglucinol stain. Let sit for a few minutes and then add a coverslip and observe with your compound microscope.

Can you see the individual fiber cells joined end to end? Sketch a few cells below.

Vascular Tissue System

The vascular tissues of plants are xylem and phloem. These tissues commonly occur in bundles that form conduction systems. The tubular shape of their cells reflects their function as conduits. Through the dead tubular cells of the xylem, water and minerals move from roots to other parts of the plant body. Through the living tubular cells of the phloem, photosynthetic products are transported from leaves to other nonphotosynthetic cells or from areas of carbohydrate storage to areas where carbohydrates are needed. In species that undergo secondary growth, xylem can occupy most of the volume of the stem or root while the phloem becomes the innermost layer of the bark.

Xylem and phloem are complex tissues. In addition to the cells that are unique to xylem and phloem, sclerenchyma fibers and parenchyma cells are often integrated into vascular tissues. We will not discuss these inclusions here.

Xylem Xylem is the principal water-conducting tissue in plants and the cells usually lack protoplasts at maturity. The protoplast's function is to form a tubular cell with a rigid, lignified secondary cell wall, and then to die, leaving the lumen as a conduit for water conduction. The thick secondary cell walls, often with spiral ridges, resist collapsing when water is pulled through the xylem by the negative pressure generated in transpiration.

Tracheids and vessel members are the cell types unique to xylem (**fig. 24.9**) that perform the function of water conduction. **Tracheids** are long, narrow spindle-shaped cells with tapered ends. **Vessel elements** are large barrel-shaped cells that join end to end to form tubes with perforated end plates between adjacent cells. Vessels transport water more efficiently than do tracheids and are thought to have evolved from tracheids. Both types of cells have thin areas in their side cell walls called **pits** where plasmodesmata passed between adjacent protoplasts when the cells were forming. The pits in mature dead cells allow lateral movements of water between adjacent conduits, an important consideration should a conduit become plugged with salts, debris, or a fungal intrusion. In woody species, wood is composed of secondary xylem that originated from cell divisions in the vascular cambium.

19 Obtain a prepared slide of macerated pine tracheids. Gymnosperms lack vessels and tracheids are the principle conducting cells in their xylem. Observe the spindle-shaped tracheids with high magnification. Look at the cell walls carefully for evidence of pits. Draw a few of the tracheids below.

Vessel elements are considered more evolutionarily recent than tracheids and are found in flowering plants (angiosperms). They are composed of short cell wall sections from different vessel elements that are joined end to end. Often cross walls are absent at the ends and water can often pass up to three meters before it encounters a pitted cross wall. This allows for rapid water movement and may be one of the reasons that angiosperms have become the dominant terrestrial plants. Angiosperm xylem typically contains both vessel elements and tracheids.

20 Obtain a macerate of a pumpkin or *Tilia* stem. Look for the vessel elements among the other cell types on the slide. Sketch and measure some of the vessel elements below.

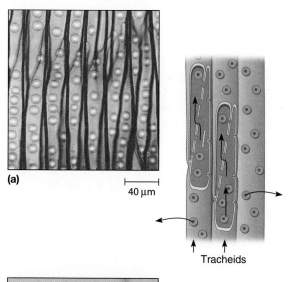

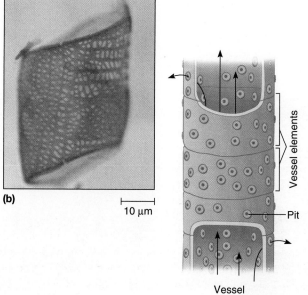

Figure 24.9 The unique cells found in xylem have stiff secondary cell walls penetrated by pits and lack protoplasts at maturity. The remaining cell wall serves as a conduit for water transfer from the roots to the leaves. (*a*) Tracheids are long narrow cells with tapered ends that overlap. Numerous pits allow water to pass both through the ends and laterally to other cells. (*b*) Vessel elements are short and wide barrel-shaped cells usually with perforated end walls.

21 Earlier in the lab, sunflower stems were placed in blue food coloring. You used these stems as a source of parenchymal cells. It is now time to use them again, but this time you will study the xylem. Since xylem is the water-conducting tissue, the blue coloring will have preferentially stained these cells. Take a 1-cm section of stem and tease it apart while looking at it through a dissecting microscope. Can you see the spiral thickenings in the vessel element cell walls? _____ What was their function?

Phloem The unique cells of this tissue function in carbohydrate conduction. Angiosperm phloem includes two unique cell types: sieve tube members and companion cells (fig. 24.10). **Sieve tube members** are cylindrical cells with thin primary cell walls that contain a living protoplast at maturity although they lack a nucleus. Sieve tube members join end to end to form long conduits. The end walls between adjacent cells, called **sieve plates,** are perforated by many pores through which enlarged plasmodesmata pass. Cytoplasmic materials flow through these bridges from one cell to the next. Each sieve tube member has lateral plasmodesmata that link its cytoplasm with a living parenchymal cell called a **companion cell.** Companion cells load and unload organic materials being transported through the sieve tube members. The lives of the two cells are tightly coupled, and when one dies, the other dies as well.

In many species, phloem has a trauma-control mechanism that prevents leakage of cytoplasm should a phloem conduit be broken by a grazing animal or accident. When a sieve tube member is injured, a slimy protein, called p-protein, surges through the adjacent cell and blocks the sieve plate pores so that no more phloem "sap" enters the cell that is broken open, thus conserving the sugars and other organics transported by the system.

22 Pumpkin stems (*Cucurbita*) contain large sieve tube elements on both the inside and outside of the xylem (bicollateral bundles). Obtain a prepared slide of a *longitudinal* section of pumpkin stem. Look at it first under medium power with your compound microscope. Look for tubular cells that are joined end to end. Globs of p-protein may be found stuck to the sieve plates. These are sieve tube members. If the angle is correct, you may be able to see perforations in the plates between adjacent cells. Associated with the sieve tube elements are small companion cells.

If you could not see sieve plates and companion cells on your slide, look at a *cross section* of a pumpkin stem. Locate the phloem and look for cells that have a honeycombed cross wall (fig. 24.10*c*).

23 Summarize your observations of vascular tissue systems by describing how you can distinguish between xylem and phloem in both cross sections and longitudinal sections.

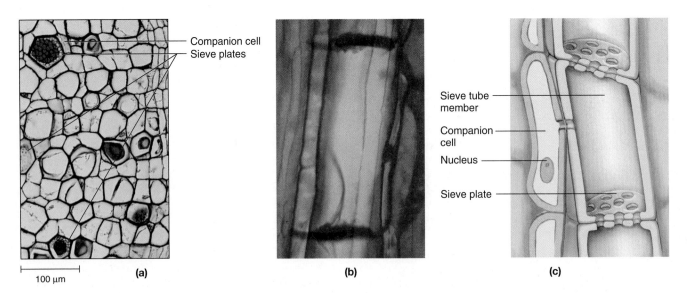

Figure 24.10 The unique cells found in phloem have thin primary cell walls with plasmodesmata connecting the living protoplasts of adjacent cells. This arrangement promotes the transfer of organic materials among the living, nonphotosynthetic cells of a plant. (*a*) Similar cells in cross section with sieve plates visible at the ends of one sieve tube member. Brown material is p-protein that plugs ruptured phloem conduits. (*b*) Sieve tube member and companion cell in longitudinal section. (*c*) Artist's interpretation of the anatomical relationship.

24 *Application of Observations* Use the cross section of the pumpkin stem as a summary slide to test your knowledge. Because all flowering plants are composed of the cell types studied in this lab, you should be able to identify them in just about any slide that you look at. On this slide, try to identify the following:

- Epidermal cells
- Parenchyma
- Collenchyma
- Sclerenchyma
- Xylem vessels
- Phloem sieve tube members
- Companion cells

If you are having trouble, see if your classmates or instructor can help you with the ones that you find difficult to identify.

Lab Summary

25 On a separate sheet of paper, answer the following questions assigned by your instructor.

1. Fill in the information required in **table 24.3**.
2. Distinguish between these three types of plant cells: parenchyma, collenchyma, and sclerenchyma.
3. During this lab, you observed a number of cells and tissues characteristic of flowering plants. These observations are evidence or data that illustrated the following list of concepts. For each statement, list the observations you made that support the statement.
   a. Protoplasts in many plant cells are interconnected.
   b. All plant cells walls are do not have the same chemical composition.
   c. Some plant cells cannot carry out photosynthesis.
   d. Certain plant cells are dead when mature and functional.
   e. At least three examples from epidermal cells can be named that support the idea that form reflects function.
   f. Parenchymal cells typically do not have a secondary cell wall.
   g. Many but not all, cell types, found in ground tissues are alive at maturity.
   h. There is more than one type of cell found in xylem.
   i. Sieve tube members are interconnected to form a pathway for transport.
   j. Vessel elements provide at least two examples of form reflecting function.

You may want to try the critical thinking questions that apply some of the knowledge you gained in doing this lab.

Internet Sources

Understanding plant cells and their requirements has led to a whole new field called plant tissue culture. It is now possible to isolate parenchymal cells from a plant and to culture them in a test tube. When a mass of cells accumulates, they can be treated with plant hormones and they will differentiate into the three types of tissues that you have studied in this lab. By manipulating hormones, it is possible to get a group of cells to actually organize themselves into a miniature plant inside a test tube. If this plant is removed and placed in soil, it grows into a mature plant. Check out this URL:

http://aggie-horticulture.tamu.edu/tisscult/tcintro.html

Alternatively, use Google and type plant tissue culture to find other sites.

Learning Biology by Writing

Because this was a discovery-based science lab rather than an experimental one, a descriptive essay is more appropriate than a report. In about 200 words, describe how different cells arise from apical meristems. Describe the different types of tissues and the associated cell types found in plants.

Critical Thinking Questions

1. A student wants to culture plant tissue so that he can asexually reproduce a fast-growing poplar tree that could be grown in plantations as a source for wood. He decides that since most of the poplar tree is secondary xylem vessel elements, he will culture isolate mature vessel elements and grow them in tissue culture. Why is this project doomed to failure from the start?
2. If plant roots are exposed to air and sunlight for a few minutes, and then the plant is placed back in the soil, it continues to wilt and then starts to recover. What is happening at the cellular/tissue level in the roots?
3. Would you expect parenchyma cells from a leaf to stain with phloroglucinol? Why? What about tracheids from an oak? Epidermal cells from a petunia?
4. When animals eat plant tissues, from what types of cells are they gaining energy and nutrients? Explain.

Table 24.3 Summary of Plant Cell Types and Characteristics

Cell Type	Found in Tissue Type	Have Primary Wall?	Have Secondary Wall and Lignin?	Cells have Protoplast?	Describe Shape	Function?
Epidermal						
Guard cells						
Trichomes						
Root hairs						
Parenchyma						
Collenchyma						
Sclereids						
Fibers						
Tracheids						
Vessel elements						
Sieve tube members						
Companion cells						

Lab Topic 25

Investigating Functional Anatomy of Vascular Plants

SUPPLIES

Resource guide available on WWW at
http://www.mhhe.com/labcentral

Equipment

Compound microscopes
Dissecting microscopes

Materials

Slides and coverslips
Prepared slides
    Basswood (*Tilia*), cross section (1 yr and 3 yr)
    Coleus shoot tip, longitudinal section
    Corn root, cross section
    Corn stem, cross section
    Corn leaf, cross section
    Medicago stem, cross section
    Eudicot leaf, cross section
    Pine needle, cross section
    Pine stem, cross section
    Raphanus root tip, longitudinal section
    Ranunculus root, cross section
Twigs of hickory or buckeye with apical bud
100-ml beaker
Razor blades and forceps
Shallow pans
Live materials
    Germinating radish seeds
    Germinating corn
    Potted, growing *Coleus* or *Zebrina*
    Sunflower plants 12" tall or larger; one well watered and other water deprived
    Various eudicot plants for leaf cross sections
    Fresh celery
    Root vegetables such as carrots, turnips, ginger, radishes, or others as available
    Sawn cross sections of large tree branches, sanded to show growth rings and bark layers

Solutions

0.5% methylene blue or food dyes
Phloroglucinol stain

STUDENT PRELAB PREPARATION

Before doing this lab, you should read the Background materials and sections of the lab topic that have been assigned by your instructor.

Use your textbook to review the definitions of the following terms:

apical meristem	phloem
bark	root cap
Casparian strip	root hair
cork	stele
cork cambium	stomata
endodermis	transpiration
eudicot	vascular cambium
herbaceous	vascular tissue
monocot	woody
pericycle	xylem

Describe in your own words the following concepts:

Primary growth in roots, stems, and leaves
Secondary growth in roots and stems
How vascular tissues are organized differently in monocots and eudicots
The transpiration–tension–cohesion theory of water movement in xylem
How guard cells regulate stomatal opening

After finishing the prelab review, write any questions you have about terms, concepts, or techniques in the margins of this lab topic. The lab experiments should help you answer these questions, or you can ask your instructor during the lab.

OBJECTIVES

1. To recognize and compare tissue organization in primary roots of monocots and eudicots
2. To compare the basis for primary and secondary growth of roots
3. To observe the differences in stem and leaf anatomy among monocots, eudicots, and conifers
4. To be able to describe the basis for primary and secondary growth in stems
5. To demonstrate the transpiration-cohesion transport theory of water movement in plants

Background

In previous labs you looked at how invasion of the terrestrial environment influenced the life cycles of plants and studied the types of cells found in plants. In this lab you will study how these types of cells are organized into functional units called tissue systems that allow vascular plants to live and grow in terrestrial environments. These tissue systems are organized into what we might call the architecture of the plant body. Just as buildings are made of basic types of building materials such as bricks, lumber, pipes, *etc.*, plants are made from the types of cells you have studied. To continue the analogy, you know that more than one building can be made from the same list of basic materials. It is how they are put together that determines the final shape and function. So it is also with plants and their cellular building blocks.

The basic parts of the vascular plant body are the root system, stem, leaves, and reproductive structures. Each has a unique arrangement of cell types which you will learn to distinguish and to correlate with function. Among flowering plants, there are two broad architectural types: monocots and eudicots. At a minimum, you need to be able to distinguish the differences between them based on their anatomy.

If we think of plant bodies in evolutionary terms, the following is a reasonable, although hypothetical, architectural analysis of why plant bodies look the way they do today. Plants most likely evolved from green algaelike ancestors. Think of the algae as resembling the chloroplast-containing parenchyma cells previously studied. These are the photosynthetic cells of the plant. Green algae require a very moist environment or they will die. How could the drying problem be solved architecturally? Encase the parenchyma cells in layers of other cells that are specialized to prevent drying. These cells need not be photosynthetic since they can get "food" from their photosynthetic neighbors. Such cooperation would allow an association of cells to function when no longer in an aqueous environment.

Photosynthesis, however, requires carbon dioxide gas from the atmosphere and produces oxygen, so the protective surface must have some openings to allow gas exchange. This would also allow some water loss, but solutions to problems always require trade-offs. These simple architectural changes would have yielded a new type of encrusting organism, maybe a few cells thick. If it were thicker, then light energy would not be available to the cells on the bottom of the stack. While forms like this might have had some momentary success, the available surface area on land is limited and low encrusting plants would compete with each other for areas to grow.

How can architecture solve the available area problem? Go three dimensional and build high-rises! Any plants that were able to grow taller than their neighbors would intercept the sunlight first, and thus another architectural change could solve a problem. However, tall structures create other problems. A structure of any height requires support. Cellular adaptations that thickened cell walls would be one response. A second would be hydrostatic support, similar to the stiffening of a fire hose when water pressure is applied.

Once a stiff, upright structure was achieved, architecture presented an opportunity to increase efficiency. Development of flat projecting structures to house the photosynthetic cells would allow plants to intercept light that was passing through space beyond the stem. However, as our hypothetical plant got taller, other problems would develop. A thin, tall, upright structure with weight at the top is unstable and will topple. Vertical supports must be anchored in the ground. Additionally, as plants got taller there was a point at which water could not diffuse the distance from the moist soil to the elevated parenchyma cells faster than it was lost by evaporation from the leaves. What were the architectural solutions?

Anchorage could be accomplished by having some cells specialize to form underground structures, but these would be dependent on above ground parts. If these structures also became highly absorptive, they would facilitate water absorption. However, anything underground could not be photosynthetic, yet energy and building materials would be required to grow and function. If a plumbing system were added to the architectural plan, then two problems would be solved. The plumbing could conduct water to the elevated photosynthetic surfaces and could transport photosynthetic products to the underground parts supplying their needs. If the plumbing structures were reinforced, then they would add to structural strength, allowing the plant to grow even taller.

Thus we can create a plausible scenario that explains the problems that were solved in the origin of plants with the development of a few mutually interdependent cell types. In today's lab, the architecture of roots, stems, and leaves will be investigated.

Stems Stems elongate by primary growth resulting from the addition of cells at the growing tip. The cells produced by the apical meristem undergo a process of differentiation to produce different primary tissue systems: dermal, vascular, and ground tissues that are architecturally arranged into the stem. Much of the support function and just about all of the transport functions are performed by the vascular tissues, xylem and phloem. In simple terms, the protoplasts of xylem tissues make elongate cells with reasonable heavy secondary cell walls and then die, leaving a hollow conduit that conducts water and minerals. Phloem tissues are composed of living cells that function in distributing photosynthetic products throughout the plant and in distributing carbohydrates from storage areas to the rest of the plant.

The support functions are legendary, the type of stuff you see in the Guiness Book of World Records. Redwood trees in California and eucalyptus trees in Australia can reach heights of a 100 meters. In the other dimension, a single banyan tree in India has a canopy (branches and leaves) that spreads over an area equal to about five football fields. It is difficult to imagine the stresses that the stems of such giants must encounter in high winds, and it speaks to the strength of the supporting xylem tissues and their root systems.

Xylem also transports water from the roots to the leaves. In a lawn grass, this is not an impressive task, but for this to happen in one of those tall eucalyptus trees, over 300 pounds of pressure per square inch must be exerted to get the water from the soil to the leaves. How does the plant accomplish this? Two mechanisms are involved: root pressure and transpiration-cohesion.

Root pressure is due to the osmotic uptake of water by a plant's root. Since root cells contain dissolved organic solutes, such as sugars and proteins, the *osmotic gradient* in wet soil goes from the soil into the root, literally forcing water into the xylem elements. It can account for raising the water column several feet.

The second and more important mechanism for transporting water involves the combined effects of **transpiration** and **water cohesion** (**fig. 25.1**). Leaves have small porelike openings called **stomata** (singular: stoma), usually on their undersurfaces. Stomata are involved in gas exchange during photosynthesis but also allow water vapor to escape by diffusion. Lost water is replaced by vapor diffusion from the xylem into the leaf, resulting in a tension on the water column in the xylem. Because of hydrogen bonding between adjacent water molecules, this tension is transmitted from the canopy to the roots by cohesion. Water, along with dissolved minerals, is pulled from roots to leaves.

Transport of organic molecules to other tissues occurs through the sieve tube members of the phloem. Carbohydrates are actively transported from the photosynthetic leaf cells into the phloem. In roots and regions of growth, these materials are removed from the phloem and used as energy sources or for making other kinds of molecules. Thus a solute concentration difference exists between the two "ends" of the phloem: high concentrations of the solute are present in the leaf, and low concentrations are in the root. Water diffuses from the xylem and enters the phloem by osmosis at the leaf end. This creates a pressure that drives the phloem fluid toward the roots where carbohydrates are removed and metabolized, and water reenters the xylem system. This mass flow results in the **translocation** of photosynthetic products. When plants break their dormancy after winter, flow of materials is reversed. Sugars from stored starches in the roots are transported to buds and cambiums where they fuel cell growth and division.

Roots Roots have three functions. First, they absorb water and minerals from the soil. Second, roots serve as storage depots for photosynthetic products and these reserves fuel metabolism over winter in the living cells in the rest of the plant and provide the energy for leafing out in the spring. The tapping of maple trees in the spring takes advantage of the maple's capacity to store carbohydrates in its roots one season and mobilize them as sugars in another. Third, and equally important, roots anchor and stabilize the stem in the soil.

Compared to stems, we know much less about roots because they are more difficult to study. Their role in ecosystems where they unlock the soil's store of minerals such as nitrate, phosphate, sulfate, and calcium cannot be overstated. Herbivores gain not only energy, essential amino acids, and vitamins from eating plants, they also gain ions necessary for their metabolism. Some scientists have observed that because of roots, animals, including humans, do not have to eat dirt to satisfy their mineral needs.

Root systems are often highly developed and account for 80% of the biomass of the plant. For example, a researcher found that a mature winter rye plant had a root system with a total length of 387 miles and a surface area of 2,554 square feet, containing 14 billion root hairs with a combined length estimated to be 6,603 miles. The most amazing fact, however, is that all of this was contained in less than two cubic feet of soil. Obviously, intimate contact occurs between soil and plant.

Mere absorption of water and minerals is not sufficient for plant survival. These essential materials must find their way from the roots to the actively metabolizing cells of the plant. As in stems, xylem and phloem are present in roots and perform this conduction function. Xylem tissues form a water-conducting conduit from near the tips of the roots to the tip of the stem. How tall was that

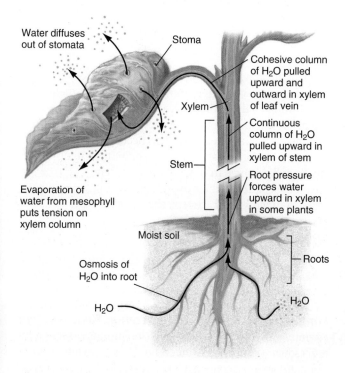

Figure 25.1 Continuous columns of water in each xylem tube extend from the roots to the mesophyll of leaves. Water is drawn upward by transpiration.

species of eucalyptus mentioned earlier? Phloem must transport photosynthetic products from the leaves to the living cells at the root tips.

Roots elongate by adding cells at the growing root tip. The cells produced by the root apical meristem undergo a process of differentiation to produce the different primary tissue systems of the root: dermal system, vascular tissue system, and ground tissue system that are architecturally arranged into the root.

Secondary Growth Most eudicots and gymnosperms exhibit some secondary growth, and in woody plants it is a substantial contributor to body mass and strength of stems and roots. **Primary growth** increases the length of the plant body by adding cells at the tips of the stem, branches, and roots. **Secondary growth** is an increase in the girth of the components of the plant body by adding cells to the periphery of the primary plant body. It is necessary because a plant that grows only in length (primary growth) will soon reach a point where a thin stem cannot support it or a thin root is not strong enough to anchor it. Those plants that lack the capacity for secondary growth are usually not very large or are aquatic where water supports them. By increasing girth, strength is added and plants can grow even larger, outcompeting neighbors for sunlight.

Secondary growth can occur because woody plants have two other meristems in addition to the apical meristems in their roots and stems. Called **cambiums,** one of these is located beneath the bark and is referred to as the **vascular cambium.** Division of it produces secondary xylem and phloem vascular tissues. The other is located in the bark and is called the **cork cambium.** Division of it produces cork cells that are components of the bark or periderm. The records for secondary growth are quite impressive. A chestnut tree in Sicily has a diameter of 18+ meters (58 feet!), yet that tree started from a single chestnut and once had a primary stem only a few millimeters across. Approximately 90% of the mass of this tree would be **secondary xylem,** the name given to xylem produced by the vascular cambium.

If you consider the geometry of secondary growth, a potential limitation is uncovered. The organization of stems and roots of woody plants can be thought of as cylinders within cylinders. If an internal cylinder expands as a result of adding cells by division of the vascular cambium, then the outer cylinders will be stressed and eventually will split longitudinally. Have you ever seen bark flaking off a tree and wonder why this happens? The explanation is secondary growth. Adding wood inside the bark stretches the bark, which gives way. You can see this as the deep longitudinal fissures in the bark of most large trees. Bark is replaced when the cork cambium produces cells that fill the voids. As secondary growth continues, the outermost layers of bark are stressed to the point that they no longer stay attached to the expanding new layers of bark and fall off.

> **LAB INSTRUCTIONS**
>
> In this lab, you will study the anatomy of roots, stems, and leaves, noting the arrangements of basic tissues. You will learn the differences between monocot and eudicot anatomy. In eudicots you will see the effects of secondary growth. At the end, you will do some experiments demonstrating water movement through xylem.

Root Structure

Whole Roots

1▶ Your lab instructor has set up demonstrations of taproots, fibrous roots, and adventitious roots. Look at each and make three quick sketches that show the differences.

The parenchyma cells of taproots store carbohydrates; some taproots are used by humans for food. Name three.

2▶ Obtain some germinating radish or corn seedlings in a petri dish and compare to **figure 25.2**. Look at them carefully through a dissecting microscope. Can you see the root cap at the apex? Note that the root hair zone begins a few millimeters back from the tip. This occurs because the cells produced by the apical meristem require time to differentiate. Root hairs are part of what tissue system _____. In the older sections of roots away from the tip, the root hairs cease to function and are replaced by other types of cells.

Root Histology

In this section, you will observe the patterns of tissue organization in both longitudinal and transverse sections of roots from eudicots and monocots.

Longitudinal Section of a Young Eudicot Root

3▶ Examine a prepared slide of a longitudinal section of the radish (*Raphanus*) under low-power magnification. Compare your specimen to the diagram in **figure 25.3**. Note how the cell shapes change as you scan away from the root tip toward the base of the root.

INVESTIGATING FUNCTIONAL ANATOMY OF VASCULAR PLANTS   **335**

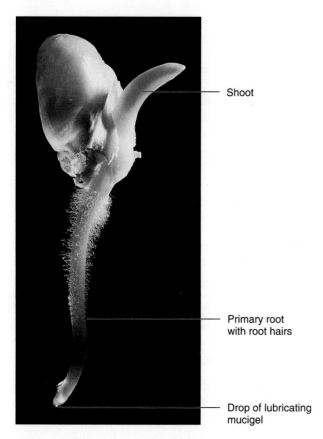

Figure 25.2 Primary root of a corn seedling. Note root hair zone.

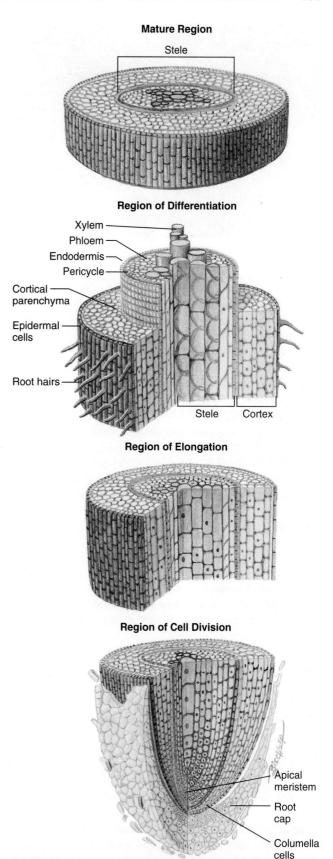

Figure 25.3 Tissue organization in a young root.

Identify the **root cap,** a thimble-like covering of cells that protects the tip of the root as it pushes through the soil (**fig. 25.4**). The cells of the root cap are constantly renewed. An elongating corn root can shed 10,000 root cap cells in a day. Once formed, root cap cells live four to nine days before they die, secreting a slimy material, called mucigel, that also lubricates the passage of the root through the soil (see fig. 25.2). An estimated 20% of the carbohydrate made in photosynthesis gets used to make mucigel. Root growth obviously contributes substantial amounts of carbohydrate to the soil where it supplies energy for the growth of soil fungi and bacteria after serving its lubrication function.

In the root tip behind the root cap is the **apical meristem,** the cells that divide during root **primary growth.** You studied mitosis in the cells from this region in onion roots in lab topic 10. The apical meristem produces some cells toward the root tip that are called **columella cells.** You should be able to see these arranged in columns passing toward the tip from a region of small densely packed cells. Older columella cells are pushed to the outside and replenish the root cap cells as they are sloughed off. Grains of starch in columella cells are thought to act as gravity sensors, directing the growth of a seedling's emerging root downward.

After cells are formed by division, they increase in size in the **region of elongation** behind the meristematic

Figure 25.4 Primary root tip. The root cap is separated, showing it is a distinct structure.

region. Cells in this region can increase in size 150-fold by water uptake. The elongation of these cells pushes the root forward through the soil up to 4 cm per day. Root growth is indeterminate, stopping only when the plant dies. It is this constant pushing forward that allows the roots to enter new soil areas and to tap the minerals dissolved in soil water.

Further up the root, you may be able to see cellular differentiation. Several types of cells should be visible. The **epidermis** is the outer cell layer of the root. As the tip of the root penetrates new soil and epidermal cells mature, **root hairs** grow out from the epidermal cells to capture previously untapped water and minerals. The **region of absorption** is marked by the root hairs. Plants may die when transplanted because the delicate root tips, with their root hairs, are torn off or dry in the moving process. Given their function would you expect root hairs to have a waxy cuticle? Why?

The roots of pines, birches, willows, and oaks often do not have well-developed root hairs. Instead, they have mycorrhizal roots in which filaments of a symbiotic fungus carry out the functions of root hairs. The fungal hyphae extend into the soil and penetrate between cells of the root cortex. They convey water and minerals into the root from the soil. (See fig. 19.9.)

In the zone of cellular differentiation, you may see a region called the **cortex** consisting of **parenchyma** cells. It extends from beneath the epidermis to a central vascular cylinder, the **stele**. The parenchyma cells of the cortex store starch and other materials in many plants. Note the loose arrangement of cells in the cortex. Water and dissolved minerals can travel through these spaces. Are the parenchyma cells normally living (with a protoplast) or dead? _____. The inner boundary of the cortex is a single layer of cells, the **endodermis.** As you will see, it is very important in regulating water and mineral absorption.

The stele is composed of all the cells inside the endodermis. **Xylem** and **phloem** tissues, which conduct materials from and to the roots, are prominent here. Parenchyma cells, also found in the stele, are living cells with relatively thin walls. They can divide to repair wounds and are physiologically active in storage. In young primary growth roots, the xylem and phloem are primary tissues derived from the apical meristem.

As the root ages and thickens, secondary xylem and phloem will develop from a vascular cambium. When this happens, the epidermis is shed and replaced by the **periderm,** a barklike covering. Because the root hairs are lost at this time, the older portions of roots with periderm do not absorb water and minerals.

Cross Sections of Primary Roots While the age sequence of cells and differentiation is best seen in longitudinal sections as you just studied, tissue arrangements are easier to understand in cross sections. In this activity you will study cross sections of primary roots from a eudicot and from a monocot. Be sure that you learn the differences in the arrangement of tissues in these two groups of flowering plants.

4 ▶ ***Eudicot Roots*** Obtain a prepared slide of a cross section of a buttercup root (*Ranunculus*), a eudicot, and look at it with low power. Identify the epidermis, cortex, and stele as you did in the longitudinal section. The bulk of young primary roots is contributed by the living parenchyma cells of the cortex.

The cylindrical stele should be readily visible at the center of the root (**fig. 25.5***a*). Examine it with the high-power objective. At the center of the stele the large, heavy-walled **xylem** tissue should be seen in an x-like configuration. Between the radiating poles of the xylem locate the **phloem.**

In some eudicots, a **vascular cambium** develops between the primary xylem and phloem. During secondary growth of eudicot roots, this cambium produces new **secondary xylem** to the inside and **secondary phloem** to the outside. Eventually, the addition of vascular tissue forms expanding concentric rings of secondary xylem and phloem. The expanding central cylinder of xylem causes tension in the surrounding primary phloem, cortex, and epidermis. These layers split longitudinally and slough off as secondary growth continues. The root also produces another cambium, the **cork cambium** (from the pericycle). Division of its cells produces a bark layer on older roots.

The outer boundary of the stele is the **pericycle** (fig. 25.5*b*) just beneath the endodermis. The pericycle is

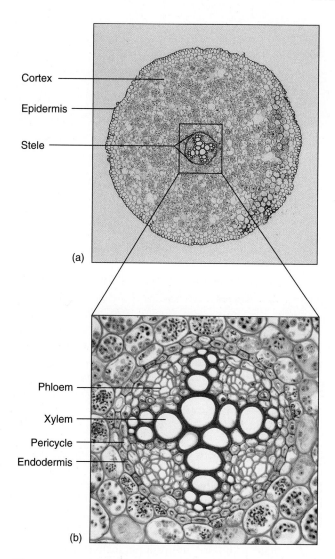

Figure 25.5 Eudicot anatomy: (a) Cross section of *Ranunculus* primary root (b) high magnification of stele showing tissue arrangement

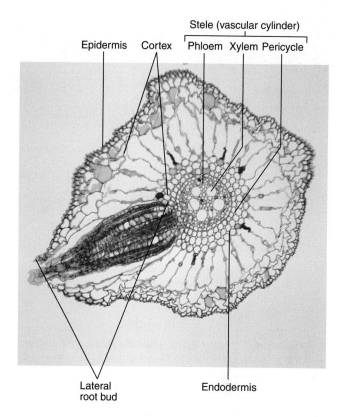

Figure 25.6 Cross section of a water hyacinth root showing formation of a branch root from pericycle.

a narrow zone of parenchyma cells that maintain their ability to divide and form a meristematic tissue.

Lateral roots arise internally from the pericycle about one inch back from the tip. The growing tissues push their way to the surface through the endodermis, cortex, and epidermis (fig. 25.6). This process differs from lateral branching in stems where the branches arise from buds located at nodes on the stem surface. Nodes and associated buds are absent in roots.

Just to the outside of the pericycle is the **endodermis**, a physiologically important boundary that regulates water and mineral absorption. With shapes like boxes, the rectangular cells of the endodermis have a cell wall on each of their six faces; four faces have a band of waxy phenolic substance called **suberin**, but two do not (fig. 25.7). The waxless walls face the outside of the root on one side and the vascular tissues on the other. Because the wax-containing walls of adjacent endodermal cells tightly abut one another, they form a water-impermeable barrier called the **Casparian strip.** It is difficult to observe, but you may be able to see it. Use high power to look for reddish dots between the adjacent cells of the endodermis.

Water and dissolved minerals entering a root can travel by two routes to the xylem in the stele. The apoplastic route is through the water-saturated cellulosic cell walls and intercellular spaces of the epidermis and cortex. This pathway stops, however, at the endodermis where the suberin bands on the abutting cell walls block the path. The second pathway, called the symplastic route, is through the living protoplasts of the epidermis, cortex, endodermis, and pericycle into the stele. The membranes of all these living cells contain transporter proteins for ions and others for water (aquaporins) and will accumulate ions and water from the apoplast. Plasmodesmata connect adjacent protoplasts. Once an ion or water molecule enters a protoplast it can travel from cell to cell without crossing membranes. By genetically regulating the types of proteins in the membranes, a plant can control its uptake of ions and water. The final step in absorption is called xylem loading. Parenchyma cells in the xylem actively transport ions from the symplast into the nonliving tracheid and vessel tubes and water osmotically follows. The absorbed materials are then transported by the xylem to other regions of the plant. As a root starts secondary growth in its aging regions and root hairs are lost from the surface, the endodermis undergoes changes. The plasmodesmata will be lost and

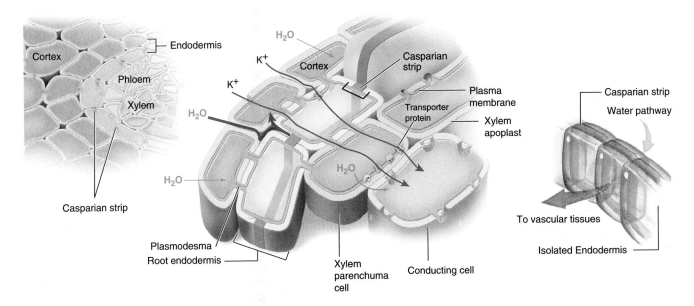

Figure 25.7 Pathways of water and dissolved mineral absorption in eudicot root. Apoplastic route shown in red only permits water to enter root to the endodermis boundary where Casparian strip prevents further passage. The symplastic route through the plasmodesmata-connected protoplasts of the epidermis, cortex, and endoderm is transfers materials to the xylem in the stele. Water and minerals, such as K^+, bind with transport proteins in the protoplasts' membranes, thus allowing the cell to regulate what enters the vascular system.

suberin will be secreted on all faces of the endodermal cells, essentially blocking any further water and mineral transport.

5 ▸ Monocot Roots Obtain a slide of a cross section of a corn (monocot) root (**fig 25.8**). Identify the outer **epidermis** and the parenchyma cells of the **cortex**. Note the intracellular spaces among the parenchyma cells. Note the general similarity of the cortex to the eudicot root but the obviously different organization of the **stele**. It is surrounded by an **endodermis** as in the eudicot root, but the **xylem** and **phloem** cells are found in clumps around the periphery of the stele rather than at its center. The large cells to the inside of the pericycle are xylem. Phloem is located between the xylem vessels. No vascular cambium is found in monocot roots. Consequently, they do not undergo secondary growth. The central area of the stele is filled with fairly uniform parenchyma cells and is called the **pith**.

6 ▸ Investigating Living Roots (Optional) If time permits, investigate the microscopic anatomy of roots by preparing freehand sections. Your instructor will have cut sections from eudicot and monocot roots and placed them in a dish of water. Your task is to examine the microscopic anatomy of a section and decide whether it came from a monocot or eudicot based on its tissue arrangement.

Remove a root tip about 2 inches long from one of the unknowns. Take a two-inch square of Parafilm and fold it in half. Place the root on the inside of the fold so that it is sandwiched by plastic. Place the preparation on a microscope slide and press on the Parafilm to hold the root in place. Cut a cross section of the root by slicing across it and the Parafilm. Discard the cutoff material. Add several drops of water to form a puddle at the newly cut edge. Now make a second cross section cut, cutting the section as thin as possible. Mount this thin section on a new slide in a drop of water and add a coverslip. Observe the cut face of the cross-sectioned root at first with the 10× objective of your compound microscope.

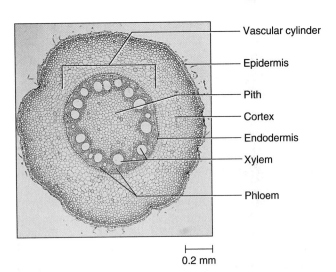

Figure 25.8 Cross section of a corn root, a monocot.

Make a sketch of the epidermis, cortex, stele, and xylem in your preparations.

Is your unknown root a monocot or eudicot? Have your instructor confirm your conclusion.

Stem Structure

Stems support the flowers, fruits, cones, and photosynthetic leaves in most plants. Conducting tissues in the stem bring water and minerals to the leaves and distribute photosynthetic products to the roots. Some stems are photosynthetic, as in cacti, and others are important in vegetative reproduction, food storage, and water storage. In this section of the lab, you will examine the basic organization of eudicot stems and a monocot stem. As with roots, you need to remember the differences in arrangement of the tissue types between monocots and eudicots.

External Eudicot Stem Antomy

7 Examine a woody twig from a hickory, buckeye, or other tree. At the tip of the stem or a branch, find the **terminal** or **apical bud.** In the bud is the **apical meristem,** the source of cells for the primary growth that elongates the stem. In most plants, the apical bud produces a hormone that inhibits development of lateral buds, a phenomenon called apical dominance. Gardeners have long recognized this phenomenon. They obtain bushy shrubs and trees by clipping off the shoot tips, which removes the inhibition of the lateral buds, allowing branches to develop. Conversely, tall plants can be "forced" by trimming lateral branches, thus removing the lateral buds and directing the plant's energy reserves to the apex of the shoot.

Note the prominent **leaf scars** where leaves were attached (fig. 25.9). If you look carefully at a leaf scar, you can see the vascular bundle scars where strands of xylem and phloem entered the petiole of the leaf. The leaves grow out from the **nodes;** the stem segments between the nodes are **internodes.** The angle formed between a leaf and the internode above it is the axil. Examine the area just above a leaf scar and you should see **lateral** or **axillary buds,** which form branches.

Small pores called **lenticels** should be visible in the bark of the internodes. These loosely organized areas of bark allow metabolically active tissues in the twig to exchange respiratory gases with the atmosphere.

Herbaceous Eudicot Stem Histology

8 Now obtain a prepared slide of a longitudinal section of a *Coleus* shoot tip. *Coleus* is a **herbaceous plant,** meaning that the stem shows limited secondary growth and the stem does not thicken very much. Annual plants such as these usually die at the end of the growing season. Using the scanning objective on your microscope, identify the terminal bud as you did for the woody stem. Note several **leaf primordia,** which surround and protect the terminal bud (fig. 25.10). As the *Coleus* matures, flower buds may develop in the terminal bud. What important group

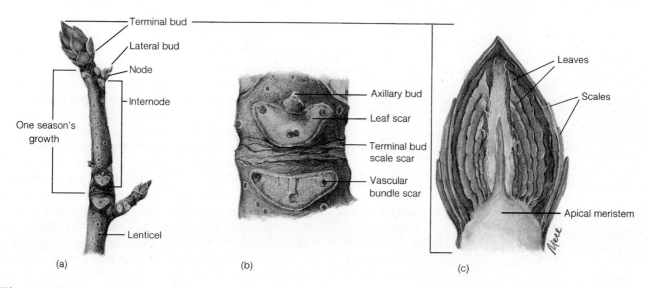

Figure 25.9 Anatomy of a woody stem: (a) external features; (b) detail of leaf scar; (c) longitudinal section of terminal bud.

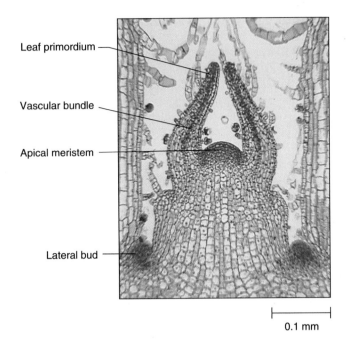

Figure 25.10 Longitudinal section of *Coleus* terminal bud.

of cells necessary for continued growth is found in the terminal bud?

Examine the terminal bud under high power and find the **apical meristem** composed of darkly staining, rapidly dividing small cells. As cells are produced at the tip and subsequently elongate as they mature, the stem undergoes **primary growth** or growth in length. Examine the cells below the meristem and note their larger size and evidence of differentiation. Some will become vascular tissues; others will give rise to parenchyma cells, leaf primordia, and lateral buds. Trace the developing vascular bundles consisting of xylem and phloem passing into each of the leaf primordia.

Note the **lateral bud** in the angle (axil) between the top surface of the leaf and stem. Lateral buds near the tip of a stem are in a state of dormancy, held in check by hormones produced by the terminal bud. As the stem elongates, the distance between the terminal and lateral buds increases and the inhibitory effect is lessened, allowing the lateral buds to develop into branches. In pines, firs and spruces this effect produces the pyramidal shape of the tree crown.

9 Now obtain a cross section of a differentiated region of *Medicago*, an herbaceous eudicot stem, and look at it under your 4× objective (**fig. 25.11**). Herbaceous stems undergo little secondary growth and are nonwoody. In young stems resulting from primary growth, the outside of the stem is covered by a single layer of living cells, the **epidermis**. The epidermis protects the underlying tissues from drying and often contains stomata. Epidermal cells on many stems form trichomes. Perhaps you have noticed the "hairy" stems of tomatoes which bear many trichomes. Waxes often are secreted by the epidermal cells to form a **cuticle**.

Just beneath the epidermis is the **cortex**, a region that contains both parenchyma and collenchyma cells. Examine one of the "ribs" of the stem and identify the **collenchyma** cells with their unevenly thickened walls. These cells are living at maturity and provide support. The bulk of the stem is **pith**, a soft matrix of living parenchyma cells. In the stems of some species the pith dies, producing a hollow stem.

The **vascular bundles** are arranged in a circle inside the cortex. Examine a single vascular bundle under high power. Xylem makes up the inward side of the vascular bundle. Most xylem cells are dead at maturity. The protoplast breaks down and the walls remain as water conduits. The larger of these are **vessels** and the smaller ones **tracheids.** The outer portion of the vascular bundle is phloem. Phloem is composed of larger **sieve tube** members and smaller **companion cells.** These cells must remain alive to function at maturity. Although difficult to see, the **vascular cambium** is a single layer of cells between the xylem and phloem (fig. 25.11). During stem secondary growth when the stem increases in diameter, cell divisions in the vascular cambium produce **secondary xylem** and **secondary phloem.** The vascular tissues of the stem are continuous with those of the leaves and roots, forming an effective transport system throughout the plant.

Woody Eudicot Stem Histology

Woody eudicots, in contrast to herbaceous eudicots, have stems that increase in girth year after year by **secondary growth.** As a plant increases in girth, its older xylem tissue is buried in the added layers and cannot be supplied with oxygen. This physiological limitation has been solved by an interesting evolutionary adaptation. Only the cells near the surface of the trunk retain their protoplasts and are alive. Those near the center die, but their cell walls remain structurally intact and provide mechanical support for the canopy of the plant.

A woody stem has three basic regions: bark, vascular cambium, and wood (**fig. 25.12**). In young twigs before secondary growth is initiated, the surface of the twig is covered by epidermis. With the onset of secondary growth, bark develops as the epidermis splits and sloughs off.

Bark is divided into an outer region, the periderm, and an inner region, the secondary phloem. The **periderm** consists of an outer **cork** layer, which prevents evaporation and protects the underlying tissues, and a **cork cambium,** which gives rise to the outer cork cells. **Secondary phloem** in angiosperms consists of sieve tubes, companion cells, sclerenchyma fibers, and storage parenchyma cells. These cells function in food conduction, storage, and support.

Beneath the phloem is the **vascular cambium.** Division of these cells produces secondary phloem to the outside and secondary xylem to the inside. Wood, the innermost region, consists entirely of **secondary xylem.** Its functions are water conduction and support.

INVESTIGATING FUNCTIONAL ANATOMY OF VASCULAR PLANTS   **341**

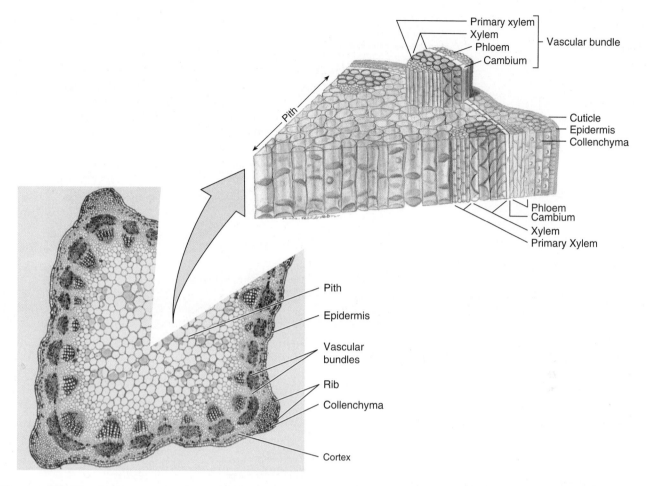

Figure 25.11   Tissue organization in a herbaceous eudicot stem: photomicrograph of a representative stem (alfalfa—*Medicago*) with artist's interpretation of tissues.

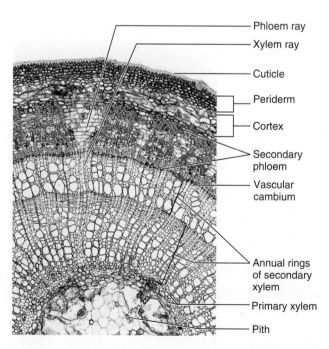

Figure 25.12   Tissue organization in a cross section of a two-year-old basswood stem, a woody eudicot.

10 Obtain a prepared slide of a cross section of a two-year-old basswood (*Tilia*) twig. Find the pith at the center of the section (fig. 25.12). As the pith ages, the protoplasts die and the cells accumulate tannins and crystals. Surrounding the pith is a narrow layer of primary xylem; both pith and primary xylem originated from the apical meristem and are the result of primary growth during the first year of life. To the outside is a relatively wide layer of **secondary xylem** laid down in annual growth rings.

Each annual ring consists of larger springwood cells and smaller summerwood cells. At one time the vascular cambium was next to the primary xylem, surrounding it as a cylinder of cells. As the vascular cambium cells divided, the daughter cell to the inside became secondary xylem, and the cell to the outside remained vascular cambium to divide again and again. Thus the cylinder of vascular cambium always enlarges and is located just outside of the xylem regardless of how many xylem cells are added in secondary growth. The vascular cambium separates the xylem from the outer phloem.

In *Tilia*, the phloem is composed of triangular sections, some of which point inward and others outward. Those that point inward, ending in sharp points, are **phloem rays** and are continuous with radial lines of

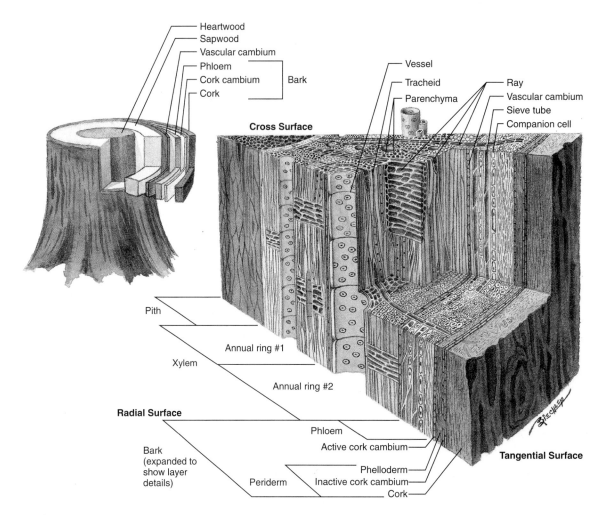

Figure 25.13 Tissue organization in a mature oak stem.

parenchyma cells, the **xylem rays,** which cross the xylem toward the pith. These cells conduct materials radially in the stem and its branches. In between the phloem rays are other triangular sections, ending in blunt points, that point outward. These consist of heavily stained **phloem fibers** and large, thin-walled **sieve tube members.** These join end to end to form **sieve tubes.** Small **companion cells** are associated with the sieve tubes.

In young eudicot woody stems, just outside the phloem is the **cortex,** a zone of loosely arranged, large parenchyma cells. The cortex is replaced by the **periderm** as the stem grows in diameter.

The periderm is divided into three layers. Adjacent to the phloem are four to six layers of thick-walled cells, the phelloderm. External to the phelloderm are the thin-walled cells of the **cork cambium,** which gives rise to phelloderm and the outermost **cork cells,** which cover the mature stem. Cork cells secrete the waterproofing compound, suberin. The bark of a woody plant consists of all the layers external to vascular cambium and thus includes the phloem and the components of the periderm.

As a tree grows in girth, what must happen to the existing bark as secondary xylem is added on the inside of the vascular cambium? _____

Flip back to lab topic 18 and look at figure 18.3, a cross section of a pine stem. List the similarities and differences you see between a woody eudicot stem and a gymnosperm stem.

Similar

Different

Structure of Wood

In the lab are polished sections of tree branches or trunks for you to study.

11 Wood is composed exclusively of years of accumulation of secondary xylem to the inside of the vascular cambium. If you look at a cross section of a tree trunk or branch two types of wood are visible (**fig. 25.13**). In the center the **heartwood** is darker in color due to the accumulation of resins, tannins and metabolic wastes. It is the older wood. Just outside the heartwood, but not including the bark, is the **sapwood**. It is lighter in color and is younger, not having accumulated the metabolic products. It will gradually darken and become heartwood as new sapwood accumulates from division of the vascular cambium.

Commercial lumber is classified as softwood or hard-wood. Softwoods are coniferous woods taken from pine, spruce, fir, or hemlock. Hardwoods are taken from eudicots, such as oak, cherry, ash, and several other species.

Look at the layers in the bark and identify the cork layer and phloem.

Monocot Stem Histology

12 Examine a prepared slide of a cross section of a corn (maize) stem. Use the 4× objective. Refer to **figure 25.14** and identify the cell types in your specimen.

Note how the monocot stem organization differs from the herbaceous eudicot stems studied earlier. The vascular bundles are not arranged in concentric circles. Instead, they are scattered throughout the ground tissue of the stem. Unlike gymnosperms and many eudicots, monocot stems do not undergo secondary growth and there is no vascular cambium. Examine a **vascular bundle** under higher magnification. (Some people think these bundles resemble a monkey's face.) The bundle is surrounded by thick-walled sclerenchyma fibers. The xylem typically includes four conspicuous vessels. The phloem is found in the area between the two larger vessels and extends from there toward the edge of the bundle. Look for sieve plates in some of the sieve tubes. Lacking a vascular cambium, monocots do not produce secondary xylem or secondary phloem.

The stem of corn is not representative of the stems of all grasses. Some grass stems are hollow, except at the nodes (**fig. 25.15**). In hollow stems, the vascular bundles are arranged as concentric circles in the walls of the stem.

In grasses, a meristem is found at the base of each internode and at the base of the leaf sheath. These are located at the base of the grass plant. When we cut grass, the tip of the leaf and stem is cut off but the meristems close to the soil are not harmed. They produce new cells and the leaf and stem "shoot up" so that the lawn must be cut again, and again, and . . .

Branches arise from the lower portions of grass stems at the soil level. Usually, the branches grow more or less horizontally, often underground, to form **rhizomes** from which shoots grow upward and fibrous roots grow downward. This type of growth is called tillering and allows a single plant to spread out from a single seed, crowding out competing plants.

Leaf Structure

Leaves are the photosynthetic organs of most plants, the exceptions being those plants with photosynthetic stems. Leaves are well adapted to the task. They are only a few cells thick and are covered by epidermal tissue that

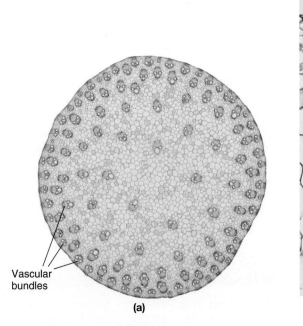
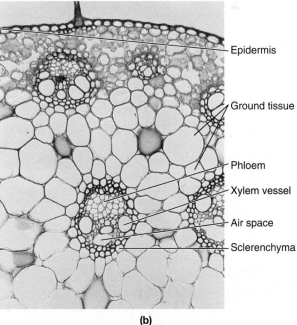

Figure 25.14 Monocot stem: (*a*) cross section of a corn, *Zea mays*, shows vascular bundles are scattered throughout parenchyma; (*b*) enlargement showing details of vascular bundles.

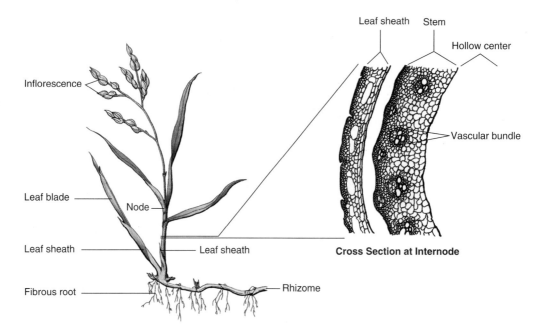

Figure 25.15 Anatomy of a grass plant. Stem is hollow in most species.

limits water loss. Leaves are supplied with vascular tissues that bring water and minerals from the roots and carry away photosynthetic products to living cells in the stem and roots.

Although chloroplasts are found in all green parts of plants, the leaves are the primary solar energy collectors, whether the plants are ferns, pines, birches, or grasses. Leaves are remarkable organs having a high surface area per unit volume. Furthermore, many plants carry and position their leaves to capture sunlight effectively. Water evaporation through leaf surfaces also creates a tension which draws water with dissolved minerals from the roots through the xylem, supplying the living cells in the above parts of the plants with needed moisture.

Many leaves display adaptations to specific habits. Leaves from dry areas are often thickened and have trichomes on their surfaces that minimize water loss. Aquatic plants with floating leaves have large air spaces in the leaf. There are often differences in the leaves of plants that are adapted to full sunlight or shade. Conifers have needles with reduced surface areas that reduce moisture loss in cold winters and dry summers.

Whole Leaves

A eudicot leaf consists of a flat **blade** and a stalk, or **petiole,** which attaches the leaf to the stem. Xylem and phloem pass from the leaf through the petiole and stem to the roots. At the point where a leaf attaches to a stem, an axillary bud is found. The bud gives rise to branches of the stem. Leaves may be **simple,** consisting of a single blade and a petiole, or **compound,** consisting of several leaflets joined to a single petiole (**fig. 25.16**).

Vascular tissues (xylem and phloem) will be visible as **veins** in the leaves. Three venation patterns are found in leaves:

1. **Parallel**—veins pass from the petiole to the tip of the blade in a more or less parallel fashion. Monocots (grasses and their relatives) have leaves with parallel venation. Many conifer needles have only a single vein. Other gymnosperms such as ginkgoes (fig. 25.16e) have dichotomous venation where veins branch into two equal parts.

2. **Net venation**—veins are not parallel. Eudicots (broadleaf plants) have leaves with net venation.

    a. **Pinnate**—a single main vein gives off smaller branch veins that run parallel to each other as in a feather.

    b. **Palmate**—several main veins radiate from where the petiole joins the blade as your fingers radiate from your palm.

13 Find examples of all three types of venation among the demonstration materials. Record below the names of the species observed and whether they have simple or compound leaves.

Leaf Histology

14 *Eudicot Leaf* Hand-cut cross sections of leaves can be easily made (**fig. 25.17**). Once you have mastered the technique, a variety of leaves can be examined and

INVESTIGATING FUNCTIONAL ANATOMY OF VASCULAR PLANTS 345

Figure 25.16 Types of leaves and venation: (a) palmately veined maple leaf; (b) simple but lobed leaf of tulip tree; (c) opposite, simple leaves of dogwood; (d) whorled leaves of bedstraw; (e) fan-shaped leaf of *Ginkgo,* with dichotomous venation; (f) palmately compound buckeye leaf; (g) pinnately compound black walnut leaf; (h) a grass leaf; (i) linear leaves of yew; (j) needles of pine.

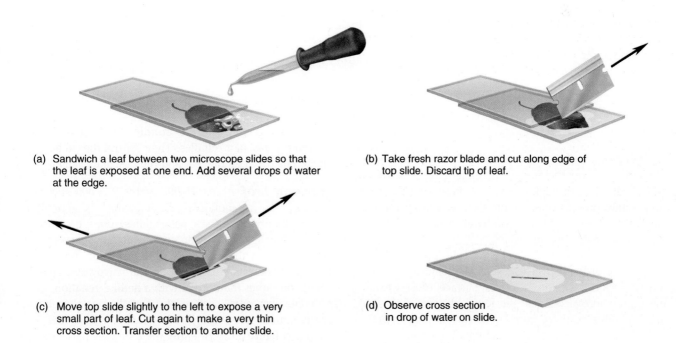

(a) Sandwich a leaf between two microscope slides so that the leaf is exposed at one end. Add several drops of water at the edge.

(b) Take fresh razor blade and cut along edge of top slide. Discard tip of leaf.

(c) Move top slide slightly to the left to expose a very small part of leaf. Cut again to make a very thin cross section. Transfer section to another slide.

(d) Observe cross section in drop of water on slide.

Figure 25.17 Two-slide method for making cross section of a leaf.

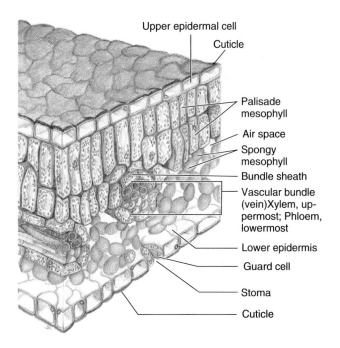

Figure 25.18 Tissues in a leaf: Artist's three-dimensional drawing of leaf organization.

compared. Try cutting sections from one of the sample plants in the lab. Mount the sections on edge in a drop of water on a slide. Add a coverslip and observe. To determine if the structure of leaves is similar in a number of different plants, the class should hand-section leaves from several different species of plants and exchange slides after completing their observations. Alternatively, obtain a slide of a cross section of a eudicot leaf. Look at it with the low-power objective of your compound microscope, switching to high power to see detail (**fig. 25.18**).

How many cells thick is the leaf?_____ Why do you think they are so thin?

In nature, leaf surfaces are usually covered by a **cuticle,** a layer of wax secreted by the underlying cells, which slows water loss through the large surface area. The slide preparation process often extracts the wax from the leaf surface but you should see it on hand-sectioned material. Is a cuticle visible on your slide?_____

Find the **epidermis,** usually a single layer of cells at the surface of the leaf. These cells lack chloroplasts and have a thick cell wall on their outer surface. Closely examine the epidermal cells on the lower surface of the leaf and find the specialized **guard cells** surrounding an opening called the **stoma** (plural: stomata). (See fig. 25.18.) Are stomata found more frequently on the upper or lower surface of a leaf?_____ Hundreds of stomata may be found in a square millimeter of leaf surface. Carbon dioxide enters a leaf through these openings and water and oxygen escape through them.

The inside of the leaf is composed of **mesophyll** tissue characterized by thin-walled parenchyma cells. They account for most of the photosynthetic activity of a plant. The mesophyll is divided into two regions. Find the **palisade** mesophyll composed of one or more layers of closely packed cells. These are near the upper surface of a eudicot leaf. Beneath the palisade cells is **spongy** mesophyll composed of irregularly shaped cells separated by a labyrinth of air-filled, intercellular spaces which allow carbon dioxide, oxygen, and water vapor diffusion. Note how these spaces are continuous with the stomata. What are the implications of this arrangement?

In which mesophyll layer do the cells contain the most chloroplasts? _____

You should also be able to see **vascular bundles** composed of xylem and phloem that transport materials to and from the mesophyll and the rest of the plant. What materials do they bring to the mesophyll?

What materials do they translocate from the mesophyll?

Note the tightly packed **bundle sheath** cells that surround the vascular bundles. They control the exchange of materials between the leaf and the vascular system.

Monocot Leaf About 70,000 of the 250,000 species of angiosperms are monocots. These include the agricultural grains, bamboos, grasses, palms, lilies, tulips, orchids, and many other species. The anatomy of monocot leaves differ from those of the eudicot you just studied (**fig. 25.19**). Monocot leaves are usually long and narrow with parallel venation rather than broad with a netlike venation.

15 Obtain a cross section of a corn leaf or alternatively hand-cut sections from living monocots, using the technique in figure 25.17. Identify the following parts: cuticle; upper and lower epidermis; stomata; mesophyll, consisting of parenchyma cells with chloroplasts; and bundle sheath cells surrounding the xylem and phloem cells in a

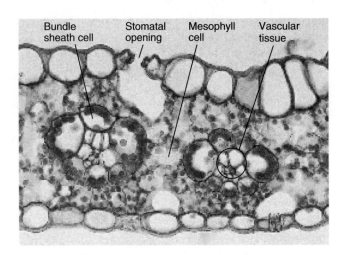

Figure 25.19 Cross section of a corn leaf as seen through a compound light microscope.

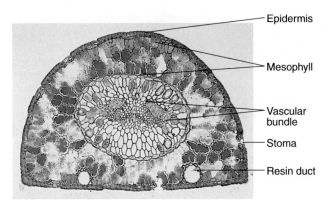

Figure 25.20 Cross section of a single pine needle.

vein. Look closely at the parenchyma. Is it organized into palisade and spongy layers as in eudicots?_____

Look closely at the upper and lower epidermis. On which surface are stomata more common?_____

16▶ **Gymnosperm Needle** Obtain a slide of a cross section of pine needle (fig. 25.20) and look at it with your compound microscope. The cuticle may not be visible because it was extracted by solvents used in the slide preparation process. Note the **epidermis,** which secretes the cuticle, and the **stomata.** What is their function?

Beneath the epidermis locate the **mesophyll,** consisting of cells that seem to fit together like pieces of a jigsaw puzzle. These are the photosynthetic cells and contain large, darkly staining chloroplasts. In the center, find one or two **vascular bundles.** Each bundle contains xylem with thick-walled tracheids and thinner-walled sieve cells in the phloem. Near the flat surface of the needle, two or more **resin ducts** should be visible. Resins seal breaks in the needles and may protect the plant from herbivores.

What similarities and differences do you observe between conifer needles and eudicot leaves?

Between conifer needles and monocot leaves?

Guard Cell Response to Osmotic Stress

The opening and closing of stomata is related to the osmotic condition of the guard cells. Guard cells change shape by taking on or losing water. Due to the orientation of cellulose microfibrils in their cell walls, they bend when they take on water, opening the stoma. When the guard cells lose water, they straighten and the stoma closes, (fig. 25.21).

17▶ These conditions can be simulated by treating a strip of leaf epidermis containing guard cells with distilled water and a 5% salt solution. Place a drop of distilled water on a slide. Obtain a *Zebrina* leaf. Fold it in half with the underside inward (fig. 25.22). As it begins to break, pull the upper half away from the break. This should produce a cellophane-thin piece of the lower epidermis. Mount it in the water drop.

Add a coverslip and observe. In half of the circle below, sketch your observations of the guard cells.

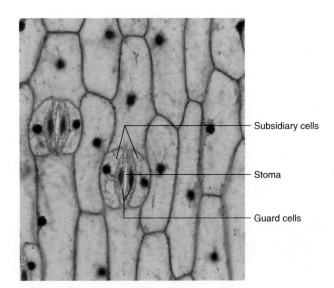

Figure 25.21 *Zebrina* epidermis. Each stoma is flanked by a pair of guard cells and subsidiary cells. Ordinary epidermal cells are the large cells.

Figure 25.22 By pinching leaf as shown, it will fracture, revealing a small, translucent edge of epidermis that can be stripped off with forceps and mounted on slide in a drop of water.

What would you hypothesize would happen if you added several drops of a 5% salt solution next to and touching the coverslip? State your hypothesis below.

Add the 5% saline by placing a drop next to the coverslip and touching a piece of blotting paper to the opposite side of the coverslip. This will wick the salt solution under the coverslip. Observe the guard cells again and sketch a few cells in the other half of the circle. How are the cells different? What caused this change in the guard cells? Must you accept or reject the hypothesis that you made?

18) It is fun to work with numbers to estimate the magnitude of an event or process. Some numbers are available regarding transpiration. A researcher counted the number of leaves on a 47-foot silver maple tree. He found that there were 177,000 leaves, with an average leaf area of 25 cm^2. If the transpiration rate per silver maple leaf is 0.01 ml/hour/cm^2, how many liters of water would be lost from a mature tree in 12 hours? _____ How many gallons move through the vascular system and out the stomata of a tree this size in a day? (A gallon equals 3.9 liters.) _____ Another researcher working with tobacco estimated there were 12,000 stomata per cm^2. Assuming the same number for maple leaves, how many stomata did the silver maple have? _____

Investigating Water Movement Mechanisms

Water Movement Due to Transpiration

In the previous sections, you studied the anatomy of roots and shoots with their vascular tissues. Now you will investigate the functions of the transport system.

During periods of active photosynthesis, the stomata of the leaves are open, allowing carbon dioxide to enter. At the same time, water escapes from the leaf and is replaced by conduction up the xylem (see fig. 25.1).

19) Obtain a fresh piece of celery stalk with leaves. Cut off and discard the bottom two inches of the stalk and quickly place the cut end of the stalk in a beaker containing a 0.5% solution of methylene blue dye. Place the celery in sunlight.

Let the stem sit in the dye for 15 minutes. Predict how far the dye will move. At the end of 15 minutes, make cross sections at 1 cm intervals up the stem. How many centimeters did the dye travel in 15 minutes? Does this value match your prediction?

Use a razor blade to make a thin section of a region that had a high dye concentration. Mount the section on a slide and look at it with your compound microscope. In what tissue is the dye located?

Testing Water Tension–Cohesion Hypothesis

Water is thought to be pulled through the xylem system by tension created in the leaves when water is lost through transpiration (fig. 25.1). If this is the case, then an experiment can be devised to test the theory. Sunflower plants can be grown in pots until they are at least a foot tall. Then they can be subjected to different watering regimens. One pot could be well watered for a week before lab, the other not watered. If the theory is correct, the water in the xylem of the unwatered plant should be under greater tension than that in the watered one.

20 To test this hypothesis, fill a beaker with a methylene blue dye solution. Take plants that have had different watering treatments and cut the stems one at a time. As quickly as you can (within a few seconds), put the cut end of the stem in the dye. After 15 seconds, remove it. If the water columns in the xylem were under tension, the dye should be sucked into the cut ends and travel different distances in the two plants.

Take the stems and cut them every 5 mm. By looking at the color of vascular bundles, you can see how far the dye has traveled.

Describe the results below.

Do the results of this experiment support or refute the water tension hypothesis for water movement in xylem? Explain.

Lab Summary

21 On a separate sheet of paper, answer the following questions as assigned by your instructor.

1. Distinguish between primary and secondary growth in plant roots and stems.
2. In **table 25.1** on the next page, make reasonably detailed drawings of cross sections of eudicot and monocot roots and stems.
3. Using both cross section and longitudinal section outline diagrams, describe the pathway for water and mineral absorption in plant roots.
4. Describe how the structure of the endodermis determines its function.
5. As a woody stem undergoes secondary growth, what will happen to cell layers located outside of the vascular cambium and phloem?
6. Draw cross sections of both a woody eudicot stem and a herbaceous stem. Indicate how they are different.
7. How would you distinguish among eudicot leaves and conifer needles based on cross sections?
8. Describe the transpiration–cohesion theory of water movement in plants. What evidence do you have that it is true? Explain.

You may want to try the critical thinking questions that apply some of the knowledge you gained in doing this lab.

INTERNET SOURCES

Use the search engine Google to locate on the World Wide Web a research report dealing with how phloem transports organic molecules. Write an abstract of the report and include the URL.

LEARNING BIOLOGY BY WRITING

Because the activities in this lab were more discovery-based science than experimental, your instructor may want you to write a summary rather than a report. In about 200 words, describe how water enters a eudicot plant and moves to the leaves. Include anatomical diagrams of the root, stem, and leaf showing the water pathway. Discuss the mechanism of movement.

CRITICAL THINKING QUESTIONS

1. Pines depend on mycorrhizal relationships. A plant nursery was sterilizing soil before planting Christmas tree seedlings and found that their trees took on average two to three years longer to reach sellable size. Why?
2. Many trees will die after their roots are flooded with water or have cars parked on them, compacting the soil. Why?
3. People often spread fertilizer under large trees to make them grow even faster. Sometimes they put the fertilizer close to the trunk, thinking it will be absorbed faster. Why is this erroneous thinking?
4. What anatomical feature of the root would allow radishes to accumulate zinc in the parenchyma tissue of the root but not in stems and leaves?
5. The tallest trees in the world are around 100 m high. Do you think this is the maximum possible height or could trees grow even higher? Explain.
6. How can information about previous climate be obtained by studying the growth rings of trees?
7. If you cut the bark of a tree down to the vascular cambium, in a circle completely around the tree, the tree dies. Why?
8. A tree grows about one foot each year. You carve your initials in the trunk of a tree that is 15 feet tall. Your initials are four feet above the ground today. How far above the ground will they be 10 years from now? Why?
9. What has happened to the epidermis, cortex, and primary phloem of a 100-year-old maple tree?

Table 25.1 Summary Anatomical Comparisons: Eudicot versus Monocot

In the spaces below, draw representative cross sections of eudicot and monocot roots, stems, and leaves. Label all regions as discussed in the corresponding sections of the lab manual.

Eudicot Root	Monocot Root
Eudicot herbaceous Stem	Monocot Stem
Eudicot leaf	Monocot leaf

Lab Topic 26

Investigating Pollination, Development, and Dispersal

Supplies

Resource guide available on WWW at http://www.mhhe.com/labcentral

Equipment
Compound microscopes
Balance
Digital camera (optional)

Materials
PBS Nature video "Sexual Encounters of the Floral Kind"
Angiosperm flower demonstration dissection
Various fleshy and dry fruits and seeds for demonstration
Soaked beans and corn kernels
Seeds (pea, bean, soybean, corn, radish, or other fast-germinating seeds)
Dark-grown 4–5-day-old oat seedlings
Light-grown 4–5-day-old oat seedlings
    Other species may also work well
Prepared Slides
    Capsella embryo development
    Pistil and ovary, longitudinal section (demonstration)
Miscellaneous
    Razor blades
    Dissecting needles
    Large petri dishes
    Large filter paper
    Parafilm
    Repeating pipette, or other dispensing device

Solutions
I_2KI stain in dropper bottles
100 μM solution of gibberellic acid
100 μM solution of abscisic acid

Student Prelab Preparation

Before doing this lab, read the Background material and sections of the lab topic assigned by your instructor.

Use your textbook to review definitions of the following terms:

abscisic acid
cotyledons
embryo sac
endosperm
fertilization
fruit
gibberellins
hypocotyl
integuments

morphogenesis
ovary
ovule
ploidy level
pollen
pollination
radicle
seed
seed coat

Be able to describe in your own words the following concepts:

Role of ovule in flowering plant reproduction
Fertilization in flowering plants
Formation of embryo in seeds
Roles of plant hormones in regulating germination of seeds

After finishing the prelab review, write any questions in the margins of the pages of this lab topic. The lab activities should help you answer these questions, or you can ask your instructor during the lab.

Objectives

1. To examine the adaptations of flowers to promote pollination
2. To examine embryo development in seeds
3. To examine seed and fruit anatomy in relation to dispersal mechanisms
4. To determine hormonal effects on germination and seedling development
5. To determine the impact of light on the growth of oat seedlings

Background

Flowering plants reproduce by two modes: asexually (vegetatively) by producing new individuals from an older individual's leaves, stems, or roots; and sexually with gametes, which give rise to an embryo contained in a seed. Both modes have their advantages.

Asexual reproduction produces progeny that can make prolonged use of the energy reserves from their parents and thus may increase the chances for survival over those of sexually produced seedlings. Because asexually produced progeny are genetically identical to the parent, they also have the same genetic advantages that allowed the parent to reproduce asexually in its environment. A genetically superior individual can thrive and multiply rapidly without risking disruption of favorable allele combinations. Asexual reproduction can yield impressive results: a single aspen tree in the Rocky Mountains is estimated to have produced by repeated root sprouting 47,000 individual trees covering 200 acres. Cattails, prairie grasses, blueberries, blackberries, cherries, and strawberries (**fig. 26.1**) are examples of other species that are asexually active reproducers. The majority of flowering plant species use asexual reproduction, although many annual plants do not.

Many named cultivars of fruits and roses are propagated only vegetatively. For example, all McIntosh apple trees in the world today are traceable back to one tree that had superior fruit qualities. Offspring were first propagated by cutting branches from the original tree and grafting them onto roots of crab apple trees. Cuttings from subsequent generations have been used to produce the present number of trees. If you planted the seeds of McIntosh apples, would you expect the tree that grew to bear exactly the same kind of fruit as its parent? Why?

While a principal advantage of asexual reproduction is genetic constancy, the main advantage of sexual reproduction is genetic variability. Offspring variability results from the meiotic production of sperm and egg with the built-in processes of crossing-over and independent assortment, and the combination of genes from two parents at fertilization. Each individual resulting from the fusion of gametes represents a unique combination of genes. If the environment changes, a genotype that was previously highly successful may no longer be so. In times of change, the inherent variability of offspring from sexual reproduction increases the probability of continued survival of a species.

Sexual reproduction in flowering plants is complex. You may want to review the last half of lab topic 18 to refresh your memory of the details. All plants have life cycles based on an alternation of generations with haploid gametophyte and diploid sporophyte stages. The sporophyte stage produces spores by meiosis. The spores develop into microscopic male or female gametophytes within the male (anthers) or female (ovules) parts of the flower. Those gametophytes produce gametes, haploid eggs and sperm, which combine at fertilization. Unlike many animals, plants are not extremely mobile, so sexual reproduction over a distance is challenging. Pollen grains are mobile male gametophytes that can travel great distances to immobile eggs held by female gametophytes in ovules, an event called pollination. Fertilization is an event distinct from pollination. Once pollen has been transferred, it must germinate, producing a pollen tube that grows through the open tissues of the female portions of the flowers to deliver sperm to the egg. After fertilization, the development of the zygote into an embryo in the ovule is supported by the sporophyte bearing the ovule. The ovule matures into a seed containing the embryo. The seed must transfer from the parent to a place suitable for growth, and germinate before the embryo can develop into a new plant.

Embryogenesis, the formation of the embryo from the zygote, accompanies the development of the seed. To achieve this multicellular stage, two developmental organizations must occur: a longitudinal apical–basal pattern and a radial pattern to produce concentrically arranged tissue systems. Over 50 genes have been identified that regulate the development of this organization, but we are still a long way from fully understanding this most fundamental mechanism of plant development.

Dispersal of offspring is another challenge. How do a plant's offspring get to locations away from their maternal parent? How do coconuts reach new islands, or weeds from Eurasia get to North America? How is it that seeds can survive for such long periods of time and then begin an intense period of development? Recently, a 2,000-year-old seed from a date palm was discovered in an archeological dig in Israel. Surprisingly, when planted, it grew into a seedling similar to present day palms despite its age. Adaptations for maintaining viability, dispersal, and signals that trigger development are important questions in plant biology.

The intent of this lab is to allow you to look at several of the morphological aspects of pollination, plant dispersal, and development. As you do so, think of these questions and others like them. It is wondering *why* and *how* that starts research projects and careers in science.

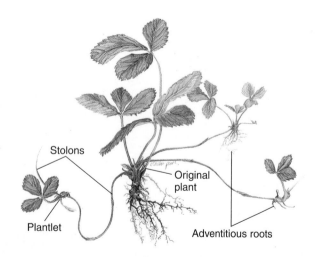

Figure 26.1 Asexual reproduction occurs by runner (stolon) formation in strawberries.

> **LAB INSTRUCTIONS**
>
> You will view a film showing plant adaptations that promote pollination and dispersal of seeds. You will then study embryo development, dissect eudicot and monocot seeds, and investigate the effects of light and hormones on seedling development.

Pollination and Seed Dispersal

Plants have evolved an array of adaptations to use mobile animals to disperse pollen and seeds, and animals have evolved complementary adaptations to the rewards that plants offer. On the plant side of the equation, the adaptations include:

- Floral structures that promote pollen transfer to animals seeking food.
- Floral colors and odors that make flowers quite apparent.
- Seed formation in tissues that are not easily detected.
- Fruits that are colorful, fragrant, and nutritious.
- Abundant fruit production to attract animals that disperse seeds.

1) To explore the adaptive strategies plants use to ensure genetic variability through sexual reproduction and to promote dispersal of offspring, you will watch the PBS Nature video called "Sexual Encounters of the Floral Kind." As you watch, look and listen for answers to the questions below. It is a good idea to take notes. If you missed a question, consult with one of your classmates to exchange information, or consult your textbook.

What are plant adaptations for wind pollination?

How do plants induce animals to transport pollen? Name several rewards that plants provide to animal pollinators.

Describe some examples of how pollinators are guided to the right spot to pick up or deliver pollen.

How do some plants attract pollinators without providing material rewards?

How do some animals cheat the system—take plant resources intended to attract pollinators without providing pollen transport services?

How is plant petal movement important in plant–pollinator interactions? Give at least two examples.

Drawing on information from previous labs, how do you think plants produce those movements?

Some plant tissue surfaces are hydrophobic and some are hydrophilic. How do plants produce those differences, and how are they important in the reproduction of some plant species?

Many flowers have both male and female parts. What pollination mechanisms reduce the likelihood of self-fertilization and increase the chance of outcrossing in such flowers? List examples.

Dispersal requires movement of seeds, not just pollen. What examples of seed dispersal mechanisms did the video provide?

Female Gametophyte

2 The film illustrated many adaptations of both plants and animals that promote cross pollination within species of flowering plants, bringing a mobile male gametophyte to the general location of a female gametophyte. The female gametophytes are in ovules contained in the ovaries usually near the base of the flower. Look at the demonstration dissections of a flower that your instructor has prepared as a review.

Look at the demonstration slide of a longitudinal section of the pistil. You can see where the pollen would be trapped on the stigma. Pollen tubes would grow through the loosely organized tissue of the stigma and style toward the ovary. In the ovary, find the **ovules.** Inside each ovule is a haploid female gametophyte stage that has produced an egg and seven other cells (see fig. 18.15). Fertilization occurs inside the ovule where the embryo also develops. As embryo development progresses, the ovule matures to become a seed.

Embryo Development

3 Obtain a slide of the fruit of the common plant shepherd's purse *(Capsella)*. A member of the mustard family, this species formerly was native only to the Mediterranean region but has spread throughout the world over several hundred years of human migrations. It is considered a noxious weed in 46 of the states in the United States of America. The small white flowers produce triangular dry fruits that resemble antique leather purses, hence the common name. It is used as a model for angiosperm embryogenesis (formation of the embryo), because each fruit is small and contains many seeds that develop in a pattern that is common to all eudicot plants.

Look at your slide with your compound microscope. You are looking at a dry type of fruit that developed from an ovary containing many ovules. Each ovule has developed into a seed and each contains an embryo at some stage of development (**fig. 26.2**). All stages may not be visible on all slides.

On your slide, locate a seed containing a mature embryo. The outer walls of the seeds, the seed coats, developed from the sporophyte integuments surrounding the ovules. When the plant was in flower, double fertilization occurred in each ovule when a pollen tube grew from the stigma through the style to the ovary and then entered the ovule though its micropyle (see figure 18.9). Pollen tubes contributed two sperm into each ovule. One fused with the two polar nuclei to form a triploid tissue called

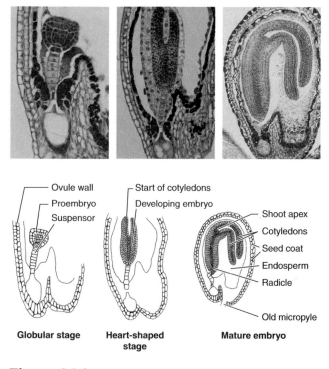

Figure 26.2 Photos and artist's interpretations of stages in the development of a *Capsella* (shepherd's purse) embryo in seeds.

endosperm and the other fused with the egg to produce a **zygote.** All other cells you now see inside the seed coats arose from these two cells by mitotic divisions.

As the endosperm cells proliferate they accumulate nutrients from the parent plant becoming rich in starches, proteins, lipids, and vitamins that will fuel the embryo's development. What is the ploidy level of the endosperm's cells? _____

The two cells that are formed by the first mitotic division of the zygote have two different fates. The one near the base, nearest to the micropyle, will divide a few times to form a filament of cells called the **suspensor** (fig. 26.2). It holds the embryo in place during the early developmental events, supplies nutrients to the embryo, and produces growth regulating plant hormones, especially gibberellins. Identify the suspensor cells in one of the seeds on your slide. As the embryo matures, the suspensor cells undergo genetically programmed cellular death and are gone by the time a mature embryo is produced.

The cell located away from the micropyle, the **apical cell,** will divide repeatedly to form a roughly spherical embryo called a proembryo or **globular stage.** What is the ploidy level of the embryo's cells? _____

Coordinate your observations with others in the lab so that you see several developmental stages: globular (proembryo), heart-shaped, mature embryo.

At first the cells of the embryo are undifferentiated and similar in appearance, but then they begin to follow different developmental pathways. Some will become meristematic tissues that will produce the tissue systems you studied in adult plants in lab topic 24. Find an example of a seed on your slide containing a globular proembryo stage. Look at the surface layer of cells and note how they are smaller and arranged on an axis parallel to the embryo's surface. This is the protoderm and will produce the dermal tissues of the embryo. Although more difficult to see, cells toward the center of the globular mass are differentiating to become ground meristem and procambium that will form the ground and vascular tissues, respectively. Cells at the top of the globular proembryo will become apical meristem cells. Their cell divisions will elongate the shoot. Cells at the base near the suspensor will become root tip meristem cells and elongate the root.

Continued differential growth and division of the embryo lead to the appearance of definable regions in the embryo. Cells on either side of the apex of the globular mass begin to grow outward, forming the **cotyledons** or seed leaves, and the remaining cells begin to elongate. The embryo at this time resembles a crude caricature of a heart. Find one of these **heart-shaped stages** among the seeds on your slide. The name *eudicot* refers to the fact that these plants have two cotyledons. *Monocots* have only one cotyledon. The cotyledons may accumulate food materials from the endosperm and become thick and fleshy as in peas or beans, where the cotyledons represent the bulk of the mature embryo. In other species, like wheat and corn, the cotyledons remain small and thin and the endosperm persists to provide starch, proteins, and lipids to the developing embryo. It is these nutrient stores that make many seeds excellent foods.

By this time two other developmental sequences are in progress. Cell divisions are becoming less common throughout the embryo and more common at the apical and root meristems, so that the embryo elongates, becoming what is called a **torpedo stage.** The procambium now begins to produce vascular tissues that you should be able to see in the center of the elongating cylindrical embryo. Look carefully at the shape of these cells compared to those on the surface of the embryo. Describe the differences below.

In a mature eudicot embryo, the **shoot apex** is seen as a small node of tissue between the cotyledons. The meristematic regions located here produce the shoot. Beneath the shoot apex and cotyledons is a short embryonic stem, the **hypocotyl.** In many plants, but not all, the hypocotyl elongates at germination. As it does, it forms a characteristic, inverted U-shape with the base in the seed and the tip attached to the cotyledons. When the curve of the U breaks the soil surface, it begins to straighten, pulling the cotyledons and shoot apex upward. This mechanism avoids pushing the shoot apex through abrasive soil, which could injure the meristematic tissues. As the hypocotyl straightens, the new shoot begins to grow from the shoot apex. It connects with an embryonic root, the **radicle.**

You have seen that plants disperse offspring in seeds as multicellular embryos, not as fertilized eggs. What advantage might this have in reproduction?

Seed Anatomy

As the embryo develops, there is a massive inflow of nutrients from the parent plant to the ovule. At first this material is contained in the endosperm, but at the heart-shaped stage it begins to transfer to the cotyledons in many, but not all, eudicots. These cotyledons enlarge as their cells store nutrients that will fuel future germination and early growth events. In monocots, the single cotyledon does not enlarge but remains thin and membranous. It will absorb materials from the endosperm during germination and pass them on to the developing plant. As these

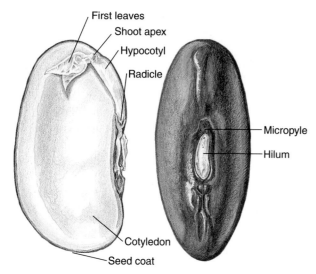

Figure 26.3   Structure of a bean seed.

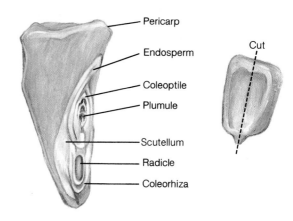

Figure 26.4   Structure of a corn kernel.

events proceed, the seed coats harden and the seed no longer takes up materials from the parent.

4) We will now look at the anatomy of whole seeds. Return your *Capsella* slide to the supply area and obtain a lima bean that has been soaked overnight to soften the seed coat and hydrate the tissues.

You are looking at a mature ovule, containing the plant embryo and stored food materials. The outer covering, the seed coat, is derived from the integuments that surrounded the ovule. Find the **hilum** that attached the ovule to the ovarian wall in the fruit (fig. 26.3; also see fig. 18.14). To one side of the hilum, the remnant micropyle should be visible as a small aperture. Peel away the seed coat to find the two **cotyledons** of the embryo. Separate the cotyledons and examine each half to see the other parts of the embryo. You should see the **shoot apex, radicle,** and the **hypocotyl**. Is the lima bean a monocot or a eudicot? _____

5) We will use corn seed as an example of a monocot seed. Obtain a soaked corn kernel, place it on a glass plate, and use a razor blade to cut it lengthwise at right angles to the broad side of the seed (fig. 26.4). Place a drop of I_2KI stain on each half. Starch in the endosperm will stain purple and the embryo will stand out against the dark background. Examine the halves to find the embryo in the base and on one side of the kernel. Note the large amount of endosperm in the kernel. A corn kernel is really a fruit consisting of a single seed to which the ovarian wall is tightly attached. This covering is called the **pericarp**.

Draw the cut surface of the corn. Label the pericarp, endosperm, shoot apex, the surrounding sheath (coleoptile), the single cotyledon (which in grains is called the scutellum), and the radicle with its enclosing sheath, the coleorhiza.

Fruits

Corn kernels, pea pods, blueberries, watermelons, and acorns are all examples of fruits because they surround the seeds and develop from the ovary of a flower.

Fleshy fruits develop from the ovary wall, the **pericarp,** which may thicken and differentiate into layers. Depending on the species, these layers may or may not be easy to distinguish. The pit of a peach is also mostly fruit. The outer "fruit flesh" layer is moist and tasty to attract animals. The peach pit is an "inner" fruit layer called the endocarp which protects a single large seed inside this hard cellulose and lignin shell. Fruit development is controlled by hormones, especially ethylene. Young fruits are usually inconspicuously green and are bitter tasting, protecting the seeds during maturation from easy detection by herbivores. When the seeds mature, the flesh softens, sugars accumulate, colors change to yellows and reds, and they become aromatic. All these adaptations attract animals that can be important agents for seed dispersal. Animals eat many fleshy fruits, ingesting the enclosed seeds in the process. The tough seed coat protects the embryo as it travels through the digestive tract. When the animal defecates, the seeds most likely are at a new location and surrounded by a pile of fertilizer!

Other species of flowering plants have other means of seed dispersal. Some fruits, such as those of maples and ash, are winged and are transported by the wind. Orchids have an almost dustlike seed that is easily airborne. Spiny fruits or seeds, such as cocklebur and beggar tick, stick to the coats of passing mammals and birds. Violets, *Trillium*, and many woodland understory plants produce seeds that have a protein and lipid-rich extension of the seed coat that attracts ants. The ants carry the seeds to their nests where they eat the seed coat outgrowth but not the seed. It may germinate in its new location in the ant's nest.

6▶ Examine several fruits, such as acorns, various edible fleshy fruits, berries, legumes, grains, and dry fruits from field plants. The fruits will have been cut in half to produce both longitudinal and cross sections. Find the mature ovules (seeds). Remove one to identify the hilum and micropyle. Look at the fleshy part to identify dermal tissues and the parenchymal cells of the pericarp. Do you think there is an endocarp? How would you tell?

Be careful when identifying fruits and seeds. What we think of as a sunflower seed is actually a fruit. The hull is the hardened ovary wall that protects the seed inside.

Horticulturist have devised a classification system for fruits that is outlined in **table 26.1**

Germination

The seed coat surrounding mature seeds is rather impermeable to water and oxygen. Furthermore, the moisture content of seeds may be quite low in the range 5 to 15%. Consequently, the embryo is in a state of suspended animation, exhibiting little, if any, metabolic activity. Before the seeds of many species will germinate, they must go through a period of dormancy. Dormancy is induced by the plant hormone abscisic acid. Low temperature exposure, fire, mechanical abrasion of the seed surface, or exposure to digestive acids and enzymes of herbivores are examples of some of the triggers that break the dormancy. As abscisic acid levels decline, water uptake, called **imbibition,** is one of the first events in breaking dormancy. As water enters, the cells rehydrate and become metabolically active, breaking down starches to glucose that is anaerobically metabolized to produce ATP. The tissues swell, exerting pressure on the seed coat, which splits, allowing oxygen to enter and metabolism becomes aerobic. Cell divisions and cell enlargement increase the size of the embryo, causing it to grow out of the seed using the food reserves stored in the cotyledons or endosperm. The root emerges first, anchoring the plant in the soil. As the shoot emerges, it begins to green, becoming photosynthetic to produce its own energy and building materials. Gibberellins are hormones that play a key role in breaking dormancy, activating genes that produce enzymes that break down the seed's food reserves, and causing the cell elongation involved in seedling emergence.

In the next two sections, you will investigate the effects of hormones and light on seedling development.

Table 26.1 Horticultural Classification of Fruits

Kind of Fruit	Description	Examples
Simple	Develop from the ovary of a flower with single pistil	
Dry		
Achene	Fruit does not split open at maturity; seed remains attached to ovarian wall by stalk	Sunflower
Follicle	Fruit splits along one edge at maturity to release seeds	Milkweed, peony
Legume	Fruit splits along both edges at maturity to release seeds	Bean pod, carob, lentil, pea pod, peanut
Nut*	Hard ovarian wall encases a single seed and does not split open at maturity	Hazelnut, hickory nut, acorn
Grain	Seed coat is permanently united with ovarian wall; does not split open at maturity	Barley, oats, rice, wheat
Samara	Ovarian wall forms "wings" that aid in dispersal	Maple, elm, ash
Fleshy		
Berry	Entire ovarian wall is fleshy	Tomato, grape, date, avocado
Drupe	Outer zones of ovarian wall become fleshy, while inner zone is hard	Cherry, peach, plum, olive
Pome	Inner zone of ovarian wall is leathery and outer is fleshy incorporating accessory structures	Apple, pear
Aggregate	Develop from single flowers with multiple pistils	Raspberry, strawberry

*Many so-called nuts are not true nuts; the peanut is a legume and walnuts, pecans, and cashews are drupes which lose the fleshy part during maturation.

Research Project: Hormone Effects on Germination and Seedlings

Plant development is influenced by external factors such as water, mineral, and light availability, and by internal regulators, especially hormones. These relatively small organic molecules play a major role in the regulation of plant growth by altering the expression of genes that lead to a change in the structure and function of the affected target cells.

Hormones are chemical signals released by some cells that direct the development and function of cells or tissues elsewhere in the organism. The plant hormones abscisic acid (ABA) and gibberellic acid (GA) affect a variety of plant processes, including seed dormancy, germination, and seedling growth. You may wish to review your textbook regarding hormonal effects on plant structures or search the library and/or Web for background information.

We can study the germination of seeds and early development of roots and shoots by placing the seeds on absorbent paper in a petri dish and adding a hormone solution to moisten the paper. That setup makes it easy to observe seeds and measure germination as well as early seedling development. At typical room temperatures, radish, pea, bean, and soy seeds will undergo enough development within a week to quantify hormonal effects. Possible outcome variables include seed water uptake, germination success, growth and branching patterns of roots and shoots, color changes, *etc.* You may be able to think of others. Additional environmental variables, like changes in temperature or light exposure, could perhaps be combined with hormone treatments.

To complete this assignment, you will:

- Write a *research proposal*, no more than two pages long. Your proposal must identify the research problem and state a reasoned hypothesis predicting the effects of ABA and/or GA on germination and development. It must present an experimental design, explain what data you will collect, and how you will analyze and interpret those data. *It must include a complete list of the materials you will need.* Your instructor may wish to view your proposal in advance to provide you with feedback on the feasibility of the experiment you designed. In actual research labs, research proposals are often subject to review before being funded and allowed to proceed.

- The following materials will be available for setting up your experiment. You may propose to use additional materials if approved by your instructor or if you supply them yourself. Viable pea, bean, radish, and/or soy bean seeds, petri dishes (maximum of four), absorbent paper, photocopies of a millimeter ruler, and hormone solutions: ABA and GA at 100 μM (micromolar) will be provided. You may choose to dilute these solutions, but higher concentrations will not be available. Pipets for measuring volumes, and a balance for weighing seeds, seedlings, or seedling parts will be available in the laboratory.

- *Set up* your approved plant hormone experiments. You will take your germination setup home with you for observation during the week. Each day, you must record observations, make measurements, and make sketches or digital photographs (cell phone) to document those observations for inclusion in your research report. Careful and complete recording of observations will enable you to document hormone effects that you might not have anticipated during the proposal phase.

- *Bring* your germination dishes and plants back to the lab after one week of observation and data collection. Balances and a camera may be available for final measurements.

- *Write up* your results in the form of a *research paper* (see appendix D).

Experimental Procedures

Prepare Germination Dishes

1. Obtain the number of petri plates and filter papers as outlined in your proposal.

2. Use a fine-tip permanent marker or pencil to label each filter paper around the edge with your name or initials, treatment, and date. Place the filter paper in the petri dishes (**fig. 26.5**)

Prepare Seeds for Germination

1. Count out the number of seeds you will be using. If you plan to weigh the seeds periodically, do so now.

2. Surface sterilize your seeds by placing them in a solution of 1% bleach for 30–60 seconds. This prevents mold and bacteria from growing on the seeds and interfering with your experiment. Do not

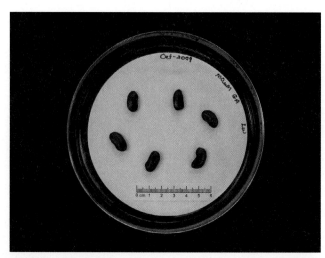

Figure 26.5 Germination dish: note labeling of filter paper, millimeter ruler, and number of seeds added to each dish.

leave seeds in the bleach for more than 60 seconds to avoid damaging the seeds themselves.

3. Rinse the seeds with distilled water.
4. Add the appropriate number and type of seeds to each germination dish.

10 Add Hormone

5. Both the GA and ABA are available at a concentration of 100 μM. If you need to prepare a dilution or a mixture of ABA and GA, prepare your dilution/mixture in a small beaker now.
6. Add hormone solution to your germination dishes. The total volume of any solution added to the dishes should be 12 to 15 ml, an amount adequate to saturate the absorbent paper.
7. If you plan to use photographs, place a ruler segment in the plate to provide a size reference. A photocopy of a ruler can easily be cut into short segments for each dish.

11 Conduct Experiment

8. Use parafilm to carefully seal the dishes to ensure you do not lose the hormone solution in transport. Take your petri dishes home and record daily observations for one week, including environmental factors such as average daily temperature and photoperiod duration. Do not place your petri dishes in full sun or they will overheat. Keep a log of your observations and measurements, and photograph or sketch your seedlings each day. If you use photographs, include a small piece of paper indicating the date and time in each photograph. If your filter paper appears to be drying out, add some water to keep the paper saturated at about the starting level.
9. After one week, bring your seedlings in their dishes back to the lab for final measurements, discussion, and disposal.
10. Your instructor may require you to submit a research paper describing your experiment and its results.

Photomorphogenesis in Seedlings

Because plants lack obvious sense organs, people often assume that plants are unable to detect changes in their environments. In fact, plants detect many aspects of the environment, and respond by altering their pattern of growth and development. Upon further consideration, this is precisely what should be expected. Unlike animals, plants are not able to move around to seek favorable environments. They must detect, and respond appropriately, to whatever environmental changes occur wherever they happen to be growing in order to survive.

12 A very clear example of response to environmental change occurs when plants are germinated in total darkness and transferred after a few days of growth into an environment in which light is present. Dark-grown seedlings are referred to as **etiolated** plants. Growth in the dark results in both **morphological** (structural) and **biochemical** differences (collectively called etiolation) compared to growth in the light (photomorphogenesis). Your instructor will have available seedlings of one or more species of flowering plants that were grown in the dark or in the light. Compare these seedlings. What differences do you observe?

13 Your instructor will provide you with a small pot of etiolated seedlings to take home with you after class. At home, place these plants in a windowsill or similarly well-lit location and be sure to water them as needed. You'll be bringing your plants back to your next class, so take good care of them! Observe your plants at various intervals after being exposed to light and record your observations in table 26.2 below.

At your next class, your instructor will have available plants that were left in the dark while your plants were in a lighted environment. Compare your plants to the dark-grown plants. What differences do you observe?

Table 26.2	Effect of Light on Morphogenesis of Etiolated Seedlings		
Time in Light	Morphology	Height	Color
0–1 hours			
6–12 hours			
24 hours			
48 hours			
72 hours			

Given that plants are autotrophic organisms, *i.e.*, they produce their food using the energy available in light through photosynthesis, why were plants kept in the dark able to grow at all?

Plants are able to detect, and respond to, a wide range of environmental stimuli in addition to light. These include gravity, salinity, touch, day length, temperature, oxygen availability, insect herbivory, and infection with pathogens. The mechanisms by which plants detect and respond to such stimuli is an area of active scientific investigation.

Lab Summary

14 On a separate piece of paper answer the following questions as assigned by your instructor.

1. What is the difference between pollination and fertilization?
2. List three different mechanisms plants use to disperse pollen. If any involve animals, explain how the plant gets the animal to help.
3. Seeds are often described as mature ovules, and fruits as mature ovaries. How can this be true? Explain your answer.
4. In your own words, describe how a zygote develops into an embryo in a maturing ovule.
5. How do the embryos of monocots and eudicots differ? Are any differences you identify likely to be important in the relative success of the two groups?
6. Why are flowers often fragrant and colorful and fruits often colorful and sweet?
7. List three different mechanisms plants use to disperse seeds. If any involve animals, explain how the plant gets the animal to help.
8. Describe the general effects of abscisic acid and gibberellins on germination and seedling development.
9. What evidence have you collected which shows that plants can detect light, even though they do not have eyes?

You may want to try some of the critical thinking questions to test your knowledge gained during this lab.

Internet Sources

Your lab observations clearly show that plants can detect whether light is present in their environment, and change their pattern of growth and development in response. Given that plants do not have eyes, how are they able to do this? Plants have at least three types of molecules called **photoreceptors** that function to detect light. These photoreceptors are phytochrome, cryptochrome, and phototropin. Use the Internet to determine the colors of light that are detected by each of the photoreceptors listed below.

Photoreceptor	Color of Light
Phytochrome	
Cryptochrome	
Phototropin	

Learning Biology by Writing

Write up the results of your experiment to determine the effects of hormones on plant development in the form of a research report (see appendix D). Be certain to state the hypothesis you tested, describe your experimental design, present your data, explain what you discovered and how the results bear on your hypotheses, and describe how your experiment and results relate to published information on the topic.

Critical Thinking Questions

1. Make a list of several foods we usually call vegetables that are actually fruits. What is the definition of a fruit you used in your list?
2. If wind and insects carry pollen from several flower species at the same time, why doesn't the pollen of one species fertilize the eggs of a second species, thus forming hybrids?
3. Many flowers have both anthers and pistils. What prevents self-pollination and later self-fertilization?
4. What is the advantage of outcrossing in sexual reproduction?
5. Many fruits are harvested before they ripen and then are treated with ethylene gas just before they are shipped to stores. Why?
6. In what ways do seedlings grown in the dark differ in form and color from seedlings grown in the light? How might those differences increase the probability of survival for a seed planted deeply under the surface of the soil?
7. Would a plant be able to grow to maturity and produce seeds if kept in total darkness? Why?

Interchapter

INVESTIGATING ANIMAL FORM AND FUNCTION

In this section of the lab manual, you are asked to dissect a fetal pig as a representative mammal. Some general instructions are given here to orient you.

VERTEBRATES

General Dissection Information

Fetal pigs are unborn fetuses taken from a sow's uterus when she is slaughtered. When the mother dies, the fetuses do as well. They are a by-product of meat preparation and are used in teaching basic mammalian anatomy. Often the circulatory system has been injected with latex so that the veins will appear blue and the arteries red.

Strong preservatives are used and can irritate your skin, eyes, and nose. Rinse your pig with tap water to remove some of the preservative to lessen irritation. If you wear contact lenses, you may want to remove them during dissection, since the preservative vapor can collect in the water behind the lens and be very irritating. If a lanolin-based hand cream is available, use it on your hands before and after dissection to prevent drying and cracking of the skin. Alternatively, your instructor may ask you to purchase rubber gloves and use them during this and subsequent dissections.

You will use this same fetal pig in several future labs to study the circulatory, excretory, and muscular systems. This means that you must do a careful dissection each time so that as many structures as possible are left undamaged and in their natural positions. Two good rules to keep in mind as you dissect your animal are: *cut as little as possible* and *never remove an organ unless you are told to do so*. If you indiscriminately cut into the pig to find a single structure without regard to other organs, you will undoubtedly ruin your animal for use in future laboratories.

Many of the instructions for dissection use anatomical terms to indicate direction and spatial relationships when the animal is alive in normal orientation. You should know the meaning of such terms as:

Anterior—situated near head or, in animals without heads, the end that moves forward.

Caudal—extending toward or located near tail.

Cephalic—extending toward or located on or near head (also cranial).

Distal—located away from the center of the body, the origin, or the point of attachment.

Dorsal—pertaining to the back as opposed to **ventral**, which pertains to the belly or lower surface.

Lateral—located toward the side; away from the median.

Median—a plane passing through a bilaterally symmetrical animal that divides it into right and left halves.

Posterior—toward the animal's hind end: opposite of anterior.

Proximal—opposite of distal.

Right-left—always in relation to the animal's right and left, not yours.

Sagittal—planes dividing an animal along the median line or parallel to the median.

1) Obtain a fetal pig and place it in a dissecting tray, ventral side up. Take two pieces of string and tie them tightly to the ankles of both right legs (the animal's right legs, not the legs to your right when the pig is lying on its back). Run the strings under the pan and tie each to the corresponding left leg. Stretch the legs to expose the ventral surface for easier dissection (**fig. I.1**). Do not pull so hard that you break internal blood vessels.

External Anatomy of Fetal Pig

Rows of **mammary glands** and the **umbilical cord** should be prominent on the ventral surface of your pig. Mammary glands and hair are two of the diagnostic characteristics of the class *Mammalia*. The umbilical cord attached the fetal pig to the placenta in the sow's uterus. Look at the cut end of the cord and note the blood vessels. They carried nutrients, wastes, and dissolved gases between the fetus and its mother's placenta.

2) Determine the sex of your pig. Identify the **anus**. In females, there will be a second opening, the **urogenital opening,** just ventral to the anus. In males, the urogenital opening is located just posterior to the umbilical cord. **Scrotal sacs** will be visible just ventral and lateral to the anus in males.

Figure I.1 shows the sequence of cuts that should be made to expose the internal organs. Make the cuts in the sequence indicated.

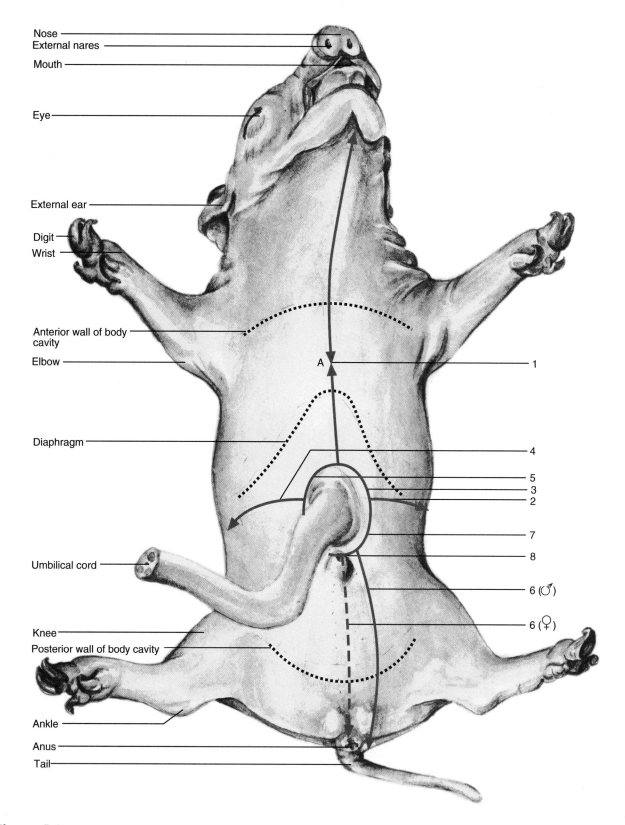

Figure I.1 External anatomy of fetal pig. Numbered lines indicate incisions to be made. Numbers show sequence of incisions.

Lab Topic 27

Investigating Digestive, Renal, and Reproductive Systems

Supplies

Resource guide available on WWW at
http://www.mhhe.com/labcentral

Equipment
Compound microscope
Dissecting microscopes

Materials
Fetal pigs
Dissecting trays
Dissecting instruments: scissors, forceps, blunt probe, razor blade, or scalpel
Thread
Prepared slides
    Mammalian small intestine, cross section
    Mammalian ovary with Graafian follicles
    Mammalian seminiferous tubules, cross section
Hydra culture
Daphnia cultures
Slides and coverslips
Pasteur pipettes with bulbs
Model of kidney and nephron

Student Prelab Preparation

Before doing this lab, read the Background material and sections of the lab topic assigned by your instructor.
    Use your textbook to review definitions of the following terms:

bladder	oviduct
Bowman's capsule	pharynx
circular muscle	seminiferous tubule
collecting duct	smooth muscle
epididymis	sphincter
esophagus	testis
glomerulus	ureter
longitudinal muscle	urethra
nephron	uterus
ovary	vas deferens

Be able to describe in your own words the following concepts:
Saccular and tubular digestive tracts
Pathway of food though mammalian digestive tract
Anatomy of *Hydra* (see fig. 20.3)
Glomerular filtration and tubular reabsorption in the kidney
Pathway sperm travel from formation to ejaculation
Pathway eggs travel from formation to uterus, and from uterus to the outer world

After finishing the prelab review, write any questions in the margins of this lab topic. The lab work should help you answer these questions or you can ask your instructor during the lab.

Objectives

1. To examine feeding behavior and the gastrovacular cavity of *Hydra*
2. To dissect the digestive system in a fetal pig and understand its regional specializations
3. To observe the histology of the small intestine and relate those observations to nutrient absorption
4. To dissect and observe the gross anatomy of mammalian excretory and reproductive systems
5. To use models of nephron structure to understand waste excretion by the kidney
6. To observe the microscopic structure of the testis and ovary and relate it to gamete production

Background

In this lab you will study the functional anatomy of three physiological systems: digestive, renal, and reproductive systems. Although they are not physiologically related, we have grouped them as one dissection because they are easily observed at one sitting. In future labs, you will reuse your fetal pig to study the circulatory, respiratory, muscle/skeletal, and nervous systems as separate dissections. Look at the Interchapter that comes just before this lab to familiarize yourself some of the general anatomical terms.
    Digestion is the enzyme catalyzed chemical breakdown of complex food materials into small molecules

that can be absorbed into the body. Chemically, it requires the hydrolysis of the covalent bonds holding together large, polymeric molecules like those you saw in lab topic 6. Proteins are hydrolyzed into amino acids and starches into sugars. These small molecules are used by the animal as either sources of energy or building units to make new molecules, such as proteins and nucleic acids. Thus, when we digest bovine proteins in a hamburger, we reuse the amino acids to make human proteins or convert the amino acids to sugars that are used as an energy source.

Digestion can occur **intracellularly** or **extracellularly.** In protozoa and sponges, the individual cells of the organism ingest food materials by pinocytosis and phagocytosis. Digestion occurs in food vacuoles inside of cells. Some endoparasites and marine invertebrates take in nutrient molecules by active transport across the body surface. Other organisms have extracellular digestion in specialized digestive organs that are either saclike or tubular. Cnidarians and flatworms have **saccular digestive tracts**; that is, food enters the mouth, passing into a chamber where enzymatic digestion and absorption occur, and nondigestible material is expelled through the same opening.

Most animals have a **tubular digestive system** between mouth and anus in which food is sequentially broken down and absorbed. This permits regional specialization and continuous processing of food. In vertebrates, different regions of the digestive tube are specialized for mechanical processing, storage, digestion, and absorption of nutrients, salts, and water from the digested material. The undigested or nondigestible residues pass on to temporary storage areas before being defecated.

Arthropods and annelids have a digestive tube that passes more or less straight through the body from head to tail. The digestive tubes in such organisms include storage areas, grinding areas, digestive areas, and absorptive areas, but the tube is not longer than the organism. In vertebrates, on the other hand, the digestive tube is many times longer than the animal's length, with much of this length devoted to absorption. After absorption, the circulatory system carries food materials to cells throughout the body.

The excretory systems of animals maintain a constant chemical state in the internal body fluids. The name *excretory system*, with its implicit emphasis on waste removal, does not suggest the three other important functions of these systems: (1) controlling water volume, (2) regulating salt concentrations, and (3) eliminating nonmetabolizable compounds and toxins absorbed from food or water.

Excretion also includes the elimination of the waste products from the metabolism of nitrogen-containing compounds. Ammonia, a compound toxic at high concentrations, is produced by all animals when they metabolize amino acids and nucleotides. Depending on whether an animal inhabits an aquatic or terrestrial environment, this toxic product is handled in different ways. Aquatic organisms generally excrete the NH_3 directly into the surrounding water, often through their gills. Terrestrial animals convert it into less toxic compounds, such as **urea** in mammals or **uric acid** in insects, reptiles, and birds. These products are collected by excretory organs and periodically voided.

Aquatic and terrestrial environments present quite different challenges to animals in terms of internal water volume regulation and regulation of salt concentrations in body fluids. The following paragraphs outline these challenges.

Many marine invertebrates lack well-developed excretory systems. They have body fluids that resemble seawater in terms of osmotic concentration and ionic composition. Active transport maintains the minor differences that exist. Furthermore, such animals usually have little tissue volume relative to total surface area. Ammonia diffuses from the body into the virtually infinite reservoir of the ocean at rates sufficient to keep internal concentration low. For freshwater invertebrates, ammonia also is lost across the body surface. Because their body fluids have more salts than the surrounding water, they constantly gain water and lose salts. Therefore, their excretory systems have become adapted to conserve salts while dumping excess water.

In other marine invertebrates and fish, the mass of tissues dictates that there be mechanisms other than diffusion to get rid of nitrogenous wastes. Some wastes and excess salts are removed by transport across the gills, where gas exchange also takes place. This mechanism is augmented by development of excretory organs called kidneys. Kidneys (the **renal** system) are organs that filter water and small molecules from blood or hemolymph, reabsorb desirable materials from the filtrate, and excrete the rest.

Terrestrial animals must always conserve water and regulate the salt concentration of their body fluids as well as excrete nitrogenous wastes. Insects, reptiles, and birds convert waste ammonia into uric acid (spiders use guanine), which is practically insoluble and requires little accompanying water when it is excreted. This is the white stuff in bird and insect droppings. Mammals convert nitrogenous waste to urea and have kidneys that can recover water or salt or both from the renal filtrate, but allow urea to be excreted.

Mammalian kidneys function by filtering the blood fluid and then selectively reabsorbing materials from the filtrate, allowing the remainder to be excreted. The mammalian kidney contains millions of capillary tufts, the **glomeruli.** They filter blood through tiny pores, so that about 20% of the plasma (the fluid portion) leaves the blood and enters the kidney tubules. Cells and large molecules like proteins are not filtered, but remain in the blood. The concentrations of water, salts, urea, sugars, amino acids, and so on in the filtrate are very similar to those in the blood. As the filtrate passes through the tubular network of the **nephrons** toward the **collecting ducts** of the urinary system, most of these materials and most of the filtered water are reabsorbed, leaving wastes, excess salts, and some water to carry the remaining materials.

INVESTIGATING DIGESTIVE, RENAL, AND REPRODUCTIVE SYSTEMS

In the adult human, about 180 liters of filtrate are produced each day, but the daily urine volume is only 0.6 to 2.0 liters. Plasma volume is about 3 liters, so the production of 180 liters of filtrate means that all of the blood plasma must be filtered through the glomeruli 60 times in 24 hours. About 99% of this filtrate, around 179 liters including most of the salts and organic molecules, is reabsorbed. The reabsorption process is obviously very important.

Because many undesirable substances are filtered by the kidneys but not reabsorbed, their concentration in urine can be 100 times their concentration in the blood. That makes it easy to understand why urine samples are often used to search for indicators of disease, toxic substances, or drug use.

In mammals the organs of the excretory and reproductive systems are closely associated, especially in males, so the two are sometimes lumped as the **urogenital system**, although they are distinctly different in their functions. The gonads produce haploid gametes by meiosis. In oogenesis, the formation of eggs in the ovary, the first stages of meiosis have occurred while the mammal was an embryo. The eggs will remain in an arrested stage until a female mammal reaches sexual maturity. Eggs are then released by the ovary into the body cavity where they are picked up by ducts of the reproductive system. As eggs descend the reproductive tract, they will finish meiosis II. The elapsed time between meiosis I and II can be years. If fertilization occurs, then the embryo implants in the uterus and completes development. In spermatogenesis, the production of sperm in the testis, meiosis does not start until sexual maturity is reached. It can then be more or less continuous throughout the life span, producing millions upon millions of cells that develop into sperm.

LAB INSTRUCTIONS

In this lab you will observe the digestive system of *Hydra* and identify the regional specialization of the mammal digestive tract and associated glands. You will use dissection to reveal the components of the mammal excretory system. Because of the close anatomical proximity of the excretory and reproductive systems, you will also study the anatomy of the mammalian reproductive system.

Feeding Behavior in *Hydra*

Hydra is a small, freshwater cnidarian related to jellyfish and sea anemones (see fig. 20.3). It lives attached to submerged rocks, leaves, and twigs. A hydra's body consists of only two layers of cells surrounding a sac-like digestive cavity. It uses tentacles to capture food and transfer it into the digestive (**gastrovascular**) cavity. Food is then digested by enzymes secreted by cells lining the cavity. Because of its small size, the cells of *Hydra* obtain nutrients from the digestive cavity by diffusion and active transport. Each body cell also exchanges oxygen and carbon dioxide directly with the surrounding water and releases cellular metabolic wastes directly into the water as well.

1) Obtain a living *Hydra* from the supply table and place it in a small dish that has been thoroughly washed so it has no chemical residues from other uses. Observe the *Hydra* with a dissecting microscope against a dark background. Sometimes a *Hydra* contracts into a small ball after it is transferred, but it should relax, attach to the glass, and resume its normal functions before long. If the *Hydra* attaches to the surface water film, sink it by dropping water on it from a dropper.

Locate the pedal disc (the animal's point of attachment), the body column, the **tentacles,** and the **hypostome,** a small raised region surrounded by the bases of the tentacles (see fig. 20.3). The **mouth** is located in the center of the hypostome but is difficult to see.

Obtain some *Daphnia,* a freshwater crustacean on which *Hydra* will feed. Small pieces of aquatic annelids such as blackworms also elicit a strong feeding response. Transfer some food organisms next to your *Hydra* with a Pasteur pipette and watch closely through a dissecting microscope. *Hydra,* like other cnidarians, is a carnivore. It traps food by stinging and paralyzing other animals with specialized epidermal cells called **cnidocytes** located on its tentacles. Each cnidocyte has a special organelle, the nematocyst, which contains a coiled tube. When stimulated, a nematocyst shoots out a barbed thread, which either entangles or pierces and poisons the prey. Seeing the *Hydra* capture food will take patience, but the time is well spent.

Record below how *Hydra* feeds. Include drawings showing how prey are transferred from the tentacles to mouth.

Once food is in the gastrovascular cavity, digestion begins. Enzymes secreted by cells lining the cavity begin **extracellular digestion.** Partially digested bits of food material can be taken up by phagocytic cells in the cavity lining, and further digestion occurs inside food vacuoles in these cells. This is **intracellular digestion.** Food absorbed by the cells lining the gastrovascular cavity supplies all cells of the body.

When you finish with the *Hydra*, return the materials to the supply area.

Mammalian Digestive System Anatomy

2 If you have not already done so, get a fetal pig and dissection tray. Read the general instructions on page 361. We will start our study of mammalian anatomy by looking first at several aspects of the digestive system and then move on to study the renal and reproductive systems.

Anatomy of the Mouth Located in the upper neck and cheek region beneath the skin are three pairs of **salivary glands: parotid, submaxillary,** and the difficult to find **sublingual.** They produce saliva containing the enzyme salivary amylase that hydrolyzes starch during chewing; mucin, a polysaccharide lubricating agent; and lysozyme, an antibacterial enzyme.

3 To view the salivary glands, remove the skin and muscle layer from one side of the face and neck, as in figure 27.1 to reveal the triangular-shaped parotid gland just posterior to the masseter muscle. Note the difference in appearance between muscle and glandular tissue. If you dissected carefully, you may find the duct that drains the gland into the mouth near the upper premolar teeth. Try to trace the duct. The other salivary glands lie beneath and below the parotid gland.

With heavy scissors, a razor blade, or scalpel, cut through the corners of the mouth and extend the cuts to a point below and caudal to the eye.

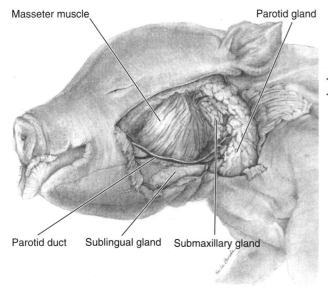

Figure 27.1 Dissected fetal pig's head showing salivary glands.

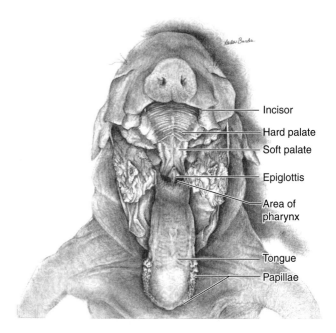

Figure 27.2 Anatomy of the fetal pig's mouth.

Open the mouth, as in **figure 27.2** and observe the **hard palate,** composed of bone covered with mucous membrane, and the **soft palate,** which is a caudal continuation of the soft tissue covering the hard palate. The oral cavity ends and the **pharynx** starts at the base of the tongue.

The pharynx is a common passageway for the digestive and respiratory tracts, as seen in **figure 27.3**. The opening to the **esophagus** may be found by passing a blunt probe (not a needle) down along the back of the pharynx on the midline. This collapsible tube connects the pharynx with the stomach. The **glottis** is the opening into the **trachea** or windpipe and lies ventral to the esophagus. It is covered by a small white tab of cartilage, the **epiglottis.** The epiglottis may be hidden from view in the throat; if so, you will have to pull it forward with forceps or a probe to see it.

4 *Alimentary Canal Anatomy* You will see the path of the esophagus to the stomach when you dissect the respiratory system. To view the rest of the alimentary tract and associated glands, use a scalpel or pair of scissors to make incisions into the abdominal cavity as indicated in fig. I.1, page 362. Cut carefully through only the skin and muscles to avoid damaging the internal organs.

The flap containing the umbilical cord will be held in place by blood vessels. Tie each end of a 15-cm piece of thread to the blood vessel about 1 cm apart. Cut the vessels between the two knots and lay this tissue flap back. Leave the thread in place so you can later trace the circulatory system.

Find the thin, transparent membranes, the **mesenteries,** which suspend and support the internal organs in the body cavity. The dark brown, multilobed **liver** should

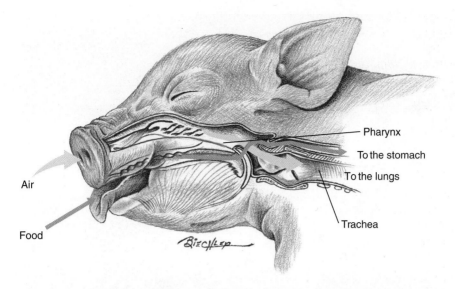

Figure 27.3 Pharyngeal crossover of digestive and respiratory systems. Food must pass over the glottis as it passes from the mouth to the esophagus. This is why we sometimes "choke" when eating and talking.

be visible caudal to the **diaphragm** (**fig. 27.4**). The liver secretes sodium bicarbonate to buffer stomach acids and bile, processes nutrients absorbed from food, and prepares some waste products for excretion. If you trace the umbilical vein from the thread to the liver, you will see a green-colored sac, the **gallbladder,** located just below the entrance of the vein into the liver. It stores **bile** produced in the liver. Bile travels from the gallbladder to the small intestine via the bile duct but the duct is quite small. Bile is an emulsifying agent that aids in digestion of fats by breaking them into very small droplets as a detergent does. These small droplets are then acted on by enzymes. What does this suggest about the likely consequences of gallbladder removal?

Under the liver is the **stomach.** Locate the point where the esophagus enters the **cardiac region** of the stomach. Gastric glands in the wall of the stomach secrete pepsinogen, hydrochloric acid, and rennin. Pepsinogen is activated by hydrochloric acid to become pepsin, which digests proteins. Rennin is an enzyme that hydrolyzes milk protein. Food leaves the stomach as a fluid suspension, **chyme.** It enters the **duodenum,** the first part of the small intestine.

Find the **pancreas,** a glandular mass lying in the angle between the curve of the stomach and duodenum. It secretes several enzymes into the duodenum that digest proteins, lipids, carbohydrates, and nucleic acids. Certain cells in the pancreas act as endocrine cells and secrete the hormones insulin and glucagon. In fact, insulin used in human diabetes therapy can be extracted from the pancreases of pigs collected at slaughterhouses. The pancreas is relatively small, but pancreatic disease (like pancreatic cancer) is more life threatening than most digestive tract diseases. Why is this so?

Although it is not part of the digestive system, identify the **spleen,** a long red-brown organ attached by mesenteries to the outer curvature of the stomach. It is made of lymphatic tissue and is important in development of immunity and the scavenging of iron from red blood cells when they break down.

5> Slit the stomach lengthwise, cutting through the cardiac and pyloric sphincters, muscles that regulate passage of material into and out of the stomach. The internal surface of the stomach is covered by gastric mucosal cells, which secrete mucus that prevents the stomach from digesting itself. When this protection fails, a peptic ulcer develops.

The small intestine is made up of three sequentially arranged regions: **duodenum, jejunum,** and **ileum.** These areas are difficult to differentiate. The small intestine is called "small" because of its diameter, not its overall size. Your small intestine makes up nearly 90% of the total length of your digestive tract. Estimate the length of the small intestine in your pig. _____ Cut out a 2-cm section of the small intestine about 5 cm posterior from the stomach, slit it open, and place it under water in a dish. Use your dissecting microscope to observe the velvet-like internal lining made up of numerous multicellular fingerlike projections called **villi.** The villi are highly vascularized, containing capillaries and lymphatics that transport the products of digestion to other parts of the body, especially the liver. Most chemical digestion and nutrient

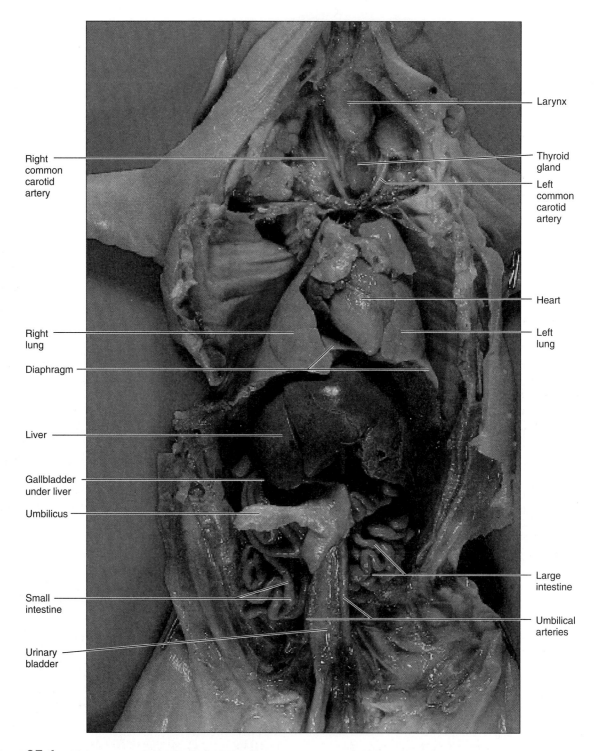

Figure 27.4 Major organs in the fetal pig. Note that the stomach and spleen are not visible.

absorption take place in the small intestine. Its small diameter, length, folds, projecting villi, and the microvilli on epithelial cells provide a large surface area for absorption.

The ileum opens into the large intestine, or **colon.** They join at an angle, forming a blind pouch, the **cecum,** which in primates and some other mammals often ends in a slender appendage, the **appendix.** What is appendicitis, and why would a ruptured appendix be life threatening?

In many herbivores, the cecum is very large and contains microorganisms that aid digestion by breaking down cellulose. The large intestine is important for reabsorption of water and sodium chloride secreted into the gut during digestion. What are the sources of that water and salt?

The **rectum** is the caudal part of the large intestine, where compacted, undigested food material is temporarily stored before being released through the **anus**. The colon of vertebrates contains large numbers of symbiotic bacteria, especially *Escherichia coli*. These bacteria produce vitamin K, which is absorbed and plays a vital role in blood clotting.

6▸ Histology of Small Intestine Put your fetal pig aside for the moment and obtain a prepared slide of a cross section of a mammalian small intestine. Examine it under scanning power with the compound microscope. Compare what you see to **figure 27.5**.

The central opening is called the **lumen** and is the space through which food passes as chyme. Using low power, observe the small fingerlike villi of the intestine's inner surface. They are covered by a layer of cells collectively called the **mucosa**. You should be able to distinguish two cell types in the intestinal mucosa: **goblet cells** and **columnar epithelial cells**. Examine them with the high-power objective. The goblet cells secrete mucus into the small intestine, serving as a lubricant for the passage of chyme. Epithelial cells are involved in absorption.

Return to the low-power objective and observe the **submucosa**, a layer of connective tissue that underlies the mucosa. Look for the blood vessels and lymphatic vessels that ramify through this layer. Sugars, amino acids, glycerides, and other components of digested food must move through the mucosal cells into the submucosa before they can enter the circulatory system and be distributed throughout the body.

To the outside of the submucosa are two smooth muscle layers: a thick **inner circular layer** and a thinner **outer longitudinal layer.** When they contract, the inner circular muscles decrease the diameter of the intestine, and the outer muscles shorten its length. These muscles contract in a wavelike motion called **peristalsis,** which pushes chyme through the digestive tract. Because pushing the chyme forward requires more force than restoring the intestine to its original length afterward, the circular muscle layer is thicker than the longitudinal muscle layer. The small intestine is covered by a connective tissue mesentary.

Figure 27.5 shows scanning electron micrographs of the three-dimensional arrangement of the small intestine. Note how the villi and microvilli increase the surface area. Microvilli are microscopic fingerlike projections on the surface of individual cells. What important process following digestion is facilitated by this increased surface area?

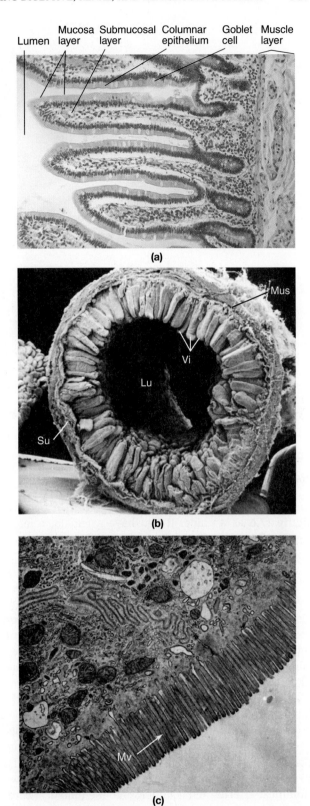

Figure 27.5 Microanatomy of intestine. (*a*) Photo taken through a light microscope of cross section of the small intestine. (*b*) Scanning electron micrograph of cross section of small intestine showing villi (Vi), lumen (Lu), submucosa layer (Su), and muscle layers (Mus). The epithelial cells on the surface of the villi have highly folded membranes, microvilli (Mv), which greatly increase the absorptive surface area of the cell layer. (*c*) a transmission electron micrograph showing the highly folded cell membrane.

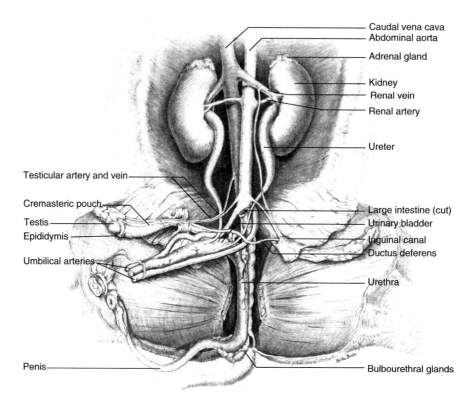

Figure 27.6 Ventral view of a male fetal pig's renal and reproductive systems.

Mammalian Excretory System

7 Return to your fetal pig. If you have a male pig, use figure 27.6 as a guide. If you have a female, refer to figure 27.7. Be sure to look at a pig of the opposite sex before you leave lab.

The kidneys are located on the dorsal wall of the abdominal cavity. Remove the **peritoneum,** the connective tissue membrane that holds them in place. Note how close the kidneys are to the **descending aorta.** The blood pressure drops very little as blood passes from the aorta into the kidneys via the **renal arteries.** A high blood pressure is essential for proper kidney function. **Renal veins** drain the kidney, returning blood to the caudal vena cava. The large diameter of the renal blood vessels indicates the importance of the renal blood filtration system.

Find the **ureter,** which originates from the medial face of the kidney. Trace it caudally to where it empties into the **urinary bladder** where urine is stored. In the fetal pig, part of the urinary bladder extends between the two **umbilical arteries** and continues into the umbilical cord, where it is called the **allantoic duct.** After birth, the duct atrophies, becoming nonfunctional.

The **urethra** proceeds caudally from the bladder to its opening; its location depends on the sex of your pig. It is difficult to observe and trace. A sphincter muscle, under voluntary control, regulates release of urine from the bladder into the urethra. Smooth muscles in the bladder wall contract to force urine out of the bladder.

Though they are not involved in excretion, note the **adrenal glands,** inconspicuous crescent-shaped masses of light-colored tissue located cranially and medially to the kidneys. Epinephrine (adrenaline) and norepinephrine, hormones produced by these glands, regulate a number of body functions, including heart rate, blood sugar levels, and arteriole constriction.

Remove one kidney. Make a longitudinal cut through the kidney with a sharp razor blade in a plane parallel with the front and back of the kidney. Study the cut surface with the dissecting microscope (**fig. 27.8**). Locate where the ureter entered the kidney and note how it subdivides into funnel-shaped **calyxes.** Urine is produced by the thousands of filtration-reabsorption units called nephrons. It flows through collecting ducts to the calyxes and passes by way of the ureter to the bladder for storage.

Nephron Structure The functional unit of the kidney is the **nephron.** In a mature pig's kidney, there may be a million of these units. Nephron structure is shown in **figure 27.9**. Blood from the renal artery flows by way of smaller arteries to tufts of capillaries called **glomeruli.** There the high blood pressure forces part of the fluid portion of the blood, minus any blood cells or large proteins, out of the capillaries. The filtrate is captured by microscopic connective tissue bags called **Bowman's capsules.** Filtrate drains from a capsule through a U-shaped tubular system where it is modified by selective absorption and secretion. Desirable molecules like amino acids, sugars, and most salts that passed through the filter are reabsorbed and

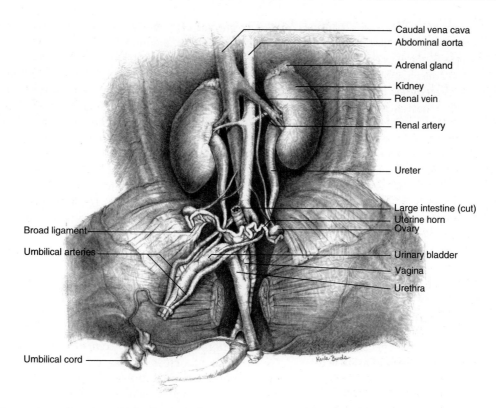

Figure 27.7 Ventral view of a female pig's renal and reproductive systems.

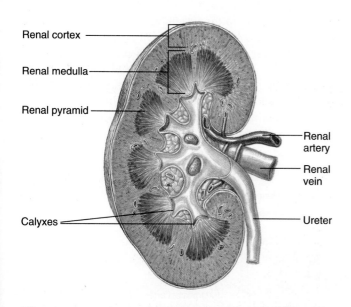

Figure 27.8 Longitudinal section of a mammal's kidney.

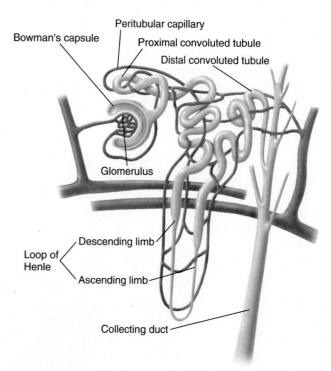

Figure 27.9 Structure of nephron and relationship to circulatory system.

water follows osmotically. Some undesirable substances are secreted into the filtrate. Hormones such as antidiuretic hormone, aldosterone, and atrial natriuretic hormone regulate the filtration, absorption, and secretion processes. Eventually, the fluids flow into collecting ducts, where additional water reabsorption can occur. Collecting ducts from the million nephrons merge together to form the ureter that carries urine from the kidney to the urinary bladder.

A whole nephron cannot be seen in a microscopic section of a kidney. The tubules and blood vessels pass in many different planes so a slice never contains a whole nephron. Your instructor may have a model of a nephron for you to view. Glomeruli, Bowman's capsules, and proximal and distal convoluted tubules of the nephron are located near the surface of the kidney, in the **renal cortex.** Loops of Henle and collecting ducts are in the middle of the kidney, the **renal medulla.** In your sectioned kidney, notice the different appearance of these regions that results.

8 If models of the kidney are available in the lab, you and a partner should trace how a drop of water can pass from the blood to the ureter.

Mammalian Reproductive System

Though not part of the excretory system, the organs of the reproductive system are located so close to the excretory system that they warrant a brief discussion here. Follow the directions according to the sex of your pig. After you have finished your dissection, explain the anatomy to someone who has a pig of the opposite sex. When finished, have them explain their dissection to you.

Male System During embryonic development, the **testes** originate in the coelom caudal to the kidneys, and then descend to lie in an external pouch, the **scrotum,** in adult pigs.

9 To find the testes in a fetal pig, locate the testicular artery and vein on one side of the animal and trace them to the **inguinal canal** (see fig. 27.6). The testis passed through this canal during its descent toward the scrotal sac. Pass a blunt probe through the inguinal canal.

The probe should help you locate a thin-walled, elongated sac extending across the ventral surface of the thighs just ventral to the tail. Cut through the skin and reveal the **cremasteric pouch,** an outpocketing of the connective tissues of the abdominal wall that contains the testis at this stage of embryo development.

Sperm are produced in the microscopic **seminiferous tubules** inside the testis. They in turn pass into the **epididymis,** visible as a tightly coiled mass of tubules along one side of the testis. The epididymis flows into the **vas deferens,** which passes from the scrotal area through the abdominal wall and into the body cavity via the inguinal canal. After entering the body cavity, the vas deferens loop over the ureter and enter the **urethra.** Paired **seminal vesicles** and single, bilobed **prostate gland** are also located at this juncture. The large **bulbourethral glands** lie on either side of the junction of the urethra with the **penis.** These three glands secrete fluids that nourish and carry the sperm during an ejaculation. The penis of a fetal pig is retracted and lies deep in the muscle layers of the groin area. The location of these structures is shown in sagittal section in **figure 27.10**.

10 *Microanatomy of Seminiferous Tubules* Obtain a slide of a cross section of a mammalian testis containing **seminiferous tubules.** Using the compound microscope, look at it first with low power. Center a tubule in the field of view and then switch to high power. Compare the specimen to **figure 27.11**.

Note that the cells grow smaller as you scan from the outer tubule wall to the inside, where mature sperm may be present in the **lumen.** The largest outer cells are special stem cells that divide by mitosis to produce other cells. Half of these cells undergo meiosis and become sperm, while the other half function as stem cells to again divide by mitosis and keep the process going. Since a single ejaculation contains hundreds of millions of sperm, rejuvenation of the stem cell population is important.

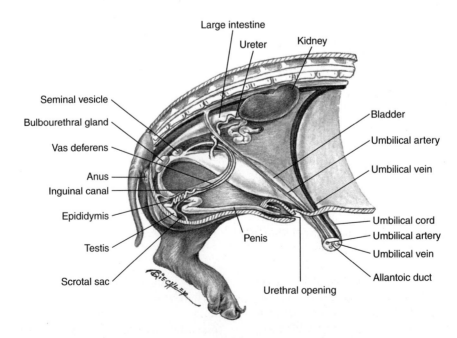

Figure 27.10 Male fetal pig's urogenital system in sagittal section view.

INVESTIGATING DIGESTIVE, RENAL, AND REPRODUCTIVE SYSTEMS   373

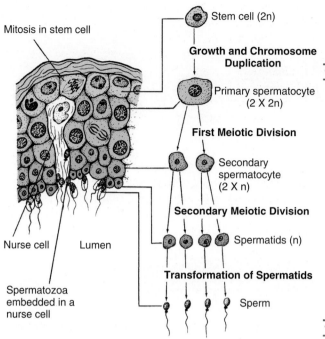

Figure 27.11 Cross section of a seminiferous tubule, showing stages in sperm development as they pass from outer wall to the lumen of tubule.

Primary spermatocytes are cells destined to enter meiosis. When these cells finish meiosis I, two **secondary spermatocytes** are produced. They rapidly undergo meiosis II to produce spermatids that mature into functional sperm. **Nurse cells** (or **Sertoli cells**) located in the seminiferous tubule walls aid in this process. Between the seminiferous tubules, **interstitial cells** are found. They produce the male hormone testosterone.

11 *Female System* The **ovaries** are small, bean-shaped organs suspended by connective tissue from the dorsal wall of the abdominal cavity near the caudal end of the kidneys (see fig. 27.7 and **27.12**). On the dorsal side of each ovary is an **oviduct** (called Fallopian tubes in humans). Eggs bursting from the ovary enter the openings of the oviducts. Cilia on cells lining the oviduct move the egg toward the uterus. Fertilization usually occurs in the oviducts. The pig **uterus** has two convoluted uterine horns where multiple embryos can imbed and develop. The uterus opens to the outside through a muscular tube, the **vagina**. The vagina and the urethra open into a common area, the urogenital sinus, in fetal pigs.

You are now finished with the fetal pig for this laboratory. Return it to the storage area as directed.

12 *Microanatomy of Ovary* Obtain a slide of a section of a mammalian ovary and look at it under low power with the compound microscope. Large, clear areas will be visible in the section, as in **figure 27.13**. These are **Graafian follicles,** structures in which egg maturation occurs. There are several hundred thousand follicles in each ovary of a female at the time of her birth. The ova in these follicles are in an arrested prophase of meiosis I. At sexual maturity, hormones from the pituitary gland cause the follicles to secrete the female hormones estrogen and progesterone.

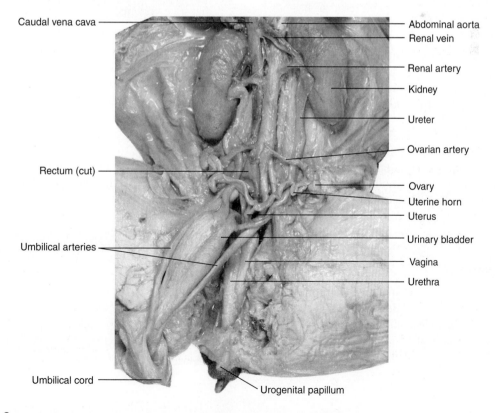

Figure 27.12 Ventral view of female renal and reproductive systems. Intestines have been removed.

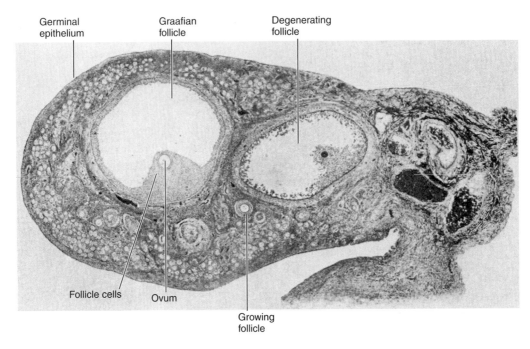

Figure 27.13 Photomicrograph of a section from a rhesus monkey's ovary.

If you are lucky, your slide will show such an enlarged follicle containing an ovum surrounded by follicle cells. During ovulation, the follicle swells until it bursts, releasing the egg. Fingerlike projections of the oviduct surround the ovary and collect the egg. The egg moves down the oviduct by the action of cilia on the tube surface. Sperm usually fertilize the egg as it passes down the oviduct. Meiosis II will take place in the egg only if a sperm fertilizes it.

Following ovulation, the collapsed follicle assumes a star-shaped appearance and fills with a yellow fluid. At this stage, it is called a **corpus luteum** and secretes progesterone. If fertilization occurs and the ovum implants in the uterus, a hormone produced by the developing placenta will sustain the corpus luteum and progesterone production. If implantation does not occur, the corpus luteum will degenerate after several days and progesterone production will decrease. Falling levels of progesterone trigger the pituitary to produce follicle-stimulating hormone (FSH), which spurs another Graafian follicle toward ovulation. The total time course for these events in a human is approximately 27 days.

Lab Summary

13 To gain an understanding of comparative anatomy, fill in tables 27.1 to 27.3. Write a brief description of each structure and its function. Include a summary statement of the organ system for each animal.

On a separate piece of paper, answer the following questions.

1. Create a flowchart that traces the pathway of food through the mammalian digestive system, and indicate to the right of the chart the functions of each organ.
2. The digestive system of *Hydra* is saclike, whereas that of mammals and many other animals is tubelike. What are the advantages of tubular digestive systems?
3. Create a flowchart showing how urine is formed in a mammal and is eventually voided.
4. Discuss why both filtration and reabsorption are both important processes in the kidney.
5. Name the structures that a sperm or an egg passes through from the time it is formed until it leaves the body.

You may want to test the knowledge gained in this lab by trying some of the critical thinking questions.

Table 27.1	*Hydra*—Digestive System Summary
Cnidocytes	
Mouth/anus	
Gastrovascular cavity	
Mesoglea	
Description of digestive process	

Table 27.2	Fetal Pig—Mammal—Digestive System Summary
Mouth	
Salivary glands	
Esophagus	
Stomach	
Small intestine	
Liver	
Pancreas	
Large intestine	
Rectum	
Anus	
Description of digestive process	

Table 27.3	Excretory System Summary
Fill in the names of structures in fetal pig that are involved in the named process.	
Filtration	
Absorption	
Secretion	
Voiding	

INTERNET SOURCES

How do the digestive tracts of herbivorous mammals differ from those of carnivorous mammals? Search the World Wide Web and record at least two differences between herbivore and carnivore digestive tracts and the URLs of your sources.

Search the World Wide Web for information on kidney stones. What are they and how are they formed? Be sure to record URLs for future reference.

LEARNING BIOLOGY BY WRITING

Write a 250-word essay describing how the surface area in the small intestine is greatly magnified compared to other regions of the gut such as the stomach or large intestine. How does this increase relate to the function of the small intestine?

CRITICAL THINKING QUESTIONS

1. Explain why the small intestine is the longest part of the mammalian digestive system.
2. Researchers interested in the physiology of the kidney made the measurements shown in **table 27.4**. On the basis of your understanding of how the kidney works, explain why some substances are found at the same concentration in the plasma and glomerular filtrate and why others are not. Also explain why most substances are found at different concentrations when the glomerular filtrate is compared to the urine.
3. Create a flowchart that traces a drop of water from the time it enters the mouth until it is voided in the urine of a mammal. You will have to list structures from the digestive, circulatory, and excretory systems.

Table 27.4	Concentrations of Selected Substances in mg per 100 ml of Fluid		
Substance	Plasma	Glomerular Filtrate	Urine
Albumin	4,500	0	0
Glucose	100	100	0
Urea	26	26	1,820
Uric acid	4	4	53
Na	330	330	297
K	16	16	192
Cl	350	350	455

Lab Topic 28

Investigating Circulatory Systems

Supplies

Resource guide available on WWW at
http://www.mhhe.com/labcentral

Equipment

Dissecting microscopes
Compound microscopes
Stethoscope
Sphygmomanometer or other blood pressure measurement device

Materials

Fetal pig
Demonstration dissection of beef, pig, or sheep heart
Dissection pans and instruments
Microscope slides
    Cross section artery
    Cross section vein
    Wright-stained human blood smear
Live guppies, goldfish, or tadpoles
Live crayfish (demonstration)
Small petri dishes
Coverslips
Absorbent cotton

Solutions

Ringers invertebrate saline
1% chlorotone (optional)

Student Prelab Preparation

Before doing this lab, read the Background material and sections of the lab topic that have been assigned by your instructor.
    You should use your textbook to review the definitions of the following terms:

aorta	lymphocyte
artery	plasma
atrium	pulmonary artery
capillary	pulmonary vein
erythrocyte	vein
hemolymph	vena cava
leukocyte	ventricle
lymphatic system	

    Describe in your own words the following concepts:
Open and closed circulatory systems
The pumping action of a four-chambered heart
The role of capillaries in exchange
General patterns of mammalian circulation to the lungs, brain, gut, and body muscles and return path to heart

    After finishing the prelab review, write any questions you have about terms, concepts, or techniques in the margins of this lab topic. The lab observations should help you answer these questions, or you can ask your instructor during the lab.

Objectives

1. To understand the pumping action of the heart and fluid flow in an open circulatory system
2. To understand the structure and function of the mammalian heart and major vessels
3. To be able to distinguish between a cross section of an artery and vein and between a red blood cell and a white blood cell
4. To observe blood flow in capillary beds of a fish
5. To measure human blood pressure

Background

Many invertebrates, such as cnidarians, flatworms, and roundworms, lack circulatory systems. Simple diffusion is sufficient for the necessary exchanges of respiratory gases, waste products, and nutrients. Larger, more complex animals require a circulatory system to supply the needs of their tissues.
    The circulatory system consists of a special internal body fluid called blood or hemolymph, a pumping system, and a vascular system consisting of tubular vessels for moving the fluid rapidly from one location to another within an animal. The circulating fluid often contains a respiratory pigment, a protein that aids in transporting oxygen and carbon dioxide between the tissues and the respiratory surface. Common respiratory pigments include

hemocyanin and hemoglobin. In vertebrates, hemoglobin is carried in red blood cells. Blood also contains cells and proteins that protect against invasion by microorganisms and proteins that are involved in clotting, the sealing of leaks. The blood-vessel system often has anatomical provisions so that the blood is brought into close contact with three exchange systems: lung or gill, where gas exchange occurs; excretory organs, where excess salts, water, and wastes are removed; and the digestive system, where nutrients are absorbed.

Circulatory systems may be either open or closed. In **open circulatory systems,** found in most molluscs and arthropods, the arterial system is not connected to the venous system through capillary beds. Instead, the small arteries simply terminate, emptying their contents into the tissue spaces, and the circulatory fluid (properly called hemolymph) directly bathes the tissues, eventually finding its way back to the heart.

In **closed circulatory systems,** found in cephalopod molluscs, annelids, and all vertebrates, the flow of blood is always within blood vessels. The arterial system is connected to the venous system by means of capillaries which have very thin walls only one cell thick. Blood entering the capillaries is under relatively high pressure, and part of the fluid portion is filtered through the capillary walls, entering the tissue spaces. On the venous side of the capillary bed, most of this fluid flows back into the capillaries due to osmosis, because the blood has retained all of its proteins that were too large to be filtered. Gaseous, waste, and nutrient exchanges between the blood and tissues occur by way of this fluid exchange as well as by diffusion (**fig. 28.1**).

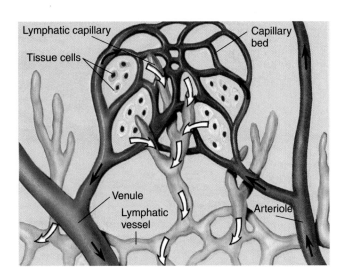

Figure 28.2 Lymphatic capillaries are interspersed with blood capillaries in tissues. Fluids that are not picked up by capillaries enter the lymphatics and are returned to circulation when lymph flows to vena cava.

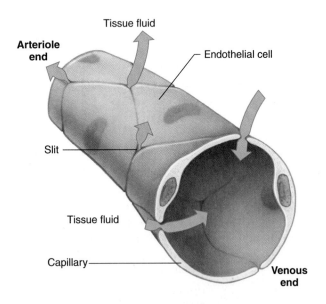

Figure 28.1 Fluid is forced out of the capillaries because of the pumping force of the heart. Pressure falls across the capillary bed due to drag and volume loss. On the venous side, fluids are drawn into the capillaries by osmosis. Excess fluids enter the lymphatic capillaries.

The capillary bed is the functional site of a closed circulatory system where essentially all chemical exchanges between the blood and surrounding tissues take place.

The **lymphatic system** consists of small open-ended lymphatic capillaries that conduct fluid into larger lymphatic ducts. Fluid that does not return to the blood capillaries enters the lymphatic capillaries from the tissue spaces (**fig. 28.2**). This fluid is collected in lymphatic ducts and returns to the venous system near the heart. Lymph flow is powered by body movements that compress lymph vessels, and is directed by one-way valves in the vessels. Thus, the lymphatic system serves as a kind of sump pump for the body, preventing fluids from accumulating in the tissue spaces. Fluids gathered in this way pass through the lymph nodes on their way to the heart where many bacteria and viruses are removed by lymphocytes in the nodes.

Vertebrate circulation is summarized in **figure 28.3**. Consider the changes that occur in the blood as it passes through the various circuits. It is more important to understand the function of the circulatory system than to know a long list of names of vessels.

Lab Instructions

You will observe the gross and microscopic features of the mammalian circulatory system and circulation in the capillary beds of a living vertebrate. You will learn how to measure the blood pressure of a human. You will also observe a living arthropod heart in order to compare open and closed circulatory systems.

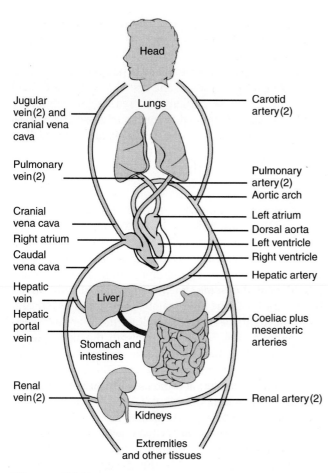

Figure 28.3 Schematic of major mammalian blood vessels and their relationship to one another. Blue denotes deoxygenated blood; red, oxygenated. Numbers indicate actual number of vessels in body.

In an open circulatory system such as this one, **hemolymph**, the circulating fluid, leaves the heart in arteries but returns in open **sinuses** instead of veins. Under a dissecting microscope, you will be able to see the fluid surrounding the heart enter it through three pairs of slit-like openings called **ostia**. When the heart relaxes, elastic ligaments expand the heart and the ostia open, allowing hemolymph to flow in. When the heart contracts, flaps of tissue inside the heart close the ostia, and hemolymph is forced out of the heart through the arteries.

You should be able to see the arteries that leave the heart, carrying hemolymph to the tail and the head regions. Other arteries lie beneath the heart. Make a diagram of how the crayfish heart works.

Open Circulatory Systems

Most molluscs, arthropods, and many other invertebrates (but not all) have open circulatory systems. Of these, the crayfish's is most easily observed. See figure 22.10 for a diagram showing the crayfish's circulatory system.

1▸ To observe the heart of a crayfish, first anesthetize an animal by packing it in crushed ice in a glass finger bowl or immersing it in a 1% chlorotone solution for about 20 minutes. Cover the abdomen with wet cotton. The dorsal part of the carapace, the exoskeleton covering the thorax, should be removed by inserting scissors under its posterior edge 1 cm to the left of the midline and cutting forward to the region of the eye. This procedure should be repeated on the right side, and the strip of exoskeleton should be carefully lifted and removed, so that none of the underlying membranes is torn.

The heart can now be seen beating in the **pericardial sinus** covered by the epidermal and pericardium membranes. These membranes can be removed to expose the heart, which should be bathed in Ringer's solution to keep it from drying.

Mammalian Circulatory System

2▸ If your fetal pig has not previously been opened, make a series of cuts as diagramed in the figure I.1, page 362. If you have followed the lab sequence in this manual, complete the opening as follows:

1. Make a longitudinal cut 1 cm to the left of the sternum from the lower ribs to the region of the forelimbs. Make a second cut parallel to the previous cut. Sever all ribs.

2. Lift up the center section, labeled (A) in figure I.1 (page 362), and cut any tissues adhering underneath. A transverse cut at the anterior end will detach this center piece, which should be discarded.

3. The heart and lungs will be easier to observe if the diaphragm is cut away from the rib cage on the animal's left side only. Cut close to the ribs.

4. In the region of the throat, remove the thymus glands, thyroid, and muscle bands, but do not cut or tear any major blood vessels.

If you have damaged or broken any of the vessels you are supposed to identify, ask someone if you can view his or her dissection to see the vessel in question.

The Heart and Major Vessels

3▶ Find the heart encased in the **pericardial sac.** Remove the sac and identify the four heart chambers. The paired **atria,** thin-walled, distensible sacs, collect blood as it returns to the heart. Their contraction aids in filling the thick-walled ventricles. The two **ventricles** are the large, muscular chambers that pump blood to the lungs (the pulmonary circulation) and to all other tissues (the systemic circulation).

Blood returning from the systemic circulation enters the right atrium (pig's right, not yours) from the cranial and caudal **vena cavae** then passes into the right ventricle. From there, it is pumped to the lungs through the **pulmonary trunk artery**, which divides into the left and right **pulmonary arteries.** The trunk is visible passing from the heart's lower right to upper left and passing between the two atria (**fig. 28.4**).

In fetal mammals, two shunts allow most blood entering the right heart to bypass the lungs, which are not functional until birth. The **foramen ovale** is an oval opening that permits blood to pass from the right atrium into the left atrium. At birth, it is normally closed by a flap of tissue that fuses with the interatrial wall, leaving an oval depression in the right atrium wall as a sign of its former presence. The **ductus arteriosus** links the pulmonary trunk artery to the aorta. Find this vessel in your animal. During the intrauterine life, most blood entering the pulmonary circuit does not pass to the lungs. Instead, it is shunted to the aorta. At birth, the shunting vessel constricts, diverting blood to the lungs. The constricted vessel fills with connective tissue to become a solid cord seen in adults as the arterial ligament.

Trace the pulmonary arteries to the lungs. Following gas exchange in the capillaries of the lungs, blood collects in the **pulmonary veins** and flows to the left atrium. These veins enter on the dorsal side of the heart and will be difficult to find. If you remove some of the lung tissue from the left side, you may be able to locate these vessels.

From the left ventricle, blood is pumped at high pressure through the **aorta** into the systemic circulation. Find the aorta. It will be partially covered by the pulmonary trunk but can be identified as the major vessel that curves 180° to the pig's left, forming the **aortic arch** (fig. 28.4).

Vessels Cranial to the Heart

4▶ **Veins** Because the venous system is generally ventral to the arterial system, it will be studied first in the congested region of the heart. Refer to **figure 28.5** and place a check next to each vein identified.

Trace the **cranial vena cava** forward from the heart to where it is formed by the union of the two very short **brachiocephalic veins.** Each of these in turn is formed by the union of the five major veins: the **internal** and **external jugular veins,** which drain the head and neck; the **cephalic vein,** which lies beneath the skin anterior to the upper forelimb and typically enters at the base of the external jugular; the **subscapular vein** from the dorsal aspect of the shoulder; and the **subclavian vein** from the shoulder and forelimb.

Trace the subclavian vein into the forelimb. In the armpit, it is known as the **axillary vein** and as the **brachial vein** in the upper forelimb.

Caudal to the union of the brachiocephalic veins, is a pair of **internal thoracic veins** that drain the chest wall. These veins were most likely cut when you opened the animal and only the stubs will be seen.

5▶ After you have revealed and identified the veins, call your instructor over. He or she will quiz you by asking you to identify one or more veins.

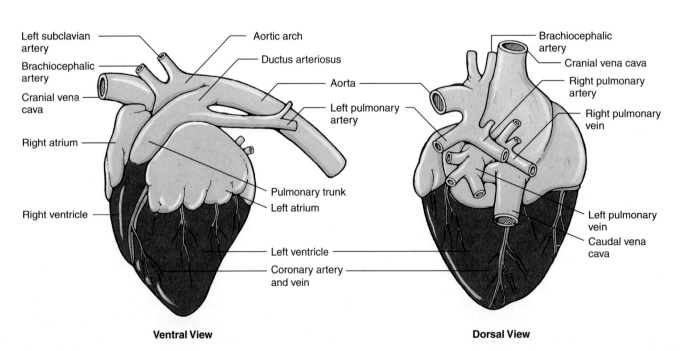

Figure 28.4 External ventral and dorsal views of the fetal pig's heart.

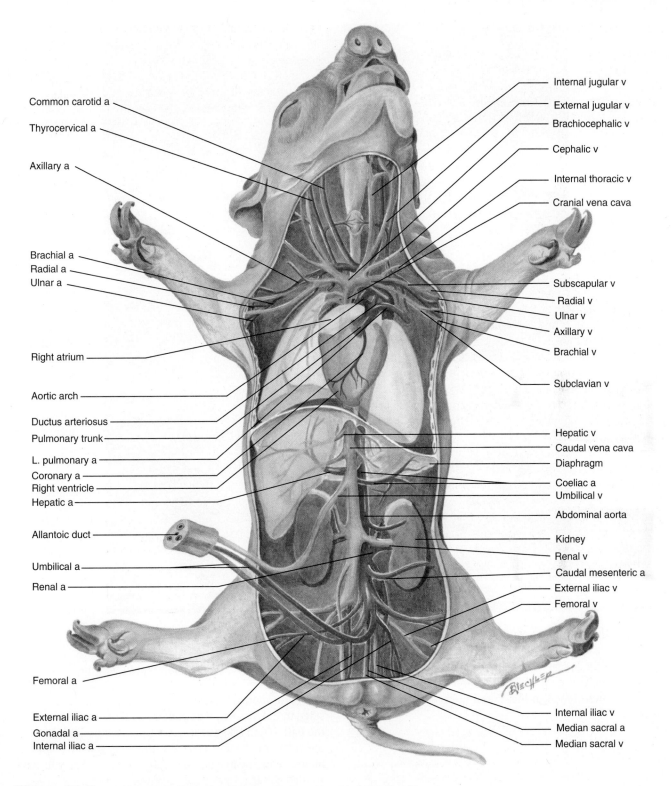

Figure 28.5 Ventral view of the fetal pig's major arteries (a) and veins (v).

Arteries Refer to figures 28.5 and **28.6** to identify the vessels. Check off the arteries as they are identified.

6 Find the aortic arch and trace it back to the heart. Note the several arteries that branch off to supply the anterior region of the animal. Find the small **coronary arteries** that arise from the base of the aorta behind the pulmonary trunk. They supply the muscles of the heart. In humans blockage of these vessels deprives the heart of oxygen, causing heart attacks.

The first major artery to branch from the aorta as it leaves the heart is the **brachiocephalic artery.** It gives rise to the two **carotid arteries,** which pass anteriorly to supply the head, especially the brain, and the **right subclavian artery,** which passes to the right forelimb.

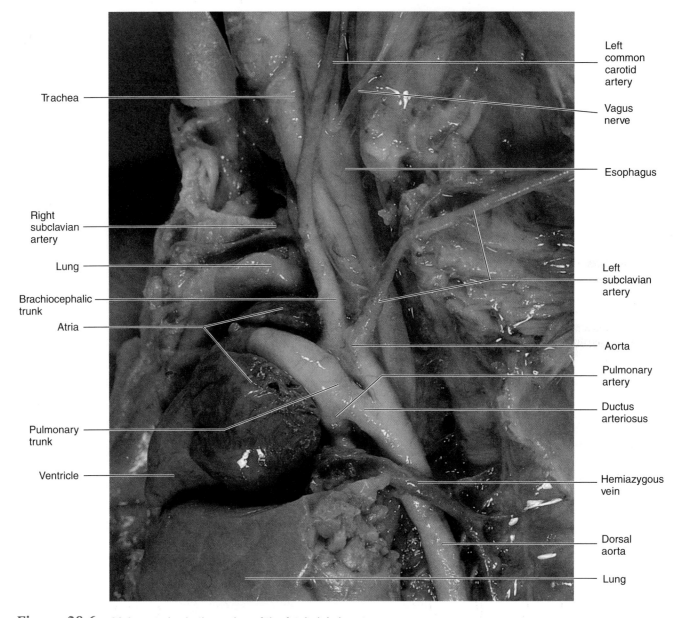

Figure 28.6 Major arteries in the region of the fetal pig's heart.

Just to the left of the brachiocephalic artery, find the **left subclavian artery** arising as a separate branch from the aortic arch. Blood in this vessel goes to which region of the body?

Trace the left subclavian artery into the forelimb, removing skin and separating muscles as necessary. In the armpit it is known as the axillary artery, and in the upper forelimb as the **brachial artery.** The subscapular artery branches from the axillary artery and supplies the shoulder muscles. The brachial artery divides in the lower forelimb, forming the radial and ulnar arteries.

Once the aorta turns posteriorward and passes along the dorsal wall of the thorax, it gives rise to intercostal arteries, which supply the walls of the chest. The aorta then passes through the diaphragm to become the **abdominal aorta.**

7 Find another student in the lab who is at the same stage in the dissection as you are. Quiz one another about the path blood takes as it flows from the forelimb through the heart to the head. You should be able to name all major veins, heart chambers, and arteries that you have studied so far.

Vessels Caudal to the Heart

Refer to figure 28.5 as an aid in identifying vessels. Place a check next to each vein and artery identified. Figure 28.3 shows you the major circulation patterns to and from this region of the body.

8 *Veins* If the heart is lifted and tilted forward, the **caudal vena cava** can be viewed at the point where it enters the right atrium. As this vein is traced caudally, several veins will be found flowing into it. After it passes through the diaphragm, the paired **hepatic veins** and single **umbilical vein** enter first. The hepatic veins drain blood from the liver. The umbilical vein carries blood from the placenta during embryonic life. It may have been cut when you opened the abdominal cavity.

If you lift and look under the lobes of the liver, you may be able to see the **hepatic portal vein,** although it is difficult to find. This vein receives blood draining from the stomach and small intestine. It will carry nutrients after birth when feeding starts. The hepatic portal vein enters the liver, where unlike other veins, it breaks down into what are called **portal capillaries.** Such capillaries receive blood from veins and then drain into veins again. The hepatic portal capillaries flow into the previously identified hepatic vein. Blood flow through the portal capillaries is slow, allowing the liver to add and remove materials and generally "condition" the chemical composition of the blood.

Return to the vena cava and follow it caudally to where the **renal veins** enter from the kidneys. The large size of the renal veins and arteries reflects the importance of the kidneys in removing undesirable materials from the blood. In the male, the spermatic veins, and in the female, the ovarian veins, enter next. On the left side, these veins may enter the renal vein first.

Below the kidneys, the vena cava splits into the **internal** and **external iliac veins** and the **median sacral vein,** a small vein that comes from the tail. The external iliac veins collect blood from the **femoral veins** in the hind legs, whereas the internal iliacs collect blood from the pelvic area.

9 *Arteries* After the aorta passes through the diaphragm and enters the abdominal cavity, a large, single **coeliac artery** arises from it at a level equal to about the cranial end of the kidneys. You will have to remove some connective tissues to obtain a full view of this artery. The coeliac artery eventually divides into three arteries supplying the stomach, spleen, and liver.

The **mesenteric artery** next arises from the aorta and supplies the pancreas, small intestine, and large intestine. The **renal arteries** are short, paired arteries supplying the kidneys. The next large arteries arising from the aorta are the **external iliacs,** which supply the hind legs with a branch to the lower back.

In the fetus, the **umbilical arteries** branch from the caudal end of the abdominal aorta and pass out through the umbilical cord. They form a capillary bed in the placenta for nutrient, gas, and waste exchange with the maternal circulatory system. Most likely these arteries were cut when you opened your pig for dissection.

10 Find another student in the lab who is at the same stage in the dissection as you are. Quiz one another about the circulation paths to the small intestine and kidney, tracing a drop of blood from the time it enters the aorta and passes to the organs until it returns to the aorta.

Look at the diagram in **figure 28.7** and add labels to the major arteries and veins that you identified in your dissection.

Internal Heart Structure

11 Study the orientation of the heart so that you can later identify it in isolation. Now, free the heart from the body by cutting through all the vessels holding it in place. Be careful to leave enough of each vessel so that they can be identified in the isolated heart. Alternative to removing the fetal pig's heart, your instructor may have a demonstration dissection of a beef heart or heart models for you to

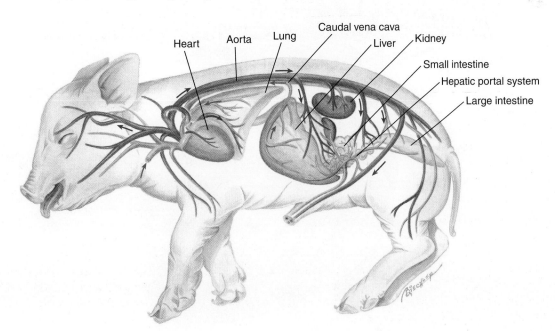

Figure 28.7 Lateral view of circulatory system in fetal pig (arteries—red, veins—blue). Label the major arteries and veins.

study. Whichever specimen you are using, orient yourself by identifying the **aorta, pulmonary artery, pulmonary vein,** and **vena cava.** Your task will be to figure out how blood flows through the heart chambers to and from the major vessels that enter it.

Place the heart in your dissecting pan, ventral side up. Make a razor cut along the pulmonary trunk down through the right ventricle. Spread the tissue open, pin it down, and remove the latex. You may wish to use a dissecting microscope to observe the open ventricle.

Identify the **semilunar valves** at the junction of the pulmonary trunk artery and ventricle. Consider how these valves work. The open flaps face into the pulmonary trunk. When the right ventricle contracts, these valves are pressed against the walls of the pulmonary trunk and blood flows by easily. When the ventricle relaxes, any backflow in the pulmonary trunk fills the valve flaps with blood and they spread out, closing the artery. These are passive valves that do not require any extra energy to operate. You may have to cover the heart with water to float the valve flaps so that you can see them (**fig. 28.8**).

Now cut through the right atrium and remove the latex and coagulated blood. The **tricuspid valves** are between the atrium and ventricle. These are also flap valves that work on the backflow principle, allowing blood to flow only one way from the atrium into the ventricle. In the ventricle, fine fibers called **chordae tendinae** are attached to the valve flaps. These cords prevent the flaps "blowing back" from the high pressures developed when this ventricle contracts.

Cut into the left atrium and ventricle as you did on the right side. Identify the **bicuspid,** also called the **mitral valve,** between the atrium and ventricle with its associated chordae tendinae. After cutting into the ventricle and cleaning it, find the **aortic semilunar valve.** Blood flow through the human heart is shown in figure 28.8. Are all of the valves of the heart flap valves?_____

12 Pair off with another student and label the major vessels shown in figure 28.7. Use the labeled diagram to describe to your partner how blood returning to the heart from the foreleg travels to the lungs. You partner should then describe to you how blood flows from the heart to the hindleg. Both of you should discuss the operation of the heart valves.

Clean your dissecting instruments and tray and return your fetal pig to the storage area.

Histology of Vessels

13 Obtain prepared slides of cross sections of arteries and veins and observe them under low power with a compound microscope. Note that arteries have thicker walls

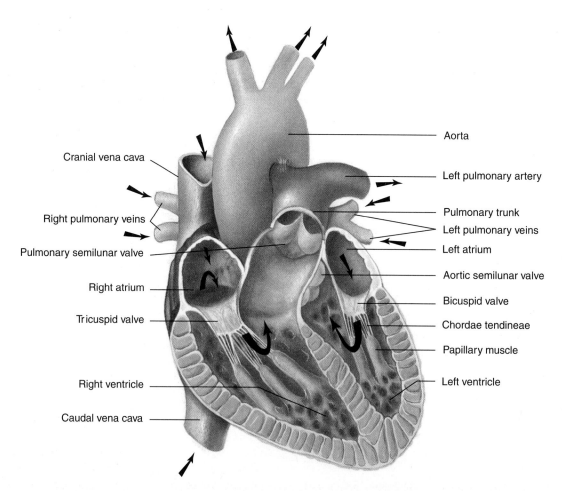

Figure 28.8 Path of blood flow through the major chambers and vessels of the human heart.

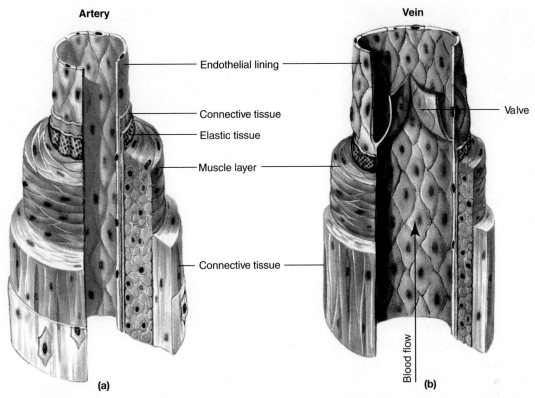

Figure 28.9 Histology of blood vessels: (a) tissues in the wall of an artery; (b) tissues in the wall of a vein; flap valves in veins prevent backflow of blood in the venous sytem.

than veins of the same size. The difference in thickness is due to the increased amounts of muscle and connective tissue in the artery. Since arteries carry blood from the heart, they operate under relatively high pressure (about 120 mm of mercury equivalent). Veins experience only one-twentieth as much pressure. Because of this pressure difference, blood flow in arteries and capillaries is unidirectional. Blood flows through veins because veins are routed in such a way that skeletal muscles press on them when the muscles contract, moving the blood along. **Valves** in the veins prevent backflow and make the passage of blood unidirectional (fig. 28.9).

Observe the tissues of a blood vessel under 10×. **Endothelial cells** are epithelial cells that line both arteries and veins; capillary walls consist of only endothelial cells. Arteries are usually smaller in diameter and have thicker layers of smooth muscle in their walls than do corresponding veins. Contraction of vascular smooth muscle, especially in the smallest arteries called arterioles, alters resistance and, consequently, the flow pattern and blood pressure.

Figure 28.10 shows the nature of the endothelial lining of blood vessels and red blood cells in an arteriole. The complexity of the microvasculature is evident in scanning electron micrographs of casts of the circulatory system. Note the capillaries and their relationship to arterioles in figure 28.11.

Blood

Human blood consists of 55% plasma and 45% cells by volume. **Plasma** is the fluid portion of the blood containing

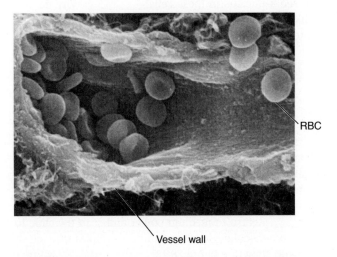

Figure 28.10 Scanning electron micrograph of a broken blood vessel, showing the smooth endothelial lining and several red blood cells (RBC).

dissolved proteins, salts, nutrients, and waste products. Several different types of cells and cell fragments are contained in blood. By far the most common (about 95% of the cells) are **erythrocytes** (red blood cells) which are red because they contain hemoglobin. During maturation red blood cells lose their nuclei and mitochondria, essentially becoming bags of hemoglobin that live for 3 to 4 months. The other 5% consists of **leukocytes** (white blood cells) which are part of the body's defense system and **platelets** that are important in blood clotting. There are several types of leukocytes. The most common, neutrophils and

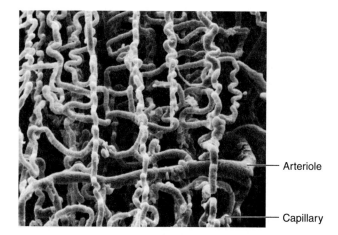

Figure 28.11 Scanning electron micrograph of a plastic cast of a capillary bed from skeletal muscle in which individual arterioles can be traced to capillaries.

lymphocytes, together representing 95% of the white blood cells, are shown in **figure 28.12**. The remaining types of cells, basophils and monocytes, represent only 5% of the white blood cells.

14▶ Get a prepared slide of a Wright-stained human blood smear from the supply area and look at it with your compound microscope under medium power.

Note the large number of round red blood cells. Can you see a nucleus in these cells?_____. Why do you think that the red blood cells are lighter in color in the center and darker at the periphery?

As you look carefully at the slide, you will see occasional cells that look different. These are leukocytes and because of the staining they have a blue/purple color. Center a leukocyte in the field of view and observe it with high power. What structure in the cell is stained? _____. Return to medium power and scan the slide to locate other leukocytes. Try to find examples of each of the types shown in figure 28.12.

Neutrophils leave the blood early in the inflammation process and become phagocytic cells consuming cell debris and bacteria. **Lymphocytes** are important in the immune response. Some are involved in cellular immunity and others secrete antibodies that neutralize foreign proteins and other macromolecules.

At high power, you may be able to see **platelets**, small cell fragments important in sealing microholes in blood vessels and in clotting. Return your slide to the supply area.

Capillary Circulation

15▶ To observe circulation in capillaries, net a small fish or tadpole from an aquarium and wrap it in dripping wet cotton, as shown in **figure 28.13**, being careful not to cover

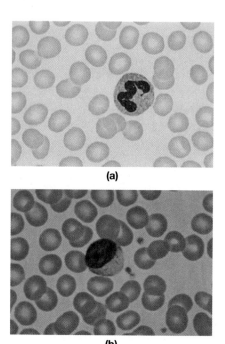

Figure 28.12 Human blood stained with Wright's stain shows red blood cells and different types of white blood cells; (a) neutrophil; (b) lymphocyte. Small dark staining objects in background of (b) are platelets.

(1) Wrap the fish (except for the head and tail) with dripping wet cotton. Place fish in half of a petri dish. Place coverslip over tail.

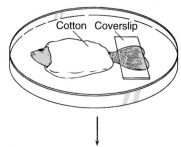

(2) Place dish on microscope so that fish's tail is over hole in stage.

(3) Examine with low- and medium-power objectives of your microscope.

Figure 28.13 Method for observing microcirculation in a fish tail.

the tail. Lay the wrapped fish in an open petri dish. Place a few drops of water on the tail, spread it, and add a coverslip. About every five minutes return the fish to water.

Place the dish on a compound microscope stage and observe the tail under scanning power. Sketch your observations, answering the following questions:

1. Can you identify capillaries, venules, and arterioles? What criteria are you using to do so?

2. Is blood flow faster in certain vessels compared to others?

3. Is blood flow continuous in all vessels? What might control this?

If dilute solutions of nicotine, caffeine, and adrenalin were available, what effects might these chemicals have on capillary circulation?

Measuring Blood Pressure and Heart Rate

16 Blood pressure is the pressure that the blood exerts on the walls of blood vessels and is usually measured in arteries. Because of the contraction cycle of the heart, pressure varies from a high (**systolic**) to a low (**diastolic**) pressure during a cycle. Pressures are reported in mm of mercury (Hg) with the systolic pressure appearing first. Thus a blood pressure reading of 120/60 means that the pressure just following maximum contraction is 120 mm of Hg and just following maximum relaxation is 60 mm of Hg. Blood pressure can be measured quickly and easily so it is commonly used as one measure of cardiovascular health. Average values for a young adult are about 120/70. The U.S. National Institutes of Health values for normal, prehypertension, and high blood pressure categories are shown in the following table.

Category	Systolic	Diastolic
Normal	< 120	< 80
Prehypertension	120–139	80–89
High blood pressure	> 140	> 90

Methods You will measure blood pressure using an instrument called a **sphygmomanometer.** It consists of an inflatable cuff that is placed around a person's arm (**fig. 28.14**). When the cuff is inflated to a pressure greater than the systolic pressure, the arteries in the upper arm collapse and blood flow to the arm stops. The pressure in the cuff, measured with a pressure gauge, is then gradually reduced until blood flow begins. The resumption of blood flow can be detected using a stethoscope to detect sounds produced by flowing blood (fig. 28.14) or electronic transducers placed on a finger tip. When cuff pressure drops below systolic pressure, blood begins to spurt through the compressed artery, and that turbulent flow produces faint tapping sounds. A pressure transducer under the cuff or on a finger can also detect the changes in blood flow. The corresponding cuff pressure is recorded as the systolic pressure. As pressure in the cuff is further reduced, sounds become louder as flow increases. When the artery is wide open and blood flow is no longer restricted, the turbulent sounds stop. The cuff pressure at which this occurs is recorded as the diastolic pressure.

Measurements Work with a partner to measure blood pressures. The subject should sit with an arm extended and resting on a table. Care should be taken to avoid moving the arm or hand during measurement. You may use a stethoscope, a commercial blood pressure monitor, or a transducer connected to a computer, to measure or record blood pressure. Your instructor will explain in more detail how to do the following procedures.

1. Wrap the cuff around the upper arm just above the elbow so the cuff's pressure gauge is over the brachial artery on the inner surface of the arm (fig. 28.14). Cuffs may have instructions printed on them—if so, follow those instructions.

2. If using a computer data acquisition system, make certain that sensors for cuff pressure and blood flow are properly connected and that you know how to start a recording.

If using a stethoscope, place the end over the pulse point located on the inside of the elbow joint.

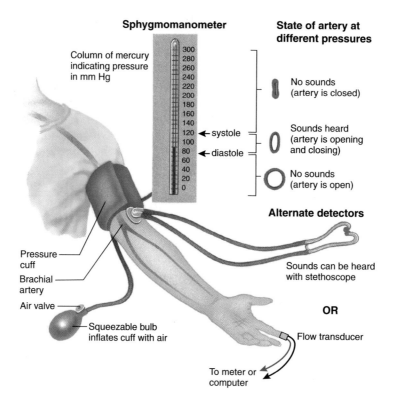

Figure 28.14 One method used in measuring blood pressure. The pressure cuff closes artery and pressure is read on mercury manometer. However there are now electronic devices to measure pressure that replace this. As pressure is released, artery opens and blood flow can be heard or sensed with transducer.

If using a finger pressure transducer, put it over the end of index finger.

3. If using a computer data acquisition system, start recording data now.

4. Inflate the cuff to about 200 mm of pressure.
   Warning: This blocks blood flow to the arm. Never keep a cuff inflated for more than about 90 seconds.

5. Most cuffs will slowly deflate automatically. If not, slowly release the pressure by opening the valve on the squeeze bulb as you listen for the first thudding sounds of blood flow or watch the electronic output from the flow transducer. Note the pressure when blood flow starts. Continue to reduce pressure until the sounds are no longer heard or flow reaches its maximum. Note the pressure reading when this happens.

6. Completely deflate the cuff, then stop recording data.

Record your results below:

Systolic pressure = _____

Diastolic pressure = _____

Now measure heart rate. If you recorded pulse pressure data, use your recorded data. Measure the difference in time between one pulse peak near the beginning of blood flow, and the tenth peak after the chosen start peak. Calculate heart rate as:

$$\text{Heart rate in beats/min} = \frac{10 \text{ beats}}{\text{Elapsed time (sec)}} \times \frac{60 \text{ sec}}{\text{min}}$$

If you did not record pressure data with a transducer and computer, you can, with a little practice, measure heart rate by counting beats. Gently place the tips of your first two fingers on the medial surface of your subject's wrist just proximal to the base of the thumb. Adjust finger position until you can feel a pulse in the underlying radial artery, then count beats for 30 seconds, and multiply by 2 to get heart rate in beats/min.

What heart rate did the subject have? _____

17. **Analysis** Class members should now write their measurements on the blackboard, creating different columns for males and females. Copy the data below.

What range of pressures and heart rates were observed in the class? Were there any differences between genders? How does blood pressure vary with heart rate?

Why doesn't blood pressure drop to zero when the heart relaxes? What is the functional importance of the elastic fibers illustrated in figure 28.9a?

Blood Pressures under Experimental Conditions

18 Assuming that you now can measure blood pressures rather quickly, try doing so under one or both of these experimental conditions, as directed by your instructor:

Laying and Standing Measure blood pressure and heart rate in a subject after reclining on a table for 2 or 3 minutes. Then have them quickly stand up and immediately measure blood pressure again. Record your results below:

Position	Systolic P (mm Hg)	Diastolic P (mm Hg)	Heart Rate (beats/min)
Reclining			
Standing			

Changes in blood pressure with posture are called orthostatic adjustments. Do these changes have any relationship to light-headedness that is sometimes felt when getting up quickly? Explain. What happens to the distribution of blood in your body when you suddenly stand up?

Rest and Exercise You already measured blood pressure and heart rate in your subject at rest. Predict how those measurements should change during exercise:

Now measure the blood pressure and heart rate immediately after 2 minutes of exercise—jogging in place. Do not remove the cuff and pressure sensors before the exercise, and start your measurement as fast as possible after exercise stops. Record your results below:

Position	Systolic P (mm Hg)	Diastolic P (mm Hg)	Heart Rate (beats/min)
Rest			
Postexercise			

Do results match your predictions? Which measures changed by the greatest percentage? What is the functional importance of these changes?

Lab Summary

19 On a separate sheet of paper, answer the following questions as assigned by your instructor.
1. Create a flowchart showing the major vessels and heart chambers that a drop of blood passes through as it returns from a pig's arm and passes to the back leg.
2. Create a flowchart showing the major vessels and heart chambers that a drop of blood passes through as it returns from the small intestine and passes to the brain.

3. List the locations and names of the valves of the heart and describe how they operate.
4. What are the structural differences between an artery and a vein? How do capillaries differ? What is the functional importance of these differences?
5. How does the open circulatory system of a crayfish differ from the closed system of a mammal? Describe how blood returns to and enters the heart of both types of animals.
6. Trace a molecule of oxygen from when it enters the pig's lungs to when it enters the tissues of the upper hind leg. Then trace a carbon dioxide molecule as it passes from the leg back to the nostril.
7. Explain what blood pressure is and what it means when a person says their blood pressure is 150/70.

You may want to try the critical thinking questions that apply some of the knowledge you gained in doing this lab.

Internet Sources

Many medical schools have extensive collections of pictures of pathological conditions. These collections are available over the WWW. Use your browser program and a search engine to locate pictures of a blood vessel with arteriosclerosis.

Compare this picture to your observations of a normal artery in the lab. Describe the differences.

Continue your search and find the functional consequences of high or low blood pressure. Identify three problems associated with each.

Learning Biology by Writing

Write an essay describing how blood flows through the heart, including a description of pressure during systole and diastole, and mechanisms that cause blood flow to be unidirectional.

Critical Thinking Questions

1. What are the roles of the lymphatic system and the venous system in returning fluid filtered through the microcirculation?
2. Because a giraffe's head is 15 feet above the ground, what circulation adaptations are necessary to allow adequate blood supply to the head?
3. If arthropods such as insects have open circulatory systems that lack veins, how does hemolymph return from the tissues to the heart?
4. Based on your lab observations, give an opinion on the following situation: A person has a blood pressure of 150/90.

Lab Topic 29

INVESTIGATING GAS EXCHANGE SYSTEMS

SUPPLIES

Resource guide available on WWW at
http://www.mhhe.com/labcentral

Equipment

Compound microscopes

Dissecting microscopes

Spirometer or pneumotachometer, recording system and associated supplies;

    ADInstruments, BioPac, IWorx and Vernier all provide suitable systems

Materials

Fetal pigs

Dissecting trays

Dissecting instruments: scissors, forceps, blunt probe, razor blade or scalpel

Prepared slides

    Mammalian lung section

Slides and coverslips

Crickets, grasshoppers, or roaches, freeze then thaw

Live goldfish and goldfish bowls

Fresh or frozen and thawed fish with head and gills intact

Petri plates with dissecting mats

STUDENT PRELAB PREPARATION

Before doing this lab, read the Background material and sections of the lab topic assigned by your instructor.

Use your textbook to review definitions of the following terms:

alveolus	pharynx
bronchiole	trachea
diaphragm	tracheole

Be able to describe in your own words the following concepts:

Pathway of air to mammalian lungs

Mechanics of mammalian breathing

After finishing the prelab review, write any questions you have in the margins of this lab topic. The lab experiments should help you answer these questions or you can ask your instructor during the lab.

OBJECTIVES

1. To observe the anatomy of an insect's tracheolar respiratory system
2. To observe the anatomy of fish respiratory systems
3. To dissect the respiratory system of a fetal pig
4. To observe the histology of mammalian lung
5. To measure human ventilation volumes

BACKGROUND

Cells of most animals are capable of aerobic metabolism, using oxygen and producing carbon dioxide (CO_2) as a waste product. Oxygen must move from the environment to every cell in the body, where it ultimately functions as a terminal electron (hydrogen) acceptor. Without oxygen, the mitochondrial cytochrome system would not operate, and cells could not carry out the Krebs cycle and other oxidations of food materials. Furthermore, CO_2, which is produced during the Krebs cycle and in other reactions, must pass from cells to the environment to maintain a consistent intracellular acid–base balance.

Aerobic metabolism produces concentration gradients favoring oxygen uptake and CO_2 efflux from cells, but diffusion is slow and limited by surface area. Consequently, only small animals with low metabolic rates can meet their gas exchange needs by diffusion across the body surface. Larger animals, however, require expanded surfaces for gas exchange. In addition, ventilation systems ensure that exchange surfaces are bathed by fresh air or water, and internal transport systems move respiratory gases between deep tissues and the exchange surface.

Protozoa, sponges, cnidarians, flatworms, and roundworms have no anatomical specializations for respiration other than shape and thickness that keep every cell near the body surface. Larger aquatic animals, such as lobsters, clams, and fish, increase the area for gas exchange through outpocketings of the body surface called gills. Surrounding water supports gills and keeps adjacent parts from sticking together. Blood or hemolymph pumped through the gills carries respiratory gases to and from distant tissues.

Dry surfaces block diffusion, so respiratory surfaces must be kept moist. In air, hydrogen bonding causes adjacent moist surfaces to adhere and surface tension causes gills to collapse, blocking gas exchange. That is why fish suffocate when removed from water, even though air contains much higher concentrations of oxygen than does water. Terrestrial animals expand surface area using inpocketings from their body surface, so that tissue anchors and pressure differences prevent respiratory spaces from collapsing. In the tracheal systems of insects and some spiders and the lungs of most terrestrial vertebrates, oxygen-rich air enters the animal by a system of tubes and is brought in close proximity to the blood or the tissues where gas exchange occurs. There are exceptions to these generalizations, of course. Many adult amphibians have lungs, but also exchange gases directly through the skin and mouth surface, and their aquatic larvae have gills. One family of salamanders commonly found in North and Middle America, the *Plethodontidae,* lacks lungs and depends entirely on cutaneous respiration.

For internal transport, the oxygen-carrying capacity of blood is increased by the presence of a respiratory pigment, such as **hemocyanin** (usually dissolved in hemolymph) or **hemoglobin** (usually in blood cells). These pigments bind loosely and reversibly with oxygen to facilitate oxygen transport from the gills or lungs to the tissues. Blood is oxygen deficient when it enters the gills or lungs. Oxygen diffuses into the blood and combines with the pigments. As the blood flows to areas where there is little oxygen, the diffusion gradient causes the oxygen to move into the tissues, where it is used by mitochondria. Conversely, high concentrations of carbon dioxide are usually present in active tissues. CO_2 moves down the diffusion gradient into the blood where it simply may dissolve or may combine chemically with the pigments. It is then carried back to the respiratory organs where it is exchanged with the environment.

LAB INSTRUCTIONS

You will study the respiratory systems of several kinds of animals and measure human ventilation.

Insect Tracheal System

The mammalian respiratory system represents only one method of gas exchange in terrestrial animals. Insects and other terrestrial arthropods make use of another, a rather remarkable **tracheal system,** which allows the exchange of gases to occur independently of the circulatory system.

In a tracheal system, respiratory gases diffuse through the body in a branching system of gas-filled tubes called **tracheoles** and **tracheae** that open to the exterior via openings called **spiracles (fig 29.1).** In insects, there is usually one pair of spiracles in each body segment. The tracheoles branch extensively, so the terminal branches are adjacent to nearly every cell in the body, and the circulatory system plays little or no role in gas transport. No special muscles circulate air in the tracheal system, but contraction of the abdominal and flight muscles can aid the exchange of air.

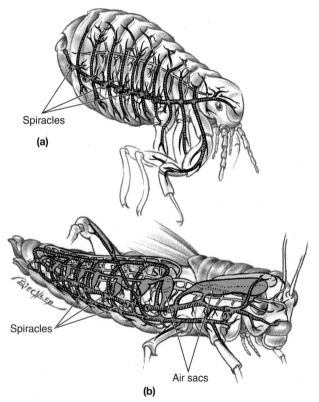
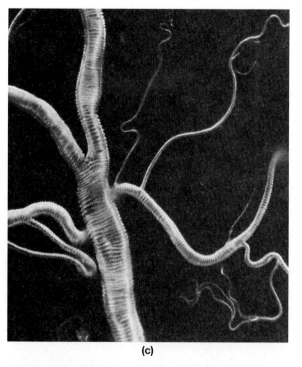

Figure 29.1 Tracheal systems. (*a*) Tracheal system of a flea penetrates all parts of its body, allowing direct gas exchange. (*b*) Large insects pump air through their tracheal systems by contracting muscles that press against air sacs. (*c*) Photomicrograph of tracheae shows reinforcing rings of chitin that prevent the tubules from collapsing.

Large insects often have tracheal expansions called **air sacs.** When compressed by body movement, they force air through parts of the system. Tracheal systems require little or no energy expenditure to maintain gas exchange, but reliance on diffusion limits maximum body size. That is why giant insects exist only in horror movies. Diffusion is much slower in water than in air, so insects are hugely successful on land, but the few aquatic insects are mostly air-breathers and confined to life near the surface.

1▶ If your instructor has not prepared a demonstration dissection of a tracheal system, obtain a large insect (hornworm larva, cricket, grasshopper, or roach) that has been euthanatized with ethyl acetate or recently frozen and thawed, and a petri dish with a dissecting mat. Locate the spiracles along the side—remove wings and legs if necessary. With fine scissors, cut through the dorsal surface close to the lateral margin but do not cut through the spiracles. (Keep the points of your scissors against the inside of the exoskeleton and take small snips so that you do not damage any interior organs.)

Run the cuts along the full length of the insect and join them together at the anterior and posterior ends. Remove the dorsal strip of cuticle and pin the insect in the dish, dorsal side up. Next flood the insect with saline solution to prevent drying and to float tissues for observation. Place the dish under a dissecting microscope and observe. You can remove yellow or white fat bodies, but be careful not to damage silvery tracheoles. Vary the lighting angle while looking for a silver-colored tube originating from a spiracle. Remove this trachea with a forceps, mount it in a drop of water on a microscope slide, and add a coverslip. Observe the trachea with your compound microscope and sketch it in the space below. Compare your observations to figure 29.1c.

Gills

In aquatic organisms, gills function in respiratory gas exchange and ion regulation. Gills are found in many different kinds of animals, including larval amphibians, fish, arthropods, molluscs, and many other aquatic invertebrates. Except in molluscs, gills originate in the embryo as featherlike outpocketings of the body wall that are amply supplied with blood. In most cases, a special mechanism exists to ventilate the gills, that is, to move water over them. In crayfish, this involves a specialized mouthpart, the **gill bailer,** which moves water forward over the gills in a sculling motion. In clams and many other animals, the gills function in both gas exchange and in filtering food particles from the water.

Fish gills are supported by a series of cartilaginous or bony arches. Each arch carries a series of **filaments,** and each filament is lined with projecting flaps of tissue called **lamellae.** Such a repeatedly divided surface provides a large area for gas exchange.

Many fish use two sets of muscles to force fresh water across their gills, one in the floor of the mouth and one in the **opercular flap,** which covers the gills (**fig. 29.2**). With mouth open, fish lower the jaw and raise the operculum to draw fresh water into the mouth and over the gills. They then close the mouth and raise the lower jaw, forcing water from the mouth over the gills and out through the gill slit. That provides a nearly continuous flow of fresh water over the gills. Blood in the gills moves in the direction opposite to the flow of water. Such a countercurrent flow maximizes exchange of oxygen.

Fast-swimming fish often use ram ventilation—they just swim with the mouth open, so swimming drives water over the gills and out through the gill slits. Active

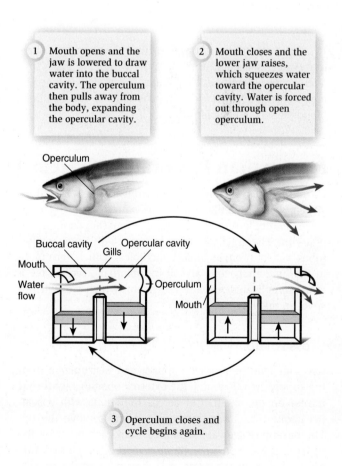

Figure 29.2 How a fish ventilates its gills.

fish like tuna and some sharks rarely stop swimming, and rely primarily on ram ventilation.

Fish that live in low-oxygen water (like goldfish) often supplement gill respiration by air breathing. They gulp air, and hold it in the mouth, digestive tract, swim bladder, or opercular cavity next to tissue with a rich blood flow. The pharyngeal lung pouches of lungfish are homologous to lungs of terrestrial vertebrates.

2 Obtain a fresh or frozen and thawed fish, or make the following observations on a demonstration prepared by your instructor. Lift one opercular flap, and look beneath to see the gills. Insert a blunt probe in the mouth, and push it past the gills and out through the opercular slit. That is the pathway that water follows over the gills. Lift the operculum, and use heavy scissors to cut forward along the edges toward the mouth, being careful not to damage the gills, until you can lift the operculum off the gills. How many gill arches are present? _____ Do any gill arches carry gill rakers, bony or cartilaginous projections opposite the soft gill filaments? _____ What might be the function of such projections?

3 Cut a small piece from one gill arch, place it in shallow water in a dish, and examine it under a dissecting microscope. Identify gill filaments and their lamellae. At higher magnification, you may be able to see capillary blood vessels in each lamella.

Mammalian Respiratory System

In amphibians, reptiles, and mammals, the respiratory system has a treelike branching pattern, starting with a single tubular trachea opening from the floor of the pharynx. The trachea branches repeatedly into smaller tubes called bronchi and bronchioles, which terminate in the lung airspaces—alveoli in mammals. Birds have a more complicated respiratory system, which we will not cover in this lab.

4 Gross Anatomy

Get your fetal pig, dissecting pan, and instruments from the supply area. Examine the external openings (**external nares**) on the snout. Cut across the snout with a scalpel about 1 to 2 cm from the end and remove the tip. The nasal passages are separated from each other by the nasal septum. You may be able to see the curved **turbinate bones** in the sinus area that increase the surface area of the passageways, and create eddy currents that, along with hairs, cilia, and mucus, help remove dust and humidify the inhaled air. The floor of the nasal passages is made up of the hard palate and the soft palate (posterior to hard palate).

Look into the pig's mouth. Behind and above the soft palate is the **nasopharynx.** It may be necessary to slit the soft palate to observe this. Air enters the nasopharynx from the posterior end of the nasal passages, then passes into the **pharynx,** through the **glottis,** and into the **larynx** and ultimately the **trachea** (see figure 27.3).

If food accidentally enters the glottis, choking results. We have all "snorted" food while talking and eating. Based on the anatomy you are observing, can you explain why?

The Heimlich maneuver can often "save" a person who has food wedged in their glottis. Your instructor may demonstrate this maneuver for you.

In the nasopharyngeal area, look for the openings of the **eustachian tubes.** They are very difficult to find. These tubes allow air pressure to equilibrate between the middle ear chamber and the atmosphere. (This is why changes in altitude cause the ears to "pop.") Throat infections often spread to the ears through the eustachian tubes.

5 Run your fingers over the pig's throat and locate the hard, round **larynx.** Make a medial incision in the skin of the throat and extend the ends of the cuts laterally, folding back the skin flaps. Repeat this procedure for the muscle layers. Use a blunt probe to separate the muscles in the midline. Be careful not to cut or tear any major blood vessels.

Just posterior to the larynx note the large mass of immune system material, the **thymus,** with two lobes flanking the underlying trachea. This is not part of the respiratory system. In young vertebrates, **lymphocytes,** an important component of the immune system mature in the thymus. Later in life, the thymus atrophies and is of little consequence.

Look between the lobes of the thymus to locate the brownish ovoid **thyroid gland.** It is not part of the respiratory system but is an important endocrine gland producing the hormone thyroxine that regulates metabolism. You can remove the thymus and thyroid to see the trachea.

Expose the trachea, and note how it is supported by rings of cartilage.

At the anterior end of the trachea is the **larynx,** or voice box. It contains folds of elastic tissue, which are stretched across the cavity. These **vocal cords** vibrate when air passes over them, producing sound, and attached muscles vary the cord tension, allowing variations in pitch. Slit the larynx longitudinally and observe the vocal cords. Continue the slit posteriorly into the trachea and observe its lining.

Dorsal to the trachea, find the esophagus, leading from the pharynx to the stomach. You may need to tear connective tissue that holds the trachea and esophagus together. Pass a blunt probe into the esophagus from the mouth to help identify it. Why is the trachea, but not the esophagus, supported by rings of cartilage?

6 If the **thorax** of your animal is not already opened, make a longitudinal cut with heavy scissors through the ribs just to the animal's right of the **sternum,** or breastbone. Always keep the lower scissor tip pointed upward against the inside of the sternum to avoid catching and cutting internal structures. Be careful! Several major blood vessels and the heart are under the sternum and should not be damaged.

The **diaphragm** is a sheet of muscle that separates the **abdominal cavity** from the **thoracic cavity.** The thoracic cavity is divided into three areas by membranes: the right and left **pleural cavities,** which surround the lungs, and the **pericardial cavity** where the heart is located. See figure 27.4 if you need help locating the diaphragm or lungs.

If the pleural membranes are removed, the **lung** structure can be seen. The trachea, when it enters the thorax, divides into two **bronchi,** which are hidden from direct view beneath the heart and blood vessels. These bronchi, in turn, divide into progressively smaller **bronchioles,** which end in clusters of microscopic air sacs called **alveoli.** Alveoli have walls only a single cell layer thick and they are covered by capillaries (**fig. 29.3**). In the alveoli, oxygen and carbon dioxide are exchanged between the erythrocytes (red blood cells) in blood and the inhaled air. Subdividing the lung volume into many thousands of tiny

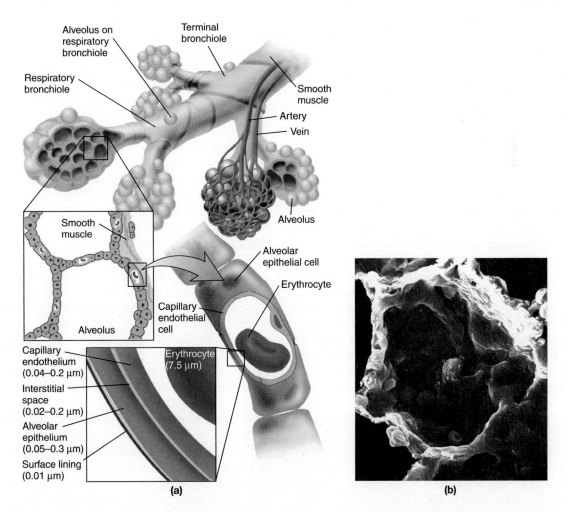

Figure 29.3 (a) Microscopic structure of lung and close relationship of alveoli and capillaries. (b) scanning electron micrograph of an alveolus, showing thickness of alveolar wall.

alveoli results in a large surface area for gas exchange. In humans, alveolar surface area equals about half the surface of a tennis court.

Bronchioles have smooth muscles in their walls. Asthma results from airway inflammation and the spasmodic contraction of bronchiolar smooth muscles. When this occurs, the bronchi are constricted preventing air flow to the alveoli. Have you ever known someone with bronchitis? Where do you think the infection was?

7 ▸ Microscopic Anatomy of Lung

Remove a piece of the lung and put it in a small bowl of water. Observe it with a dissecting microscope and find alveoli and bronchioles.

Also look at the demonstration slide of a section across several alveoli. How many cell thicknesses separate the air in a mammalian lung from the red blood cells in the capillaries?_____. The inset in figure 29.3 gives you the distances in micrometers between an erythrocyte in a capillary and the air in an alveolus. What is the diffusion distance for a molecule of oxygen in the alveolus to an erythrocyte and its hemoglobin?_____.

Figure 29.3 contains a scanning electron micrograph of an alveolus and surrounding capillaries. The alveoli form an interconnecting system of chambers, and macrophages in the alveoli scavenge for microorganisms and particulate material.

Lung Ventilation Mechanism

Air enters the lungs as a result of contraction of the **diaphragm** aided by intercostals muscles (fig. 29.4). When the dome-shaped diaphragm contracts, it flattens out, pushing down against the abdominal contents and increasing the volume of the thorax. Air pressure in the lungs drops, sucking air in through the trachea and bronchi and inflating the lungs until lung air pressure rises to equal atmospheric pressure. For a deeper breath, the **external intercostal** (between the ribs) muscles also contract, raising the rib cage and further increasing the volume of the thoracic cavity. When the diaphragm and external intercostal muscles relax, the rib cage drops and the diaphragm rises compressing the lungs and driving air out of the "respiratory tree." Figure 29.4 illustrates these mechanics in humans. Forced expiration can occur by contracting the **internal intercostal muscles,** which pull the rib cage down. In addition, contraction of the abdominal muscles force the organs in the abdominal cavity against the diaphragm, driving it upward.

8 ▸
Note the convex shape of the diaphragm in your pig and imagine how contraction of the diaphragm increases the volume of the thoracic cavity. Pull the diaphragm

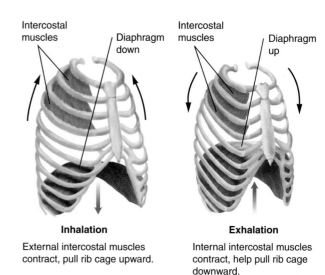

Figure 29.4 Inspiration and expiration mechanics. Contraction of intercostal muscles lifts rib cage and contraction of diaphragm muscles flattens diaphragm: both enlarge chest cavity and cause its pressure to drop. During resting expiration, elastic recoil of lung tissues, together with dropping of the rib cage and recoil of diaphragm because of organ pressure, decrease size of chest cavity.

posteriorly to mimic the effect of muscle contraction and note the expansion of the chest cavity.

You are now finished with the fetal pig for this lab. Return it to the storage area according to the directions given by your lab instructor.

Measuring Human Respiratory Volumes

Ventilation in mammals is tidal, consisting of alternating inhalations and exhalations. Volume and frequency of breaths are adjusted to match metabolic demand. Respiratory volumes vary with a person's size, age, gender, and physical condition. The total volume of the respiratory system can be divided into two major subdivisions, the vital capacity and the residual volume.

The **vital capacity** is the respiratory system volume that can be filled or emptied by voluntary breathing. It can be subdivided into the tidal volume, inspiratory reserve volume, and expiratory reserve volume. The **tidal volume** is the volume of gas inhaled or exhaled during normal tidal breathing, averaging about 500 ml in adult humans. The **inspiratory reserve** is additional volume that can be inhaled with maximal effort, and the **expiratory reserve** is the additional volume that can be forced out by maximal effort after tidal exhalation.

The **residual volume** is the volume of gas remaining in the lung after a maximal exhalation.

Not all the air in a breath reaches the alveoli. Gas exchange in the airways (the trachea, bronchi, and bronchioles) is negligible because they have thick walls, limited surface area, and low blood flow, so they constitute the anatomic **dead space.** The dead space remains filled with used air after an exhalation, so inhaled volume must exceed the dead space volume before any fresh air is delivered to the lung.

Dead space doesn't fit in to the subdivision of respiratory capacity described above. At the end of exhalation, the dead space is part of the residual volume. At the end of inhalation, the dead space is filled with freshly inhaled air, so at that point it represents part of the tidal volume. Average dead space in adult humans is about 150 ml.

Ventilation rates and ventilation volumes, including tidal volume, inspiratory reserve, and expiratory reserve can be measured relatively easily using a spirometer (measures breath volumes as a function of time) or pneumotachometer (measures breathing flow rates as a function of time). Other lung capacity variables can be calculated from these measurements. The **inspiratory capacity** is the sum of tidal volume and inspiratory reserve. The sum of the expiratory reserve volume and the residual volume gives the **functional residual capacity**. The **vital capacity** is the sum of tidal volume, inspiratory reserve, and expiratory reserve volumes; and **total lung capacity** is equal to vital capacity plus residual volume. Notice that variables called capacities are the sum of two or more volumes, but variables called volumes cannot be separated into functionally different parts. These concepts are summarized in **figure 29.5**.

9 *Methods* In this section of the exercise, you will measure ventilation volumes using a mechanical or computer-based data acquisition system. Your instructor will provide directions on how to set up and calibrate your system to collect and record data like those illustrated in figure 29.5. Work in pairs, with one person acting as the subject and the other managing the recording system and giving instructions. Your experimental subject will breathe into the system through a mouthpiece, while wearing a noseclip to prevent air exchange via the nostrils. Each subject should use a sterile mouthpiece, and the system should include a filter to keep it clean. If you are the subject, remember that accurate measurement of ventilation capacities requires maximal effort to accurately measure reserve volumes.

- Your subject should sit or stand upright, and focus on activity across the room, rather than watching the computer screen or thinking about his or her breathing.
- Ask your subject to start breathing through the mouthpiece and record normal (tidal) breathing for at least 60 seconds.
- After 60 seconds, ask the subject to inhale as deeply as possible, then exhale as deeply as possible. **Encourage s/he to try hard!** After this maximal breath, they should resume normal tidal breathing.
- After breathing has returned to normal, ask your subject to **inhale as deeply as possible** then return to normal tidal breathing.

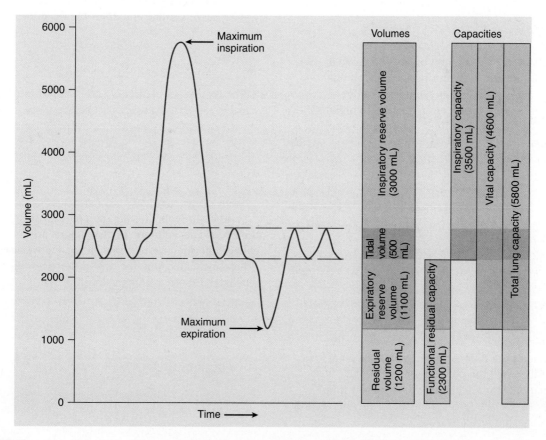

Figure 29.5 Volumes and capacities of human lungs. To understand the different lung volumes, try to mimic the graph in your breathing. Start by breathing normally, then breathe in as deeply as possible, but exhaling only to the point that you again breathe normally. After breathing normally again do a forced exhalation.

- When breathing has returned to normal after the forced inhalation, ask the subject to **exhale as deeply as possible,** then return to normal tidal breathing. After a few breaths, they can stop breathing into the spirometer. Continue recording for a few seconds after they release the mouthpiece, then stop recording and examine the data. Use your results to calculate the values below. Some computer-based systems will calculate some of these variables for you.

Analysis

Breathing Frequency. Count the number of breaths in 60 seconds to calculate breathing frequency in breaths/minute and record:
_____.

Tidal Volume. Average the volumes of six tidal breaths from your experiment to estimate tidal volume, and record that volume in the space below.

Minute Volume. The amount of gas you move in and out of your lungs during 1 minute of tidal breathing, is your minute volume equal to tidal volume × breathing frequency, in breaths/min. Calculate your subject's minute volume below.

Inspiratory Reserve. Calculate inspiratory reserve volume as the difference between a maximal inhalation and the preceding normal tidal inhalation. Use the highest value (why?) from either of the two forced inhalations on your record, and enter the result below.

Expiratory Reserve. Calculate expiratory reserve volume as the difference between a maximal exhalation and the preceding normal tidal exhalation. Use the highest value from either of the two forced exhalations on your record and enter the result below.

Vital Capacity. Calculate vital capacity by adding tidal volume, inspiratory reserve volume, and expiratory reserve volume and enter the result below.

How do your results compare with those in figure 29.5? The volumes in that figure are averages from college-age men in reasonably good physical condition. What might account for any differences between your results and those average values?

11 *Optional Activity: Dead Space and Changes in Tidal Breathing* When you breathe through the system you are adding a substantial dead space to your normal gas exchange system. Dead space is that part of the respiratory tree where no gas exchange occurs. Should that have any impact on your breathing? If so, what ventilation parameters are most likely to change, and why? Make some predictions. How is more or less dead space likely to affect your normal tidal ventilation, and why? Why aren't snorkels 20 feet long, so you can swim a lot further below the surface and still breathe air? Ask your subject how s/he thinks s/he was affected by breathing through the spirometer.

If time permits, you can design an experiment to test your predictions. How could you change the amount of dead space the spirometer adds to your respiratory tree?

12 *Summarize Observations* To gain an understanding of comparative anatomy, fill in **tables 29.1** to **29.3**. Write a brief description of each structure and its function. Include a summary statement of the organ system for each animal.

Lab Summary

13 On a separate sheet of paper, answer the following questions as assigned by your instructor.

1. Outline the pathway of air from the nostrils to the alveoli in a mammal.
2. Describe how the diaphragm and rib cage function in moving air into and out of a mammal's lungs.
3. Contrast how cells in a mammal receive oxygen with how cells in an insect receive oxygen.
4. Explain what is meant by the term vital capacity of the lungs. How is it different from total capacity?

You may want to try the critical thinking questions that apply some of the knowledge you gained in doing this lab.

Table 29.1	Insects—Gas Exchange System Summary
Spiracles	
Trachea	
Tracheoles	
Tissues	
Air sacs	
Description of ventilation process	

Table 29.2	Fish—Gas Exchange System Summary
Mouth	
Gill arches	
Gill filaments and lamellae	
Tissues	
Operculum	
Description of ventilation process	

Table 29.3 Fetal Pig—Mammal—Gas Exchange System Summary

External nares	
Nasopharynx	
Larynx	
Trachea	
Bronchi	
Bronchioles	
Alveoli	
Blood	
Diaphragm	
Rib cage muscles	
Description of ventilation process	

INTERNET SOURCES

The World Wide Web has become an important resource for those who work in the medical field. You can explore this resource by using Google to search for information on "pulmonary pathology." When the list of citations is returned, look for those that have images of black lung disease, silicosis, and asbestos accumulation. From your observations of the photos, write a short paragraph describing how these diseases affect the lung. Be sure to include the URL for future reference.

Use the WWW to find a description of a fish that can breath air. Write a paragraph describing the site used for gas exchange during air breathing, and the ecological circumstances under which air breathing occurs. Cite the URL for future reference.

LEARNING BIOLOGY BY WRITING

Write an essay comparing how an insect's tissues receive oxygen to how a mammal's tissues do. Although the essay is comparative it should mention all the structures that you observed in lab.

CRITICAL THINKING QUESTIONS

1. Consider air and water as respiratory media. Which contains a greater concentration of oxygen? Which takes more energy to move over the respiratory surface? Why can't mammals survive breathing water?

2. What effect does the giraffe's long neck have on respiratory gas exchange? (Consider volume of trachea.) How do you think a giraffe's tidal volume differs from that of a cow with the same body mass?

3. Reptiles lack a diaphragm muscle. How do they ventilate their lungs?

4. This is the GRAND SYNTHESIS question, designed to test your knowledge of the three anatomical systems studied so far in the fetal pig. Make a list of all the structures you have seen in you dissections that an oxygen molecule would pass through sequentially from the time it enters the nostril, until it combines with hydrogen to form water in a cell in the upper arm, and then goes on in the water molecule to pass out of the pig's body in its urine.

Lab Topic 30

Investigating the Properties of Muscle and Skeletal Systems

Supplies

Resource guide available on WWW at
http://www.mhhe.com/labcentral

Equipment

Compound microscopes

System for stimulating and recording muscle contractions. ADInstruments, BioPac, and iWorx all provide suitable systems.

Computer

Materials

Fetal pig, for optional muscle anatomy
Millipedes, spiders, crabs, or insects
Bat, snake, or fish skeletons
Bird skeleton
Human skeleton
Frog skeleton
Fresh beef long bones split longitudinally
Dry long bones
Prepared slides
    Skeletal muscle, longitudinal
    Smooth muscle section
    Cardiac muscle, longitudinal
    Bone
    Earthworm, cross section

Student Prelab Preparation

Before doing this lab, you should read the Background material and sections of the lab topic that have been assigned by your instructor.

Use your textbook to review the definitions of the following terms:

antagonistic muscles
appendicular skeleton
axial skeleton
endoskeleton
exoskeleton
extensor
flexor
hydrostatic skeleton
muscle recruitment
muscle twitch
pelvic and pectoral girdles
sarcomere
smooth, cardiac, and skeletal muscle
summation
temporal
threshold

Describe in your own words the following concepts:

Sliding filament theory of muscle contraction
How a nerve impulse causes a muscle to contract
Basic structure of bone

After finishing the prelab review, write any questions you have about terms, concepts, or techniques in the margins of this lab topic. The lab experiments should help you answer these questions, or you can ask your instructor during the lab.

Objectives

1. To compare the microanatomy of smooth, skeletal, and cardiac muscle
2. To identify the surface muscles in the fetal pig
3. To investigate mechanisms that alter force produced during muscle contraction
4. To compare hydrostatic skeletons, exoskeletons and endoskeletons
5. To observe the microanatomy of bone
6. To compare the structural and functional differences among endoskeletons of several vertebrates

Background

Simple small organisms, such as bacteria, algae, protozoa, and sponges, are capable of movement without muscle systems. They use cilia or flagella to move through their aqueous environments. All eumetazoan animals, whether aquatic or terrestrial, depend on muscle systems for movement and on nervous systems to control that movement. Muscles are closely associated with the skeletal system, which converts muscular contraction into effective movement of body parts and locomotion.

Skeletal muscles are nearly always arranged as **antagonistic** pairs. Contraction of one member of the pair causes an action, while contraction of the second member restores the body to its original position. In earthworms, the longitudinal and circular muscles in the body wall are antagonists. Contraction of the circular muscles elongates the segments by forcing the fluid of the hydrostatic skeleton forward and back in each segment, while contraction of the longitudinal muscles shortens the segments, leading to an increase in girth. Neuronal control usually ensures that the contraction of one muscle is accompanied by the relaxation of its antagonist.

Figure 30.1 shows the antagonistic arrangement of human muscles associated with an internal skeleton. Straightening a joint like your elbow is extension and bending it is flexion. Muscles that increase the angle between two bones at a joint are called **extensors. Flexor** muscles decrease the angle between two bones and are antagonistic to extensors. The **origin** of a muscle is the end of the muscle nearest to the central axis of the body, and is usually attached to a bone that does not move much when the muscle contracts. The opposite end, furthest from the body axis, is the **insertion,** and attaches to a bone that moves when the muscle contracts. The belly of the muscle is the region between origin and insertion. **Tendons** are specialized structures of dense connective tissue that attach muscles to the skeleton.

About 80% of the mammalian body mass is muscle. Most of this is skeletal muscle, sometimes called striated muscle or voluntary muscle. Muscle proteins, mainly actin and myosin, occupy most of the cytoplasm of the muscle cell, and are arranged in regular arrays that make the muscle look banded (striated) under a microscope. Each skeletal muscle is composed of many cylindrical, parallel muscle cells.

Two other types of muscle occur in vertebrates. **Cardiac muscle** is found in the heart. **Smooth muscle** is found in the walls of tubular organs including the digestive tract, blood vessels, urinary and reproductive systems, and in the iris of the eye. It is called smooth because its actin and myosin fibers are not arranged in the regular arrays that give skeletal and cardiac muscle a striated appearance.

An integral part of any animal's locomotory system is its skeleton. While the word *skeleton* usually brings to mind the bony **endoskeleton** of vertebrates, other types of skeletons occur in animals. Earthworms and other soft-bodied invertebrates use **hydrostatic skeletons** in which muscle contraction pressurizes internal fluid compartments to change shape. Nematodes, arthropods, molluscs, corals, and others have **exoskeletons,** hard outer coverings that provide support and sites for muscle attachment. Skeletons often serve other functions including protection of soft tissue, calcium and phosphate storage, and housing marrow where blood cells and immune cells are produced.

Skeletons are often the only parts of animals preserved as fossils. Luckily, skeletons exemplify the biological principle that form reflects function. They provide information about size, locomotion, feeding, protection, and size and location of muscles. Careful study of a skeleton permits inferences about how an animal, even if it is extinct, lived on a day-to-day basis.

LAB INSTRUCTIONS

You will study the anatomy and properties of muscle and compare several types of skeletal systems.

Muscular System

Microscopic Anatomy of Muscle

Obtain prepared slides of smooth muscle, cardiac muscle, and a longitudinal section of skeletal muscle. Compare the tissues on your slide to **figure 30.2**.

1 ▶ *Smooth Muscle* Observe the smooth muscle slide first. Note the shape of individual cells, the presence of nuclei, and the absence of filament organization in the cytoplasm. These muscle cells are spindle shaped, and their actin and myosin are not organized in regular arrays. Smooth muscle is innervated by the autonomic nervous system and is not under voluntary control. It contracts much more slowly than skeletal muscle and can maintain contraction for long periods. In some smooth muscles, adjacent cells are linked by gap junctions and contractions are propagated along the length of the organ, as in the peristaltic movement of chyme through the intestine. Smooth muscle cells in the digestive tract may contract rhythmically

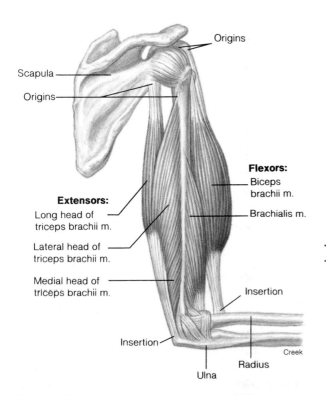

Figure 30.1 Antagonistic muscle arrangements in human limb. Contraction of human biceps and branchialis flexes the forearm, while contraction of the triceps extends the forearm.

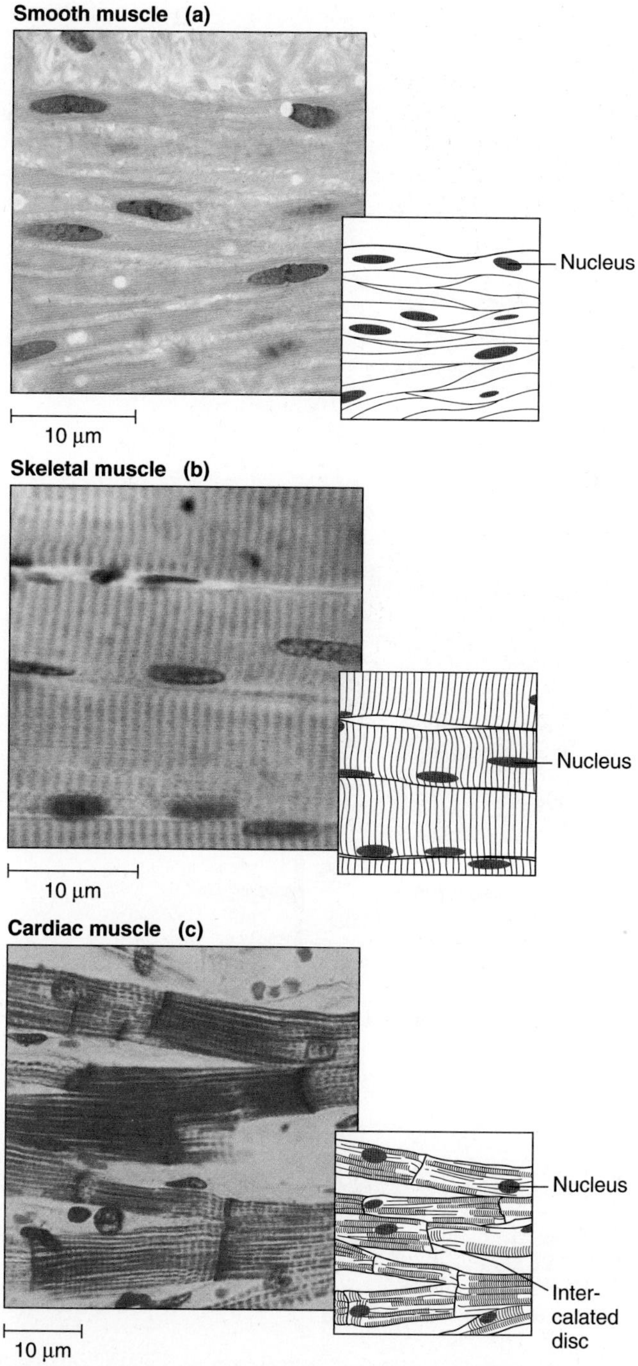

2. Skeletal Muscle

Now examine a slide of skeletal muscle under the medium-power objective. You should be able to see cross striations in the cells. This is why skeletal muscle is sometimes called striated muscle. Contraction of skeletal muscle can be controlled by conscious thought, so it is also called voluntary muscle. Note the substantial differences from smooth muscle. Identify the individual cylindrical fibers that run the length of the whole muscle. Find a very thin area of the section that is only one fiber thick and examine it under high power.

On the periphery of each fiber, you will see several darkly stained nuclei. Each skeletal muscle fiber is **multinucleated** because it is formed by the fusion of numerous cells, each with its own nucleus, during embryonic development. Thus, a muscle fiber is really a syncytium of several smaller cells. Mitochondria are also present in this peripheral cytoplasmic area but are not visible at lower magnifications. The central area of the cell consists of many parallel fibers called **myofilaments** running lengthwise in the cell, but they are difficult to see. Draw two or three of these cells.

To understand how your drawing relates to the structure of a whole muscle, such as the biceps in your upper arm, look at **figure 30.3**. What most people call a muscle contains thousands of **muscle fibers,** each a multinucleate cylindrical cell, arranged parallel to one another and surrounded by a sheath of connective tissue. Within a single muscle fiber are many **myofibrils,** which consist of contractile units called **sarcomeres** joined end to end at protein structures called Z discs. On microscope slides, the Z discs are viewed from the side and are often called Z lines. Each sarcomere contains two types of filamentous proteins. **Actin** is the protein in thin filaments and **myosin** is the protein in thick filaments. The actin filaments at each end are anchored in the Z discs. The myosin filaments are suspended between and surrounded by actin filaments. It is the interdigitation of these filaments and the areas of overlap that create the alternating light and dark banding patterns that you saw on your slide of skeletal muscle (fig. 30.2b).

During contraction, the thin filaments and Z discs are pulled toward the middle shortening the sarcomeres and the muscle cell. Because this occurs simultaneously in each sarcomere within a fiber, muscle cells can quickly shorten by 20 to 40%.

Muscle contraction is initiated in vertebrate skeletal muscle when a nerve impulse arrives at the **neuromuscular junction** (see figure 31.4). It causes the release of

Figure 30.2 Three types of muscle. (a) Smooth muscle has spindle-shaped cells, each with one nucleus (b) Skeletal muscle cells are cylindrical and have many nuclei. (c) Cardiac muscle is unique to the heart. Its cells join at connections called intercalated discs. The cylindrical cells branch. Cardiac and skeletal muscle cells are striated due to the orderly arrangement of contractile proteins.

without nervous system input. Sketch two or three examples of smooth muscle cells.

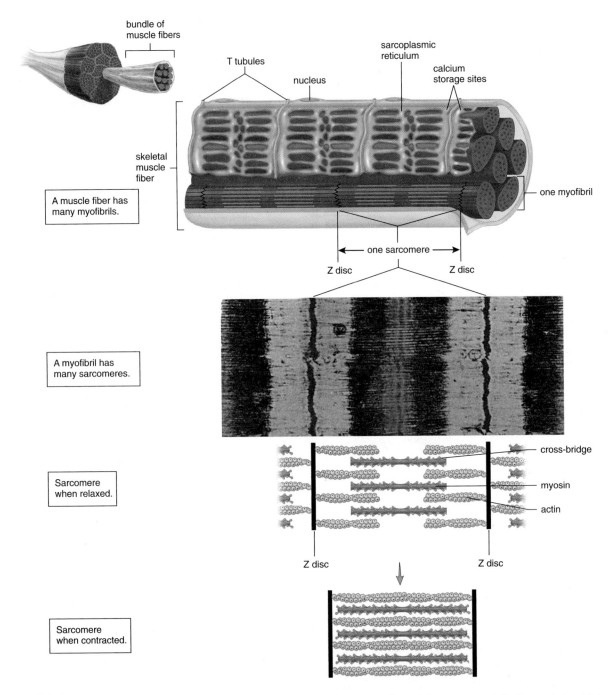

Figure 30.3 Skeletal muscle structure and function. A muscle fiber or cell contains many myofibrils, organized into contractile units called sarcomeres. During contraction, each sarcomere shortens as the actin filaments and Z discs are pulled toward the center of the sarcomere, shortening the overall length of the muscle cell.

a chemical called acetylcholine that depolarizes the muscle cell membrane and triggers an action potential in the muscle cell. The action potential moves over the surface of the muscle cell, including the T tubule system. Membrane depolarization causes the sarcoplasmic reticulum (specialized endoplasmic reticulum) to release calcium ions into the sarcomere. The calcium ions allow myosin to interact with actin, and shorten the muscle. A single twitch contraction lasts from a few milliseconds to tens of milliseconds, depending on the muscle. When nerve stimulation ends, the muscle cell repolarizes, the sarcoplasmic reticulum quickly reabsorbs calcium, myosin's interaction with actin is blocked, and the muscle relaxes.

3) *Cardiac Muscle* As the name implies, cardiac muscle is found only in the heart. Cardiac muscle is autorhythmic, producing regular contractions on its own, though neural input can modify the frequency and force of contractions. The contraction of heart muscle is not under voluntary control. Cardiac muscle cells are linked by gap junctions in the intercalated discs that hold adjacent cells together. Consequently when cells in the pacemaker region contract, a wave of contraction sweeps through the whole heart.

Use your compound microscope to look at a slide of cardiac muscle. Would you say that it is more similar to smooth muscle or to striated muscle? Why?

Use your 40× objective and look closely at one muscle fiber. Move the slide so that you can follow a single fiber for some distance. You should find two things in cardiac muscle that are different from what you saw in striated muscle. What are these differences?

The actin and myosin components in cardiac muscle are found in interdigitating parallel arrays, as in skeletal muscle. The muscle cells are shorter than in skeletal muscle, often branch, and join end to end by means of special tight junctions, called **intercalated discs.** These electrically couple the cells and allow contraction to spread from cell to cell independent of the nervous system.

Fetal Pig Superficial Muscles (Optional)

This dissection may be done by students or by the instructor as a demonstration, depending on the time available and the emphasis of the course on gross anatomy.

4 If your fetal pig is not already skinned, skin it—being careful not to tear away muscle as you remove the skin. Use a probe or finger to separate the two. Under the skin, there may be adipose (fat) tissue, which should be removed to reveal the muscles. When the skin is removed, dry the carcass with paper towels to improve viewing. In young fetal pigs, the muscles are not fully developed and are often tightly connected to the skin so that they are torn during skinning. Identification can be difficult.

5 Starting at the head, identify the major muscles indicated in **figure 30.4.**

1. The **latissimus dorsi** is a broad muscle running obliquely across the lateral thorax. It originates on the vertebrae and inserts on the proximal end of the humerus. The latissimus dorsi extends the foreleg and moves it closer to the body.

2. The **trapezius** originates on the occipital bone of the skull and from the first ten vertebrae and inserts on the scapula or shoulder blade. This broad muscle draws the scapula medially. When this muscle contracts, how does the front leg move?

3. The **brachiocephalic** muscle extends obliquely as a flat belt from the back of the skull (the occipital

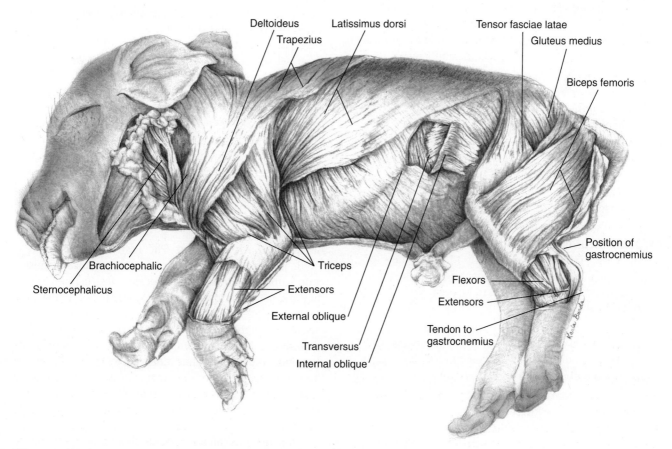

Figure 30.4 The major muscles of the fetal pig.

bone) to the foreleg (distal end of the humerus). It moves the leg anteriorly.

4. The **sternocephalic** is a long muscle below the brachiocephalic muscle. It controls the flexing of the head. It originates on the sternum and inserts on the mastoid process of the skull by means of a long tendon. Remove the parotid salivary gland to see the tendon.

5. The **brachialis** is a small muscle located in the angle formed by the flexed foreleg. It originates on the humerus and inserts on the ulna. What do you think its function is?

6. The **deltoid** is a broad shoulder muscle originating on the scapula and inserting on the humerus. It aids in flexing the humerus.

7. The **extensors** are several muscles in the lower foreleg that extend and rotate the wrist and digits.

8. The **triceps** (brachii) is a large muscle making up practically the entire outer surface of the forelimb. It originates on the humerus and inserts on the proximal end of the ulna. Triceps contraction extends the forelimb.

9. The **external oblique** muscles make up the outer wall of the abdomen and run obliquely from the ribs to a ventral **longitudinal ligament** along the ventral midline. Contraction constricts the abdomen.

10. The **internal obliques** lie just under the external oblique muscle and run almost at right angles to them. Contraction of these fibers also results in abdominal constriction.

11. The **tensor fasciae latae** is the most cranial of the thigh muscles. It originates on the pelvis and continues ventrally as a thin, triangular muscle until it becomes a sheet of connective tissue called the **fasciae latae,** which inserts on the kneecap and extends the leg.

12. The **gluteus medius** is a thick muscle covered by the tensor fasciae latae in the hip region. If the overlying muscle is removed, its origin on the hip and insertion on the femur can be seen. What action does it cause?

13. The **biceps femoris** is a large muscle making up most of the back half of the thigh. It originates on the pelvis and inserts on the lower femur and upper part of the tibia.

14. The **gastrocnemius** is the large muscle of the calf, originating on the lower end of the femur and attaching to the heel via the Achilles tendon. Its action extends the foot. The **soleus** muscle lies close to the gastrocnemius and has a similar function.

15. The **digital flexors** and **extensors** originate on the tibia and fibula and insert on the metatarsals. What are their functions?

16. The **peroneus** muscles have origins and insertions similar to digital muscles. They are involved in moving the whole foot.

Physiology of Muscle

This part of the lab may be performed by the instructor as a demonstration or by students working in groups of three or more, depending on the time and equipment available.

Overview of Experiment You will investigate how skeletal muscles respond to stimuli that differ in strength and frequency to develop insight into how you control the force of muscle contraction from gentle to forceful movements.

You will use four major pieces of equipment—a stimulator, a force transducer, an interface unit, and a computer running special software. A **stimulator** produces pulses of electrical current that can vary in amplitude, duration (how long the pulse lasts), and frequency (number of pulses per second). Stimulators used in this experiment are designed so that the electrical current they deliver is small and safe. The **transducer** is a device that converts the force developed by the movement of a finger into an electrical signal that can be recorded. The **interface unit** regulates the stimulator, converts the transducer output to numbers, and sends the information to your computer. The **software** allows your computer to control the stimulator, graphically display the results, and make quantitative measurements rapidly.

Equipment from different manufacturers can be used, so detailed instructions for any one set of equipment are not given here. Your instructor will provide supplemental materials that will instruct you on how to use your equipment to carry out the following experiments.

Locating a Motor Point The first step is to locate a motor point that you can stimulate to produce a muscle contraction. **Motor points** are not special structures, rather they are just locations on the skin where nerves controlling skeletal muscles lie near the surface. An electrical current passing through the skin to a motor point will depolarize neurons, stimulating them to produce action potentials that signal the muscles they innervate to contract. You will locate a motor point for the median nerve, which includes axons of motor neurons that control the *flexor digitorum superficialis*, a muscle responsible for fast flexing of the fingers.

6 ▶ One member of each group should volunteer to be a test subject. Others can assume different jobs: run the computer software, record important comments, read directions, *etc.* Anyone who has a cardiac pacemaker or suffers from neurological or cardiac disorders should not volunteer. If the volunteer feels major discomfort during the exercises, stop and consult your instructor.

To locate a motor point, you will place a pair of stimulating electrodes on the skin surface, moving the electrodes until you see the fingers twitch. The median nerve runs near the midline of the medial surface of the arm. Place a tiny amount of electrode cream on the metal surface of each stimulating electrode. Have your subject sit with one forearm on the table, palm up, and hold the metal electrodes in contact with the skin midway between wrist and elbow (**figure 30.5a**). Align the two electrodes parallel to the long axis of the arm.

To start, set your stimulator to deliver a series of pulses at a current of 10 mA and a 200 μsec duration at 1-second intervals. Set your recording system to record a measurement at least every 5 msec (sampling frequency of 200 Hz). Your instructor may provide a file with these settings already programmed, or may suggest other settings more appropriate for your equipment.

Turn on the stimulator. If the subject feels unacceptable discomfort, they can immediately stop stimulation by removing the stimulator electrodes. Stimulation in most places gives rise to little discomfort. A painful sensation in the forearm or hand, away from the site of stimulation (toward the fingers) means a sensory nerve is being stimulated—the sensation will probably stop if you move the electrodes to a different spot.

The electrical stimuli should produce twitches of the fingers or thumb if you are near a motor point. Pick the electrode up and move it from place to place until you locate the position giving the largest twitches of one of the three longest fingers. Mark that electrode position on the subject's skin with a marker or ballpoint pen.

If you see the thumb or little finger twitch, you are probably stimulating the ulnar nerve, not the median nerve. Move the stimulating electrodes more toward the thumb side of the arm to locate a median nerve motor point. If no twitch occurs and the subject cannot feel the stimuli, check to make certain the electrodes are properly connected and the stimulator is turned on. More electrode cream or slightly higher stimulus current (mA) might help.

Determining Threshold and Observing Motor Unit Recruitment

7 Nerve cells are excitable cells. When subjected to an electrical current, ions move across the cell membrane causing a change in the membrane potential that is proportional to the current. If the change in potential exceeds a threshold value, an action potential is triggered in the neuron. You will determine this threshold for the nerve innervating the muscles that move the fingers. What do you hypothesize will happen when the stimulating current exceeds this threshold?

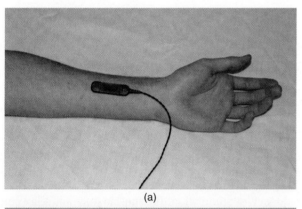

(a)

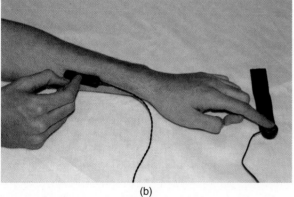

(b)

Figure 30.5 Stimulating and measuring muscle contraction. a) Move the stimulating electrode across lower surface of arm to locate a motor point. b) When measuring muscle contraction, hold electrode on motor point and place finger on pulse pressure transducer that is anchored to table with double-sided tape.

In this experiment you will gradually increase stimulus amplitude to determine the **threshold,** the smallest stimulus (measured in milliamps, mA) that produces a detectable response, and then examine the effect of increasing stimulus amplitude above that threshold. You should set up your system to record both the time (and amplitude, if possible) of each stimulus and the force produced by any resulting muscle twitch. Some systems may not record stimulus amplitude, so be careful to write down those data if necessary.

You will use a force transducer to measure muscle response. If you use a transducer like that you used to detect finger pulse in your blood pressure experiment in lab topic 28, ask your subject to rest the forearm on the table, palm down, then tape the transducer to the table top so your subject can comfortably rest the fingertip that moved the most in your motor point test on the transducer. If you use a pistol-grip transducer, the subject should hold the transducer between fingertips and the base of the thumb. Test your recording system to make certain you see a response when the subject pushes very lightly on the transducer with the fingertip.

Set your stimulator to deliver a sequence of four 200 μsec pulses at 1-second intervals, then set the amplitude of the stimuli to zero mA or the lowest possible setting. Ask your subject to hold the stimulator electrodes over their motor point, and record the four stimuli and their result. What should happen? _____

Continue recording, as you gradually increase the stimulus amplitude in small (1 mA) steps, until your subject produces detectable finger twitches after each stimulus. Record or write down each stimulus amplitude that you use. The smallest stimulus at which your subject produces detectable twitches is the **threshold** current required to depolarize at least one motor neuron in the median nerve enough to produce an action potential. Record that threshold here, including units of measurement: _____

Continue to increase stimulus amplitude in small steps, and watch what happens to muscle force. Stop the stimulation and the recording when your subject becomes uncomfortable, when muscle force stops changing, or when you reach the current limit of your stimulator. Save your recording.

Analysis How did muscle twitch force change as stimulus amplitude increased? Illustrate your answer by constructing a graph showing contraction force as a function of stimulus amplitude on your computer or on the graph paper at the end of this lab topic.

Why does contraction force reach a maximum, even though stimulus amplitude may continue to increase?

Now return to your recording, and zoom in to examine one contraction in more detail. Did the muscle contraction start or peak at the same time as the stimulus? How long was the delay (called latency) before contraction started? _____ How long before it peaked? _____ What do you think accounts for that delay?

Compare your threshold value with those recorded by other groups around you. What might account for any variation between groups?

Each skeletal muscle is composed of many motor units. Each **motor unit** consists of a group of muscle cells that are all controlled by a single motor neuron, and the muscle cells in each motor unit contract in an all-or-none fashion. As additional motor units are stimulated the force of contraction increases (a process known as **recruitment**). Can you use this information to account for your results?

Temporal Summation So far you have used a low frequency (1 cycle per second, or 1 Hertz, abbreviated Hz) of stimulation. Now, you will examine the effects of changing the frequency of stimulation. With increasing frequency of stimulation, the muscle may not relax completely before the next stimulus arrives. As a result, a new contraction begins in a muscle fiber that is already partly **8** contracted. What effect do you think this might have on the force of contraction of that muscle? State this as a hypothesis:

Why is there a threshold? Why does contraction force increase with stimulus amplitude beyond this threshold?

To test your hypothesis use the same setup as before. Be sure to use the stimulus intensity that gave the maximum contraction. In this experiment, you will vary the frequency of stimulation delivered to the motor point.

Start by verifying the setup and recording the effect of four stimuli at 1-second intervals. Now increase the frequency of stimulation from 1 Hz to 5, 7, 10, 15, 20, and if possible, 40 Hz, following the directions for the equipment that you are using. Record the muscle contraction as you increase frequency. At 1 Hz, it takes 4 seconds to deliver 4 stimuli. How long do four stimuli take at 20 Hz? _____

Analysis Save a copy of your results, then use them to answer the following questions:

9. What happened to muscle force as stimulus frequency increased? Illustrate your answer with a graph showing how force changes with stimulus frequency, using your computer or the graph paper at the end of this lab topic.

The increase in contraction force when stimulus frequency increases so that the muscle is stimulated again before it can fully relax is called **temporal summation.** At very high frequencies, muscle does not relax at all, producing a sustained contraction called **tetany.** In life, your nervous system often stimulates skeletal muscles with bursts of stimuli at high frequency, to take advantage of the resulting increase in force.

At what frequency of stimulation did temporal summation begin? _____

Tetany is a sustained maximal contraction. Some transducer systems respond only to changes in force, and return to baseline if force is sustained, so a tetanic contraction might appear as a single peak rather than a sustained maximal force.

At what frequency of stimulation did the muscle go into a state of tetany, with no relaxation between stimuli? _____

In percentage terms, how much greater was the force developed during tetanic contraction than during a single twitch contraction at the same stimulus amplitude? _____

What feature of your experimental design suggests that the mechanism of temporal summation is not just recruitment of additional motor units?

Temporal summation results from two mechanisms. First, Z-disc proteins and the tendons that anchor muscles to bones are elastic, and unless the contraction is sustained long enough to fully stretch those elastic elements, the maximum force produced by contraction will not act on the bones. Second, rapidly repeated stimuli increase the concentration of free calcium in the sarcomeres, and permit more myosin to bind and pull actin filaments.

Skeletal Systems

Hydrostatic Skeletons

Many soft-bodied invertebrates, such as cnidarians and annelids, do not appear to have specialized, differentiated skeletal systems. However, close examination reveals that the body wall muscles act on the incompressible fluids in the body cavity and intracellular spaces to facilitate very effective movement and support. Such skeletal systems are called **hydrostatic skeletons.**

10. Obtain a microscope slide of a cross section of an earthworm. Look at it first under the low-power objective and note the general anatomy of the animal by comparing the slide to figure 21.9.

The external surface of the earthworm is covered by a highly flexible noncellular **cuticle** composed of the protein collagen that is secreted by the underlying cells of the dermis. The proteins of the cuticle protect the worm and help maintain its form.

Beneath the cuticle are two layers of muscles: an outer circular set and a featherlike inner longitudinal set. Since the longitudinal set runs parallel to the long axis of the body, it will appear in cross section in the slide. Also note the body cavity, which is filled with fluid in live animals.

What happens to the worm's shape when the circular muscles contract? (Consider the effects of both muscles and internal fluids.)

What happens to the worm's shape when the longitudinal muscles contract? (Consider the effects of both muscles and internal fluids.)

The body cavity of the earthworm is divided into compartments, corresponding to each segment, by cross walls called **septa.** These prevent fluids from moving from one end of the worm to the other and allow earthworms to contract some segments while lengthening others.

Although they are not part of the hydrostatic skeleton, you should note the spinelike **setae** in the body wall.

Muscles attached to each seta extend or retract them. When extended they project into the soil and anchor the worm.

Describe how you think an earthworm could crawl forward using its setae, circular muscles, and longitudinal muscles.

Exoskeletons

Exoskeletons are characteristic of several animal phyla. In addition to supporting the animal, exoskeletons prevent water loss, protect the organs, and serve as points of attachment for muscles. Animals in the phylum **Arthropoda,** which includes crayfish, insects, spiders, and millipedes, have a well-developed exoskeleton composed of a complex polysaccharide called **chitin.** Several examples of these animals should be available in the laboratory.

11 Look at an arthropod through your dissecting microscope and note the segmentation of the exoskeleton. In insects, the **head** and **thorax** are composed of several fused body segments. The thorax bears the walking legs and wings. The segments should be clearly visible in the **abdomen.**

An insect's exoskeleton consists of hardened plates called **sclerites.** In the head and thorax, the sclerites are rigidly joined to one another along **suture** lines. In the abdomen, the dorsal and ventral sclerites are enlarged and joined to one another by **pleural membranes,** which correspond to thin, reduced lateral sclerites. Similar membranous areas are found between adjacent dorsal or ventral sclerites. These thin, flexible areas allow the animal to expand, contract, and curl its abdomen. Flexible membranous areas are also found between the joints of the legs. Muscles attach to the inside of the exoskeleton.

Endoskeletons

An endoskeleton is an internal supporting system of hardened material. They are found in vertebrates and cephalopod molluscs. Cartilage and bone make up the

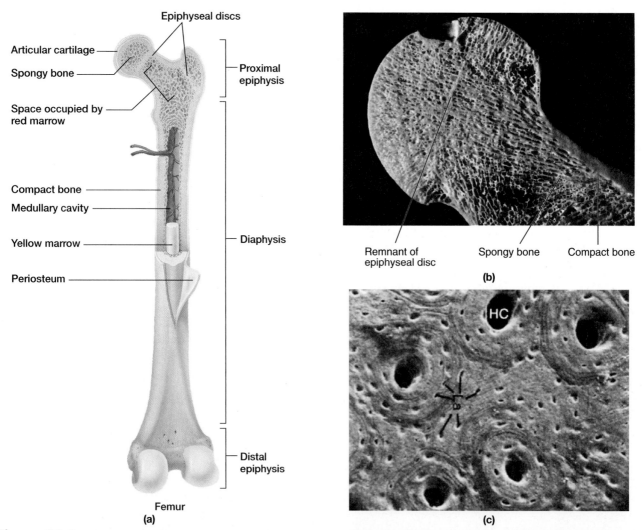

Figure 30.6 Structure of a bone: (*a*) a bone in partial longitudinal section; (*b*) micrograph of cancellous (spongy) bone; (*c*) scanning electron micrograph of Haversian canal (HC) systems in compact bone. Canals allow blood vessels to pass into bone. Bone cells are located in the lacunae (La).

vertebrate endoskeleton. You will start your study of endoskeletons by examining the structure of bone.

12▶ Bone Structure Examine a fresh beef femur that has been cut in half down its long axis. The central shaft of a long bone is called the **diaphysis,** while the enlarged ends that articulate with adjacent bones are the bone's **epiphyses** (fig. 30.6). Though difficult to observe in older bones, a narrow zone of cartilage, called the **epiphyseal disc** or growth plate of the bone, separates the epiphysis from the diaphysis.

The rounded projection from the upper end of the femur (fig. 30.6a) is called the **head,** and it articulates with the **acetabulum** of the pelvis, a depression where all three pelvic bones intersect. **Condyles** (rounded, articular prominences) are located on the lateral and medial sides of the distal end of the femur. With which bone(s) do they articulate?

Find the **periosteum,** the tough connective tissue layer surrounding the diaphysis. The surfaces of the condyles are covered by a glassy smooth **articular cartilage,** which, along with synovial fluid, reduces friction between articulating bones at the joints. Residues from the ligament that held the joint together may also be visible.

On the outer surface of the bone, find the narrow ridges called crests and the small, round, elevated tubercles, where muscles attach by tendons to the bones. The surface of the bone is perforated by openings called **foramina** (singular **foramen**) that allow blood vessels and nerves to pass into the center of the bone.

The center of the bone is a hollow chamber called the **medullary cavity.** It is filled with **marrow.** Red marrow is where new red blood cells are formed, but in older animals much of the marrow is filled with yellow fat and no longer produces blood cells.

Compare the structure of the bone in the diaphysis with that in epiphysis. Compact bone is very dense, whereas spongy bone has many interior cross braces. Spongy bone is less dense than compact bone.

Is one type of bone more common in one area compared to the other? _____

13▶ Microanatomy of Bone Obtain a prepared slide of ground bone and examine it under low power. Using figure 30.6c as a guide, find a **Haversian canal.** These canals allow nerves, blood vessels, and lymphatics to penetrate bone. Note the black chambers, called **lacunae,** arranged in concentric circles around the canal. **Osteocytes,** mature bone cells, live in these spaces and are responsible for depositing and removing calcium salts. Their activity is regulated by hormones. Several **canaliculi** radiate from the lacunae. These canal networks allow osteocytes to extend their cytoplasm and form intimate contact with the bone matrix.

Comparative Vertebrate Endoskeletons

The vertebrate skeleton has two components: the **axial skeleton,** consisting of the **skull** and **vertebral column;** and the **appendicular skeleton,** which includes the **pelvic** and **pectoral girdles** as well as the bones of the appendages.

14▶ Examine the skeletons of at least three vertebrates, ideally including a frog, a bird, and a human (figs. 30.7–30.9).

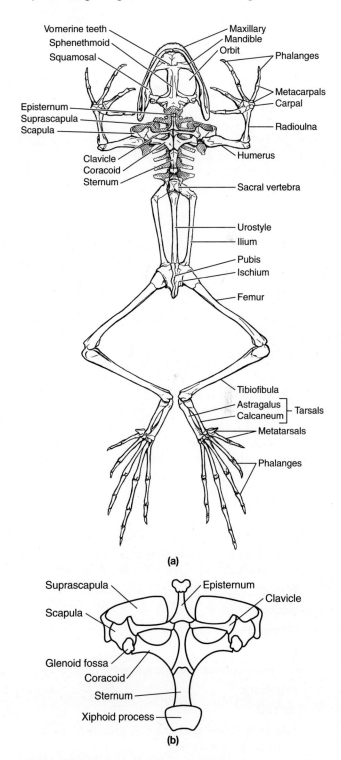

Figure 30.7 (a) Skeleton of frog (ventral view), including (b) ventral view of pectoral girdle.

412 LAB TOPIC 30

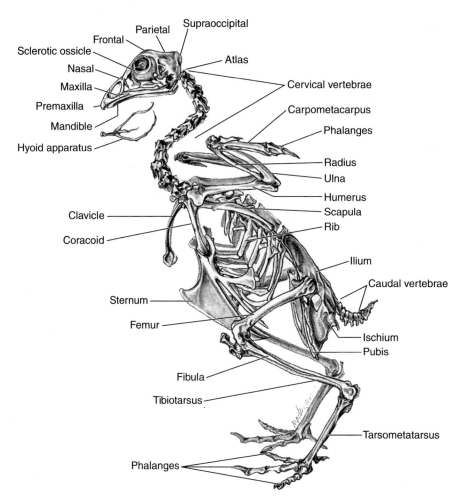

Figure 30.8 Skeleton of a bird.

The task before you is to determine how these skeletons are similar and how they are different. For the differences, you should offer explanations of how these differences correlate with locomotion or life style. For each skeleton you should be able to quickly identify the axial and appendicular skeletons. Then you should be able to find and identify the bones listed below.

Axial skeleton

Skull

cranium (fused bones encasing brain and sense organs) consisting of the following major bones:

frontal

parietal

temporal

occipital

facial bones consisting of the following major bones:

zygomatic

maxilla

mandible (lower jaw), the only movable bone in the skull of mammals

Vertebral column, which supports and protects the spinal cord. Cartilaginous discs are found between the vertebrae in living animals. The vertebral column is divided into five regions:

cervical vertebrae (neck)

thoracic vertebrae (on which the ribs articulate)

lumbar vertebrae (abdominal region)

sacral vertebrae (enclosed by pelvic girdle)

caudal vertebrae or tail (coccyx in human)

Ribs (attached and floating)

Sternum

Appendicular skeleton

Pectoral girdle consisting of:

clavicle (collarbone)

scapula (shoulder blade)

coracoid

Forelimbs consisting of:

humerus (long bone of upper limb)

radius (long bone of lower limb, which forms a pivot joint with the ulna and is also part of the hinge joint of the elbow)

ulna (other long bone of lower arm, forming hinge joint at elbow)

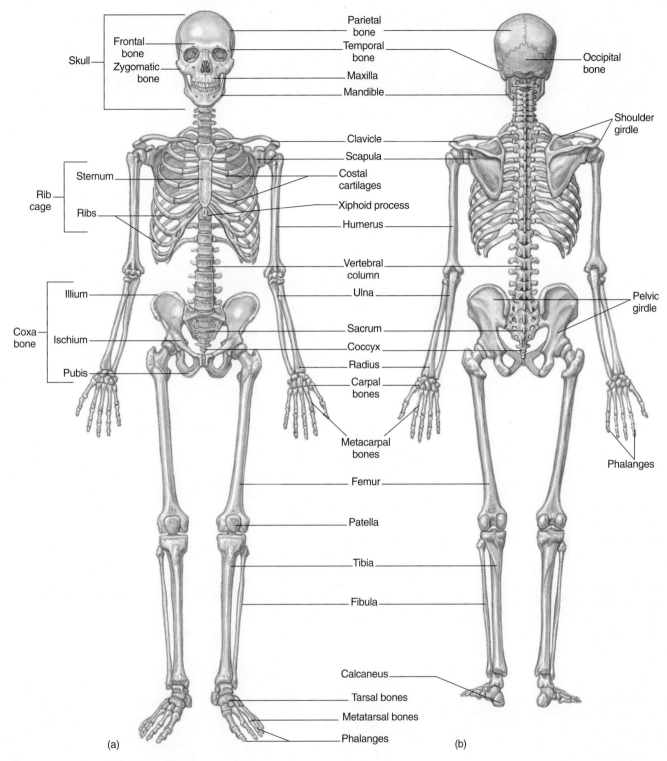

Figure 30.9 Skeleton of a human.

    carpals (small bones of wrist)
    metacarpals (bones of palm)
    phalanges (finger bones)
  Pelvic girdle consisting of:
    ilium
    ischium
    pubis

  Hind limbs consisting of:
    femur (long bone of thigh)
    patella (kneecap)
    tibia (larger of two long bones of lower limb)
    fibula (smaller long bone of lower limb)
    tarsals (bones of ankle)

calcaneus (heel bone)

metatarsals (slender foot bones)

phalanges (toe bones)

Students in the lab should divide into two teams to do comparative studies. Half should look at the pectoral girdle and forelimb skeleton of a frog, bird, and human. Half should look at the pelvic girdle and hind limb skeleton of the same animals. Discuss among yourselves the similarities and differences that you see.

Name the bones common to all three species.

How do the shapes and relative sizes of homologous appendicular skeletal bones differ among these three species?

How are those differences related to mode of locomotion of the animal?

Choose a spokesperson for the group to present your results. In a whole lab discussion, present the group results.

Lab Summary

15 On a separate sheet of paper, answer the following questions as assigned by your instructor.

1. What evidence do you have that smooth, cardiac, and skeletal muscle cells are different? Give examples of where you would find each in your body.
2. Describe the all-or-none response in muscle cells. How is it possible for us to have graded responses in our muscles if muscle cells respond in an all-or-none manner to stimuli?
3. Both recruitment and temporal summation lead to greater force of muscle contraction. Explain what is happening in both of these phenomena.
4. Describe the structure of a bone, such as the humerus, at the macroscopic and microscopic levels.
5. Describe the major common features of skeletons found in all vertebrates.
6. "Form reflects function" is a statement that applies to the skeletal system. What evidence can you provide from your lab work that the form of the vertebrate skeleton relates to locomotion in a frog compared to a human? In a bird compared to a human?

You may want to try the critical thinking questions that apply some of the knowledge you gained in doing this lab.

Internet Sources

Use an internet search engine to find answers to these questions.

1. Skeletal muscle fibers are not all the same, but differ in contraction speed and energy sources. Search the Internet to find descriptions of at least two different muscle fiber types, and the distribution and functional importance of those differences.
2. One class of drugs used to treat high blood pressure is calcium channel blockers. Why do calcium channel blockers help reduce blood pressure?
3. Vitamin D is a micronutrient we all require. Search on vitamin D to identify its importance in maintaining your skeletal system, and find the symptoms and name of an illness that result from vitamin D deficiency.

Learning Biology by Writing

Write a lab report that describes your experiments with human muscles. It should have three sections in the results:

Determining threshold

Demonstrating motor unit recruitment

Demonstrating temporal summation and tetany

The discussion section should describe how recruitment and temporal summation are important in graded responses such as are used in sports or playing a musical instrument.

Critical Thinking Questions

1. About 99% of the body's calcium is found as calcium phosphate salts in bone tissue. The calcium in plasma is in an ionic form, Ca^{++}. Although the level of Ca^{++} in the blood and tissues is closely regulated by parathyroid hormone and calcitonin, imbalances in Ca^{++} can occur. What effect would an elevated Ca^{++} concentration have on muscle activity? What effect would a Ca^{++} deficiency have on muscle activity?
2. As can be seen in figure 30.8, the skeleton of a bird is modified for flight. Aside from the size and arrangement of the bones in the skeleton, what other feature of its bones might affect the ability of a bird to fly?
3. How do scientists deduce the appearance of hominids or dinosaurs when all they have to study are a few bone fragments?

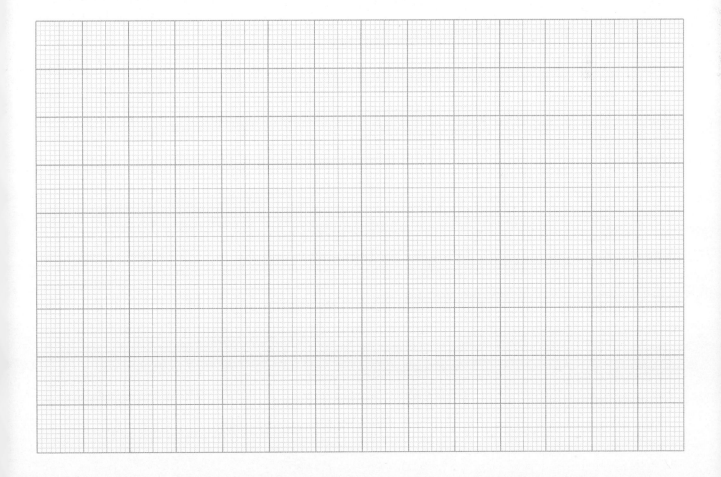

Lab Topic 31

Investigating Nervous and Sensory Systems

Supplies

Resource guide available on WWW at
http://www.mhhe.com/labcentral

Equipment

Compound microscopes
Dissecting microscopes
Slide or video projector

Materials

Fetal pig
Pig brain, sagittal section
Whole pig brains
Prepared slides
  Mammalian nerve cord with dorsal and ventral roots, cross section
  Neurons in smear of bovine spinal cord
  Muscle with nerve end plate
  Mammalian retina (demonstration slide)
  Mammalian cochlea, cross section
Models of human eye and ear
Patellar reflex hammers
Slides, computer images or transparencies project individual red, blue, and green shapes against a black background, with each color to be followed by a white slide or transparency.

Student Prelab Preparation

Before doing this lab, read the Background material and sections of the lab topic assigned by your instructor.
Use your textbook to review definitions of the following terms:

axon	medulla oblongata
cerebellum	myelin sheath
cerebrum	organ of Corti
cochlea	retina
cranial nerves	semicircular canals
dendrite	spinal nerves
gray matter	white matter

Be able to describe in your own words the following concepts:
How a nerve cell functions
Major regions of the mammalian brain
Reflex arc
How the eye "sees" light
How the ear "hears" sound

After finishing the prelab review, write any questions in the margins of this lab topic. The lab experiments should help you answer these questions, or you can ask your instructor during the lab.

Objectives

1. To observe the microscopic anatomy of nerve cells, neuromuscular junctions, and the spinal cord
2. To learn the gross anatomy of the mammalian central nervous system, eye, and ear
3. To determine experimentally some of the properties of cones
4. To use models to review how the ear functions

Background

Multicellular animals must coordinate activities of different specialized tissues to operate as an integrated whole. The nervous system uses electrical signals to coordinate rapid responses to changes in the environment. The endocrine system uses soluble chemicals circulated in body fluids to regulate longer-term responses. The functions of these two systems complement one another, so biologists often link the two as the neuroendocrine system. In this lab, you will study the nervous system and two related sensory systems.

The nervous system has three functions: (1) to receive signals from the environment and from within the body through sense organs; (2) to process the information received, which can involve integration, modulation, learning, and memory; and (3) to cause a response in appropriate muscles or glands. Sensory receptors are specialized cells that detect physical and chemical changes. The function of receptor cells is to transduce an environmental signal, like light, movement, chemical concentration, or temperature into a change in the cell membrane ion permeability, leading to a voltage change across the cell membrane. If the voltage change is of sufficient magnitude, the sensory cell or an adjacent nerve cell produces an **action potential,** an electrical signal that travels from the sensing zone to the central nervous system.

The central nervous system consists of networks of nerve cells (**neurons**) which are involved in decision making. A decision may involve the simple passage of action potentials from an afferent sensory fiber to an efferent motor (effector) fiber, assuring that perception results in a response. In other cases, many interneurons in the nervous system become involved, tempering the inputs with memory or reasoning. The response resulting from this interaction of nerve signals may differ according to conditions. For example, a person frightened by a loud sound will gasp, but if the person is under water, higher nerve centers inhibit the gasp response.

To gain insight into how nervous systems operate, biologists often study the simpler systems of invertebrates. Neurons are generally similar in form and function no matter what type of animal they are taken from. The differences between simple and complex nervous systems involve primarily the organization of the central nervous systems.

Fast communication is obviously important, and one major difference between vertebrates and other animals is the mechanisms they use to increase neuron conduction velocity. The invertebrate solution is giant fibers, cells with diameters as large as 1 mm, because conduction velocity goes up with diameter. Giant fibers are easy to study, and much of what we know about neuron function came from the study of squid giant fibers. Vertebrates do not use giant fibers to increase conduction velocity, but instead insulate neurons with wrappings of accessory cell membranes. Smaller fiber size permits vertebrates to pack more complex neural systems in less space, in much the same way that decreases in size of electronic components led to smaller and more powerful computers.

One of the simplest nervous systems is found in the radially symmetrical cnidarians, like *Hydra*. The **nerve net** in these organisms is not organized into tracts or centers (**fig. 31.1***a*). Simple bilaterally symmetrical animals like flatworms have tracts of neurons running the length of the organism, with the anterior end serving as a coordinating center or brain (fig. 31.1*b*). Selection for rapid response to changes encountered in forward motion led to the evolutionary trend of **cephalization,** in which sense organs and the coordinating centers are concentrated at the animal's anterior. In more complex animals, the size and intricacy of the coordinating centers increase.

The central nervous systems of annelids and arthropods consist of two ventral nerve cords with interspersed **ganglia,** or groups of cell bodies. The cords consist mainly of axons, cytoplasmic extensions of ganglia neurons, which rapidly transmit action potentials, whereas the ganglia serve as processing centers (fig. 31.1*c*). In arthropods, the ganglia are clustered in the head and thorax regions.

In the vertebrates, it is obvious that the brain dominates the nervous system. It is estimated that our brains contain between 10 and 100 billion neurons, excitable cells capable of producing action potentials. Those neurons make over 100 trillion points of contact with each other. Only a few million are motor or sensory neurons. The remaining 99+% are interneurons, connecting only to other neurons and are involved in processing incoming information and making "decisions." The nervous system also contains abundant **glial cells,** which provide support, nutrition, and insulation for neurons.

The nervous system receives information about the external and internal environment and the position of an animal in its environment from sensory cells. Sensory cells

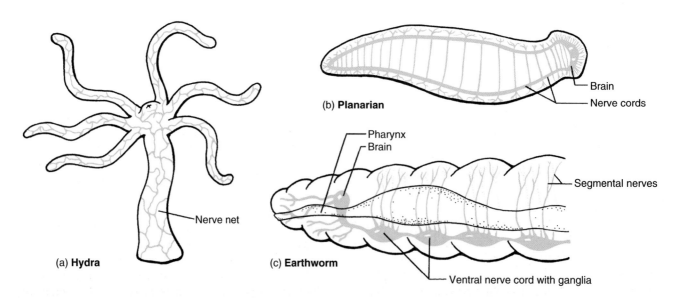

Figure 31.1 Nervous systems in invertebrates: (*a*) nerve net in *Hydra* lacks central aggregation of cells; (*b*) flatworms have neurons concentrated in longitudinal nerve cords with cross connectives; (*c*) ventral central nervous system of earthworm shows more extensive cephalization and has integrating ganglia in each segment.

are often organized into larger units, the sensory organs. The mammalian eye includes not only light-sensitive rod and cone cells, but also a focusable lens, a light-regulating iris, a set of skeletal muscles that move the eye, a tear gland, and a protective eyelid. The proper functioning of a sense depends not only on its sensory and neuronal cells, but on the anatomical structures that collect, amplify, and direct environmental signals to the receptor cells.

Senses are "hard-wired" into the organism. You feel cold because temperature receptors in the skin or hypothalamus are stimulated, and the affected neurons carry impulses to specific areas of the brain. You see something because light has stimulated individual cells in a particular pattern, and these cells each stimulate different neurons that carry nerve impulses to the brain. The brain filters and integrates those signals, and produces a pattern of brain cell activity that you interpret as an image. Sensory cells do not perceive; they receive. Perception is an integrative process in which nerve impulses from receptors, not environmental stimuli, are processed. The perception of a face, especially a familiar face, is actually a comparison of new neuronal inputs with memory of previous neuronal inputs.

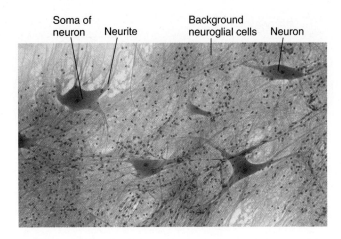

Figure 31.2 Neurons in a smear from a bovine spinal cord.

LAB INSTRUCTIONS

You will study the structure and function of major components of the mammalian central nervous system and of nerves, the eyes, and the ear.

Histology of the Nervous System

Structure of Neurons

1▶ The neuron is a basic cell of the nervous system. Their structure reflects their functions, which are integration and rapid conduction of information. Obtain a slide of neurons from a spinal cord and look at it through a compound microscope on low to medium power, comparing what you see to figure 31.2.

Find a large, angular-shaped neuron. Its cell body is called the **soma,** and the cytoplasmic extensions are axons or dendrites. The many small cells surrounding the neurons are glial cells. **Axons** are cytoplasmic extensions that conduct action potentials away from the soma. **Dendrites** conduct impulses toward the soma. It is difficult to distinguish between axons and dendrites on a slide, because only small segments of individual axons or dendrites are visible at one time. Consider a tall basketball player whose axons extend from neuronal soma in the base of the spine to the toes, a distance of more than a meter! It would be impossible to see the entire cell. The term **neurite** is used to describe any cytoplasmic extension, be it axon or dendrite.

Cross Section of Spinal Cord

2▶ Obtain a slide of a spinal cord cross section through a **dorsal root ganglion.** Observe it at first under low power with a compound microscope or with a dissecting microscope against a white background.

Note the general organization and compare it with figure 31.3. The nerve bundles leaving and entering the spinal cord from the sides are called the dorsal and ventral roots and merge to form the spinal nerves. A pair of spinal nerves leaves the spinal cord from each vertebral segment. The structures we call nerves are not neurons, but are collections of axons or dendrites bundled together that carry signals from cells in one location to distant cells. Sensory nerves always enter the top (dorsal root) of the spinal cord. The dorsal root has a swollen area compared to the ventral root. This is called the **dorsal root ganglion** and contains the soma of many sensory neurons. Motor nerves contain axons of motor neurons and always leave the bottom (ventral root) of the cord. There are no ventral root ganglia because the soma of these axons are in the gray matter within the spinal cord. You should be able to see tracts of axons leading from the gray matter toward the ventral root in your section.

Note that the nerve cord is hollow, containing a **central canal,** a vestige from its embryonic development. A dorsal hollow nerve cord is a synapomorphy, a derived character that is shared by all chordates. The central canal is connected with the ventricle spaces of the brain.

Observe the peripheral white matter surrounding the central gray matter using a high-power objective. The **white matter** consists of axons or dendrites that have been cut in cross section and supporting glial cells, so that they appear as a white ring with the dark axon at the center. These axons are myelinated, meaning each is surrounded not only by its own cell membrane, but also by a spiral wrapping of layers of membrane from an adjacent glial cell called a Schwann cell. That **myelin sheath** electrically insulates an axon, and increases the speed of action potential transmission. Myelin sheaths

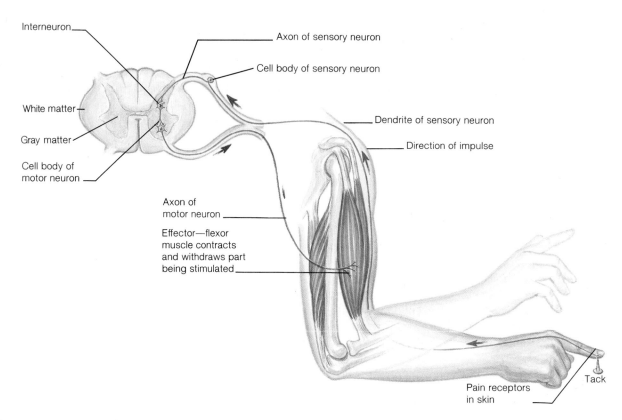

Figure 31.3 Cross section of spinal cord showing reflex arc components.

reflect light, so areas where they are abundant (white matter and nerves) look white. Spinal cord axons extend up and down the spinal cord and out into the peripheral nervous system. When the spinal cord is severed in an accident, it is impossible for the parts of thousands of neurons to reconnect, resulting in permanent numbness and paralysis.

Examine the central **gray matter** and find the soma in this area. They include interneurons that process and relay information, motor neurons that send axons via ventral roots to skeletal muscles, and glial cells. Because cell bodies are not wrapped in myelin, these areas absorb light and appear dark.

3 *Spinal Reflex Experiment* The simplest example of neural integration and control is a stretch reflex. A stretch reflex involves just two neurons (the pain reflex illustrated in figure 31.3 is a little more complicated, and includes an interneuron). If a skeletal muscle is suddenly stretched by some unanticipated change, a sensory cell in that muscle responds by sending action potentials to the spinal cord. They stimulate a motor neuron in the spinal cord to signal muscle cells in that muscle to contract, and return the stretched muscle to its original length. Other branches of the sensory cell can have an inhibitory effect on the contraction of antagonistic muscles. Stretch reflexes are important in maintaining position, especially on unstable surfaces. They are fast, no thought is required, and we are usually not aware of them. If we choose, we can easily block a stretch reflex by commands from higher centers in the central nervous system (CNS).

To demonstrate the stretch reflex, ask a volunteer to sit with her or his upper leg fully supported and the lower leg and foot dangling free. Tap their patellar tendon, just below the kneecap, with a rubber hammer or the edge of your hand. The tendon tap stretches the tendon and the relaxed muscle. The stretch is transient, and gone by the time the muscle can respond, so the reflex contraction of the quadriceps muscle in the thigh produces a "knee jerk." Unlike most movements, spinal stretch reflexes like this do not involve any conscious thought, so the term "knee-jerk reaction" has entered popular English as an idiom meaning any action that seems to be a fast response to some stimulus without any thought. Testing the knee-jerk reaction is, however, useful. It shows that your peripheral nervous system and spinal motor neurons are working normally.

Neuromuscular Junctions

4 Now examine the prepared slide showing neuromuscular junctions where a motor neuron innervates muscle fibers. The axon of the motor neuron coming from a ventral root of a spinal nerve divides into fine branches each of which terminate as a **motor end plate** on the surface of a different muscle fiber (**fig. 31.4**). Nerve impulses arriving at the end plates cause the release of acetylcholine, a neurotransmitter chemical that diffuses across each neuromuscular junction, triggering an action potential and contraction in

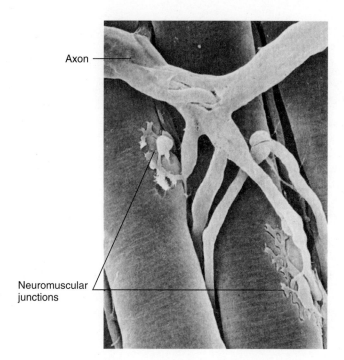

Figure 31.4 Scanning electron micrograph showing an axon ending at the neuromuscular junction on a muscle cell. Action potentials arriving here cause the nerve endings to secrete the neurotransmitter acetylcholine that triggers muscle contraction.

all the muscle cells innervated by that neuron. Sketch and label the parts of a motor unit visible in your slide.

Mammalian Central Nervous System

The mammalian nervous system can be subdivided to reflect differences in structure and function. The **central nervous system (CNS)** includes the brain and spinal cord, and is the site of most integration and control decisions. The **peripheral nervous system (PNS)** includes all nervous tissue outside the CNS. The PNS includes the **sensory nervous system,** afferent neurons, which carry information from peripheral sensory cells to the CNS, and the **efferent system,** neurons that carry commands from the CNS to muscles and glands.

The efferent peripheral nervous system is further subdivided. The **somatic nervous system** consists of axons of motor neurons that control skeletal muscles and use acetylcholine as their neurotransmitter at the neuromuscular junctions you just observed. Because we can easily control skeletal muscles with conscious thought, the somatic system is sometimes known as the voluntary nervous system. The somas of the somatic nervous system neurons are located in the CNS. The **autonomic nervous system** controls glands, smooth muscles, and modulates cardiac muscle activity, and includes neurons whose somas lie outside the CNS. It can be divided into the sympathetic and parasympathetic nervous systems.

The **sympathetic nervous system** has clusters of neurons in ganglia just outside the spinal cord that send long axons to target tissues, and usually stimulate changes that prepare animals for action. The flight or fight response to emergency conditions includes a suite of changes (increased cardiac output and blood flow to muscles, elevated blood glucose, dilated airways and pupils, inhibition of digestion) mediated by the sympathetic nervous system.

The **parasympathetic nervous system** has somas that lie in ganglia far from the CNS but close to the tissues they control. Parasympathetic system activity slows heart rate and increases maintenance activities like digestion.

The distinction between sympathetic and parasympathetic systems is important. Individual tissues, like the heart, often have both sympathetic and parasympathetic controls, which have opposing effects on tissue function. Because the two control systems use different chemicals as neurotransmitters to affect target tissues, different drugs can be used to stimulate or inhibit one without affecting the other. For example, drugs called beta blockers reduce heart rate by blocking one class of sympathetic synapses, while the drug atropine increases heart rate by blocking parasympathetic synapses.

5) We will now look at the gross anatomy of the mammalian nervous system. Get your fetal pig and a dissecting pan from the supply area. Remove a strip of skin about two inches wide from the area over the backbone. Lay the pig on its stomach. Expose the **spine** by removing the muscles that cover the vertebrae. Since it is time consuming for each student to expose the entire spine, groups in the laboratory should expose only short sections of about five to eight vertebrae each. Coordinate your efforts so that representative parts along the entire spinal cord are exposed. You can then look at one another's dissections to see the entire spine.

Spinal Cord

6) Once the vertebrae are exposed, take a sharp scalpel and gradually remove the cartilaginous spines and neural arches of several vertebrae to reveal the spinal cord (**fig. 31.5**). The spinal cord will be surrounded by three

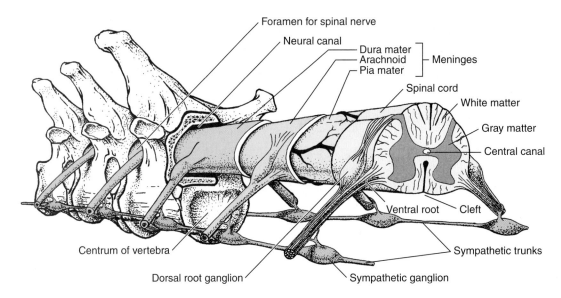

Figure 31.5 The relationship of the spinal cord, spinal nerves, and sympathetic nervous system to the vertebrae and meninges.

membranes, or **meninges.** The meninges constitute part of a barrier that separates the CNS from other tissues, and helps protect it against infection and toxins. Slit the meninges to observe the spinal cord.

Spinal nerves branch laterally from the cord. Each nerve begins as the union of several fiber tracts from the spinal cord, including sensory fibers from the dorsal root and motor neuron axons from the ventral root. The best way to see the segmental arrangement of spinal nerves is to turn your animal ventral side up and look at the dorsal wall of the body cavity. Each spinal nerve consists of thousands of axons or dendrites extending from different neurons in the spinal cord and dorsal root ganglion. They are bundled by connective tissue into a large "cable." As a nerve proceeds away from the spine toward the periphery, it branches, and specific neurons follow certain branches to particular muscles or sensory structures.

The cranial end of the spinal cord gradually enlarges to become the **medulla oblongata.** This region of the brain contains the cardiac and respiratory centers as well as numerous sensory and motor nerve tracts that transmit impulses to and from the higher brain centers.

Autonomic Nervous System

7 To see part of the sympathetic nervous system, look inside at the dorsal wall of the thoracic and abdominal cavities on either side of the spinal column. The **sympathetic nerve trunks** run parallel to the spine. The bulges at regular intervals along the trunks are **sympathetic ganglia** containing the soma of sympathetic neurons.

Some parasympathetic fibers leave the spinal cord in the sacral region, but they are difficult to identify in gross dissection. Most of the parasympathetic fibers to the thorax and abdomen leave the CNS from the brain, mainly via the vagus nerves.

Brain Anatomy

Dissection of a fetal pig brain is challenging; your instructor may substitute a demonstration or a larger brain from an adult pig or sheep.

8 If you are to do the dissection, expose the skull by making a longitudinal cut from the base of the snout to the base of the skull. Make lateral cuts from the ends of the first cut to the angle of the jaws at the anterior end and to the level of the ears at the caudal end. Remove the muscle layers. Make a shallow longitudinal cut in the skull and then lateral cuts from this incision at 2 cm intervals. Break off chips of the skull and gradually expose the brain. At the base of the skull, the tough **occipital** bone will have to be dissected out separately. The spinal cord passes through this bone.

9 The brain is surrounded by three meninges as was the spinal cord. Remove the membranes and observe the gross features of the brain (**fig. 31.6**). A longitudinal fissure separates the right and left hemispheres of the **cerebrum.** Higher-order functions, such as memory, intelligence, and perception, are associated with this part of the brain. Caudal to the cerebrum is the smaller **cerebellum,** which coordinates motor activity and equilibrium. The **medulla oblongata** lies ventral and caudal to the cerebellum and, as you saw earlier, is continuous with the spinal cord.

The cerebrum has a convoluted surface. The ridges are called gyri and the valleys, sulci. The outer part of the cerebrum, the **cortex,** is made up of **gray matter** composed of nerve cell bodies and supporting glial cells. The inner part of the cerebrum is made up of **white matter** composed of axons encased in an insulating myelin sheath. This arrangement is the opposite of the spinal cord, where myelinated nerve fibers were on the outside (white matter) and cell bodies (gray matter) were on the inside.

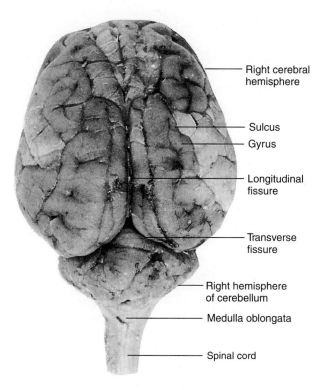

Figure 31.6 Dorsal view of the fetal pig's brain.

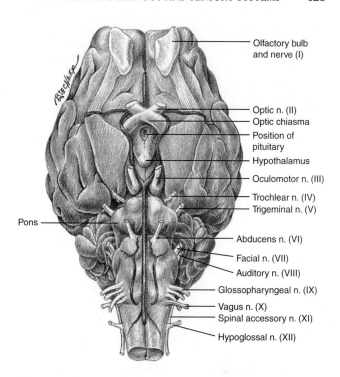

Figure 31.8 Ventral view of the fetal pig's brain with cranial nerves indicated by standard numbering system.

The transverse fissure separates the cerebrum from the cerebellum. There are three parts to the cerebellum, the two lateral hemispheres and a central vermis. If the caudal edge of the cerebellum is raised, a thin vascular membrane called the posterior choroid plexus may be observed covering the medulla. The space beneath the choroid plexus is the **fourth ventricle** of four ventricles, or cavities, found in the brain. **Cerebrospinal fluid** created by filtration from the capillaries located in the ventricles passes into the spinal cord spaces. This fluid carries nutrients to the cells and protects the nervous system from mechanical shock. The ventricles are best seen in a sagittal section of the brain (**fig. 31.7**).

10▶ Remove the brain from the skull (or obtain demonstration material) and orient it with the ventral surface up (**fig. 31.8**). At the anterior end of the cerebrum, find the **olfactory bulbs.** In the midportion of the cerebrum, two large nerve trunks, the **optic tracts,** cross to form the **optic chiasma.** Just posterior to the crossover is the **pituitary gland,** an important link between the nervous and endocrine systems, but it is usually broken off when the brain is pulled from the cranium. When looking at the ventral surface of the brain, a conspicuous oval structure, the **hypothalamus,** will be found dorsal to the pituitary. The hypothalamus integrates many autonomic functions, such as sleep, temperature regulation, appetite, and water balance. Posterior to this is a wide transverse group of fibers, the **pons,** which serves as a passageway for neurons running from the medulla to higher centers.

Twelve pairs of **cranial nerves** carry information directly into or out of the brain (fig. 31.8), including sensory information concerning sight, sound, taste, equilibrium, and touch. Motor fibers to the head region and parasympathetic fibers also exit via these nerves.

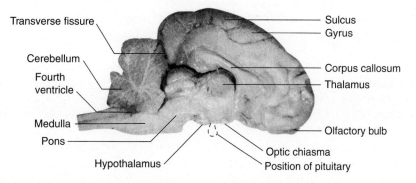

Figure 31.7 Sagittal section of the fetal pig's brain.

11 *Summarize Observations* In **table 31.1** write a brief description of each structure and a summary statement of its function.

Mammalian Sensory Systems

Eye Functional Anatomy

Dissection of larger beef or sheep eyes may be substituted for those of a fetal pig. If doing so, skip the next paragraph and proceed to arrow 13.

12 To remove an eye from the fetal pig, make an incision that extends from the external corner of the eye completely around the eye, removing the upper and lower lids. A thin mucous membrane, the **conjunctiva,** covers the eye and folds back to line the undersurface of the lids, preventing foreign material from entering the socket. Now remove the connective tissue, the muscle, and part of the bony orbit surrounding the eye. Push the eye to the side and up and down, noting the seven thin strips of the ocular muscles that originate on the skull and insert on the eye. Cut these muscles as far as possible from the eye. Draw the eye out of the orbit and snip the optic nerve. Once the eye is free, note the muscle mass that surrounds the eyeball and controls its orientation.

13 *Gross Anatomy of the Eye* With a sharp razor blade, cut longitudinally through the eye to one side of the **optic nerve** and place the halves in a dish of water. Identify the three layers making up the wall of the eye: the outermost white **sclera;** the **choroid,** dark in pigs, cattle, and sheep, which are primarily diurnal; and the innermost **retina** (**fig. 31.9**). The sclera and choroid block any light from entering from the back or side of the eye. The retina houses the light-sensory cells, rods, and cones, and a network of associated neurons. The front surface of the eye is covered by a transparent layer of connective tissue

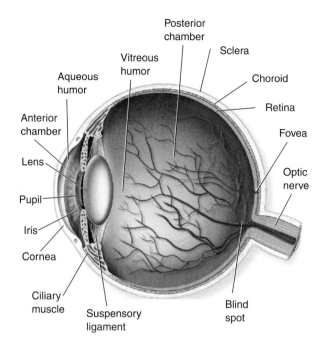

Figure 31.9 Anatomy of mammalian eye in longitudinal section.

called the **cornea.** The large white structure within the eye at the front is the **lens.** The lens is normally crystal clear, but preservative chemicals cause it to turn white. The ciliary muscles change the shape of the lens to focus images of objects at different distances on the retina. The colored material surrounding the lens is the **iris,** and the central opening in the iris is the **pupil.** Muscles in the iris adjust the diameter of the pupil to control the amount of light reaching the retina.

The large posterior space of the eye is filled with a gelatinous fluid, the **vitreous humor,** whereas the small chamber anterior to the lens is filled with **aqueous humor.** The sensory cells of the retina lie behind several

Table 31.1	Summary Description of Nervous System Components in Fetal Pig
Spinal nerves	
Meninges	
Sympathetic trunks	
Cerebrum	
Cerebellum	
Medulla oblongata	
Pons	
Hypothalamus	
Ventricles	
Cranial nerves	

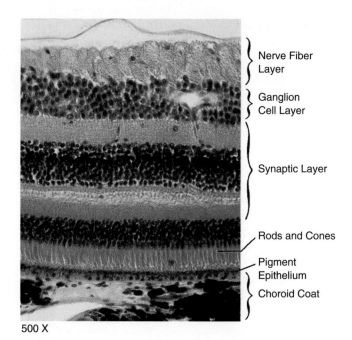

Figure 31.10 Section of back wall of eye showing cell layers of retina. The back of the eye is at the bottom and the front, where light enters, is at the top.

layers of neurons, as shown in figure 31.10. Where the axons of those neurons pass out of the back of the eye as the optic nerve, there is no room for sensory cells. This area is called the **blind spot**. The small, yellowish spot in the center of the retina is called the **macula lutea**. The depression in the center of this spot is the **fovea**.

14▶ Histology of Retina In the human retina, there are two kinds of light-sensitive cells: **rods** and **cones**. These cells are on the retinal surface facing away from the vitreous humor, toward the choroid. Look at the demonstration slide of the mammalian retina and identify the following layers: choroid, rods and cones, and nerve cell layers.

Resolution, Sensitivity, and Color Vision Eyes, like cameras, collect three kinds of information about light: direction, intensity, and color. **Resolution** or **acuity** measures the ability to detect differences in direction of incoming light and thus see detail. **Sensitivity** is the ability to discriminate between small differences in intensity, and allows us to see in dim light. **Color perception** requires that we distinguish between different wavelengths of light. Neither eyes nor cameras can maximize all three at the same time, and understanding the necessary compromises helps us understand the structure of our retina.

The rod cell system emphasizes sensitivity. Rod cells respond to light of any color, so they use all available light to work even in dim light, but cannot distinguish different colors. Information from neighboring rod cells is sent to the same ganglion cell (a retinal neuron). That **convergence** adds together light information gathered from an area larger than one cell. Convergence improves sensitivity to dim light, but sacrifices direction information—there is no way to tell where on a group of rod cells incoming light fell.

The cone cell system, in contrast, emphasizes color information and resolution. Three types of cone cells each respond to a narrow range of wavelengths, so by comparing the response of different cone cells, we discriminate colors. Of course, that reduces sensitivity, because each cone cell can only detect incoming light in its wavelength range of absorption. Convergence in the cone system is minimal; a ganglion cell may receive input from only a single cone. That maximizes resolution, because our brain can tell exactly which cone was illuminated, but also reduces sensitivity. The **fovea** has a very high density of cones, and is the area producing the best color vision and greatest visual acuity.

The trade-off between resolution and sensitivity can also affect choroid structure. Our black choroid, like the inside of a camera, minimizes reflection of any light that passes through the retina, and prevents uncertainty about the direction from which that light came. For nocturnal or deep-water animals that never see bright light, sensitivity is more important, and they have choroids that reflect any light that went through the retina without being absorbed back toward the retina, so sensory cells have a second chance to detect it. Reflective choroids produce the "eye-shine" you see when you shine a light on the eyes of many nocturnal animals. Many mammals that are nocturnal have few or no cones in their retinas so they are color blind in daylight. That is why deer hunters can wear fluorescent orange.

When light strikes a receptor cell, it triggers a photochemical reaction that leads to neuronal firing. After each photon capture, it takes a brief time for the sensory protein (opsin) in the receptor to regain its sensitivity to light. After viewing a bright object, if the eye is quickly shut or turned to a dark wall, you see a positive (bright) **afterimage** because the chemical changes that stimulate firing of visual neurons take time to decay. After a second or two, however, if you look at a light background, a negative (dark) afterimage of the bright object will appear. The photopigment proteins of cells that were brightly illuminated remain bleached and are less sensitive to the light coming from the light background. We can use this phenomenon to gain some insight into how the color vision system of the eye works.

Afterimages Experiment According to the Young-Helmholtz theory of color vision, humans have three types of cone cells, which respond respectively to red, green, or blue light, the same three colors your computer monitor uses in different mixes to display millions of colors. All other colors are perceived as the brain interprets impulses coming from a mix of these receptors. If an object of one color is viewed at high light intensity, the cones for that color are bleached, but cones sensitive to other colors are not. When you look at a white area, afterimage of the colored object will be "seen" in the complementary color for a few seconds until the cones recover.

15) Your instructor will demonstrate afterimages to you using a projection system. A small shape, colored bright red, on a black background will be projected onto a screen in the darkened lab room. You should stare intensely at the center of the shape for 20 seconds or so without shifting your gaze, then the slide will be changed to a uniform white. White is not a wavelength, but is how we perceive light composed of a mixture of red, green, and blue wavelengths at about the same intensity. Continue staring at the screen after the change is made. What do you "see"? What color was the afterimage?

According to the Young-Helmholtz theory, which set of cones were bleached by staring at the red image? _____

Which sets were not bleached? _____

Which sets of cones were more sensitive to the white light after the red shape was replaced by a uniform white background?

Now predict what should happen if you repeat the experiment using a blue shape, rather than a red shape:

Do the experiment to test your prediction. Write the results below:

Repeat the experiment using a green shape. Record your results below.

Explain the colors seen in these afterimages using the Young-Helmholtz theory.

Mutations that affect cone pigments (proteins) can result in various kinds of color blindness. The most common color blindness in humans is difficulty is discriminating between colors red and green. Under the Young-Helmholtz theory, what sorts of changes in cone wavelength sensitivity might be responsible? How would red–green color blindness affect the results of the afterimage experiment?

If a class member is red–green color blind, and willing to share his or her perceptions, compare the afterimage results with your predictions.

16) *Summarize Observations* Discuss the following situation with your lab partner, using your understanding of rods and cones to provide an explanation. You immediately walk into a darkened movie theater (no film on the screen) after having been outside in strong sunlight. You cannot see where you are stepping or any seats at first, but then your sight slowly returns. Explain what happened and is happening to your pupil, rods, and cones.

Ear Functional Anatomy

The mammalian ear houses a collection of mechanoreceptors. Receptors in the cochlea detect sound vibrations and receptors in the semicircular canals detect changes in head position and acceleration/deceleration. The inner ear is difficult to dissect because it is embedded in the bony skull, but the anatomy is not difficult to visualize from models. For this section, study **figure 31.11** and models of the ear provided in the laboratory.

17) Sound waves are collected by the external ear and pass down the auditory canal, causing the eardrum (**tympanic membrane**) to vibrate. The movement of the eardrum is transmitted by the three **auditory ossicles** of the middle ear to a flexible membrane covering the **oval window** of the **cochlea** or inner ear. Sound energy is amplified in this process for two reasons: the area of the tympanic membrane

INVESTIGATING NERVOUS AND SENSORY SYSTEMS   427

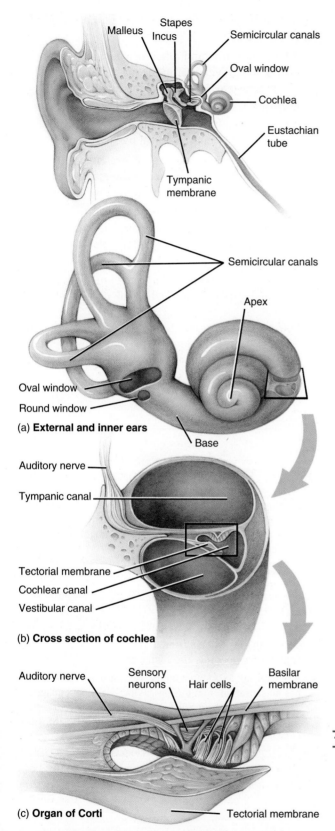

Figure 31.11 Anatomy of the human ear: (a) the relationship between the external and inner ears, and a cross section of the cochlea, showing the canal and membrane organization; (b) magnified view of the cochlear canal showing organ of Corti and the relationship of hair cells to the tectorial membrane; (c) the organ of Corti in magnified view.

is about 30 times larger than the oval window membrane; and the mechanical leverage systems of the ossicles, the **malleus, incus,** and **stapes,** greatly amplify any movement. The back-and-forth movement of the oval window causes the fluid in the cochlea to move and it is this movement that we interpret as sound.

The cochlea is divided lengthwise into three canals by soft-tissue walls. Pressure waves generated at the oval window travel up the **vestibular canal** to the apex of the cochlea and down the **tympanic canal.** The **round window** at the base of the tympanic canal serves as a pressure-release valve, allowing the incompressible fluid of the cochlea to move back and forth. Thus, these anatomical structures of the ear convert the movement of air molecules, or sound, into the movement of the fluid (perilymph) in the inner ear.

The pressure waves cause the thin, longitudinal **basilar membrane** to vibrate. **Sensory hair cells** on this membrane in an area known as the **organ of Corti** (fig. 31.11c) are displaced and brush against a stiff, overhanging structure called the **tectorial membrane.** Hair cells do not produce hairs, but have specialized projections (sensory cilia) that look hairlike under a microscope. Distortion of the projections causes changes in membrane electrical potential (generator potentials), which in turn cause cochlear neurons to produce action potentials that travel along the auditory nerve to the brain.

Ears can extract three kinds of information from sound waves. **Loudness,** or amplitude, affects how much the basilar membrane moves, and is encoded by how frequently nerve impulses are generated in auditory neurons. Frequency or **pitch** is detected in a different way. High-frequency sounds cause greater movement and stimulate hair cells near the base of the cochlea and low-frequency sounds stimulate cells near the apex. Separate neurons lead from hair cells at different points along the cochlea to the brain. The location at which receptors are stimulated most thus indicates pitch. To detect **direction** of sounds, we compare the loudness and arrival time of sound waves between the two ears. It takes about 2 msec for sound to travel the distance from one ear to the other, so if a sound stimulates the left ear 2 msec after it stimulates the right ear, it must have come from the right.

18 *Histology of Inner Ear* To see the anatomy of the sound receptors, obtain a slide of a cross section of the cochlea. First look at the slide under the scanning objective using figure 31.11b and c to orient yourself. Switch to the medium- then high-power objective to observe the hair cells and their relationship to the tectorial membrane in the organ of Corti. The slide preparation process often folds tectorial membrane away from the basilar membrane, so your slide may look different from the figures.

Semicircular Canals The semicircular canals of your inner ear respond not to sound, but to angular acceleration. They consist of three curved, fluid-filled tubes perpendicular to each other. Each has hair cells in the walls. When you change position, the walls accelerate faster than the fluid, and hair cells are distorted, leading to action potentials. Each tube is most sensitive to acceleration in one plane, so comparing results from all three allows your brain to distinguish up–down, right–left, and front–back accelerations and decelerations.

19. Summarize Observations Pair off with another student and quiz one another about how the ear functions. Use the model and explain to one another how sound passes into the inner ear and is transduced into a nerve impulse.

Lab Summary

20. On a separate sheet of paper, answer the following questions as assigned by your instructor.

1. Describe the organizational structure of nervous systems in the animal kingdom as you proceed from Cnidaria to Vertebrata.
2. How does the shape of a neuron differ from other animal cells you have studied this semester, and what is the functional importance of those shape differences?
3. What is the structural difference between white matter and gray matter in the mammalian central nervous system, and what is the functional importance of that structural difference?
4. Name the major anatomical features of the brain as viewed whole from the dorsal perspective.
5. Describe how light entering the eye is regulated and focused on the retina. Describe the response of rods and cones to different colors of light.
6. Explain the nature of afterimages that are seen after looking at a bright light. Why are afterimages a different color than the actual images?
7. Trace the pathway of sound energy as it enters the auditory canal and passes into the inner ear. How are pitch and loudness encoded by the nervous system?

You may want to try the critical thinking questions that apply some of the knowledge you gained in doing this lab.

INTERNET SOURCES

1. Neuromuscular junctions are fascinating anatomical sites where the nervous system meets the muscular system. These are still the topics of much research. Use a search engine to search the World Wide Web to discover the type of research that is being conducted now. Jump to at least three of the sites returned in the list and summarize the research that is being done. Include the URLs for future reference.
2. How can you remember the names of all 12 pairs of cranial nerves in order? Use the Internet to locate a way to help you to remember their order.
3. According to the text, there is a blind spot in your retina, but if you look out at the world, you cannot perceive such a blind spot in your visual field. Use the Internet to find a way of showing that the blind spot really exists. Use it to see if you really have a blind spot, and propose an explanation for why you cannot normally perceive its existence.

CRITICAL THINKING QUESTIONS

1. Hearing impairment can result from mechanical failure of parts of the ear, or from sensory cell damage, or from auditory nerve damage. Which of these might be easily corrected by a hearing aid? Why?
2. If you examined the retinas of various animals with a microscope, what differences would you expect to find between animals that are active at night versus those active during the day?
3. As humans age, the lens of the eye often hardens and the ciliary muscles will not bend it as readily as when we are young. How does this affect what can be seen?
4. A person suffers a spinal injury. He can move his legs but has no sensation from the legs. What has been damaged and what is the prognosis for recovery?

LEARNING BIOLOGY BY WRITING

Write a 150-word essay on how a sensory nerve impulse travels from a muscle spindle in the calf to the central nervous system and back to a muscle cell in the leg via a motor neuron. Discuss the ways in which action potentials, synapses, and conduction velocities are involved.

Lab Topic 32

Investigating Early Events in Animal Development

SUPPLIES

Resource guide available on WWW at www.mhhe.com/labcentral

Equipment
Dissecting microscopes
Compound microscopes
Refrigerator

Materials
Prepared slides
    Sea star development from egg to larval stage
    Early cleavage stages of *Cerebratulus* sp.
    Frog blastula, ls
    Frog gastrula, ls
    Frog yolk plug, ls
    Frog late neural groove, cs
    16 hr chick blastoderm (wm)
    18 hr chick (cs)
    33 hr chick (wm)
    72 hr chick (wm)
Live sea urchins in reproductive condition
Photographs, models, or plastic whole mounts of 24- and 48-hour stages in the development of a chick
Depression slides
Slides and coverslips
Thermometers
Dropper bottles
Pasteur pipettes
Syringes
Miscellaneous beakers

Solutions
Seawater
0.5 M KCl
3.5% NaCl

STUDENT PRELAB PREPARATION

Before doing this lab, read the Background material and sections of the lab topic that have been assigned by the instructor.

You should use your textbook to review the definitions of the following terms:

archenteron	gastrula
blastomere	mesoderm
blastopore	neurula
blastula	radial cleavage
ectoderm	spiral cleavage
endoderm	zygote

Describe in your own words the following concepts:
Cleavage of the zygote to produce a blastula
Significance of gastrula formation

After finishing the prelab review, write any questions you have about terms, concepts, or techniques in the margins of this lab topic. The lab experiments should help you answer these questions, or you can ask your instructor during the lab.

OBJECTIVES

1. To observe early developmental stages of sea stars
2. To compare radial to spiral cleavage
3. To observe fertilization and early cleavage in sea urchins
4. To observe blastula, gastrula, and neurula stages in frog development
5. To identify and describe the early events in chick development

BACKGROUND

Development is one of the truly amazing processes in both plants and animals. Cells from different parents fuse to produce a deceptively simple-looking zygote that, in turn, undergoes a series of cell divisions to produce a multicellular adult. The adult is made up of hundreds of different kinds of cells that perform myriad highly integrated and coordinated functions. For most organisms, although the stages of development have been outlined, the underlying mechanisms that control development remain an enigma. It is known that the developmental program is encoded in the genetic material, but no one yet understands exactly how certain genes are activated and others are repressed at just the right times during development.

An animal's embryonic development can be divided into four major stages: (1) **fertilization,** when the sperm penetrates the egg and is soon followed by syngamy, the fusion of the egg and sperm nuclei; (2) **cleavage,** the mitotic divisions that partition the large cytoplasm of the fertilized egg into smaller cells; (3) **gastrulation,** a morphogenetic cellular movement that produces an embryo with three layers of cells; and (4) **organogenesis,** the process whereby specific organs develop from the primary germ layers.

The specifics of development differ among animals, but there are many similarities among species in general development. As development progresses, all embryos become more complex as new tissues and organs appear and the embryo becomes capable of performing new functions. This increase in complexity is called **differentiation.**

However, the developing organism is not simply a random assortment of new cell types and organs; these new features always have specific spatial relationships with existing cells and with those that will form later. This development of form is called **morphogenesis** and comes about as a result of cell growth, movement and differentiation as the embryo uses yolk materials for a source of energy and for chemical building units. During morphogenesis, the cells must "communicate" to inform each other of their location during migration movements and to trigger the differential use of genetic information.

The physical characteristics of eggs are related to the environment in which an animal lives, the place where the embryo develops, and the stages in the life cycle of the species. Eggs of animals that live in aquatic environments generally have gelatinous coats, which protect the egg from physical and bacterial injury, and moderate amounts of yolk, which supply the embryo with sufficient energy to achieve a developmental stage that allows food gathering.

Land animals that release their eggs for development outside of the mother have eggs covered by shells that provide protection against physical injury and desiccation. However, such a protective device is not without its problems. Fertilization must occur before the shell is formed because the sperm cannot penetrate the hard structure. This means that fertilization must be internal in the female because the shell is produced by cells in her reproductive tract, not by the zygote. The embryo inside a shelled egg must exchange O_2 and CO_2 with the environment, must have sufficient energy and raw materials to develop to an advanced stage, and must have a means of disposing of nitrogenous wastes. Inside such eggs of vertebrates, four extraembryonic membranes grow out from the embryo, surround it, and function in gas exchange, waste storage, and nutrient procurement.

In mammals, since the developing embryo is carried inside the mother's uterus, no outer shell or jelly coats are necessary. When the egg implants in the uterus, extraembryonic membranes surround the embryo and also form the placenta, a highly vascularized organ that brings extraembryonic capillaries close to the mother's capillaries where nutrients and wastes are exchanged.

In animals, the way in which an egg undergoes early development is strongly influenced by the amount and distribution of yolk in the egg. Eggs of some species have little yolk and it is uniformly distributed in the egg's cytoplasm, as in the starfish and sea urchin you will study in this lab. Others have a large amount of yolk that displaces the active cytoplasm to one pole of the spherical egg. Such eggs are characteristic of fish, amphibians, reptiles, and birds.

LAB INSTRUCTIONS

You will observe developmental patterns in an echinoderm, a marine worm, a frog, and a bird, which have very different kinds of eggs. You will study the early developmental patterns of each organism to see the effects of egg type on development.

Early Sea Star Development

The sea star is used to illustrate early development because the events of cleavage and gastrulation are especially easy to see. More complex development patterns in other organisms may be compared to this simple pattern.

Fertilization and Cleavage

1 Obtain a prepared slide containing early developmental stages of the sea star. Scan the slide with your compound microscope and locate the stages described in this section and shown in **figure 32.1**. These slides are usually *thicker* than most slides you have studied. *Do not use high power,* or you may push the objective through the coverslip while focusing.

An **unfertilized egg** contains a single egg nucleus and equally distributed yolk surrounded only by a cell membrane.

When a sperm penetrates the egg, a **fertilization membrane** forms as materials stored beneath the cell membrane are released. This new membrane may be visible as a "halo" above the surface of fertilized eggs on your slide. It acts as a barrier that prevents the penetration of additional sperm. Why is this important? Think of the genetic consequences of polyspermy: the fusion of more than one sperm with an egg.

It is interesting to note that in all animals it is only the sperm nucleus that enters the egg. This means that the male contributes only genes to the next generation. All of the organelles in the new generation are derived from the female. This includes mitochondria that contain a small amount of DNA, setting up what is called

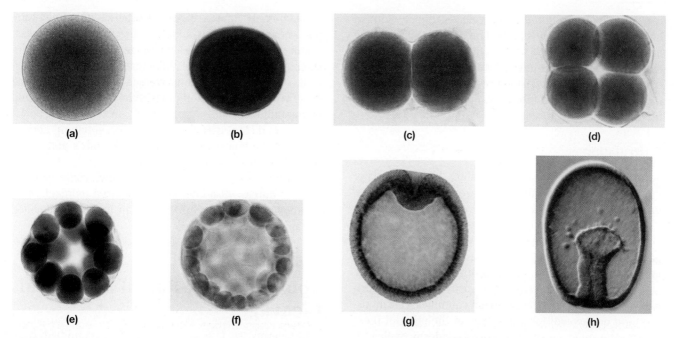

Figure 32.1 Stages in echinoderm (sea star) development: (*a*) unfertilized egg; (*b*) zygote surrounded by fertilization membrane; (*c*) two-cell stage (blastomeres); (*d*) four-cell stage; (*e*) morula; (*f*) blastula; (*g*) early gastrula; (*h*) late gastrula.

maternal inheritance. Once inside the egg, the haploid sperm nucleus migrates through the cytoplasm. It fuses with the haploid egg nucleus, in an event called **syngamy** that forms the diploid nucleus of the **zygote**, the first cell of the new generation. Find examples of unfertilized and fertilized eggs on the slide.

For several hours following fertilization, the zygote undergoes a series of rapid mitotic cell divisions without any intervening periods of growth. This **cleavage** process involves the duplication of chromosomes followed immediately by mitosis and cytokinesis, followed again by chromosome duplication, followed by mitosis, and so on. The result is that the large uninucleate fertilized egg is divided into smaller and smaller cells, each with a single nucleus. If the type of cell division is mitosis, do the nuclei of the developing embryo contain genetic information identical to or different from the zygote's nucleus?

The cleavage divisions are usually synchronized in all cells, so that the embryo goes through a series of stages in which it contains at first 1 then 2, 4, 8, 16, 32, 64 cells, and so on. Around the 32-cell stage, the number of cells becomes difficult to count, and the developing embryo is referred to as a **morula,** meaning mulberry-like.

Find examples of 2-, 4-, 8-, and 16-cell stages on your slide. Note that the size of the developing embryo does not increase as cell number increases. The existing material of the embryo is merely partitioned by cleavage into smaller cells.

2 If ocular micrometers are available on your microscopes, you can test this hypothesis.

The cells in the 1- through 16-cell stages are spherical. The volume of a sphere equals $4/3\ \pi r^3$. If you measure the diameter of cells in each stage, you can calculate the volume of cytoplasm in each stage by:

Zygote volume = $4/3\ \pi r^3$ = _____

16-cell volume = $16(4/3\ \pi r^3)$ = _____

Within an acceptable margin of error are the volumes equivalent? Do you accept or reject this hypothesis?

The developmental process requires energy. Where does this come from? Would you predict any effect on cytoplasmic volume as a result of using energy?

As cleavage continues, the cells eventually form a hollow ball of cells that is ciliated on its outer surface. The beating of the cilia allows the embryo to swim out of the enveloping fertilization membrane. This stage is called a **blastula**, the cells are **blastomeres**, and the central cavity is the **blastocoel**. Find a blastula on your slide and identify these areas.

Gastrulation

Several hours after blastula formation, a second major morphogenic event occurs. Cells at one end of the blastula undergo rapid growth and move into the blastocoel. This infolding is called **invagination** or **gastrulation** and results in a two-layered embryonic stage called the **gastrula** (fig. 32.1g and *h*). The gastrula then elongates, forming a tube-within-a-tube cylindrical body.

The two layers of cells in the gastrula are called **primary germ layers**. As development continues, certain tissues and organs will form from one or the other of these layers. The inner layer, called the **endoderm**, will form the lining of the digestive system and digestive glands. The outer layer is the **ectoderm** and will form the skin and the nervous system of the adult. The midgastrula and late gastrula stages can be identified by the elongation of the gastrula and changes in the shape of the endoderm tube. The elongated tube is called the **archenteron**, meaning primitive gut, and its opening to the outside is called the **blastopore**. It will become the anus of the sea star.

The end of the archenteron away from the blastopore eventually develops two lateral pouches that will grow outward and pinch off, forming a third germ layer called the **mesoderm**. The start of this process is shown in figure 32.1*h*. Muscles, connective tissues, and gonads will develop from the cells in this third layer. Find a late gastrula stage on your slide and sketch it in the circle. Label all the structures indicated by boldface terms in this paragraph.

In the transition from blastula to gastrula, a lot has happened. The basic shape of an animal has appeared. Most animals have a digestive tube-within-a-body-tube plan of organization, if you neglect the appendages. Think of the bodies minus appendages for an earthworm, insect, and human to test this idea. In addition, the gastrula stage is the first time we see distinct tissues, and these will further develop to yield the many tissues found in an adult.

Beyond the gastrula stage, differentiation continues. Most animals, except for reptiles, birds and mammals, have a free living larval stage. A larva is capable of feeding itself and continues to grow, using external energy sources. At a later stage, the larva reorganizes its tissues to create the adult body plan in a process called metamorphosis.

Comparison of Radial and Spiral Cleavage

Echinoderms, such as the sea star you just studied, have what is called **holoblastic** cleavage. Holoblastic refers to the fact that when the cells (blastomeres) divide, the daughter cells are usually of equal size and completely separate from one another. Such division processes are characteristic of eggs in which the yolk is uniformly distributed. In contrast to holoblastic cleavage, animals that have eggs with large amounts of yolk often have **meroblastic** cleavage. In this case, the daughter cells resulting from a division are often unequal in size with the larger cells containing greater amounts of yolk. You will see examples of meroblastic cleavage when you study frog and bird development.

Echinoderms also have what is described as **radial** cleavage (**fig. 32.2a**). Each successive mitotic spindle and cleavage plane forms at right angles to the previous one so that the cells are arranged on radii extending from the center of the cell mass. In contrast to radial cleavage, many invertebrates have **spiral** cleavage. In fact, this basic difference correlates with the two clades of higher animals, the Protostomes (spiral cleavage) and Deuterostomes (radial cleavage). (See lab topics 21–23.) Figure 32.2 compares radial to spiral cleavage. The differences result from the angle at which the spindles form during early cleavage. In radial cleavage the spindles always form at right angles to the spindles seen in the previous cleavage. In spiral cleavage, after the four-cell stage, the mitotic spindles form at oblique angles to the previous one so that the cleavage planes are also displaced to oblique angles.

3▶ To observe evidence of spiral cleavage, look at a slide of the early developmental stages of *Cerebratulus* sp., a marine ribbon worm. Find an eight-cell stage where you are looking down on the organism from the top, a polar view. Note the positions of the top four blastomeres relative to the positions of the lower four. The top blastomeres lie over the grooves between the lower blastomeres. Sketch the stage.

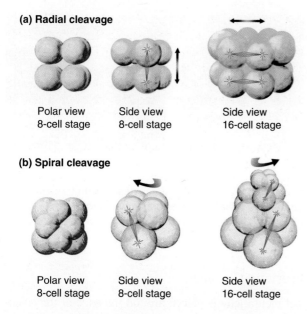

Figure 32.2 Comparison of radial and spiral cleavage. (*a*) In radial cleavage in the deuterostome clade, the spindles form either parallel or at right angles to the axis running through the poles of the cells. Consequently, cells sit directly on top of cells underneath. (*b*) In spiral cleavage in the protostome clade after the four-cell stage, the mitotic spindles form at oblique angles to the polar axis of the cells. Because cytokinesis always occurs across the center of the spindle, the oblique spindles result in displacement of the newly forming cells. Arrows indicate the direction of displacement.

Look again at the eight-cell stages of both the sea star and *Cerebratulus*. Are each of the cells in each embryo the same size? Which of the organisms has holoblastic cleavage?

Experimental Embryology with Sea Urchins (Optional)

Sea urchins, as well as sea stars, are members of the phylum Echinodermata and show the same early developmental patterns. Sea urchins are found in both the Atlantic and Pacific oceans. The breeding cycles are such that the West Coast species are fertile in the fall and winter, whereas the East Coast species are fertile from April to September. This makes sea urchins ideal for teaching laboratories because animals in breeding condition are available throughout the year.

Sea urchins in breeding condition can be induced to release gametes by injection of KCl that causes the gonads to release either eggs or sperm. The separately collected eggs and sperm can be mixed on a microscope slide so that fertilization and development can be directly observed.

Obtaining Gametes

Because it is difficult to tell the difference between the sexes in sea urchins, several sea urchins may have to be used to obtain both eggs and sperm. One male and one female, however, provide enough gametes for a whole lab.

5) To induce shedding, inject three to four animals with 2 ml of 0.5 M KCl through the soft tissue surrounding the mouth as shown in **figure 32.3**. Place the animals in a dry bowl with the **oral** (mouth) surface down and watch the upper **aboral** surface for the release of gametes through the genital pores. Males release a white suspension of sperm, and the females a colored suspension of eggs. Collect the eggs and sperm as follows (see also fig. 32.3).

4) Change slides and look again at the slide of sea star developmental stages. Find an eight-cell stage in polar view and note the relative positions of the upper and lower blastomeres. They lie directly on top of one another with the cells and grooves in register. Sketch the stage below.

   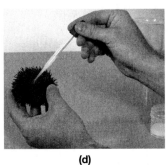

(a) (b) (c) (d)

Figure 32.3 Technique for collecting sea urchin gametes. (a) Inject 0.5 M KCl into soft tissues near mouth to stimulate gamete release. (b) Place animals oral side down in dish and watch for gamete release from aboral surface. (c) Place female over beaker of cold seawater with aboral surface down in water. Eggs will sink to bottom. (d) Sperm should be removed from male's aboral surface and stored undiluted in a dropper bottle.

Females should be placed over a small beaker filled with cold seawater. Put the aboral surface down and immersed in the water. When shed, the eggs will sink to the bottom of the beaker. To facilitate union with sperm later, these eggs should be washed three times with seawater. Swirl the water in the beaker, allow the eggs to settle and decant the supernatant. The eggs will remain viable for two to three days if stored in the refrigerator in seawater.

Sperm can be collected by holding a male over a sterile petri plate with the oral surface up so that the sperm drip into the plate. They may also be collected as in figure 32.3d. They can be transferred to a cold, clean dropper bottle for storage up to 24 hours. No seawater should be added until the sperm are to be used. At that time, add three to four drops of sperm to 25 ml of cold seawater to create a suspension that will remain active for 20 to 30 minutes.

Observing Fertilization

Fertilization events are temperature sensitive and rarely occur above 23°C. If your lab room is hot, cool all slides and suspensions in a refrigerator before starting the experiment.

6 To observe fertilization, take a depression slide and add one to two drops of seawater containing 10 to 30 eggs. If depression slides are not available, add a few washed grains of sand to a drop of seawater containing the eggs on a normal slide. This will keep the coverslip elevated. Examine these eggs under the 10× objective on the compound microscope and try to locate the egg nucleus. The yolk is concentrated in one hemisphere of the egg called the **vegetal hemisphere.** It is usually on the bottom as the egg floats. The other hemisphere is called the **animal hemisphere.**

After adjusting the light intensity for good viewing, add a drop of the diluted sperm suspension to the slide while watching through your microscope. Carefully observe the gametes. You will probably not see the single sperm that fertilizes the egg. When a single sperm fuses with the egg, it will release a **fertilization membrane,** which you should be able to observe. Your textbook should provide some of the details on the development of the fertilization membrane.

Observing Cleavage

If petroleum jelly is placed around the edge of the coverslip, the slide can be kept and later developmental stages observed during the next 24 hours. (Be sure the temperature does not exceed 23°C.) Refer to the following schedule of developmental events to see the approximate timing for different stages.

Sperm contacts egg	0 minutes
Fertilization membrane forms	2 minutes
1st cleavage (2-cell stage)	1 hour
2nd cleavage (4-cell stage)	1 hour, 30 minutes
4th cleavage (16-cell stage)	2 hours, 30 minutes
Blastula (about 1,000 cells)	7 to 8 hours
Gastrula	12 to 15 hours
Pluteus larva	2 days

7 While waiting for the first cleavage to occur, obtain a regular microscope slide and add a drop of diluted sperm suspension. Add a coverslip and observe the slide under high magnification (or oil immersion) with your compound microscope. Sketch what you see.

8 If you do extend your observation time to view the developmental stages, keep a journal using the blank space at the end of this lab topic. Enter the elapsed time from fertilization for all observations.

Amphibian Development

Frogs must return to water to breed, although there are exceptions—one of the more bizarre being frogs that ingest their eggs, incubating them in the mouth or even in the stomach where digestive enzymes are inhibited. Eggs are released by the female as a male grasps her from above and behind and releases sperm. Fertilization is external and the eggs develop into an aquatic larval stage (tadpole) that feeds and grows before metamorphosing into an adult.

Cleavage

Frog eggs contain more yolk than starfish eggs. It is concentrated in one-half of the egg and, being heavier, this half always floats downward in water. The yolk-rich **vegetal hemisphere** is opposite the darkly pigmented **animal hemisphere** (fig. 32.4).

As in the starfish, the fertilized egg undergoes cleavage, dividing first into 2 cells, then 4, 8, and so on, until a blastula formed (fig. 32.4). Cleavage differs in the frog, however, in that the resulting cells are not equal in size. The greater amount of yolk in the frog's egg displaces the planes of cell division as well as slowing it so that those cells in the vegetal hemisphere are larger.

9 Obtain a prepared microscope slide with a section of a frog blastula on it and check it out under low power. Note the smaller well-defined cells in the animal hemisphere and the larger faintly outlined cells in the vegetal. Nonetheless, the blastula stage is similar in organization to that of a starfish and is a hollow ball of cells. Identify the **blastocoel**, animal hemisphere, and vegetal hemisphere.

Frog gastrulation differs from that in the starfish. In the starfish, the archenteron formed by a tube invaginating on one side of the developing embryo. In frogs, it is more complicated because the large yolk-filled cells modify the process. Nonetheless, a blastopore and archenteron do form a primitive gut.

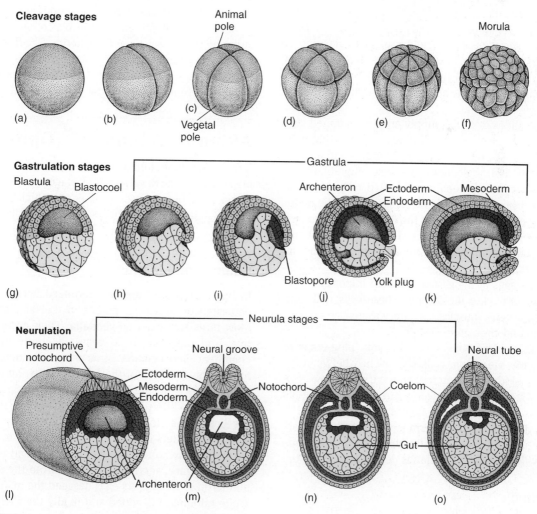

Figure 32.4 Stages in development of frog egg.

Invagination (involution) of cells occurs at a region called the **gray crescent.** This lightly pigmented area develops at fertilization. When a sperm fuses with an egg at fertilization, most of the dark pigment of the animal hemisphere shifts toward the sperm entrance point, much like a cap shifting on our head. The area that is uncovered retains a small amount of pigment and is called the gray crescent. Thus, an event as simple as sperm penetration establishes a major axis in what seems to be a relatively uniform sphere except for the hemispheric differences.

Gastrulation

At the end of the blastula stage and the beginning of the gastrula stage, cells above the gray crescent begin to migrate inward between other cells. This begins as a small slit which develops into an arc and then a circle. Here is an analogy to help you visualize this: If you took a spoon and pressed it against the side of a very soft rubber balloon, as the spoon passed inward it would gradually be covered by soft wall material coming from the stretching surface of the balloon. If the spoon were invisible, you would see a slit appear on the surface and balloon wall material would start migrating into the slit, forming a flattened tube extending from the surface into the balloon. The slit is the **blastopore** you easily saw in the starfish and its upper surface is called the **dorsal lip** of blastopore. At this important point, the major axes of the embryo are established. The blastopore establishes the anterior-posterior axis and the dorsal lip forms the dorsal–ventral axis.

10▶ Get two slides from the supply area, a sectioned gastrula stage and a sectioned yolk plug stage. Starting with the gastrula stage and using figure 32.4 as a guide, find the blastopore and archenteron. Because amphibians are deuterostomes, the blastopore is destined to become the anus. The cells lining the archenteron are called the **endoderm** and will form the lining of the digestive tract as well as various digestive glands. The cells on the external surface are **ectoderm** and will become the outer skin and the nervous system.

In later stages of the gastrula, a third important layer of cells will form, the **mesoderm.** These cells will form the skeletal system, muscles, excretory system, reproductive system, and circulatory system.

11▶ Look at a microscope slide of a yolk plug stage to see these early mesoderm cells sandwiched between the dorsal cells of the archenteron and the overlying ectoderm (fig. 32.4k). As these events are occurring, the old blastocoel is eventually obliterated as it fills with cells. At this point, the three primary germ layers have been established, the major axes of the embryo are laid down, and the tube-within-a-tube body plan is developing.

Neurulation

Following gastrulation in frogs, the next significant event is the development of the nervous system, through a process called **neurulation.** The embryo elongates and a groove can be seen running from anterior to posterior on the dorsal surface. The ectodermal cells folding inward on the surface will form the nervous system.

12▶ From the supply area, get a prepared slide of a cross section of a neural groove stage. The slide and figure 32.4 show you how the nervous system forms. On the dorsal surface of the section, two ridges of ectoderm (neural folds) should be seen, defining the **neural groove** between them (fig. 32.4m).

The formation of the groove is induced on the dorsal surface by chemical signals from the underlying mesoderm. By this stage, it has increased in size and some of its cells have formed a longitudinal skeletal element called the notochord. At a later stage, it will be replaced by the vertebral column. At this stage, it induces the ectodermal cells above it to fold inward. As the fold deepens, the upper edges fuse so that a tubular group of cells is internalized from the surface. These cells comprise the nerve cord. Its anterior end will enlarge to become the brain.

From the hollow ball of cells, the blastula, you have traced how the general body plan of a tadpole larva is laid down. You have seen how the gastrula stage forms the tube-within-a-tube body plan where the inner tube is the gut and the outer one becomes the body wall. The tube is polarized so that the end at the blastopore becomes the anus and the opposite end the anterior. The three germ layers of cells give rise to the tissues of the body. You traced how the nerve cord is formed from the ectoderm. Time does not permit the further tracing of organ development.

Avian Development (Optional)

Because of time limitations, it will not be possible to observe all of the developmental stages of the chick. Three stages have been chosen to illustrate chick development: 16-, 18- and 24-hour embryos. You will look at prepared slides.

Gametes and Mating

In birds, eggs are internally fertilized before the shell is deposited by accessory glands in the female's oviducts. Most birds lack external genitalia. Their excretory, digestive, and reproductive systems all empty into a single chamber called a **cloaca.** Birds mate when the male briefly mounts on the back of the female while she elevates her tail feathers exposing her cloaca opening. Sperm are produced in two internal testes in the male. The male presses the lips of his cloaca to hers and passes sperm to her. The sperm enters the female's reproductive tract from her cloaca and are stored. As eggs are produced and released from the ovary, sperm are released from the storage area and fertilize the egg. As eggs descend the oviduct, the egg white (albumen) is added and finally the shell (**fig 32.5**). Some birds, especially ducks and geese, do have penises. In 2001, scientists found that one Argentine lake duck has

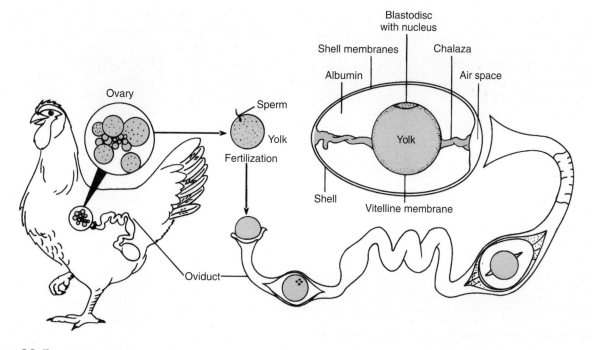

Figure 32.5 Egg development in chicken. Albumin (egg white) and shell develop as egg passes through oviduct. Embryo is at gastrula stage when egg is laid.

a penis over 42 cm long (almost the same as body length). Subsequent research indicated more modest lengths were the norm, but nonetheless large.

Because the eggs of birds (also fish and reptiles) contain large amounts of yolk, the nucleus and cytoplasm are confined to a small disc-shaped area called the **blastodisc**, which sits on top of the yolk. Fertilization occurs at the blastodisc, and cleavage divisions are confined to this area. The large yolk does not divide as in the starfish; the developing chick embryo "rides" atop the yolk during development (**fig. 32.6**). When slides are prepared, the yolk is removed and only the blastodisc area is on the slide.

Chick Cleavage and Gastulation

13 Obtain a whole mount slide of a 16-hour old chick embryo. You will be looking at the entire blastodisc after it has undergone several cell divisions. As the cleavage divisions occur in the blastodisc, a flat layer of cells called the **blastoderm** is produced. As the cleavage continues, the blastoderm becomes several cell layers thick with a fluid-filled space, the **subgerminal space**, developing between it and the yolk. The blastula is formed when the cell layers of the blastoderm separate into an upper **epiblast** layer and a lower **hypoblast** layer (fig. 32.6). The space between them is the avian equivalent of the **blastocoel**. Only the epiblast cells will form the tissues of embryo.

As development continues, cells in the center of the epiblast migrate down from the surface and away from the center area in a process that is equivalent to gastrulation. These movements result in the recognizable **primitive streak** stage (fig. 32.7). Find the primitive streak on your slide. Epiblast cells migrating into the primitive

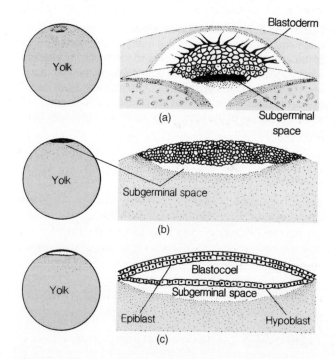

Figure 32.6 Side view of developing chick embryo. (*a*) Blastodisc confined to top of yolk. (*b*) Several cell layers accumulate. (*c*) A cavity appears in cell mass when blastula forms.

streak delaminate and enter the blastocoel. Some remain there to become mesoderm. Others displace hypoblast cells to become endoderm. Cells remaining in the epiblast become ectoderm.

14 Now get a slide of a cross section through the blastoderm for an 18-hour chick. Compare what you see to

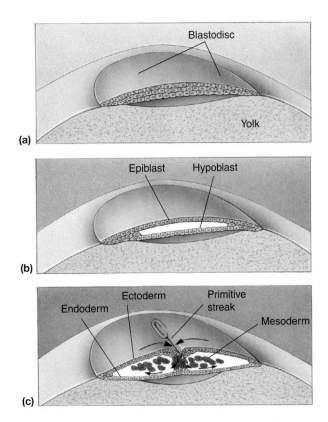

Figure 32.7 Formation of primary germ layers in avian egg. (a) The blastodisc is a mass of cells found on top of the yolk; (b) The cell mass splits to form a cavity yielding a blastula-like stage where the upper layer is the epiblast and the lower is the hypoblast; (c) Cell migrate down into the cavity from the epiblast at the primitive streak and along the edges of the disc, forming the three primary germ layers as indicated.

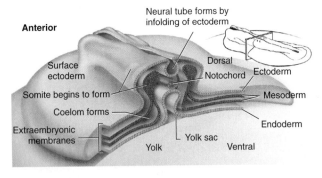

Figure 32.8 Organ formation from germ layers.

figure 32.8. By this time most of the cellular migrations are complete and new structures are beginning to appear from the primary germ layers. Ectoderm on the outside will produce the skin and feathers as well as the nervous system; mesodermal layers in the blastocoel will form the muscles and bones and other connective tissues; and endoderm will form a digestive tube. A **neural tube** may be developing from the infolding of the ectoderm. It will become the nerve cord. A notochord should be visible below it. It will be replaced by the vertebral column.

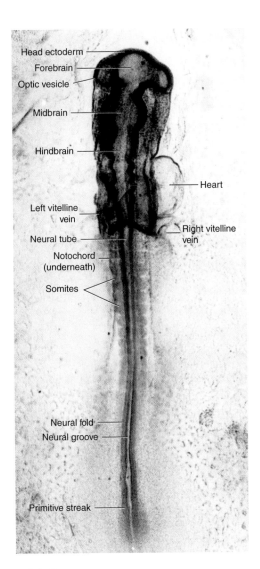

Figure 32.9 Dorsal view of 33-hour chick.

Organogenesis

15 Study photographs or whole mount microscope slides of 33-hour and 72-hour chick embryos. Note the developing nervous system and the anterior-posterior axis.

By 33 hours, the relatively unorganized structure of the blastoderm is gone, replaced by an obviously animal-like embryo (**fig. 32.9**). The **neural tube** runs the length of the embryo and its anterior portion is beginning to differentiate into regions of the brain. A **heart** has formed. **Somites,** segments of body muscle, are beginning to appear.

By 72 hours, note that the anterior half of the chick has rotated and lies on its left side, while the posterior half remains dorsal side up (**fig. 32.10**). Considerable differences may also be noted in the nervous and circulatory systems. The neural tube has differentiated into a five-part brain and a nerve cord. **Optic cups** are developing from the optic vesicle, and the optic lens is starting to form. An auditory vesicle should be visible. The heart has changed from a muscular tube to a two-chambered organ, and aortic arches with associated gill slits have developed. Note the vitelline arteries and veins. What are their functions?

INVESTIGATING EARLY EVENTS IN ANIMAL DEVELOPMENT 439

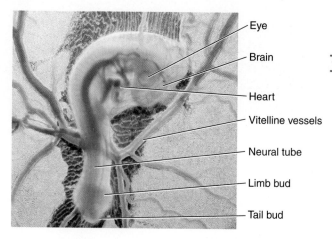

Figure 32.10 Whole mount of a 72-hour chick embryo.

Lab Summary

16 On a separate sheet of paper, answer the following questions assigned by your instructor.

1. Describe how a fertilized sea star egg develops into a blastula.
2. Why is the gastrula considered an important stage in development? How does the gastrula stage reflect the general adult body plan of most animals?
3. Describe the events that occur when a sea urchin's egg is fertilized.
4. Compare and contrast radial and spiral cleavage.
5. Compare gastrulation in a starfish to that in a frog.
6. Describe how the nerve cord forms in a frog from ectoderm.
7. Describe the formation of a blastula and gastrula in a bird egg.

You may want to try the critical thinking questions that apply some of the knowledge you gained in doing this lab.

Internet Sources

Check the WWW for current research on early developmental stages in animals. Use the search engine Google at www.google.com and enter the phrase research on blastula or research on gastrula. Scan the sites returned and write a summary of one that is of most interest to you.

Learning Biology by Writing

Compare and contrast radial cleavage with spiral cleavage. Describe how the gastrula stage forms from a blastula and discuss the significance of the gastrula stage, *i.e.*, why is it important. Discuss how events in the gastrula and following stages differ in protostomes and deuterostomes.

Critical Thinking Questions

1. All bilaterally symmetrical animals have three axes: anterior/posterior; left/right; dorsal/ventral. Describe when in development these axes originate.
2. Reflect on the general life cycles of a bird and an invertebrate such as a starfish. Recognize that many adult birds are smaller than starfish. Why are bird eggs always larger than invertebrate eggs? Consider in your answer the energy budgets for development and the role of larval free-living stages.
3. Many invertebrates produce several thousand eggs during their lifetimes. Why is the world not overrun by invertebrates?
4. The eggs of most terrestrial animals are surrounded by shells while those of most aquatic species are not. Why do you think this is so? In species with shells would you expect that fertilization is internal or external? Why?

Lab Topic 33

Estimating Population Size and Growth

Supplies

Resource guide available on WWW at
http://www.mhhe.com/labcentral

Equipment

Balances, with 0.01 g sensitivity
Computers connected to Internet

Materials

Wooden pegs or applicator sticks
Small plastic bags
4-m string
40-cm string
About 200 beads per student lab group (all one color, in coffee cans)
Beads of another color, 50 per group
Markers
Means of generating pairs of random numbers

If lab done outside
  Maps of quadrat sampling area
  Resource books on identification of common lawn weeds in area
  Identified samples of lawn plants in lab for reference

If lab done inside
  Table top covered by 1-m² paper ruled in 10-cm grid
  Beads to scatter on table, count and weigh
  175 beads one color
  250 beads another color
  Boards 1 m × 5 cm to edge grid

Student Prelab Preparation

Read the Background material and sections of the lab topic that have been scheduled by your instructor.
  You should use your textbook to review the definitions of the following terms:

density   population
dispersion   population density
exponential   quadrat
mark and recapture

Describe in your own words the following concepts:
Population growth curves
Exponential growth
Logistic growth
Carrying capacity

After finishing the prelab review, write any questions you have about terms, concepts, or techniques in the margins of this lab topic. The lab experiments should help you answer these questions, or you can ask your instructor during the lab.

Objectives

1. To use quadrat sampling techniques to estimate the size of nonmobile populations
2. To simulate mark-and-recapture methods for estimating the size of large, mobile populations
3. To plot logistic growth curves, using real data
4. To determine what species are rare and endangered in your state

Background

Organisms do not usually exist as isolated individuals in nature but are parts of larger biological units called **populations.** These reproductive and evolutionary units consist of the members of a single species residing in a defined geographical area, for example, in a small park, in a mountain range, or on a continent. A population of clover in a lawn may be easy to recognize because of its compactness, whereas a widespread population of moose in northern Minnesota may be less apparent. Organisms within each population are interlinked by reproductive gene flow. Natural selection operates on individuals within these populations to shape a population's adaptation to an environment over several generations. Populations are the fundamental units of species on which the mechanisms of evolution operate.

Biologists study populations for several reasons: to understand gene flow, selection, adaptation mechanisms and to manage the populations. Management addresses the problem of the relationship between humankind and the environment. In some cases, management involves controlling the size of populations of noxious organisms, such as rabid skunks, mosquitoes, or parasites of humans. In other cases, the goal may be to increase populations of beneficial organisms, such as lumber-producing trees, edible fish, or game species. To manage populations, it is necessary to know the basic biology of the organisms involved as well as to understand the physical and biological environment.

Studies of populations start with such basic questions as: How many individuals are in the population? How is the population distributed in the study area? Is the population increasing or decreasing in size?

To answer these questions, biologists must use sampling procedures that allow them to estimate the number of individuals in the population distributed in a certain space, that is, the **population density.** Rarely is it practical to count the entire population. In this lab, you will use two often-employed techniques of estimation: **quadrat sampling** for nonmobile animals and plants and **mark-and-recapture techniques** for mobile animals. With the passage of the National Environmental Protection Act in 1970, these techniques have been, and are, frequently used by biologists to determine the environmental impact of changes in ecosystems.

Both techniques are based on random sampling statistical procedures; care must be taken to assure that randomness occurs when these techniques are applied to an actual biological problem. Sampling plots must be randomly located and representative of the study areas, and animals must be randomly captured and released so that they can freely mix with the total population. If randomness is not realized, the sampling procedure will lead to erroneous estimates. Sampling procedures repeated over time allow the investigator to determine whether a population is growing or declining.

Estimation of population size alone, however, is not sufficient for making management decisions. Additional information is required, such as what resources are needed by and available to the population, how other organisms influence the population, and what the population age structure is. The question of resources involves the physical environment: for example, water quality, soil type, temperature, or available nesting sites. The influence of other organisms on a population affects food availability, parasitism, and predation. Population structure refers to age distributions in the population. A stand of white pine trees could consist of all young, all old, or a mixture of young and old trees. Harvest practices and replanting programs would depend on the population structure in a particular timber tract. Similar problems exist in managing deer herds and coho salmon and in assessing the effects of agricultural pesticides on harmful insects.

Populations of organisms, whether they be bacteria in culture, algae in a pond, fungi in the soil, oak trees, or frogs, have the potential to increase in numbers exponentially, if no factors in their environment are limiting. However, this potential is rarely realized. Pacific salmon lay thousands of eggs that are fertilized, but we are not overrun by salmon. Why? Many die from natural causes: disease, predation, limited food supplies, and competition with other members of their species. Most species maintain rather constant population levels, although they may fluctuate around an average value from year to year as density-dependent factors reduce the population one year but allow an increase the next year only to have density-dependent factors again reduce it.

Exponential growth occurs most often in newly available environments or when a population has been dramatically reduced. If you build a farm pond where there are few algae to start or kill off the algae in an existing pond by adding a chemical that is then removed, the algae will begin to increase exponentially in numbers. Their birthrate (= cell divisions) will greatly exceed the death rate in the population. At some point, the competition among algae will increase as their density increases—for example, during the day for light or carbon dioxide, and at night for oxygen. As a result, death rate starts to increase and birth rate falls off and population growth slows. At some point the two rates become equal and the population establishes an equilibrium with its environment. In this example, if algae population size is plotted as a function of time, the early stages of population growth, before density-dependent factors became significant, would be an **exponential growth curve** or J-shaped curve. When density-dependent factors started to come into play, the exponential curve would change into a sigmoid or S-shaped curve called a **logistic growth curve.** The number of individuals in the population at equilibrium is called the **carrying capacity** for that environment.

Many biologists feel that population ecology is the most demanding and comprehensive field of biology because the investigator must understand and deal with environmental variables, metabolic efficiency, physiological reactions, hereditary mechanisms, and the interactions of organisms.

LAB INSTRUCTIONS

You will learn two estimation techniques used by population ecologists: quadrat sampling for nonmobile plants and the mark-and-recapture technique for mobile animals. You will also investigate the phenomenon of exponential growth in populations and will explore what species are considered threatened and endangered in your state.

Quadrat Sampling

The structure and composition of populations of nonmobile organisms, such as terrestrial plant communities or sessile animals on a coral reef, can be sampled by a number of techniques. The most common procedure involves using randomly located plots called **quadrats.** Quadrat sampling varies in terms of the size, number, shape, and arrangement of the sample plots, all of which depend on the information sought and the nature of the populations or communities being studied.

Technique

The size of a plot is determined by the size and density of the organisms being sampled. It should be large enough to include a number of individuals but small enough to allow easy separation and counting of individuals. Obviously, plots of different sizes would be used to estimate the number of trees in a forest in contrast to the number of cattails in a swamp.

For the sampling procedures to be statistically valid, quadrats must be chosen randomly within the study site. To do this, visualize the study site as a piece of graph paper on which you want to randomly locate small quadrats. A baseline is laid out along one side of the study area. A pair of random numbers is then chosen that corresponds (1) to distance along the baseline and (2) to perpendicular distance from the baseline to the center of the sampling quadrat. The units of length will depend on the size of the study area and may be in feet, meters, paces, or any other appropriate linear measure.

Inside Quadrats Because weather might not allow you to perform quadrat sampling in the field, the technique can be simulated in the laboratory. If the indoor simulation is to be used, your instructor will have 1 m^2 of paper taped to a table top. It will be ruled into a 10-cm grid. How many 10-cm squares are in 1 m^2?_____

1) One edge of the paper will have the 10-cm rows numbered 1 through 10. An adjacent edge will have the columns lettered A through J. The location of each 10-cm square is given by a letter-number combination. For example, quadrat H8 is found at the intersection of column H with row 8.

Randomly scattered on the grid will be 200 beads of one color and 50 of another. The colors represent juveniles and adults of the same species. Some squares will have boxes taped to them to exclude beads. These areas are uninhabitable by the species (such as bare rock) and should be subtracted from the total area available. Imagine that the bead distribution represents trees in a forest and that you are looking down on it or that you have taken an aerial photo on which a grid was superimposed.

Your task is to estimate the total population size in the study area and the immature population size by quadrat sampling.

Student pairs will each be assigned three random coordinates, such as A5, D2, and F7. Count the number of beads of each color in each assigned square and calculate an average per square. Record below.
   Average for total _____
   Average for immatures _____

2) Place these estimates on the blackboard and calculate a grand average across all lab groups. Record below.

Class average, total_____
Class average, immature_____
Is there a significant difference between the average total population per square based on your data and the average of the class data?_____
Why do you think this is so?

Repeat this analysis for the immatures.

3) Determine the total number of inhabitable squares in your study area and multiply your average estimates per square by this number to get your estimation of entire population in the study area. Repeat for the number of immatures. Record below.
   Estimated entire population = _____
   Estimated immatures = _____

4) How many individuals were actually in the study area _____

What percentage were immatures? _____
Calculate the percent error for the estimations based on your data and those based on the class data, using the relationship:

$$\% \text{ error} = \frac{(\text{estimate} - \text{actual}) \times 100}{\text{actual}}$$

My % error = _____
Class % error = _____
Use statistical sampling theory to explain why there is a difference between population estimations when you use your data versus class data.

Why would you not simply count all of the individuals in a natural population to eliminate error?

Use the actual numbers of beads in the population to answer this question. If all adults were to die and only the juveniles produce offspring during the next year, is the population expanding or contracting? (Assume each pair of juveniles will produce three offspring and that the number of males equals the number of females.)

stem to be a grass "plant" but realize they will be connected to other plants by underground stems. Convert the count per small quadrat to the count per large quadrat by multiplying by 100.

Note any unusual features in your study area. Are there any ant hills? Is the quadrat located on a path? Under a tree? In the open? Is there drainage into the area? Are plants uniformly distributed or in clumps?

Based on your data, how many immatures were estimated to be in the population? _____ If you were managing a forest, would you have made the same decision about harvesting (adults die) based on your estimations as those based on the actual population counts?

Outside Quadrats (Optional) If you are to do this lab outside, your instructor will provide you with a map of an area on your campus that will serve as your study area. You will also be given a set of random coordinates, a piece of string 40 cm long and one 4 m long and five pegs. Study the map so that you understand where the study area is and the orientation of the baseline.

Take your lab manual, a pencil, and your strings to the study area. Using your coordinates, pace out the distances along the baseline and perpendicular from it to locate the center of your sampling quadrat. Place a peg at this point. Tie the ends of the 4-m string together to form a loop. Using the other four pegs, stretch the string as a square around the center peg (the area inside the string will be 1 m²).

6 Count the number of nongrass plants contained within your quadrat. Keep separate counts for each species. If you cannot identify what kind of plants they are, make up a temporary name and then check resource books available in the laboratory. (Specimens can be collected and placed in plastic bags to prevent drying for later identification.) Take a sheet of paper and make a table similar to **table 33.1**. You will record each species found on a new line.

If, by chance, you have a quadrat with thousands of clovers in it, you will not be able to accurately count them in the time provided. If this is the case, switch to the 40-cm string and lay out a smaller square around the center peg. Count the clover within this area and record the count separately.

7 To determine the density of grass in the lawn, use the 40-cm string to lay out the quadrat around the center peg or the upper-right corner. Count and record the number of grass plants in this quadrat. Consider any aboveground

8 Remove one of each type of broadleaf plant from the lawn by cutting it off at the soil level. Collect several dozen grass plants in the same way.

Analysis After you have completed your counts, notes, and collection, return to the lab. Put your data on the blackboard, including brief summarizations of your notes. After all lab groups have reported, the data should be discussed to see if any should be rejected because of atypical localized situations, such as chemical spills, trench excavation, or other isolated interferences. Combine the acceptable data and record them as a class summary in **table 33.2**.

9 Calculate the mean and standard deviation for the number of each type of plant per square meter. (Refer to lab topic 5 or appendix C for a description of how to calculate standard deviations.)

Mean = _____

S.D. = _____

A **hectare** is a unit in the metric system equivalent to 2.47 acres or 10,000 m². This would be equivalent to an area of 100 meters by 100 meters. How many square meters were contained in the total study area? Remember that length times width equals area. How many hectares are in the study area?

Study area = _____ m²

= _____ ha

Using your measurements of grass and broadleaf weed density, how many of each type of plant would be found in a hectare? Record your results in your copy of table 33.1.

How many of each species are contained in the entire study area? Record your answer in your copy of table 33.1.

10 Weigh the total mass of each type of plant you collected. Calculate the average weight of each type of plant. Since you know the number of plants in a hectare and the weight of each type of plant, calculate the aboveground wet biomass in a hectare of lawn.

Biomass/ha = _____

Table 33.1	Number of Nongrass Plants in Quadrat		
Species	Number Counted	Estimated Number Per Hectare	Estimated Number in Entires Study Area

Table 33.2	Class Results from Quadrat Analysis		
Group	Area Size	No. of Grass Plants	No. of Other Plants (Specify)
1			
2			
3			
4			
5			
6			
7			
8			
9			
10			

Mark and Recapture

In the field, it is difficult to measure the size of a population of randomly dispersed, mobile animals, but estimations can be obtained by mark-and-recapture methods. In these methods, living individuals are trapped or collected, marked, and released at the site of capture. A record is kept of the number marked. At a later time that allows for dispersal, animals are again collected from the population. Some of these will be marked and some will be unmarked due to immigrations and emigrations of individuals to and from the collection sites.

Assuming the animals are randomly dispersed, the frequency of marked and unmarked animals in the second collection will allow you to estimate the total population size, using the following proportion (Lincoln-Peterson method): *The number of animals marked and released is to the total number of animals in the population as the number of marked animals recaptured is to the total number of animals recaptured.* Or,

$$\frac{M}{N} = \frac{R}{C}$$

solving for N

$$N = \frac{MC}{R}$$

where

- N = total number of individuals in population
- M = number of animals marked and released (any number but for this lab 25, 50, or 100 depending on your assignment by the instructor)
- C = total number of animals caught in second sample (any number but for this lab 40 or 80 depending on your assignment)
- R = number of marked animals caught in second sample (recaptured)

When using the mark-and-recapture technique in natural situations, several assumptions are made. These are:

1. Capturing the animals the first time does not influence whether they will be caught a second time;
2. Marking does not influence behavior and chances of a second capture;
3. All animals in the population are equally catchable;
4. Sample sizes are large enough relative to the total population to be statistically valid;
5. Animals are not migrating into or out of the population;
6. Traps are randomly spread through the population's territory.

These assumptions are not always met. For example, some animals prefer to feed on bait in traps rather than on natural food while others do not. Or, traps might provide shelter for animals that they prefer compared to natural shelters. Thus, one needs to carefully analyze any situation where traps are used.

Technique

In the lab are several cans containing about 250 beads of one color. These beads are to simulate animal populations. Each lab group will be given a can and will be asked to estimate the population size, using mark-and-recapture methods. Each group will use a slightly different sampling regime. The results of the different groups will be compared to determine what sampling, procedures are optimum.

11 Two groups should each mark 25 beads in their population, while two others should mark 50 and two others, 100. Do this by removing the assigned number of beads from the can and replacing them by adding the same number of different color beads to the can. By comparing results with other groups, you will determine if the size of the marked sample influences accuracy. One group from each of the above pairs should recapture 40 beads from

Table 33.3 Results from Recaptures for Sampling Parameters Assigned to You

Record parameters used: no. to be marked = _____ no. to be recaught = _____

	Trial 1	Trial 2	Trial 3	Average
Number Marked				
Number Unmarked				

its population, while the other group will recapture 80. This will test the effect of the recapture sample size on accuracy.

Once you are sure of your group's assignment for the number of beads to be marked and the number to be recaptured, start the experiment. Remove the assigned number from the can and add "marked" individuals to the population. Close the can and shake well to ensure random mixing. Without looking, withdraw a bead and tally its color on a piece of scrap paper. Stir the beads, and withdraw another, adding to the tally. Continue sampling, one bead at a time, until you withdraw the number of beads corresponding to your assigned recapture sample size. After every five draws put the top back on the can and shake it to ensure random mixing.

Count how many beads in the recapture sample are marked and how many unmarked. Record the counts in **table 33.3**.

Return the beads to the can and repeat the recapture sampling. Record the results. Calculate an average value for your samples.

Analysis

12 Using the average values and the Lincoln-Peterson formula, estimate the population size in your can.

Now count all the beads (both colors) in your can and compare the actual population size to the estimated population size.

Estimated number in population _____

Actual number in population _____

Calculate the % error in your estimation:

$$\% \text{ error} = \frac{(\text{estimate} - \text{actual}) \times 100}{\text{actual}}$$

% error = _____

Lab groups should now share their results by placing their % error values on the board in a facsimile of **table 33.4**. Record class values in table 33.4.

13 Take these class data and plot them on graph paper with the absolute value (no plus or minus sign) of % error (y axis) as a function of marked sample size (x axis). Two

Table 33.4 % Error of Different Sample Sizes

	Recaptured Sample Size	
Number Marked	40	80
25		
50		
100		

lines will be drawn on the graph: one for the recapture sample of 40 and the other for 80. If both of these lines are extrapolated to zero percent error, how many individuals must be marked before you will get correct estimates?

_____

What fraction of the total population would this number be? _____

Describe the relationship between % error and marked sample size.

Describe the relationship between % error and recepture sample size.

If you were optimizing effort and accuracy, what sampling strategy would be best? Why?

Population Growth

An endangered species is a species of fish, plant, or other wildlife that is in danger of becoming extinct throughout a significant portion of its range. A threatened species is any species likely to become endangered in the foreseeable future. In 1963, surveys of bald eagle populations indicated that the number of breeding pairs in the lower

48 states was 487, down from the U.S. Fisheries and Wildlife Service's estimate of 50,000 pairs in 1782 when the bald eagle was adopted as our national emblem. What caused this drastic decline?

No doubt habitat loss and disturbance of nesting sites contributed to the decline, but a more pervasive factor was also at work. DDT, an insecticide developed during WWII, was widely used for insect control on crops and in cities where mosquitoes and noxious insects were a problem. So much was being used, that it started to show up in samples of ice from Antarctica where it had never been applied, indicating that it had entered the global ecosystem, although at very low levels. In aquatic ecosystems, algae accumulate DDT because it dissolves more readily in their fatty membranes than it does water. Minute amounts in water move into the membranes and are concentrated. Algae, however, are not directly affected. Algae with their DDT burden are, in turn, eaten by various aquatic invertebrates and fish eat both as well as each other. This results in a biological amplification as the DDT accumulates in the membranes of the fish to much higher levels than were found in the water, algae, or invertebrates. Eagles feed on fish and through their food were getting high dosages of the pesticide. It did not directly affect the adults, but did affect their reproductive ability. Female eagles were laying eggs with thin shells as a result of the DDT effect on their shell glands. When they sat on the thin-shelled eggs, the eggs cracked; obviously resulting in reduced populations. In 1972, Federal legislation banned all further use of DDT in the United States. A year later, Congress passed the Endangered Species Act of 1973. It listed the eagle as an endangered species. Together these legislative efforts protected habitat and stopped any new DDT from entering the environment. In 2006, bald eagles were removed from the endangered and threatened listing when surveys indicated the population had reached 9,789 breeding pairs: a fantastic recovery in 42 years.

14) Table 33.5 lists data from the U.S. Fisheries and Wildlife Service for the recovery of bald eagle populations. Surveys were not conducted in all years. The table lists breeding pairs, so the population level is at least twice as many.

To see the nature of the recovery, plot the data on the graph paper at the end of this lab topic or use a computer spreadsheet program. Choose an appropriate scale and label all axes. Draw a smooth curve to fit the data points or use the *Trendline* option in a spreadsheet program to fit a curve to the data points.

How would you describe this curve? Is it a J-shaped exponential curve or a S-shaped logistic growth curve? _____

15) Use your graph to determine how many years it took for the bald eagle breeding pairs to double from:

1,000 to 2,000 _____

2,000 to 4,000 _____

4,000 to 8,000 _____

Table 33.5 Breeding Pairs of Bald Eagles in Lower 48 States

Year	No. of Pairs
1963	487
1974	791
1981	1,188
1984	1,757
1986	1,875
1987	2,238
1988	2,475
1989	2,680
1990	3,035
1991	3,399
1992	3,749
1993	4,015
1994	4,449
1995	4,712
1996	5,094
1997	5,295
1998	5,748
1999	6,404
2000	6,471
2005	7,066
2006	9,789

Source: U.S. Fisheries and Wildlife Service, 2009.

Over the last 30 years or so, what has been the average doubling time of the population?
_____

Is the doubling time increasing or decreasing over the study period? _____

16) Use the average doubling time to see if you can theoretically replicate the actual population growth curve of breeding pairs observed for bald eagles. We will start the analysis in 1974, allowing 11 years from 1963 for DDT to degrade in the environment. Complete the following worktable, by adding your doubling time to the starting year 1974, then add it again to the new entry, and so on.

Year	No. of Pairs
1974	791
____	1,582
____	3,164
____	6,328
____	12,656

Plot this data on the same graph as the original data to see if it is a reasonably accurate prediction. Is it? _____

Assuming that the population continued to grow at this rate, approximately how many years would it take for the population to come back to the 1782 level of 50,000 breeding pairs?

_____

The assumptions used to make this prediction are not realistic for two reasons. First, the habitat available to bald eagles has changed dramatically since 1782. Cities, harbors, and vacation resorts have cleared away roosting trees near water where bald eagles once fished and the fish populations (food) are drastically reduced. Consequently, the carrying capacity of the environment for bald eagles is not what it once was. Second, as any population in exponential growth nears the carrying capacity of its environment population density factors come into play. Population growth begins to slow as either birthrates decline or death rates increase. This results in an S-shaped curve called a logistic growth curve. When the carrying capacity is reached, birthrate and death rate are equal. Whenever an individual dies another is born and population numbers remain constant. We cannot predict what the environmental carrying capacity for bald eagles is today. Your analyses of the bald eagle data show that the population is in exponential growth and that appears as if the carrying capacity has not been reached.

Exploring Threatened and Endangered Species Lists

17. Each state publishes a list of threatened and endangered species, which will be investigated in this portion of the lab. Work with a partner and use a computer with Internet access. Open Google and type in the terms "threatened and endangered species" followed by the name of your state. In the long list that is returned, look for sites provided by state environmental agencies. Open a site that gives a listing of animals that are threatened or endangered in your state and choose one for further research. Record its common name and scientific name:

Common _____
Scientific _____

Note: Scientific names are always written with first letter of first name as a capital and first letter of second name in lowercase, e.g. *Homo sapiens.*

Once each pair decides on a species, write its common name on the board. No two groups should research the same animal. The bald eagle is off-limits for all groups.

Once an animal is identified, spend some time looking for a photograph (use Google Image Search) to share with the class. Either write down the URL or open a word processor and cut and paste the URL into it with a short description. This photo will be shared with the rest of the class.

Now do research on the species of interest, using Google to locate other sites that discuss the biology and conservation of the species. Enter the scientific name, not the common name, as you search to eliminate many sites of lesser scientific value. The object of your Internet research is to find information that you can share with your classmates during a presentation at the end of the lab period.

Below are a few questions that indicate the types of information you should be seeking. Make notes, but do not copy sentences, as you go along to help you organize a verbal presentation of your results. Bookmark the best sites, should you want to return and check something after leaving a site.

- What is some basic information about the biology of your animal?
- In what kind of habitat does this animal live and what is its ecological role?
- What does it eat?
- Does it have any major predators, parasites, or symbionts?
- How often does it breed and how many offspring are produced each time?
- Is the decline in population size due to decreases in birthrate or increases in mortality?
- Is the animal a migrant or permanent resident of your state?
- What is the animal's current population size in your state and what was its maximum size in the past?
- Is this animal broadly dispersed in your state or restricted to just one or two counties?
- Is this animal threatened only in your state, or more broadly across the country?
- What has caused the population to decline to the point that it is listed as a threatened and endangered species (habitat loss, pollution, disease, management practices, *etc.*)?
- Describe any specific programs in place to restore the population levels and any data describing the outcomes to date.

- What is the nature of any studies on the species that have been published in science journals in the last three years?
- Describe the economic or ecological value of this species.

Allow some time after you finish your research to organize your thoughts for a brief presentation to the class that will allow them to understand why the species you chose is listed as a threatened and endangered species, what the ecology of the species is, and what is being done or can be done to bring it back.

Lab Summary

18 On a separate sheet of paper, answer the following questions as assigned by your instructor.

1. Under what conditions should quadrat sampling be used instead of mark-and-recapture methods?
2. What are the underlying assumptions of the mark-and-recapture technique and what suggestions would you make to improve the accuracy?
3. Why is random sampling important in the techniques used to estimate the sizes of populations?
4. Distinguish between arithmetic and exponential increases.
5. Name five species that are threatened or endangered in your state and what has caused the populations to decline.

You may want to try the critical thinking questions that apply some of the knowledge you gained in doing this lab.

Internet Sources

Mark and recapture is a standard technique for estimating the population sizes of animals. Use a search engine to locate information on the WWW for this technique. Connect to three URLs and read what they have to say about this technique. Summarize in a paragraph what you read, and record the URLs for future reference.

Learning Biology by Writing

Write an essay describing why quadrat sampling or mark and recapture would not be good techniques for surveying populations of bald eagles.

Critical Thinking Questions

1. List several sources of error in estimating population sizes by the quadrat sampling technique and indicate whether these would lead to an overestimation or an underestimation of population size.
2. Population growth results from a balance between the rate of birth and the rate of death in the population. For a hypothetical species of tree in a national forest, make a long list of natural factors that would influence the "birth rate." Do the same for those factors that influence the death rate.
3. The world human population during the twentieth century increased exponentially. Speculate on the logical outcome if this rate of growth continues. How can the rate of growth be reduced? If the practices that you suggest were implemented, would there be a resulting ethical dilemma?

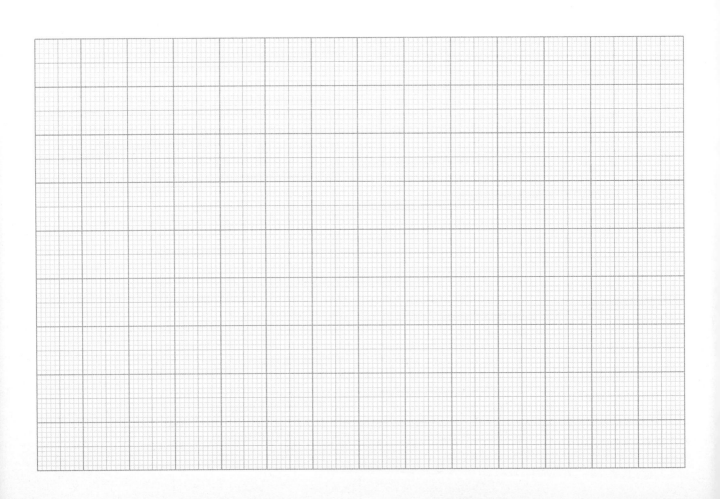

Appendix A

Significant Figures and Rounding

In the laboratory, students often ask how precise they should be when doing laboratory calculations. A few simple explanations are given here to guide you in these quantitative aspects of laboratory biology.

What Are Significant Figures?

Significant figures are defined as *the necessary number of figures required to express the result of a measurement so that only the last digit in the number is in doubt.* For example, if you have a ruler that is calibrated only in centimeters and find that a pine needle is between 9 and 10 cm long, how do you record the length?

The definition of significant figures tells you that you should estimate the additional fraction of a centimeter in tenths of a centimeter and add it to 9 cm, thus indicating that the last digit is only an estimate. You would never write this additional fraction as hundredths or thousandths of a centimeter because it would imply a precision that did not exist in your measuring instrument.

However, suppose you have a ruler calibrated in millimeters and measure the same pine needle, finding that the needle is between 93 and 94 mm long. You should then estimate the additional fraction in tenths of a millimeter and add it to 93 mm which would be 9.3 cm plus the estimate.

Memorize and use this rule throughout the course: *When recording measurements, include all of the digits you are sure of plus an estimate to the nearest tenth of the next smaller digit.*

Doing Arithmetic with Significant Figures

Other rules apply to calculations, especially when doing multiplication or division with an electronic calculator. Such devices provide an answer that is carried to an almost unbelievable number of decimal places, often giving a false sense of precision. To guide you, several situations you will encounter are discussed in the following paragraphs.

When converting measurements from one set of units to another in the metric system, be sure not to introduce greater precision than exists in the original number. For example, if you have estimated that something is 4.3 cm long and wish to convert it to millimeters, the correct answer is 43 mm, not 43.0 mm because the number of centimeters was known only with precision to a tenth of a centimeter and not a hundredth of a centimeter as 43.0 mm implies.

When performing additions or subtractions, the answer should contain no more decimal places than the number with the least number of digits following the decimal place. Thus, 7.2°C subtracted from 7.663°C yields a correct answer of 0.5°C not 0.463°C. If the first number had been known with a precision of 7.200°C, then the latter answer would have been correct.

When performing multiplication with numbers having different levels of significant figures, express the answer using the number of significant figures associated with the least precise value. For example, if you wish to calculate the weight of 10.1 ml of water and you are told the density of water is 0.9976 g/ml, you would multiply the density times the volume to obtain the weight. However, the correct answer would be 10.1 g, not 10.07576 g. Because the water volume measurement is known only to three significant figures, reporting the more precise value is not justified given the precision of the water volume measurement.

When performing division, the same rule applies. For example, let's say you weighed a small sample of water and found it was 3.2 g. Your balance had a digital readout and did not measure with any more precision. If you had to convert that weight into a volume in milliliters, you would divide the density of water (0.998 g/ml) into the weight, yielding an answer of 3.2076984 ml on an electronic calculator. However, this greatly exceeds the precision of the measuring device. Only if you had a balance that could measure with a precision of 0.0000001 g could you accept your calculator's answer. The correct answer is 3.2 ml.

What Is Rounding?

The last example introduces the concept of rounding to the appropriate number of significant figures. The rules governing this are straightforward. You should not change the value of the last significant digit if the digit following it is less than five. Therefore, 3.449 would round off to 3.4 if two significant figures were required. If the value of the following number is greater than five, increase the last significant digit by one. Therefore, 88.643 would round off to 89 if two significant figures were required.

There is some disagreement among scientists and statisticians as to what to do when the following number is exactly five, as in 724.5, and three significant numbers are required. Some will always round the last significant figure up (in this case 725), but others claim that this will introduce a significant bias to the work. To eliminate this problem, they would flip a coin (or use another random event generator) every time exactly five is encountered, rounding up when heads was obtained and leaving the last significant digit unchanged when tails was obtained. Recognize, however, that if the number were 724.51 or greater, the last significant digit would always be rounded up to 725.

Examples of Rounding

49.5149 rounded to 5 significant figures is 49.515 ($= 4.9515 \times 10$)

49.5149 rounded to 4 significant figures is 49.51 ($=4.951 \times 10$)

49.5149 rounded to 3 significant figures is 49.5 ($=4.95 \times 10$)

49.5149 rounded to 2 significant figures is 50 ($=5.0 \times 10$)

49.5149 rounded to 1 significant figure is 50 ($=5 \times 10$)

Appendix B

Making Graphs

Graphs are used to summarize data—to show the relationship between two variables. Graphs are easier to interpret and remember than are numbers in a table and are used extensively in science. You should get in the habit of making graphs of experimental data, and you should be able to interpret graphs quickly to grasp a scientific principle.

In using this lab manual, you will be asked to make two kinds of graphs—line graphs and histograms. **Line graphs** show the relationship between two variables, such as amount of oxygen consumed by a tadpole over an extended period of time (**fig. B.1**). **Histograms** are bar graphs and are usually used to represent data in which measurements are repeated, such as the values obtained when an object is weighed several times (see figure B.3).

Line Graphs

When you make line graphs, always follow these rules.

1. Decide which variable is the dependent variable and which is the independent variable. The **dependent variable** is the variable you know as a result of making experimental measurements. The **independent variable** does not change as a result of the dependent variable but changes independently of it. In figure B.1, time does not change as a result of oxygen consumption. Therefore, time is the independent variable and the amount of oxygen consumed (which is dependent on time) is the dependent variable.

2. Always place the independent variable on the x-axis (the horizontal one), and the dependent variable on the y-axis (the vertical one).

3. *Always label* the axes with a few words describing the variable, and *always put the units* of the variable in parentheses after the variable description (fig. B.1).

4. Choose an appropriate *scale* for the dependent and independent variables so that the highest value of each will fit on the graph paper.

5. Plot the data set (the values of y for particular values of x). Make the plotted points dark enough to be seen. Always use pencil not pen in case you need to erase. If two or more data sets are to be plotted on the same coordinates, use different plotting symbols for each data set ($\cdot$, $\times$, $\odot$, $\otimes$, $\square$, etc.).

6. Draw *smooth curves* or *straight lines* to fit the values plotted for any one data set. **Do not connect the points** with short lines. A smooth curve through a set of points is a visual way of averaging out variability in data. Do not extrapolate beyond a data set unless you are using it as a prediction technique, because you do not know from your experiment whether the relationship holds beyond the range tested.

7. Every graph should have a legend, a title sentence, explaining what the graph is about.

Derivative Graphs

Four experiments are summarized in figure B.1, demonstrating that oxygen consumption in tadpoles changes with temperature. What you might miss, however, is that oxygen consumption increases only from 15° to 25°C and then declines at 30°C, perhaps because the heat is killing the animals. The problem to address here is how to clearly show this response, so that a reader gets the point at first glance. A derivative graph can be used to present the same data in a more summarized form. It is a line graph where the dependent variable is derived from a data set as follows.

For linear functions (and you can see that oxygen uptake at any one temperature is linear in figure B.1), the derivative is the slope of the line. So, to make a derivative plot of the oxygen consumption data, you first plot the data to check linearity (fig. B.1) and then calculate the slopes of the four lines.

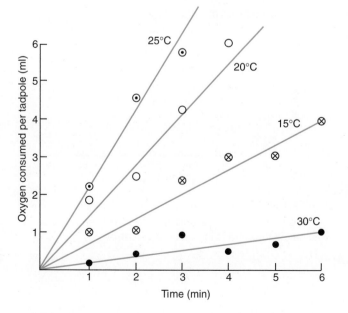

Figure B.1 Oxygen consumption by a tadpole at four temperatures.

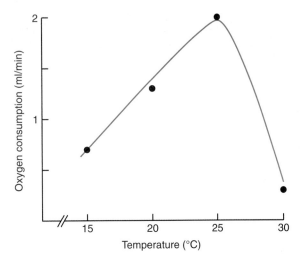

Figure B.2 Derivative plot summarizes data from figure B.1 and clearly shows how rate of oxygen consumption rises with temperature only to 25°C before declining. Futhermore, it suggests that additional measurements should be made between 25° and 30°C to better define temperature of which maximum oxygen consumption occurs.

The slope of a straight line is the change in the y-value for each unit change in the x-value. For the tadpole data this would have the units ml of oxygen consumed per minute. When this was done for the data in figure B.1, the following values were derived:

Temperature, °C	Slope, ml O_2/min
15	0.7
20	1.3
25	2.0
30	0.3

This derived data can now be replotted as shown in **figure B.2**, where the dependent variable on the y-axis is the slope (rate of oxygen consumption per minute) and the x-axis is the independent variable (temperature) that yielded the set of slopes. This plot clearly shows that the rate of oxygen consumption rises to a maximum and then quickly falls off beyond that. A reader will not miss the point.

Histograms

Histograms are bar graphs that summarize count data. In making histograms, the count data are always on the y axis. The categories in which the data fall are on the x axis. For example, the data used to draw **figure B.3** are:

Class Results from a Series of Weighings of the Same Sample (in Grams)				
61.0	60.0	60.0	59.8	58.0
61.5	61.5	61.0	60.9	58.0
59.0	60.0	60.2	61.7	60.6
59.7	59.0	60.3	63.0	60.4
62.0	59.0	60.7	58.5	59.0

Obviously there is variability in the data but it is difficult to interpret. By preparing a histogram (fig. B.3) a "picture" of the variability is created that is easy to understand. To make a histogram (also known as bar graph), the x axis was laid out with a range of 58 to 64 so that all values would be included. The values were marked on the graph as lightly penciled Xs, one in each square of the graph paper for each observation. After all data were plotted, bars were then drawn to show the frequencies of measurements. On histograms, it is a good idea to show the average value across all measurements with an arrow. In calculating an average, remember the significant figure rule (appendix A).

A final note about making graphs. Be neat! Remember most people do not trust sloppy work. Always print labels and use a sharp pencil, not a pen. Use a ruler to draw straight lines and a drafting template called a French curve to draw curved lines. Alternatively, many computer programs have various graphing options and can neatly prepare graphs. However, the same rules apply as when preparing graphs by hand.

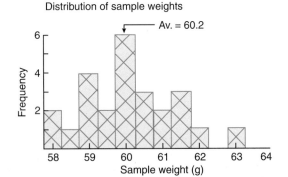

Figure B.3 Histogram of a series of weights obtained by weighing the same sample several times. An average was calculated and is indicated by an arrow.

Appendix C

Simple Statistics

Quantitative data may be expressed in two forms, **count data** or **measurement data**. Count data are discontinuous variables that always consist of whole numbers. They are derived by counting how the results from an experiment fall into certain categories; for example, in a genetic cross between two heterozygotes, you expect to obtain a genotypic ratio of 1:2:1. Measurement data are continuous variables obtained by using some measuring instrument. The precision of the measurement depends on the fineness of the scale on the instrument; for example, a ruler calibrated in centimeters is not as precise as one calibrated in millimeters. Accuracy differs from precision in that it depends not on the scale used but on the calibration against a standard and the proper reading of the scale.

Whenever measurements are made, there are potential sources of error. Instruments and the humans who read them make random errors. If the instrument is properly calibrated and if the person who is reading it is careful, the percent error is small and will be randomly distributed around the true measurement.

In some cases, another source of error may be introduced. Bias occurs when an instrument is improperly calibrated or when the operator makes a consistent error in reading or sampling. For example, if a watch that is five minutes slow is used to measure the time of sunset for several days, the data will reflect a consistent bias, showing sunset as occurring five minutes earlier than it really did. Similarly, if a balance, spectrophotometer, or pipette is improperly calibrated, it will consistently yield a biased estimation either over or under the true value; and if the operator of the device misreads the instrument, additional bias enters. No statistical procedure can correct for bias; there is no substitute for proper calibration for accuracy and care in making measurements for precision.

Assuming that all sources of bias have been ruled out, there is still the problem of dealing with random fluctuations in measurement and in the properties (size, weight, color, and so on) of samples. In biological research, this is especially important since variation is the rule rather than the exception. For example, white pine needles are approximately 4 inches long at maturity, but in nature, the length of the needles varies due to genetic and environmental differences. A biologist must constantly be aware that biological variability and bias influence the results of experiments and any analysis should include procedures to minimize the effects.

Dealing with Measurement Data

When several measurements are made by an individual or by a class, most would agree that it is best to use the **mean** or **average** of those measurements to estimate the measured value. However, determining the average alone may mean that important information concerning variability is ignored. Look at **table C.1**, which contains two sets of data: the average temperature in degrees Celsius at 7:00 P.M. each day in September for two different geographic locations.

The average temperature for each location can be calculated by adding the readings for that location and dividing by the total number of readings (N):

$$\text{average temperature} = \frac{\Sigma \text{ Readings}}{N}$$

where Σ equals "sum of"

For both data sets, the averages are the same (15°C). Based on the averages alone, one might conclude that the two locations have similar climates. However, by simply scanning the table, one can see that location A has a more variable temperature than location B. Such temperature variations, especially those below 0°C, may have a tremendous effect on organisms; many plants and small animals may die at subzero temperatures. Therefore, reporting only the average temperature from this set of data does not convey crucial information on variability.

The **range** of values can convey some of this information. Location A had a mean temperature of 15°C with a range from −5 to 35, while location B's mean temperature was 15°C with a range of 10 to 20. Unfortunately, the range of values has a limited usefulness because it does not indicate how often a given temperature occurs. For example, if another location (location C) had 15 days at −5°C and 15 days at 35°C, it would have the same mean and range as location A, but the climate would be harsher.

To overcome the limitations of range, the data could be plotted in frequency histograms, with the x-axis showing the daily temperature and the y-axis showing the frequency of that temperature, or how often it occurs. The data for all three locations are plotted in **figure C.1**. It is now obvious, looking at the plotted data, that these three locations have quite different climates even though the mean temperatures are the same and the ranges overlap for two of the three. However, this method of reporting is cumbersome.

Table C.1	September Temperatures in °C at 7:00 P.M. for Two Locations	
Day	Location A	Location B
1	0	15
2	0	20
3	5	20
4	10	20
5	15	20
6	15	15
7	20	10
8	25	10
9	30	10
10	35	10
11	30	10
12	35	15
13	30	20
14	30	13
15	25	17
16	25	18
17	20	19
18	20	20
19	19	10
20	15	11
21	15	12
22	11	13
23	10	17
24	10	15
25	5	10
26	5	20
27	0	10
28	0	10
29	−5	20
30	−5	20

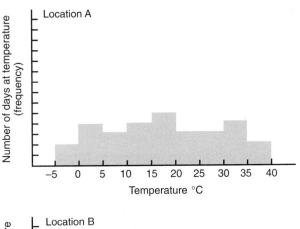
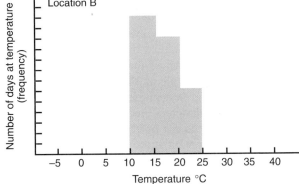
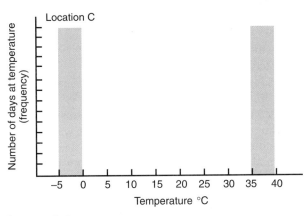

Figure C.1 Frequency histograms of temperatures at three locations each having the same mean temperature of 15°C. Note differences in variation.

An efficient way to report information about variability and its frequency is to calculate the **standard deviation** from the mean. This value is calculated by using the difference between each observation and the average. If a group of measurements is distributed randomly and symmetrically about the average, then the standard deviation defines a range of measurements in which 68% of the observed values are expected to fall.

Figure C.2 shows that distributions can differ in three ways: the means may be different, the standard deviations may be different, or both the means and the standard deviations may be different. Obviously, when the standard deviation is large, the variability is great, and when the variability is small, the standard deviation is small.

The following formula is used to calculate standard deviation:

$$\text{standard deviation} = \pm\sqrt{\frac{\Sigma(x_o - \bar{x})^2}{n - 1}}$$

where

$\sqrt{}$ = square root
Σ = sum of
x_o = an observed value
$\bar{x}$ = average of all observed values
n = number of observed values

The computation of standard deviation is best performed by setting up a calculation sheet as in **table C.2**. The numbers there are the ones from location A in the earlier

SIMPLE STATISTICS   457

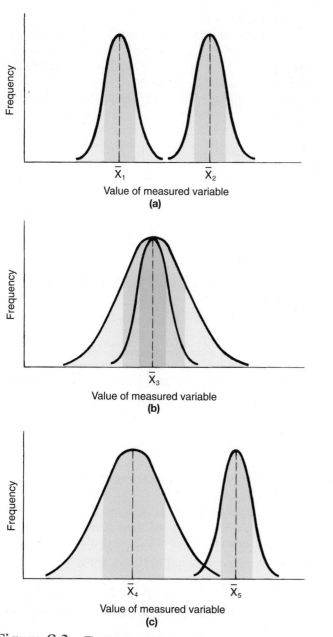

Figure C.2 Three comparisons of normal curves that differ in their means and standard deviations. On normal curves, the mean ($\bar{x}$) coincides with the mode, or the most frequently occurring value, and also with the median (the value at which 50% of all values are higher and 50% are lower). The dark shaded area corresponds to plus or minus one standard deviation about the mean, and will always include 68% of all the observations. The lighter shaded area is plus or minus two standard deviations and includes 95% of the observations. In (a), the curves differ in their means but the standard deviations are the same. In (b), the means are the same but the standard deviations are different. In (c) both the means and standard deviations differ.

example. The value for x_o is the temperature observation on each day, and $\bar{x}$ is the average temperature for the month, 15°C.

The mean and the standard deviation for this set of data are 15 ± 12°C. This means that the average temperature of 15°C plus or minus 12°C (3°C to 27°C) represents

Table C.2	Calculation Sheet for Standard Deviation using Location A Data		
n	x_o	$(x_o - \bar{x})$	$(x_o - \bar{x})^2$
1	0	−15	225
2	0	−15	225
3	5	−10	100
4	10	−5	25
5	15	0	0
6	15	0	0
7	20	5	25
8	25	10	100
9	30	15	225
10	35	20	400
11	30	15	225
12	35	20	400
13	30	15	225
14	30	15	225
15	25	10	100
16	25	10	100
17	20	5	25
18	20	5	25
19	19	4	16
20	15	0	0
21	15	0	0
22	11	−4	16
23	10	−5	25
24	10	−5	25
25	5	−10	100
26	5	−10	100
27	0	−15	225
28	0	−15	225
29	−5	−20	400
30	−5	−20	400
	Σ = 450		Σ = 4182

$$\bar{x} = \frac{\Sigma x_o}{n} = \frac{450}{30} = 15$$

$$SD = \pm \sqrt{\frac{4182}{30-1}} = \pm 12$$

68% of the temperatures you would expect to measure in this location during the same month that these temperatures were measured. If the same figures are calculated for the data set from location B, the results are 15 ± 4°C (not 4.2 because of significant figure rule; see appendix A). The mean and the standard deviation together convey a better impression of the similarities and differences between the two sets of data than the means alone.

In teaching laboratories, each student usually makes only one or two measurements of a biological phenomenon. The concept of standard deviation is not appropriate for so few readings; a range and average are more suitable.

However, if the class combines its readings so that there are 20 or 30 measurements, it is better to calculate the average and the standard deviation. Form the habit of reporting averages with some estimate of the variability when presenting data.

Comparing Count Data: Dealing with Variability

Often in experiments, theories are used to predict the results of experiments and data are recorded as counts or frequencies in categories. For example, Mendelian genetics can be used to predict the progeny from a cross between two heterozygotes. In cases where there is clear dominance and recessiveness, we would use our understanding of genetics to predict that offspring would occur in a ratio of 3:1. What happens when the actual results from a genetic cross of this type do not exactly fit the predictions? First, we would review the data for any obvious errors of technique or arithmetic and reject the data based on identified errors. If the data still do not conform exactly to predictions, we must decide whether the variations are within the range expected by chance.

When variations from the expected are small, we usually just round off and say that there is a good fit between the data and the expected. However, this is somewhat arbitrary and can become a problem as deviations from the expected become large. When are deviations no longer small and insignificant in comparison to the expected results (i.e., the theory is not a good predictor)?

Statisticians use a statistical test called the **chi-square** (χ^2) **goodness of fit test** to help them decide how much variation is acceptable. This test creates a number, the χ^2, which summarizes the differences between the data and what was expected. If that number is large, then the theory may be wrong. If it is small, the theory is a reasonable predictor of the results. The steps in using this test follow.

Scientific Hypothesis

The χ^2 test is always used to test a hypothesis known as the **null hypothesis (H_o)**. It is based on the scientific theory known as the **model,** which is the basis for predicting the results. A null hypothesis proposes that there is no difference between the results **actually obtained** and those **predicted** from a model within an acceptable range of variation.

A null hypothesis is paired with an **alternative hypothesis (H_a),** which proposes an alternative explanation of the results not based on the same scientific model as was the null hypothesis. The acronym "HoHa" emphasizes the coupling of hypotheses in statistical testing.

If a statistical test indicates that the results of an experiment do not fit the expected, then the null hypothesis must be rejected and the alternative hypothesis is true by inference: the data variation is greater than that expected by chance, and a factor other than that tested is influencing the results, implying that the model is not correct.

Rejection of H_o does not prove H_a! This is a common mistake in the use of statistics. Rejection of H_o only rejects the model on which H_o is based.

Let us suppose that you are conducting breeding experiments. You are looking at a dihybrid cross involving dominant and recessive traits located on autosomes. Mendel's principle of independent assortment predicts off-spring in the phenotypic ratio of 9:3:3:1. When the experiments were finished, the following results were obtained:

150 phenotypes dominant for both traits

60 phenotypes dominant for the first trait and recessive for the second

67 phenotypes recessive for the first trait and dominant for the second

23 phenotypes recessive for both traits

How would you determine whether these results were an acceptable variation from the expected 9:3:3:1 ratio? The actual ratio in this case is close to 7:3:3:1. Has the experiment failed to show Mendelian inheritance, or are the results simply within the limits of chance variability?

First, a null hypothesis should be formulated and then an alternative, mutually exclusive, hypothesis proposed. For the above experiment, these would be:

H_o: There is no significant difference between the results obtained and those predicted by the Mendelian principle of independent assortment.

H_a: Independent assortment does not predict the outcomes of this experiment.

Note how H_o and H_a are mutually exclusive and both cannot be true.

Testing the Null Hypothesis

Because we are dealing with count (frequency) data, which should conform to those predicted by a model, the chi-square goodness of fit test can be used to compare the actual and predicted results.

First, a frequency table is created (**table C.3**). The entries on the first line are the observed results from an experimental genetic cross in the lab. Those on the second line are the expected results obtained by multiplying the total by the fractions expected in each category ($9/16 \times 300$, $3/16 \times 300$, etc.).

The summary chi-square statistic is calculated using the following formula:

$$\chi^2 = \Sigma^n \frac{(O_i - E_i)^2}{E_i}$$

Table C.3 — Frequency Table of Observed and Expected Results from Hypothetical Fruit Fly Cross

	Dom-Dom	Dom-Rec	Rec-Dom	Rec-Rec	Total
Observed (O)	150	60	67	23	300
Expected (E)	169	56	56	19	300

where

Σ indicates "sum of"

O_i is the observed frequency in class i

E_i is the expected frequency in class i

n is the number of experimental classes (in this case 4)

χ^2 is the Greek letter *chi*, squared

It should be clear from inspecting this formula that the value of χ^2 will be 0 when there is perfect agreement between the observed and expected results, whereas the χ^2 value will be large when the difference between observed and expected results is large.

To calculate χ^2 for the data in table C.3, the following steps are required:

$$\chi^2 = \frac{(150-169)^2}{169} + \frac{(60-56)^2}{56} + \frac{(67-56)^2}{56} + \frac{(23-19)^2}{19}$$

$$= \frac{(-19)^2}{169} + \frac{(+4)^2}{56} + \frac{(+11)^2}{56} + \frac{(+4)^2}{19}$$

$$= 2.14 + 0.29 + 2.16 + 0.84$$

$$\chi^2 = 5.43$$

Therefore, for this experiment, the variability from the expected result is now summarized as a single number, $\chi^2 = 5.43$.

Making a Decision about the Null Hypothesis

To determine if a value of 5.43 is large and indicates poor agreement between the actual and predicted results, the calculated χ^2 value must be compared to a **critical χ^2 value** obtained from a table of standard critical values (**table C.4**). These values represent acceptable levels of variability obtained in random experiments of similar design.

To use table C.4, you must know a parameter called the **degrees of freedom (d.f.)** for the experiment; it is always numerically equal to one less than the number of classes in the outcomes from the experiment. In our example, the experiment yielded data in four classes. Therefore, the degrees of freedom would be 3.

In addition to knowing the degrees of freedom for an experiment, you must also decide on a **confidence level**, a percentage between 0 and 99.99, which indicates the confidence that you wish to have in making a decision to reject the null hypothesis. Most scientists use a 95% confidence level. Any other confidence level can be used and several (but not all) are given in table C.4.

Table C.4 — Critical Values for χ^2 at Different Confidence Levels and Degrees of Freedom (d.f.)

d.f	$\chi^2.90$	$\chi^2.95$	$\chi^2.98$	$\chi^2.99$	$\chi^2.999$
1	2.7	3.8	5.4	6.6	10.8
2	4.6	6.0	7.8	9.2	13.8
3	6.3	7.8	9.8	11.3	16.3
4	7.8	9.5	11.7	13.3	18.5
5	9.2	11.1	13.4	15.1	20.5
6	10.6	12.6	15.0	16.8	22.5
7	12.0	14.1	16.6	18.5	24.3
8	13.4	15.5	18.2	20.1	26.1
9	14.7	16.9	19.7	21.7	27.9
10	16.0	18.3	21.2	23.2	29.6

If you now read table C.4 by looking across the row corresponding to 3 degrees of freedom to the column for 95% confidence, you see the value 7.8. This critical value means that 95% of the χ^2 values for experiments with 3 degrees of freedom in which the H_o is true fall below 7.8 due to chance variation and that only 5% will be above that value due to chance.

The next step is to compare the calculated χ^2 value to the critical value for χ^2. When this is done, two outcomes are possible:

1. The calculated value is less than or equal to the critical value. This means that the variation in the results is of the type expected by chance in 95% of the experiments of a similar design; the null hypothesis cannot be rejected.

2. The calculated χ^2 value is greater than the critical value. This means that results would occur only 5% of the time by chance. Stated another way, these results should not be accepted as fitting the model; the null hypothesis should be rejected.

When these decision-making rules are applied to our calculated χ^2 value, 5.43 is obviously less than 7.8. Therefore, although our ratio was not 9:3:3:1, it is within an acceptable limit of variation and we cannot reject the H_o. A model based on the Mendelian principle of independent assortment predicts the results of the experiment even though the numbers are not exactly what was expected. Note, however, that you have not proven your H_o; you simply failed to reject it with 95% confidence in your decision. This is more than a subtle difference because it indicates that scientific knowledge is probabilistic and not absolute.

A Hypothetical Alternative

For illustration, let us suppose that another group got different results. When they calculated their χ^2 value, it was 9.4. If they went through the same decision-making steps that we just did, they would reject the null hypothesis: Mendelian genetics did not have predictive power for their experiment. In rejecting the null hypothesis, they can have 95% confidence that they are not making a mistake by rejecting what is actually a true null hypothesis.

Those who have studied table C.4 might suggest a change in strategy here. If this second group changed its confidence level to 97.5%, the critical value becomes 9.8, which is greater than the calculated value of 9.4, thus keeping the group from rejecting the null hypothesis. This illustrates that by choosing a higher value for a confidence level at a constant number of degrees of freedom, the critical value will be larger, allowing one to accept almost any null hypothesis. Is this not arbitrary, the very situation we sought to avoid by invoking this statistical test?

The solution to this dilemma is found in what is really tested by the chi-square goodness of fit test. The comparison of calculated and critical χ^2 values is done to attempt to falsify the H_o, to reject it at a predetermined confidence level. If H_o cannot be rejected, then by default it is accepted. *You have not proven H_o, you have simply failed to disprove it.* The confidence level that is stated before each test represents the confidence you have in rejecting the null hypothesis, not the confidence you have in accepting it. As you increase the confidence level, you decrease the likelihood of rejecting H_o.

Statisticians often speak of type I and type II errors in statistical testing. A **type I error** is the probability that you will reject a true null hypothesis. A **type II error** is the probability that you will accept a false null hypothesis. As you increase the confidence level, you reduce the probability that you will reject a true null hypothesis (type I error), but you increase the probability that you will accept a false one (type II error). The confidence level of 95% is used by convention because it represents a compromise between the probabilities of making type I versus type II errors. The basis of this conservatism in science is the recognition that it is better to reject a true hypothesis than it is to accept a false one. Once a confidence level is stated for an experiment, it should not be changed according to the whim of the experimenter who wants a model to have predictive power. Those who change the confidence level run the risk of accepting fiction as fact.

Appendix D
Writing Reports and Scientific Papers

Verbal communication is temporal and easily forgotten, but written reports exist for long periods and yield long-term benefits for the author and others. Gregor Mendel's work is a perfect example. When Mendel finished his research, he gave a verbal presentation to a scientific meeting, but few understood it. That was in 1872. Fortunately, he also wrote a paper and published it. About 30 years later that paper was read by scientists who realized its significance and Mendel has since been given deserved credit for founding the modern study of genetics.

Scientific research is a group activity. Individual scientists make observations or perform experiments to test hypotheses about natural phenomena. A lab report or scientific paper can be a source of information about data or a vehicle of persuasion; when it is published, it is available to other scientists for review. If the results and conclusions stand up to critical review, they become part of the accepted body of scientific knowledge unless later disproved.

Most scientific work is first published in the form of papers, meaning relatively short reports (2 to 20 pages or so) in periodicals known as scientific journals. Such papers are normally carefully reviewed by the editorial staff of the journal to make certain they meet standards for content, style, and accuracy, and are often subject to additional peer review by other scientists working in the field. That peer review process and the format distinguish scientific research publications from news releases, newspaper and magazine articles, and advertisements.

In scientific writing, clarity is more important than elegance. Simple declarative sentences and accurate use of words improve communication. Brevity reduces publication costs and saves readers' time, so flowery language, unnecessary adjectives, and repetition must be avoided. Technical terms and abbreviations must be defined the first time they are used in a paper. Experimental methods must be described so that other individuals can repeat observations or experiments to confirm results, an essential part of the self-checking nature of scientific discovery.

In scientific writing, it is essential to document sources of facts and ideas, so they can be verified by others and because the use of text, ideas, data, or figures produced by others without crediting the source constitutes plagiarism and intellectual dishonesty.

There are a number of style manuals that provide detailed directions for writing scientific papers. Some are listed in further readings at the end of this section.

Format
A scientific report usually includes the following sections:

1. Title
2. Abstract
3. Introduction
4. Materials and methods
5. Results
6. Discussion
7. Literature cited

There is general agreement among scientists that each section of the report should contain specific types of information.

Title
Titles should be brief and inform a reader about the factual content of the report. Scientific titles are usually not designed to catch the reader's fancy, but to convey information. A good title is straightforward and uses keywords that researchers of related topics will recognize.

Abstract
Most scientific papers begin with a brief abstract. An abstract is a concise (often only 100–200 words) summary of the purpose of the report, the data obtained, and the author's major conclusions. The abstract should allow a reader to judge whether it would serve his or her purposes to read the entire report. Though the abstract may follow the title in a finished paper, abstracts are best written only after the rest of the paper has been completed.

Introduction
The introduction outlines the scientific purpose of the paper and the hypotheses tested, giving the reader sufficient background to understand the rest of the report. Care should be taken to limit the background to whatever is pertinent to the investigation being reported. A good introduction answers such questions as:

What knowledge already exists about this subject? The answer to this question briefly reviews what is known about the topic, including existing data and the history of ideas about those data, and identifies conflicts or gaps in existing knowledge addressed in the new research.

What is the purpose of the study? State the nature of new data being provided and specific hypotheses being tested. Hypotheses should include both direction and magnitude of predicted effects, and a summary of the logic that led you to propose a particular hypothesis..

Materials and Methods

This section describes materials and methods in enough detail to enable a reader to understand what the author did, and enable another scientist to repeat the work if necessary. Identify experimental organisms by scientific name (in italics): chemicals by name, concentration, and source if not widely available; and describe instrumentation used. Describe your experimental design, including sample sizes, treatments, and methods of analysis. This includes statistical methods that you use to describe variation in data or decide whether a difference between treatment groups is significant. The amount of detail requires judgment. The model and supplier of equipment with special capabilities or new to researchers in the field might be important, but naming the brand of micropipetter you use or paper towel you used is unnecessary. The challenge is to allow a reader to understand the the experiments without wasting space or the reader's time. When procedures from a lab book or a published report are followed exactly, simply cite the work, noting that details can be found there. Illustrations of specialized apparatus can be valuable.

Generally, this section attempts to answer the following questions:

What materials were used?

How were they used?

Where and when was the work done? (Very important for field studies.)

How were the results analyzed?

Results

The results should present the experimental and descriptive data from your research. Many authors organize and write the results section before the rest of the report. Your results section should identify significant observations ("treatment with drug x increased survivorship by 47%"), but is not the place to discuss the implications of those observations. Be certain that your results address each of the hypotheses you put forward in the introduction.

Results must include both written text describing your findings and supporting figures, including tables, graphs, and illustrations or photographs as needed. Use the text to identify significant trends or differences between groups. Avoid redundancy—do not include the same data in the form of both a table and a graph. Graphs often enable readers to understand relationships much more easily than do tables of data.

All figures and tables must have descriptive titles and include a legend explaining any symbols, abbreviations, or special methods used. Figures and tables should be self-explanatory so a reader can understand them without referring to the text. All columns and rows in tables and axes in figures must be labeled. See appendix B for graphing instructions.

Number figures and tables consecutively and be certain to refer in the text to each figure or table by number, so the reader can understand their importance in your analysis. If there is no reason to refer to a figure in your text, it does not belong in your report. Here are two ways to reference data in a figure:

1. Figure 1 shows that the activity decreased after five minutes.
2. The activity decreased after five minutes (fig. 1).

Discussion

In the discussion, identify the significance of important observations and explain the logic that allows you to accept or reject your hypotheses. Show how your results are related to published data and the work of others. This requires that you locate, read, and discuss published descriptions of related research by others.

Your discussion should address each hypothesis you present, and describe the biological or physical mechanisms that might be responsible for any patterns you observe. Be as thoughtful and creative as you can—draw together and integrate different parts of your data and knowledge and try to look for relationships, causes, and effects.

The discussion is usually the most important part of a scientific report, and should never just rehash the results. Emphasize interpretation of your data, and explain how they relate to existing theory and knowledge. Identify new hypotheses suggested by your data. Suggestions for the improvement of techniques or experimental design, or future experiments that might clarify areas of doubt or expand the significance of your results can be included.

Literature Cited

Any source you used to obtain a particular fact or idea that is not widely accepted, or text or figures that you used, must be cited *in the text* where that material is used, not just listed at the end of the report. When citing references in the text, do not use footnotes. Most scientific journals refer to sources using the author's last name and date of publication. Here are two ways to cite a source using that form:

1. Haussmann *et al.* (2005) investigated how immune function changes with age in birds.
2. Immune response declines slowly with age in long-lived species (Haussmann *et al.*, 2005).

When citing papers that have two authors, list both names. When three or more authors are involved, the Latin *et al. (et alia)* meaning "and others" is usually used. The Haussmann *et al.* paper cited above really is shorthand way of listing seven authors that worked together. When this paper is listed in the literature cited section at the end of the report, all seven authors' last names and initials would be spelled out.

The **literature cited** section provides complete references (including names of all authors) for all sources cited in your report. It is not the same as a bibliography, which simply lists references regardless of whether they were actually cited in the paper. The listing is usually alphabetized by the last names of the authors. Different journals require different formats for citing literature. The format that includes the most information is illustrated below:

For journal articles:

Fox, J. W. 2000. Nest-building behavior of the catbird, *Dumetella carolinensis. Journal of Ecology* 47: 113–17.

For books:

Watson, J. D. with A. Berry (2003). DNA: The Secret of Life. New York: Random House.

For a chapter written by a named author(s) in books, edited by a third person(s):

Halpern, M. 1992. Nasal chemical senses in reptiles: structure and function. pp. 422–523. In: *Hormones, Brain and Behavior. Biology of the Reptilia,* Volume 18, *Physiology* E. C. Gans and D. Crewes (eds). Chicago: University of Chicago Press.

Citing information from the Internet is challenging, because of the variety and transitory nature of Web sites. Most Internet sources lack any peer review or other quality control, and must be treated with caution as sources of fact. Citation of any article or book that is also available in published form should always include the information listed for books and journals above. Citation of any electronic source should include as much as possible of the following: author's last name and initials or responsible organization's name; date of publication on the Internet, including revision dates; title of the electronic document; the URL; and the date on which you accessed it. Always check that the URL you provide is complete, correct, and will work another Web browser. Examples are:

Soltis, Pam, Doug Soltis, and Christine Edwards. 2005. Angiosperms. Flowering Plants. Version 03 June 2005. http://tolweb.org/Angiosperms/20646/2005.06.03 in The Tree of Life Web Project, http://tolweb.org/. Accessed 7/1/2009.

Wikipedia contributors. Protein structure [Internet]. Wikipedia, The Free Encyclopedia; modified June 23, 2009. http://en.wikipedia.org/wiki/Protein_structure. Accessed 6/28/2009.

General Comments on Style

1. Use correct English, and always check grammar and spelling. Each sentence must have a subject and a verb.
2. Divide text into paragraphs to improve clarity, and make certain each paragraph includes a topic sentence. Do not run unrelated sentences together. A section of a report is not one long paragraph.
3. Use the first person and active voice, in spite of what your high school science teacher may have taught you. If you did something, write; "In the light, the plant produced oxygen," not "In the light, oxygen was produced by the plant." The active voice is clearer and briefer.
4. All scientific names of organisms (genus and species) must be italicized. (Underline if italics are not available.)
5. Use the metric system of measurements, and always specify the units of measurement. Abbreviations of units are used without a following period.
6. Numbers should be written as numerals for 10 and above or when associated with measurements; for example, 6 mm or 2 g but two explanations or six factors. When one list includes numbers 10 and above, all numbers in the list may be numerals; for example, 17 sunfish, 13 bass, and 2 trout. Never start a sentence with numerals, instead, spell any number beginning a sentence.
7. Use the past tense to describe completed actions and results. Be consistent in the use of tense throughout a paragraph.
8. Avoid the use of slang and contractions.
9. Be aware that the word data is plural while datum is singular. This affects the choice of a correct verb—"data are. . . ," not "data is. . . ." The word species is used both as a singular and as a plural.
10. Be certain antecedents are clear when you use pronouns. For example, in the statement, "Sometimes squirrels are in walnut trees but they are hard to find," does "they" refer to squirrels or trees?

After writing a report, take a break, then reread it to search for lack of precision and for ambiguity. Each sentence should present a clear message. The following examples illustrate lack of precision:

"The sample was incubated in mixture A minus B plus C." Does the mixture lack both B and C or lack B and contain C?

The title "Protection against Carcinogenesis by Antioxidants" leaves a reader wondering whether antioxidants protect from or cause cancer.

The only way to prevent such errors is to read and think about what you write.

Further Readings

CSE Style Manual Committee. 2006. *Scientific Style and Format: The CSE Manual for Authors, Editors, and Publishers.* 7th ed. Reston, Va: Council of Science Editors and Rockefeller University Press.

McMillan, V. E. 2006. *Writing Papers in the Biological Sciences.* 4th ed. New York: Bedford/St. Martin's Press.

Patrias, K. 2001. *National Library of Medicine Recommended Formats for Bibliographic Citation Supplement: Internet Formats.* Bethesda, MD: U. S. Department of Health and Human Services, National Institutes of Health. Available online at http://www.nlm.nih.gov/pubs/formats/internet2001.pdf.

Pechenik, J. A. 2007. *A Short Guide to Writing About Biology.* 6th ed. White Plains: Longman/Prentice Hall.

Photo Credits

Lab Topic 2
Figure 2.1: Courtesy Leica, Inc.; **Figure 2.2:** Courtesy of the Olympus Corporation, Lake Success, NY.

Lab Topic 3
Figure 3.4, Figure 3.5: © Ed Reschke; **Figure 3.6:** © Inga Spence/Visuals Unlimited; **Figure 3.7:** Courtesy of Joseph Viles, Iowa State University; **Figure 3.8:** © E. H. Newcomb & W. P. Wergin/Biological Photo Service.

Lab Topic 4
Figure 4.3: © M. Abbey/Visuals Unlimited.

Lab Topic 5
Figure 5.3: Linda Westgate; **Figure 5.5a:** Courtesy of Bausch & Lomb; **Figure 5.5b:** Warren Dolphin; **Figure 5.5c:** Courtesy of Bausch & Lomb; **Figure 5.5d:** Warren Dolphin.

Lab Topic 6
Figure 6.8: © CNRI/SPL/Photo Researchers, Inc.

Lab Topic 7
Figure 7.2: Created by Warren Dolphin using Jmol (http://www.jmol.org/) and pdb files available from the National Center for Biotechnology Information.

Lab Topic 9
Figure 9.1 (top): © Bruce Coleman, Inc.; **Figure 9.1 (middle):** © Michael Eichelberger/Visuals Unlimited, Inc.; **Figure 9.1 (bottom):** © Visuals Unlimited, Inc.

Lab Topic 10
Figure 10.1: © Dr. Don Fawcett/Photo Researchers, Inc.; **Figure 10.2 (all):** © Ed Reschke; **Figure 10.3 (all):** © The McGraw-Hill Companies, Inc./Kingsley Stern, photographer

Lab Topic 11
Figure 11.2: © Cabisco/Visuals Unlimited; **Figure 11.3b:** © Fred Hostler/Visuals Unlimited; **Figure 11.4:** © Cabisco/Phototake.

Lab Topic 13
Figure 13.1: © K. G. Marti/Visuals Unlimited; **Figure 13.2:** © David Dresser/Huntington Potter/Tiepin/Getty.

Lab Topic 15
Figure 15.1, Figure 15.8: © Steve Alexander and Dennis Strete.

Lab Topic 16
Figure 16.1: © David M. Phillips/Photo Researchers, Inc.; **Figure 16.3:** © Carolina Biological/Visuals Unlimited; **Figure 16.4a:** © Carolina Biological/Visuals Unlimited; **Figure 16.4b:** Courtesy of Dr. Yuuji Tsukii, Hosei University; **Figure 16.4c:** © Nalco Chemical Co./Fundamental Photographs; **Figure 16.4d:** © Roland Birke/Peter Arnold Inc.; **Figure 16.4e, Figure 16.4f:** © Bruce J. Russell/BioMedia Associates; **Figure 16.5:** © John D. Cunningham/Visuals Unlimited; **Figure 16.6 (all):** © Kingsley R. Stern; **Figure 16.7 (a–c):** © Carolina Biological/Visuals Unlimited; **Figure 16.7d:** © Cabisco/Visuals Unlimited; **Figure 16.11:** © E. R. Degginger/Photo Researchers, Inc.; **Figure 16.12a:** © Eric V. Grave/Photo Researchers, Inc.; **Figure 16.12b:** © Wim Van Egmond; **Figure 16.12 (c–d):** Courtesy of Dr. Yuuji Tsukii, Hosei Univeristy; **Figure 16.13:** © Bill Beatty/Visuals Unlimited; **Figure 16.14:** Courtesy Barry Leadbeater.

Lab Topic 17
Figure 17.3a: © Heather Angel; **Figure 17.3b:** © Cabisco/Phototake; **Figure 17.4, Figure 17.7a:** © Runk/Schoenberger/Grant Heilman; **Figure 17.7b:** © David S. Addison/Visuals Unlimited; **Figure 17.7c:** © Ed Reschke; **Figure 17.8a:** John D. Cunningham/Visuals Unlimited; **Figure 17.8b:** © John D. Cunningham/Visuals Unlimited; **Figure 17.8c:** © S. Elms/Visuals Unlimited.

Lab Topic 18
Figure 18.2a: © George Loun/Visuals Unlimited; **Figure 18.2b:** © Runk/Schoenberger/Grant Heilman; **Figure 18.2c:** © Science/Visuals Unlimited; **Figure 18.3:** © Jack M. Bostrack/Visuals Unlimited; **Figure 18.4a:** © Doug Sokell/Visuals Unlimited; **Figure 18.4b:** © Dr. John Cunningham/Visuals Unlimited; **Figure 18.4c:** © George J. Wilder/Visuals Unlimited; **Figure 18.6 (a&b):** © Cabisco/Visuals Unlimited; **Figure 18.11:** © Cabisco, Visuals Unlimited; **Figure 18.13:** © Ed Reschke; **Figure 18.14a:** © Cabisco/Visuals Unlimited; **Figure 18.15:** © Kingsley R. Stern.

Lab Topic 19
Figure 19.1a: © Kingsley R. Stern; **Figure 19.1b:** © Biophoto Associates/Photo Researchers, Inc.; **Figure 19.3b:** © Cabisco/Phototake; **Figure 19.3c:** © Ed Reschke; **Figure 19.4:** © Joyce Photographics/Photo Researchers, Inc.; **Figure 19.5b:** © James Richardson/Visuals Unlimited; **Figure 19.5c:** © Ed Degginger/Color-Pic, Inc.; **Figure 19.5d:** © Doug Sherman/Geofile; **Figure 19.6a:** © William Ormerod/Visuals Unlimited; **Figure 19.6b:** © Richard Thom/Visuals Unlimited; **Figure 19.6c:** © Bill Keough/Visuals Unlimited; **Figure 19.7b:** © Biophoto Associates; **Figure 19.8(a-c):** Courtesy James T. Colbert; **Figure 19.8e:** © V. Ahmadjian/Visuals Unlimited.

Lab Topic 20
Figure 20.7: © Carolina Biological Supply/Phototake; **Figure 20.10c:** Courtesy Dr. Fred Whittaker; **Figure 20.10d:** © Ed Reschke.

Lab Topic 21
Figure 21.2a: Courtesy of David Barnes; **Figure 21.2b:** © Dr. John Cunningham/Visuals Unlimited; **Figure 21.3a:** © Runk/Schoenberger/Grant Heilman; **Figure 21.3b:** © Kjell Sandved/Visuals Unlimited; **Figure 21.3c:** © Robert & Linda Mitchell; **Figure 21.4a:** Courtesy of University of Saskatchewan Archives; **Figure 21.6 (a&b):** Courtesy of Dr. Clyde Herreid and the Department of Biological Sciences/University at Buffalo, SUNY; **Figure 21.7b:** © David M. Dennis/Animals, Animals; **Figure 21.13:** © John D. Cunningham/Visuals Unlimited.

Lab Topic 22

Figure 22.5b: Linda Westgate and David Vleck; **Figure 22.9:** © Ken Taylor.

Lab Topic 23

Figure 23.1: © William Van Egmond/Visuals Unlimited; **Figure 23.2a:** © Daniel Gotshall/Visuals Unlimited; **Figure 23.2b:** © Carl Roessler; **Figure 23.2c:** © Bill Ober/Visuals Unlimited; **Figure 23.2d:** © Daniel Gotshall/Visuals Unlimited; **Figure 23.2e, Figure 23.4b:** © Michael DiSpezio; **Figure 23.5 (b&c):** © John D. Cunningham/Visuals Unlimited.

Lab Topic 24

Figure 24.1: © Harry Horner, Iowa State University; **Figure 24.2, Figure 24.3:** © Biophoto Associates/Photo Researchers, Inc.; **Figure 24.5c:** Courtesy Eva Frei and R.D. Preston; **Figure 24.6:** © John D. Cunningham/Visuals Unlimited; **Figure 24.7, Figure 24.8a:** © Biophoto Associates/Photo Researchers, Inc.; **Figure 24.8 (b&c):** © Michael W. Clayton; **Figure 24.8d:** © Kingsley R. Stern; **Figure 24.9a:** © John D. Cunningham/Visuals Unlimited; **Figure 24.9b:** D. T. Webb, University of Hawaii; **Figure 24.10a:** © Randy Moore/Visuals Unlimited; **Figure 24.10b:** © George J. Wilder/Visuals Unlimited.

Lab Topic 25

Figure 25.2: © Jeremy Burgess/SPL/Photo Researchers, Inc.; **Figure 25.4:** © BioPhoto; **Figure 25.5a:** © Lynwood M. Chace/Photo Researchers, Inc.; **Figure 25.5b:** © Ed Reschke; **Figure 25.6:** © Cabisco/Phototake; **Figure 25.8:** © John D. Cunningham/Visuals Unlimited; **Figure 25.10:** © Jack M. Bostrack/Visuals Unlimited; **Figure 25.11:** © Cabisco/Phototake; **Figure 25.12:** © Ed Reschke; **Figure 25.14a:** © Cabisco/Phototake; **Figure 25.14b:** © Runk/Schoenberger/Grant Heilman Photography; **Figure 25.19:** © BioPhot; **Figure 25.20:** © John D. Cunningham/Visuals Unlimited; **Figure 25.21:** © BioPhot.

Lab Topic 26

Figure 26.2 (all): © Cabisco/Visuals Unlimited; **Figure 26.5:** Linda M. Westgate.

Lab Topic 27

Figure 27.4: © Ken Taylor; **Figure 27.5a:** © Lester V. Bergman/Corbis; **Figure 27.5b:** © Dr. Richard Kessel & Dr. Randy Kardon/Visuals Unlimited, Inc.; **Figure 27.5c:** © Keith Porter; **Figure 27.12:** Courtesy of William Radke; **Figure 27.13:** From W. Bloom and D. S. Fawcett, *A Textbook of Histology*, 10th edition, Fig. 33.2, p. 859. Copyright © 1975 Chapman & Hall.

Lab Topic 28

Figure 28.6: © Ken Taylor; **Figure 28.10:** © Warren Rosenberg/Biological Photo Service; **Figure 28.11:** From R. G. Kessel and R. H. Kardon, *Tissues and Organs: A Text-Atlas of Scanning Electron Microscopy*, 1979, W. H. Freeman and Company; **Figure 28.12 (a&b):** © Ed. Reschke.

Lab Topic 29

Figure 29.1c: © Thomas Eisner, Cornell University; **Figure 29.4b:** Courtesy Gregory J. Highison, provided by Frank N. Low.

Lab Topic 30

Figure 30.2 (a&b): © Ed Reschke; **Figure 30.2c:** © Manfred Kage/Peter Arnold; **Figure 30.3:** © Janzo Desaki, Ehime University School of Medicine, Shingenobu, Japan; **Figure 30.5:** David Vleck; **Figure 30.6b:** Courtesy John W. Hole; **Figure 30.6c:** From R. G. Kessel and R. H. Kardon, *Tissues and Organs: A Text-Atlas of Scanning Electron Microscopy*, 1979, W. H. Freeman and Company.

Lab Topic 31

Figure 31.2: © Biodisc/Visuals Unlimited; **Figure 31.4:** Courtesy of Hugh E. Huxley; **Figure 31.6, Figure 31.7:** Courtesy Theron O. Odlaug; **Figure 31.10:** © Professor Harold Benson.

Lab Topic 32

Figure 32.1 (all), Figure 32.3 (all), Figure 32.9: © Cabisco/Visuals Unlimited; **Figure 32.10:** Oxford Scientific Films.